Marine Fisheries Management

Marine Fisheries Management

Edited by **Simon Oakenfold**

Callisto Reference
New York

Published by Callisto Reference,
106 Park Avenue, Suite 200,
New York, NY 10016, USA
www.callistoreference.com

Marine Fisheries Management
Edited by Simon Oakenfold

International Standard Book Number: 978-1-63239-755-3 (Hardback)

Printed in China.

Contents

Preface

The main aim of this book is to educate learners and enhance their research focus by presenting diverse topics covering this vast field. This is an advanced book which compiles significant studies by distinguished experts. This book addresses successive solutions to the challenges arising in the area of application, along with it; the book provides scope for future developments.

Marine fisheries management refers to the practice of using fisheries science in order to protect the marine ecosystem, especially fishes. It aims to device methods to sustainably harvest fishes along with minimizing exploitation of the natural resources involved in the process. This book is compiled in such a manner, that it will provide in-depth knowledge about the theory and practice of marine fisheries management. It presents this complex subject in the most comprehensive and easy to understand language. Students, researchers, marine biologists, aquaculturists, and all other associated with this area will find this text helpful. It will prove to be a beneficial source of reference for readers.

It was a great honour to edit this book, though there were challenges, as it involved a lot of communication and networking between me and the editorial team. However, the end result was this all-inclusive book covering diverse themes in the field.

Finally, it is important to acknowledge the efforts of the contributors for their excellent chapters, through which a wide variety of issues have been addressed. I would also like to thank my colleagues for their valuable feedback during the making of this book.

Editor

1

A Surplus Production Model Considering Movements between Two Areas using Spatiotemporal Differences in CPUE: Application to Sea Ravens *Hemitripterus villosus* off Fukushima as a Practical Marine Protected Area after the Nuclear Accident

Yasutoki Shibata*
Tohoku National Fisheries Research Institute, Fisheries Research Agency, Samemachi, Aomori 031 0841, Japan

Manabu Yamada and Toshihiro Wada
Fukushima Fisheries Experiment Station, Iwaki, Fukushima 970 0316, Japan

Masaki Itou
Tohoku National Fisheries Research Institute, Fisheries Research Agency, Samemachi, Aomori 031 0841, Japan

Harumi Yamada
Seikai National Fisheries Research Institute, Fisheries Research Agency, Taira-machi, Nagasaki 851 2213, Japan

Tadahiro Sohtome, Takashi Iwasaki, Tooru Sakuma, and Takuji Mizuno
Fukushima Fisheries Experiment Station, Iwaki, Fukushima 970 0316, Japan

Abstract

Although the number of marine protected areas (MPAs) for stock management has increased, movements or differences in population structure of a target species between an MPA and surrounding fishing areas have rarely been considered in stock biomass estimations. We developed a surplus production model considering seasonal movements between two areas; the model was applied to Sea Ravens *Hemitripterus villosus* off Fukushima, where almost all fishing has been prohibited since the 2011 accident at the Fukushima Dai-ichi Nuclear Power Plant. We predicted future biomass by using CPUE data from coastal gill-net fishing and offshore bottom trawl fishing in 2000 to 2009. The model reflected the seasonal coastal–offshore movements of Sea Ravens well, and it predicted increasing Sea Raven biomass in both areas, which was validated by the CPUEs observed after 2010—including those for trial bottom trawl fishing that occurred within limited offshore areas after the accident. Our results indicate that the newly developed model incorporating seasonal movements of Sea Ravens is feasible and that the waters off Fukushima have effectively been

Subject editor: Anne Hollowed, Alaska Fisheries Science Center, Seattle

*Corresponding author: shibatayas@affrc.go.jp

serving as an MPA since the nuclear accident. We also demonstrated the model's applicability for estimating the optimal fishing effort and designing a new MPA for stock management that considers seasonal movements.

Spatially based effort management, such as no-take areas or rotating closures, has been widely recognized as a feasible stock management method for conserving biodiversity, ecosystem services, habitats, and endangered species as well as for increasing biomass (Agardy 1994; Allison et al. 1998; Jones 2001; Halpern and Warner 2002; Field et al. 2006; Lester et al. 2009). Marine protected areas (MPAs) are one of the management methods currently used. They are defined as areas with legal boundaries in which fishing is prohibited for all or a subset of a species, or where particular fishing gears are disallowed with some degree of permanence (Field et al. 2006). In line with global targets agreed upon by the Convention on Biological Diversity, the number of MPAs is increasing rapidly (Edgar et al. 2014).

Virtual population analysis and surplus production models have been widely used to assess fish abundance (Hilborn and Walters 1992; Coggins et al. 2006). In general, these models assume that the assessed population has a high diffusivity, which means that the effects of local recruitment events and regional fisheries spread rapidly throughout the entire stock (Field et al. 2006). However, the establishment of an MPA is expected to generate heterogeneous patterns of abundance, age structure, and size structure inside and outside of the MPA, especially in less-migratory species with strong local retention mechanisms (Guénette et al. 1998; Holland 2002; Punt and Methot 2004; Hart 2006). Consequently, the development of stock assessment models that can incorporate spatial differences in population structure and movement patterns in waters within and outside of MPAs may be necessary in some special cases (Field et al. 2006).

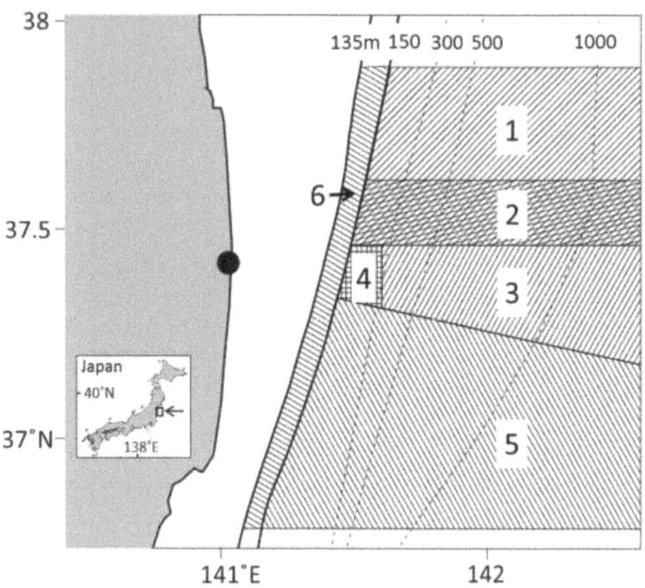

FIGURE 1. Trial fishing area off Fukushima, Japan. Numbers in shaded areas show the order of opening dates for trial fishing (see Table 1); numbers above the dotted lines indicate depth. The black circle shows the location of the Fukushima Dai-ichi Nuclear Power Plant.

Recently, a spatially explicit age-structured model and a spatially explicit surplus production model were developed to address fish movement between MPAs and the surrounding fishing areas (Punt and Methot 2004; Pincin and Wilberg 2012). These studies indicated that the relative error in estimated biomass increased if the data were combined across MPAs and fishing areas (i.e., heterogeneity among areas was not considered). However, the negative impact on estimates was relatively low when differences in population structure due to fish movements were accounted for in the assessment. Spatially explicit age-structured models require age data, and they are usually applied in data-moderate to data-rich situations (Punt and Methot 2004). In the case of spatially explicit surplus production models, movement between an MPA and fishing areas is only based on a single, fixed movement parameter through time. This type of model is well suited for a data-poor situation in which only CPUE data are available.

We attempted to develop a spatially explicit surplus production model and apply it to a local population of Sea Ravens *Hemitripterus villosus* in the waters off Fukushima, Japan, where almost all fishing (except for offshore purse-seine and stick-held dip-net fishing) has been practically banned since the Great East Japan Earthquake and tsunami and the subsequent accident at the Fukushima Dai-ichi Nuclear Power Plant (FDNPP) in March 2011 (Wada et al. 2013).

In June 2012, trial fishing operations (bottom trawl only) commenced in a limited area offshore (depth > 150 m) of northern Fukushima (Wada et al. 2013). Since then, the target areas for trial bottom trawl fishing have been expanded southward (Figure 1; Table 1). However, as of December 2014, a shallow coastal area has still not been opened to fishing. Some

TABLE 1. Opening dates for trial bottom trawl fishing in the waters offshore of Fukushima, Japan. See Figure 1 for the location of each area.

Area	Opening date	Boundary depth (m)
1	Jun 18, 2012	150
2	Oct 19, 2012	150
3	Feb 18, 2013	150
4	May 24, 2013	150
5	Aug 28, 2013	150
6	Dec 25, 2013	135

specimens of demersal fish species (including Sea Ravens) caught in this area during monitoring surveys conducted after April 2011 have shown radioactive cesium (^{134}Cs and ^{137}Cs) concentrations above the Japanese standard limit for foodstuffs (100 becquerels/kg wet weight). The total effort of the trial bottom trawl fishing operations in 2012 was estimated to be about 2.5% of the mean total effort that occurred from 2007 to 2009 (Yamada et al. 2014). This suggests that the waters off Fukushima have effectively acted as an MPA since the FDNPP accident.

The aim of this study was to develop a new surplus production model considering seasonal movements and different fishing pressures between two areas: the no-take area and the limited fishing area off Fukushima. The model was applied to Sea Raven populations off Fukushima by using CPUEs from 2000 to 2009 and was also used to predict the future biomass of Sea Ravens. We chose the Sea Raven as a suitable candidate because previous studies have reported that they exhibit clear seasonal movements from coastal to offshore areas in association with their reproductive cycles (Munehara 1996; Antonenko et al. 2010). The feasibility of the model was evaluated by comparing the predicted and observed CPUEs before (i.e., January 2010–March 2011) and after (i.e., September–October 2012) the nuclear accident at FDNPP based on data that were not included in the estimation step. Results of the model were used to evaluate whether the waters off Fukushima have been effectively serving as an MPA since the accident. We also demonstrate the model's applicability for estimating the optimum fishing effort and for designing a new MPA based on consideration of seasonal movements.

METHODS

Study area.—The waters off Fukushima, where the cold Oyashio Current meets the warm Kuroshio Current (Shimizu et al. 2001), constitute a fertile fishing area. Many commercially important demersal fish stocks (e.g., Pacific Cod *Gadus macrocephalus*, Barfin Flounder *Verasper moseri*, Willowy Flounder *Tanakius kitaharai*, and Spotted Halibut *Verasper variegatus*) are distributed in this region and display some degree of seasonal movements between coastal and offshore areas (Narimatsu 2006; Narimatsu et al. 2007; Wada et al. 2012; Kayaba et al. 2014). Demersal fishes off Fukushima were fished by various commercial methods, such as gill-net fishing and bottom trawl fishing, prior to the FDNPP accident.

At the instruction of the Fukushima Prefectural Government, the Fukushima Fisheries Experimental Station has conducted monitoring surveys for radioactive iodine (^{131}I) and Cs concentrations in Fukushima's marine products since April 7, 2011 (Wada et al. 2013). Whether or not fish can be landed via the trial fishing is determined based on the results of the monitoring surveys. The first trial fishing area was opened on June 18, 2012, and five other areas were opened sequentially thereafter (Figure 1; Table 1).

Data set used for analysis.—Off Fukushima, 82.9% of the Sea Raven landings from 2000 to 2010 were caught by coastal gill-net fishing and offshore bottom trawl fishing. The waters off Fukushima were divided into coastal and offshore areas based on the operated depths of gill-net fishing and bottom trawl fishing. The location of each depth observation (Appendix, Figure A.1) was obtained from daily fishing logbooks. Some outliers or typographical errors were included in the logbooks; therefore, we removed the upper and lower 2.5th percentiles of the observed depth. The shallowest and greatest depths were calculated as 9.0 and 69.0 m for gill-net fishing and as 87.0 and 576.0 m for bottom trawl fishing. The average of the deepest observed gill net and the shallowest observed bottom trawl ([69.0 + 87.0]/2 = 78.0 m) was adopted as the boundary between gill-net and bottom trawl fishing areas. Here, we refer to areas shallower than 78.0-m depth as coastal areas, and we refer to waters deeper than 78.0 m as offshore areas. The fishing logbooks were recorded by a chief fisherman on each vessel and included data on fishing dates, depth, longitude, latitude, net deployment and net hauling times for each fishing operation, gill-net length and mesh size (for gill-net fishing), and maximum catch (nearest 0.1 kg) of 15 species for each operation. The number of vessels that cooperated to record the fishing logbooks differed among years. On average, 48 gill-net vessels (range = 44–52) and 11 bottom trawl vessels (range = 10–12) per year cooperated to record the logbook data during 2000–2010. The total number of registered vessels in Fukushima Prefecture was approximately 600 for gill-net fishing and approximately 40 for bottom trawl fishing, thus indicating that the cooperating vessels constituted 8.0% of all gill-net vessels and 27.5% of all bottom trawl vessels. Since the FDNPP accident, all bottom trawl vessels participating in the trial fishing operations have been recording logbook data. However, no data were available for gill-net fishing after the accident because there were no trial gill-net fishing operations for demersal fishes as of December 2014.

The tonnage of bottom trawl fishing vessels ranged from 17.0 to 47.6 metric tons, and the engine power of those vessels ranged from 502 to 670 kW. The reported cod-end mesh size used on the bottom trawl nets ranged from 5.0 to 9.0 cm, and the headrope length ranged from 28.3 to 37.4 m (Hirata 2000; Inoue and Honda 2002). Two types of otter boards were used: monoplane and biplane. The tonnage of gill-net fishing vessels ranged from 1.30 to 6.98 metric tons, and engine power ranged from 281 to 540 kW. The gill-net mesh size ranged from 9.59 to 16.7 cm, but the mean gill-net mesh size did not affect the monthly gill-net fishing CPUE (Figure A.2). Because there was little difference between observed and unobserved vessels, the vessels that cooperated to record the fishing logbooks were assumed to be representative of all gill-net and bottom trawl fishing vessels.

Monthly Sea Raven catch (C; metric tons/month) for all gill-net fishing and bottom trawl fishing from January 2000 to

March 2011 was obtained from the annual report of fishery statistics in Fukushima (Fishery Office of the Fukushima Prefectural Government 2013). We obtained the logbook-recorded catch from the trial bottom trawl fishing that was conducted in September and October 2012. Effort from trial bottom trawling was obtained for June and September–December 2012. From 2000 to 2010, average catch per month was 1,020 kg (CV = 0.83) for gill-net fishing and 2,770 kg (CV = 0.98) for bottom trawl fishing. Total gill-net length per month and total bottom trawl towing hours per month as calculated from the fishing logbooks were used as the effort values (*E*). In other words, each monthly value of *E* was calculated from 8.0% of the gill-net vessels and 27.5% of the bottom trawl vessels, whereas data on monthly catch was from all of the vessels.

The total gill-net length or total bottom trawl towing hours were summed for each month. Because the unit of effort differed and data were obtained from different fishing gears, we transformed the monthly effort to values from zero to 1.0 by dividing by the monthly maximum effort values for 2000–2009 as follows:

$$E_s^{'}(t) = \frac{E_s(t)}{\max[E_s(t)]}, \tag{1}$$

$$E_d^{'}(t) = \frac{E_d(t)}{\max[E_d(t)]}, \tag{2}$$

where *s* represents the shallow coastal fishing area (<78.0 m) for gill-net fishing and *d* represents the deep offshore area (≥78.0 m) for bottom trawl fishing. From the distribution of operated depth, we assumed that gill-net fishing was not carried out in the offshore fishing area and likewise that bottom trawl fishing did not occur in the coastal fishing area (Figure A.1), although at most 2.5% of the fishing activity by each vessel might involve crossing the boundary. The symbol *t* represents a cumulative month from 2000 to 2009; therefore, $t =$ 1 represents January 2000 and $t = 120$ represents December 2009. The CPUEs of gill-net fishing (I_s) and bottom trawl fishing (I_d) were then calculated as

$$I_s(t) = \frac{C_s(t)}{E_s^{'}(t)}, \tag{3}$$

$$I_d(t) = \frac{C_d(t)}{E_d^{'}(t)}. \tag{4}$$

There was no effort in July and August for bottom trawl fishing, which was prohibited during those months. Records for depths greater than 314 m were excluded from the calculation because 314 m is the recorded deepest depth of the Sea Raven catch off Fukushima.

Population model.—We developed a surplus production model that included a monthly rate of Sea Raven movement, X_m ($-1 \le X_m \le 1$), between the two fishing areas,

$$B_s^{'}(t) = B_s(t) + rB_s(t)\left[1 - \frac{B_s(t)}{K_s}\right] - q_s E_s^{'}(t) B_s(t), \tag{5}$$

$$B_d^{'}(t) = B_d(t) + rB_d(t)\left[1 - \frac{B_d(t)}{K_d}\right] - q_d E_d^{'}(t) B_d(t), \tag{6}$$

$$\begin{cases} B_s(t+1) = B_s^{'}(t) + X_m B_d^{'}(t) \\ B_d(t+1) = B_d^{'}(t) - X_m B_d^{'}(t) \end{cases} \text{if } 0 \le X_m \le 1, \tag{7}$$

$$\begin{cases} B_s(t+1) = B_s^{'}(t) + X_m B_s^{'}(t) \\ B_d(t+1) = B_d^{'}(t) - X_m B_s^{'}(t) \end{cases} \text{if } -1 \le X_m < 0, \tag{8}$$

where $B^{'}$ is biomass (metric tons) prior to movement, and *m* represents each month corresponding to *t* (i.e., $m = 1$ [January], . . . , 12 [December]). For example, X_m with $t = 2$ and $t =$ 14 shows the same movement rate (X_2) in February 2000 and in February 2001. Sea Ravens move from the offshore area to the coastal area if X_m is larger than zero (equation 7); they will move from the coastal area to the offshore area if X_m is less than zero (equation 8). Estimated parameters include *r* (growth rate), *K* (carrying capacity in metric tons; $K \ge B$), and *q* (fishing efficiency).

Observation model.—We assumed that the natural logarithm of observed monthly CPUEs for gill-net fishing and bottom trawl fishing followed a normal distribution (N[μ, σ^2]), with a mean value (μ) that was the natural logarithm of the estimated CPUE ($\log_e[q_s B_s]$ and $\log_e[q_d B_d]$) from the model,

$$\log_e[I_s(t)] \sim \mathrm{N}\left(\log_e[q_s B_s(t)], \sigma_s^2\right), \tag{9}$$

$$\log_e[I_d(t)] \sim \mathrm{N}\left(\log_e[q_d B_d(t)], \sigma_d^2\right). \tag{10}$$

The variance (σ^2) was assumed to be different for each gear (σ_s^2 and σ_d^2).

Parameter estimation.—The parameters were estimated by using the Markov chain–Monte Carlo method under the framework of Bayesian estimation. The software R version 3.1.0 (R Development Core Team 2014), the R package R2WinBUGS (Sturtz et al. 2005), and the Gibbs sampling program WinBUGS (Spiegelhalter et al. 2003) were used to estimate posterior distributions. The first 500,000 samples were not used due to the potential for them to be affected by initial values that were randomly selected from their prior distributions. The last 1,000,000 samples were used to obtain posterior distributions. Some parameters were sampled in the logarithmic scale to avoid cancellation of significant digits or to increase the efficiency of searching. Prior distributions of parameters were set as follows:

$$\begin{cases} \sigma_s^{-2}, \sigma_d^{-2} \sim \text{Gamma}(10^{-6}, 10^{-6}) \\ \log_e(q_s), \log_e(q_d) \sim U(-10, 0) \\ \log_e[B_s(1)], \log_e[B_d(1)] \sim U(1, 5) \\ \log_e(K_s), \log_e(K_d) \sim U(3, 10) \\ \log_e(r) \sim U(-6, -1) \\ X_m \sim U(-1, 1) \end{cases} . \quad (11)$$

In this case, the CPUEs from 2000 to 2009 were used to estimate the parameters. The developed model was evaluated by applying it to predict biomass before the nuclear disaster at FDNPP (January 2010–March 2011) and biomass after the disaster (September–October 2012).

Model evaluation and prediction of future biomass.—By using estimated posterior distributions, we predicted future biomass of Sea Ravens from 2010 to 2015 under two scenarios. In scenario 1, the observed effort for each gear was continued at the same level as 2010. In scenario 2, the observed effort for each gear after the FDNPP accident was used (i.e., only the effort from the trial bottom trawl operation). In scenario 2, the predicted and observed CPUEs before the nuclear disaster (January 2010–March 2011) and after the disaster (September–October 2012) were compared to evaluate the predictive performance of the model. We randomly selected samples of each parameter from the posterior distributions, and future biomass from 2010 to 2015 was then predicted for each scenario. We repeated this prediction 10,000 times to calculate the median and 95% credible intervals for predicted biomass.

Although the developed model was able to estimate CPUE in the defined offshore area ($q_d B_d[t]$), it was unable to estimate the CPUE for the trial bottom trawl fishing area during 2012. Therefore, a corrected value of predicted biomass in the area deeper than 150 m was needed to compare the predicted and observed CPUEs from trial bottom trawl fishing. To correct the predicted biomass, we calculated a correction factor (CF) by using the monthly CPUEs from the daily bottom trawl fishing logbooks for 2000–2010. Generally, the CF provides good correction when CPUE is proportional to biomass,

$$CF_{m,z} = \frac{\sum_y C_{d,m,z,y} \Big/ \sum_y E'_{d,m,z,y}}{\sum_y C_{d,m,y} \Big/ \sum_y E'_{d,m,y}} = \frac{I_{d,m,z}}{I_{d,m}} = \frac{q_d B_{d,m,z}}{q_d B_{d,m}} = \frac{B_{d,m,z}}{B_{d,m}}, \quad (12)$$

where y represents the year ($y = 2000, \ldots, 2010$); $I_{d,m,z}$ is the CPUE of bottom trawl fishing that occurred deeper than depth z during month m; and $I_{d,m}$ is the bottom trawl fishing CPUE calculated from all depth ranges greater than 78.0 m. This equation shows the ratio of CPUE equal to that of biomass. Therefore, $CF_{m,z}$ is the ratio of biomass in the area deeper than 78.0 m and depth z during month m. The 95% confidence interval of $CF_{m,z}$ was calculated by the bootstrap method. We randomly chose 11 years from 2000 to 2010 (with replacement) and calculated $CF^h_{m,z}$ by using the hth bootstrap sample. This step was repeated 1,000 times (i.e., $h = 1, \ldots, 1{,}000$), and the confidence interval of $CF_{m,z}$ was calculated.

For the evaluation step, mean $CF_{m,150}$ was multiplied by the corresponding monthly predicted biomass. The corrected predicted biomass in the area deeper than 150 m ($B_d[t]CF_{m,150}$) was then obtained and used to predict CPUE for September and October 2012. The value of $CF_{m,z}$ was also used to examine the effect of a simulated MPA (see below).

Estimating the optimum fishing effort.—Two different simulations were carried out to estimate the optimum levels of effort for gill-net fishing and bottom trawl fishing and to show the effects of an MPA. The period of simulations was set at 100 years, which was long enough to confirm whether predicted catch and biomass (see equations 13–17 and equation 20 below) decreased if effort was too high. For simplicity, effort for each gear type was fixed through time (i.e., $E'_s[t] = E'_s[t+1]$; and $E'_d[t] = E'_d[t+1]$), and the estimated median values of the parameters were used in calculations.

First, we simulated future biomass to investigate the optimal fishing effort for maximizing total catch by varying the levels of gill-net fishing effort and bottom trawl fishing effort from 0 to 1. We calculated a predicted total catch for 100 years (TC) and the predicted biomass 100 years later (PB) as

$$TC_s = \sum_{t=133}^{1332} C_s(t), \quad (13)$$

$$TC_d = \sum_{t=133}^{1332} C_d(t), \quad (14)$$

$$PB_s = \frac{1}{12} \sum_{t=1321}^{1332} B_s(t), \quad (15)$$

$$PB_d = \frac{1}{12} \sum_{t=1321}^{1332} B_d(t), \quad (16)$$

where $t = 133$ represents January 2011; $t = 1321$ represents January 2111; and $t = 1332$ represents December 2111. The mean gill-net fishing effort (ME'_s) and the mean bottom trawl fishing effort (ME'_d) before the nuclear disaster were calculated as follows:

$$ME'_s = \frac{1}{132} \sum_{t=1}^{132} E'_s(t), \quad (17)$$

$$ME'_d = \frac{1}{132} \sum_{t=1}^{132} E'_d(t). \quad (18)$$

The gill-net effort and bottom trawl effort that maximized the value of $TC_s + TC_d$ were defined as the optimal effort levels, and the ME' values before the disaster were compared.

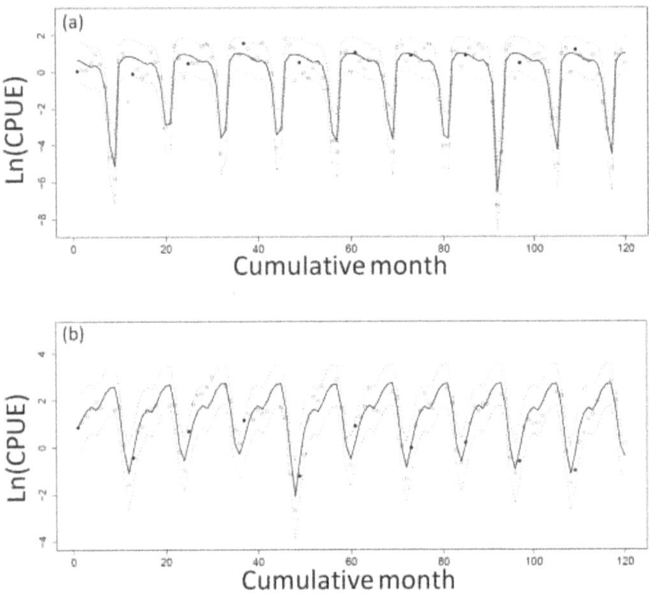

FIGURE 2. Estimated median CPUEs (solid line; $\log_e$ transformed) and 95% credible intervals (dotted lines) for Sea Ravens captured by **(a)** gill-net fishing and **(b)** bottom trawl fishing off Fukushima, Japan (open circles = observed monthly CPUEs; black circles = CPUEs observed in January of each year). There are no CPUEs for bottom trawl fishing in July or August because bottom trawling was prohibited during those months.

Estimating the effects of a simulated marine protected area.—We also calculated the total catch and biomass by varying the boundary depth and $E_d^{'}(t)$. The boundary depth determines the area where trial bottom trawl fishing can occur. In other words, a change in the boundary depth is equivalent to a change in the size of an MPA. We used the mean $\mathrm{CF}_{m,z}$ to simulate the trial fishing catch in the area deeper than the boundary depth z as

$$C_{d,z}(t) = q_d E_d^{'}(t) B_d(t) \mathrm{CF}_{m,z}, \tag{19}$$

where $B_d(t)\mathrm{CF}_{m,z}$ is the corrected biomass in the area deeper than depth z (i.e., biomass outside of the MPA) and $C_{d,z}$ is the expected catch for bottom trawl fishing conducted in that area. In this simulation, the range of $E_d^{'}(t)$ was set at 0 to 5, and the boundary depth was set at 78.0 to 180 m. The gill-net fishing effort was set at zero through time. We calculated the total catch in the area deeper than the boundary depth z ($\mathrm{TC}_{d,z}$) by using $C_{d,z}$ as follows:

$$\mathrm{TC}_{d,z} = \sum_{t=133}^{1332} C_{d,z}(t). \tag{20}$$

RESULTS

The natural logarithms of estimated monthly CPUEs for both gill-net fishing and bottom trawl fishing are shown in Figure 2a and 2b, respectively. Almost all of the observed CPUEs for gill-net fishing (93.3%) and bottom trawl fishing (97.0%) were included within the 95% credible intervals of the estimated CPUEs. The estimated cyclic variation in CPUE was consistent with the cyclic variation in estimated Sea Raven biomass. The medians of the posterior distribution of biomass at each time point changed cyclically between the coastal and offshore areas (Figure 3). When biomass in the offshore area was high, biomass in the coastal area was low (e.g., in September) and vice versa (e.g., in December).

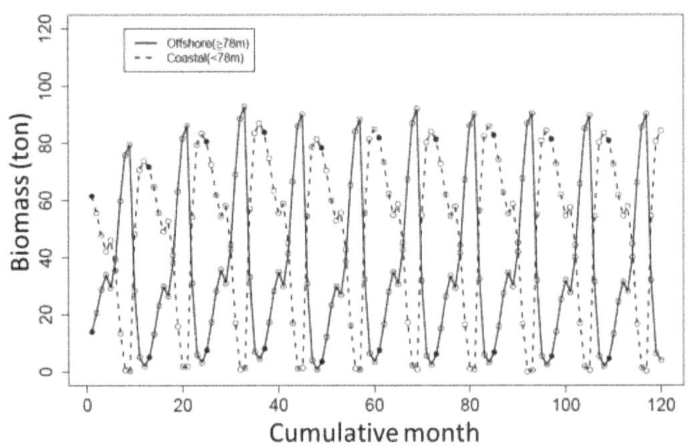

FIGURE 3. Estimated median biomass of Sea Ravens (metric tons) obtained from the model developed in this study. Biomass in the coastal area (<78.0 m; dotted line) and biomass in the offshore area (≥78.0 m; solid line) are depicted (open circles = monthly biomass; black circles = biomass in January of each year).

Monthly movement rates (Figure 4) reflected the spatiotemporal variation in Sea Raven biomass between the coastal and offshore areas. The estimated posterior distributions of movement rate showed that Sea Ravens stayed in the coastal area from approximately January through April (Figure 4). The fish then started to move from the coastal area to the offshore area from May to August. In September, Sea Ravens began to move back into the coastal area from the offshore

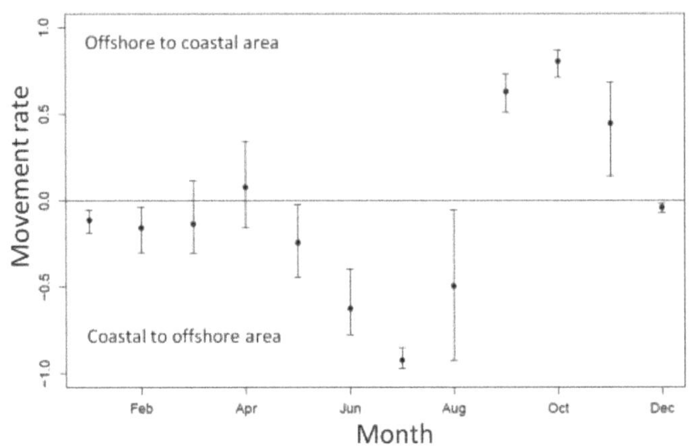

FIGURE 4. Estimated posterior distributions of Sea Raven movement rates off Fukushima (black circle = median; ±95% credible interval). Positive values indicate fish movement from the offshore area to the coastal area; negative values indicate movement from the coastal area to the offshore area.

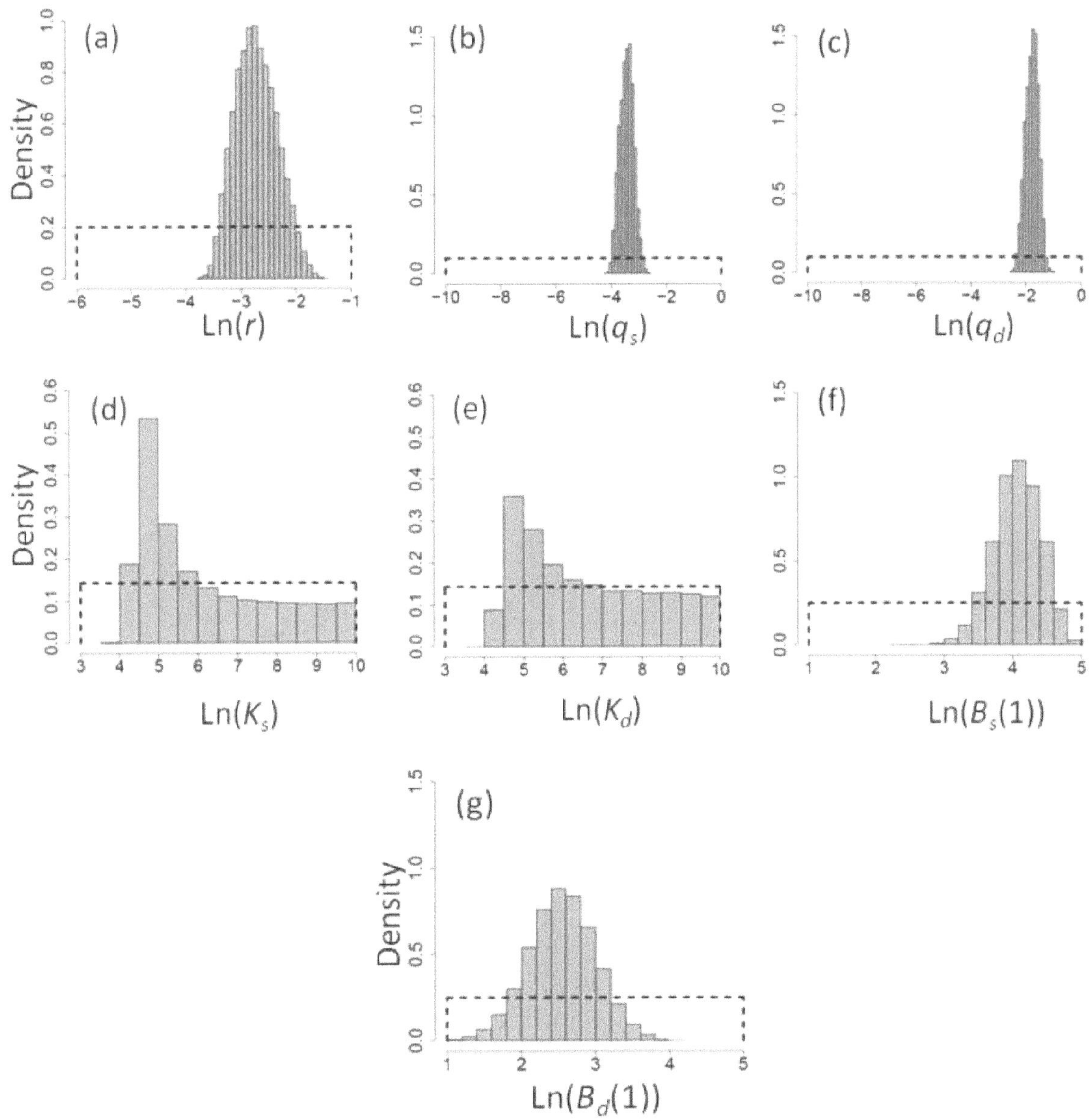

FIGURE 5. Histograms of estimated posterior distributions for the natural logarithms of **(a)** Sea Raven intrinsic growth rate (r); **(b)** gill-net fishing efficiency (q_s); **(c)** bottom trawl fishing efficiency (q_d); **(d)** carrying capacity of the coastal area (<78.0 m; K_s); **(e)** carrying capacity of the offshore area (≥78.0 m; K_d); **(f)** biomass for the coastal area in January 2000 (B_s[1]); **(g)** biomass for the offshore area in January 2000 (B_d[1]). Dotted lines indicate the prior distributions.

area. Medians and 95% credible intervals for the other parameters are shown in Figure 5a–g and Table 2.

The median of the average yearly biomass (i.e., the mean monthly biomass from January to December) calculated for scenario 1 was smaller than that for scenario 2 in both the coastal area and the offshore area (Figure 6). This result indicates that Sea Raven biomass will increase in the no-take area and in the trial fishing area off Fukushima if fishing effort only occurs at the trial levels.

Almost all of the observed monthly CPUEs for gill-net fishing (86.7%) and bottom trawl fishing (73.3%) were included in the 95% credible intervals of predicted CPUEs (Figure 7). During 2012, catch of Sea Ravens was only recorded in September and October, consistent with the result that the fish were mainly distributed in the offshore area during those months (Figure 3).

Estimated Optimum Fishing Effort to Maximize the Total Catch

The sum of TC_s and TC_d was maximized at 8,090 metric tons when $E_s^{'}(t)$ was 0.20 and when $E_d^{'}(t)$ was 1.00 (Figure 8a). The calculated $ME_s^{'}$ and $ME_d^{'}$ were 0.79 and 0.59, respectively, with a summed TC of 904 metric tons. This indicates that the mean gill-net fishing effort was too high and that 7,186 metric tons would be lost during this period. The sum of PB_s and PB_d was

TABLE 2. Estimated median, 2.5% quantile, and 97.5% quantile of the posterior distributions for each parameter (subscript s represents the shallow coastal fishing area [<78.0 m] for gill-net fishing; subscript d represents the deep offshore area [≥78.0 m] for bottom trawl fishing; r = growth rate; q = fishing efficiency; K = carrying capacity; $B[1]$ = biomass in January 2000).

Parameter	2.5%	Median	97.5%
$\log_e(r)$	−3.40	−2.74	−1.94
$\log_e(q_s)$	−3.93	−3.46	−2.97
$\log_e(q_d)$	−2.26	−1.77	−1.35
$\log_e(K_s)$	4.31	5.48	9.74
$\log_e(K_d)$	4.41	6.23	9.77
$\log_e(B_s[1])$	3.35	4.07	4.68
$\log_e(B_d[1])$	1.65	2.54	3.42

1.13 metric tons when the mean effort levels ($ME_s^{'}$ and $ME_d^{'}$) were continued, whereas the optimized effort values yielded a summed PB of 330 metric tons (Figure 8b). Sea Raven catch and biomass both increased under the optimized levels of gill-net and bottom trawl fishing effort compared with the mean fishing effort levels that were observed before the disaster.

Effects of the Boundary Depth on Total Catch and Biomass

The calculated mean $CF_{m,z}$ is shown in Figure 9, and the confidence intervals of mean $CF_{m,z}$ as calculated by the bootstrap method are presented in Figure A.3. In some cases, the calculated $TC_{d,z}$ decreased when $E_d^{'}(t)$ was high (Figure 10a). For example, the maximum $TC_{d,z}$ was 8,071 metric tons when (1) PB_s and PB_d were 98.2 and 119 metric tons, respectively; (2) the boundary depth was set at 78.0 m; and (3) $E_d^{'}(t)$ was set at 2.20 (Figure 10a–c). However, $TC_{d,z}$ decreased and Sea Raven biomass approached zero when $E_d^{'}(t)$ was greater than 2.20. The sum of PB_s and PB_d was 332 metric tons when $E_d^{'}(t)$ was set at 1.0, the maximum effort of bottom trawl fishing. This was greater than the mean biomass for the period 2000–2009 (85.6 metric tons).

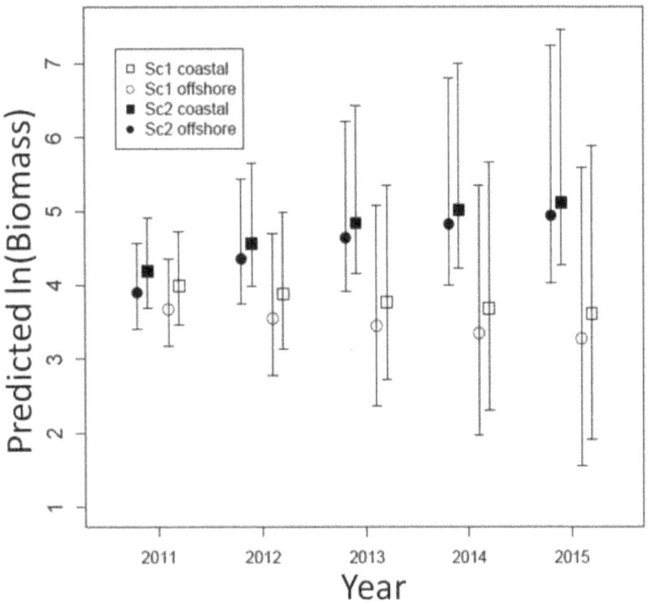

FIGURE 6. Medians (±95% credible interval) of the average predicted Sea Raven biomass ($\log_e$ transformed) per year in the coastal area (squares) and offshore area (circles; open symbols = scenario 1, in which the observed effort for gill-net fishing and bottom trawling was continued at the same levels as 2010; shaded symbols = scenario 2, in which the observed effort after the FDNPP accident was used [i.e., trial bottom trawl fishing only]).

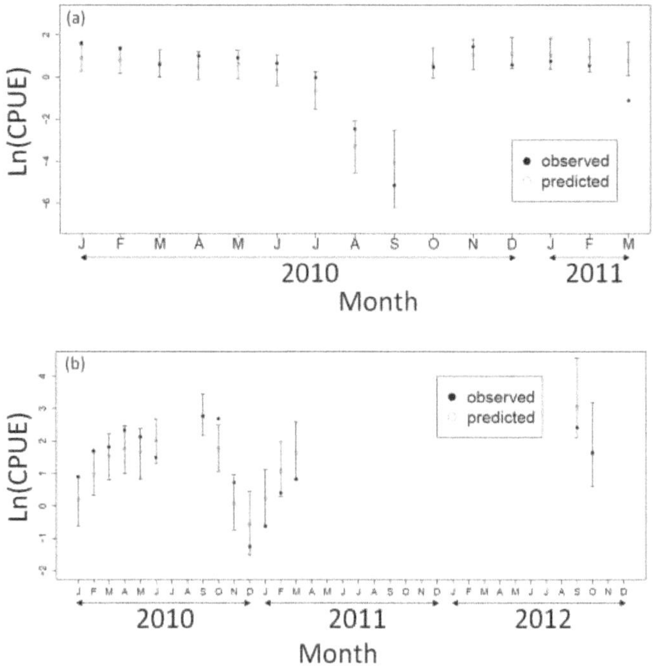

FIGURE 7. Medians of the predicted CPUE (open circles; $\log_e$ transformed, ±95% credible interval) and observed CPUE (black circles; $\log_e$ transformed) for Sea Ravens captured by **(a)** gill-net fishing (January 2010–March 2011) and **(b)** bottom trawl fishing (January 2010–October 2012).

DISCUSSION

Waters off Fukushima as an Effective Marine Protected Area

We demonstrated that spatiotemporal changes in Sea Raven biomass caused by seasonal movements and different fishing pressures between the two areas can be estimated using the developed surplus production model. In addition, biomass after the FDNPP accident can be predicted with this model. The results suggest that the predicted Sea Raven biomass in the coastal and offshore areas increased when fishing effort remained at levels occurring since the FDNPP accident (Figure 6). Moreover, the high predictive ability of the model was demonstrated by the fact that almost all of the observed CPUE values were within the 95% confidence intervals of the predicted CPUEs (Figure 7).

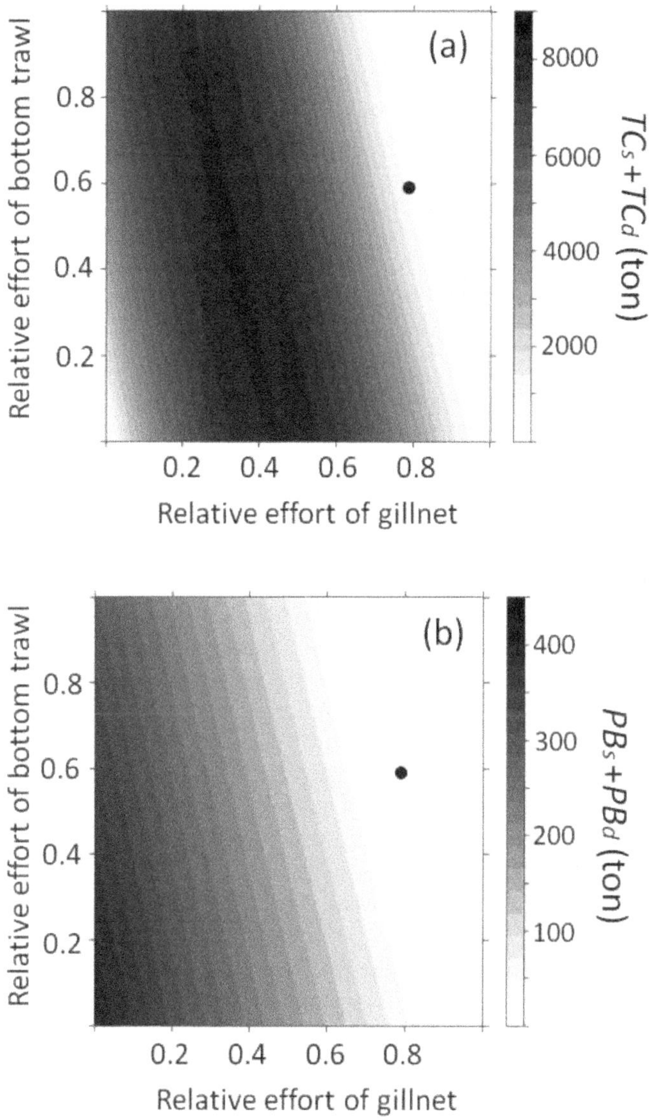

FIGURE 8. Sum of **(a)** the predicted total Sea Raven catch (metric tons) for 100 years of gill-net fishing (TC_s) and bottom trawl fishing (TC_d) and **(b)** the predicted Sea Raven biomass (metric tons) 100 years later (2111) in the coastal area (PB_s) and offshore area (PB_d) off Fukushima, Japan. The *x*- and *y*-axes show the gill-net fishing effort ($E_s^{'}$) and bottom trawl fishing effort ($E_d^{'}$). The black circle indicates the mean gill-net fishing effort ($ME_s^{'}$) or mean bottom trawl fishing effort ($ME_d^{'}$) before the nuclear disaster.

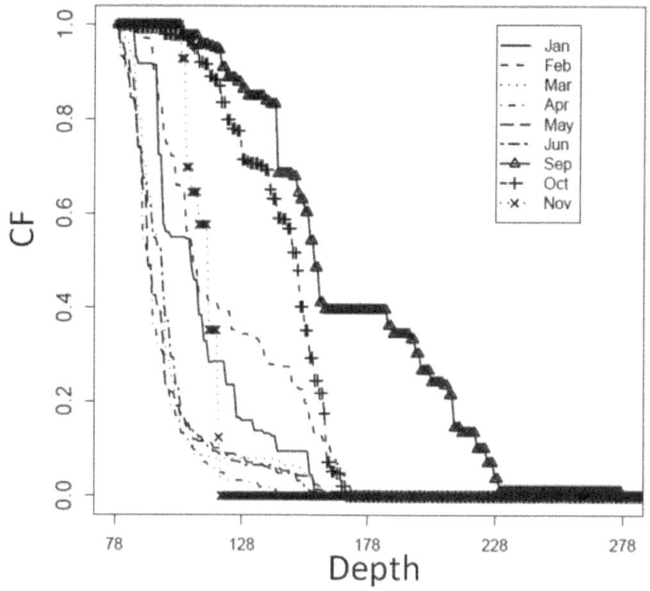

FIGURE 9. Calculated mean value of the correction factor ($CF_{m,z}$) for each month from 2000 to 2010 (see Methods). Note that there was no CF for July or August because bottom trawl fishing was prohibited during those months. There was no CF for December because Sea Ravens were not recorded in the fishing logbooks for December of 2000–2010.

Results showed that the sum of PB_s and PB_d was larger than the mean biomass before the disaster when $E_d^{'}(t)$ was set at 1.0 and when boundary depth was set at 78.0 m (Figure 10). This finding indicates that Sea Raven biomass will increase and that the waters off Fukushima can serve as an effective MPA for Sea Ravens even if the bottom trawl fishing area and effort are completely restored.

A recent study proposed five key features of a successful MPA: no take, well enforced, old (>10 years), large size (>100 km^2), and isolated from fishing areas by deep water or sand (Edgar et al. 2014). Under its current status, the Fukushima coast has at least two of those key features (well enforced and large), and the coastal area is a no-take zone. Our present results support the area's potential to be a successful MPA. The mean CPUE of exploited species from the trial fishing in 2012 was about three times the mean CPUE per year from 2007 to 2009 (Yamada et al. 2014), indicating that the effective MPA has not only influenced Sea Raven biomass but also the biomass levels of other commercial species.

General Applications of the Model

The model developed here is not limited to biomass estimation between a no-take area and a fishing area; it is also useful for estimating optimal fishing pressure under a stock management strategy by varying the gear-specific effort levels, as we have shown. By applying our model to other exploited species, we could simulate various fishery management strategy scenarios. For example, the model may be applicable in estimating biomass and optimal fishing pressure in cases where the seasonal migration of a stock occurs inside and outside of the exclusive economic zone.

Furthermore, our model will be useful in evaluating the effect of an MPA on catch and biomass before MPA establishment if data on CPUEs at the candidate site and surrounding fishing areas are available. In fact, we used our model to show the expected catch and biomass in a simulated MPA (Figure 10). However, at least a one-dimensional relative spatial distribution of the target species is needed to estimate the

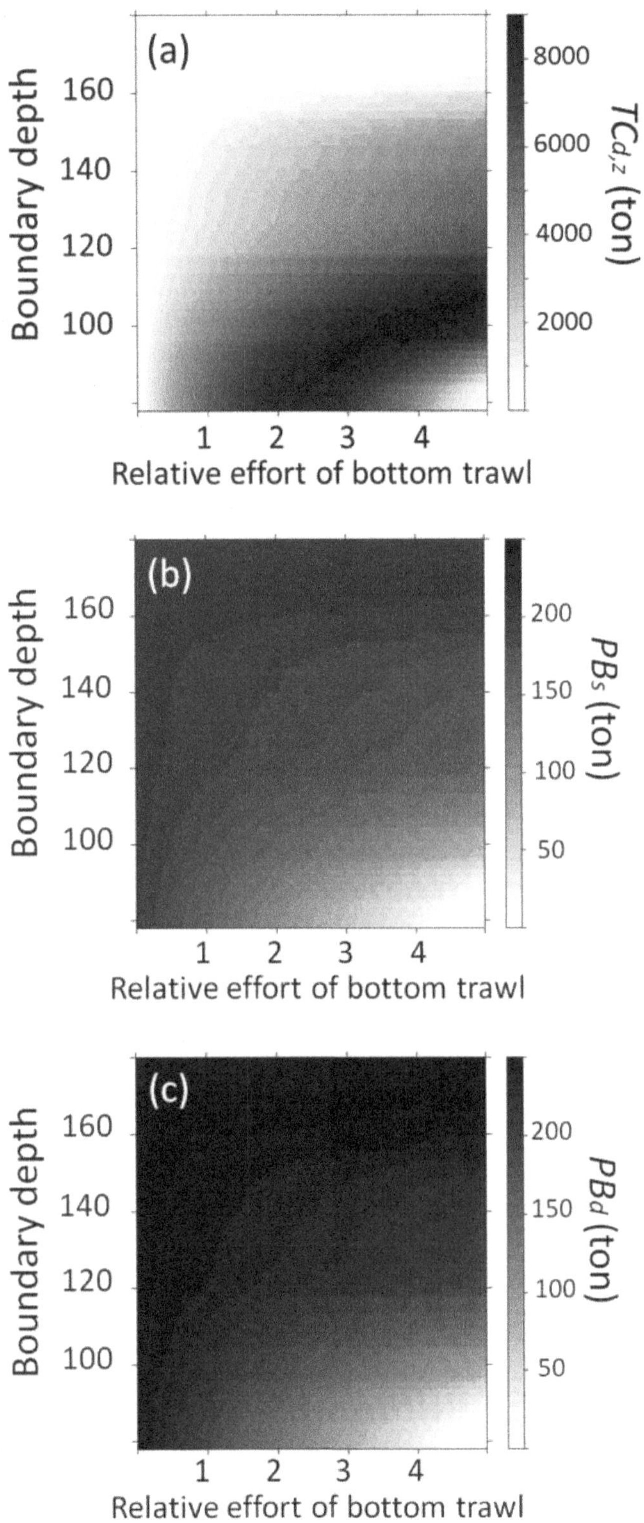

FIGURE 10. (**a**) Predicted total catch (metric tons) of Sea Ravens by bottom trawl fishing for 100 years ($TC_{d,z}$) with different boundary depths; (**b**) predicted Sea Raven biomass (metric tons) 100 years later in the coastal area (PB_s); and (**c**) predicted biomass 100 years later in the offshore area (PB_d). The *x*-axis shows the bottom trawl fishing effort ($E_d^{'}$).

effect of different MPA sizes. Although we used a CF to obtain the relative spatial distribution of Sea Ravens (Figure 9) against a certain depth, depth-specific CPUEs are usually difficult to obtain. A statistical model (e.g., a generalized linear model [GLM]) for estimating spatial distribution (Guisan and Thuiller 2005) would be useful if related covariates are available as explanatory variables.

Required Data and Model Assumptions

The model we developed is applicable to species that move between two areas if spatiotemporal CPUEs are available from both areas. Although we used monthly CPUEs, the temporal and spatial scales can be changed to match a migrating type of target species.

The efficiencies of gill-net fishing and bottom trawl fishing were estimated from the developed model. Some researchers may standardize CPUEs so as to estimate the difference in efficiency between gill-net and bottom trawl fishing with a GLM. Estimated efficiencies from GLMs are usually relative efficiencies among gears (e.g., Campbell 2004); however, our model gave estimates of absolute efficiency when the unit of effort and biomass was given. Moreover, although both types of model can estimate the effect of gear on CPUE, the present model was able to estimate the incremental effect of gill-net fishing effort on the CPUE of bottom trawl fishing and vice versa. In terms of effort control for stock management, the model we developed is useful.

We used separate values of *K* for the coastal area and the offshore area because coastal waters usually display higher productivity than offshore waters. Moreover, the two study areas differed in size. If there are no environmental differences between two areas, setting a single *K*-value in the model and allocating it to the areas based on the ratio of their sizes might be useful for reducing the number of parameters.

The posterior distribution of *K* had longer tails than the distributions of the other parameters (Figure 5). In a previous study, Hilborn and Walters (1992) reported that a CPUE determined under conditions of high biomass and low fishing effort is necessary for the estimation of *K*. However, the mean gill-net fishing effort before the FDNPP disaster was more than four times the optimal effort. This might explain why the posterior distribution of *K* had long tails.

Although conventional tagging data have been widely used to examine population size and mortality, exploitation, growth, and movement rates in fisheries (Kurota et al. 2009), two assumptions must be met: (1) the reporting rate must be known and (2) fish behavior must not be impacted by tagging (Sibert et al. 1999; Pine et al. 2003). Our study suggests that given a sufficient contrast in CPUE between two areas, tagging is not required to estimate biomass, fishing efficiency, growth, movement rate, and *K*. Eliminating the need for tagging data is one advantage of the present model, which is easily implemented for new data sets.

Conclusions

Our model predicted that Sea Raven biomass off Fukushima will increase in the future if fishing effort is kept at trial levels. The results revealed that the waters off Fukushima have effectively acted as an MPA since the FDNPP accident. This model will also be useful for other studies of migrating species—for example, to estimate the optimal fishing pressure for a stock that migrates seasonally within and outside of the exclusive economic zone or to evaluate effects on biomass at a candidate site before its establishment as a new MPA.

ACKNOWLEDGMENTS

We are grateful to all those who contributed to data collection. We thank Hiroyuki Matsuda, Takashi Matsuishi, Miyako Naya, and two anonymous reviewers for useful comments on a draft of the manuscript. This work was supported by a Grant-in-Aid (25082C) from the Agriculture, Forestry, and Fisheries Research Council to T.M.

REFERENCES

Agardy, T. M. 1994. Advances in marine conservation: the role of marine protected areas. Trends in Ecology and Evolution 9:267–270.

Allison, G. W., J. Lubchenco, and M. H. Carr. 1998. Marine reserves are necessary but not sufficient for marine conservation. Ecological Applications 8(Supplement):S79–S92.

Antonenko, D., O. Pushchina, S. Solomatov, and P. Kalchugin. 2010. Distribution and some features of biology of Sea Raven *Hemitripterus villosus* (Hemitripteridae) in the northwestern part of the Sea of Japan. Journal of Ichthyology 50:640–649.

Campbell, R. A. 2004. CPUE Standardisation and the construction of indices of stock abundance in a spatially varying fishery using general linear models. Fisheries Research 70:209–227.

Coggins, L. G. Jr., W. E. Pine III, C. J. Walters, and S. J. Martell. 2006. Age-structured mark–recapture analysis: a virtual-population-analysis-based model for analyzing age-structured capture–recapture data. North American Journal of Fisheries Management 26:201–205.

Edgar, G. J., R. D. Stuart-Smith, T. J. Willis, S. Kininmonth, S. C. Baker, S. Banks, N. S. Barrett, M. A. Becerro, A. T. Bernard, and J. Berkhout. 2014. Global conservation outcomes depend on marine protected areas with five key features. Nature 506:216–220.

Field, J. C., A. E. Punt, R. D. Methot, and C. J. Thomson. 2006. Does MPA mean "major problem for assessments"? Considering the consequences of place-based management systems. Fish and Fisheries 7:284–302.

Fishery Office of the Fukushima Prefectural Government. 2013. Annual report of fishery statistics in Fukushima. Available: http://www.pref.fukushima.lg.jp/sec/36035e/suisanka-toukei-top.html. (July 2015.) (In Japanese.)

Guénette, S., T. Lauck, and C. Clark. 1998. Marine reserves: from Beverton and Holt to the present. Reviews in Fish Biology and Fisheries 8:251–272.

Guisan, A., and W. Thuiller. 2005. Predicting species distribution: offering more than simple habitat models. Ecology Letters 8:993–1009.

Halpern, B. S., and R. R. Warner. 2002. Marine reserves have rapid and lasting effects. Ecology Letters 5:361–366.

Hart, D. R. 2006. When do marine reserves increase fishery yield? Canadian Journal of Fisheries and Aquatic Sciences 63:1445–1449.

Hilborn, R., and C. J. Walters. 1992. Quantitative fisheries stock assessment: choice, dynamics and uncertainty. Springer, New York.

Hirata, T. 2000. Structure of existing offshore trawl in the Soma Haragama region of Fukushima Prefecture. Bulletin of the Fukushima Prefecture Fisheries Experiment Station 9:83–89. (In Japanese.)

Holland, D. S. 2002. Integrating marine protected areas into models for fishery assessment and management. Natural Resource Modeling 15:369–386.

Inoue, Y., and T. Honda. 2002. Haragama design of coastal selective trawl net (HD-STN). Technical Report of the National Research Institute of Fisheries Engineering 24:27–30. (In Japanese.)

Jones, P. J. 2001. Marine protected area strategies: issues, divergences and the search for middle ground. Reviews in Fish Biology and Fisheries 11:197–216.

Kayaba, T., T. Wada, K. Kamiyama, O. Murakami, H. Yoshida, S. Sawaguchi, T. Ichikawa, Y. Fujinami, and S. Fukuda. 2014. Gonadal maturation and spawning migration of stocked female Barfin Flounder *Verasper moseri* off the Pacific coast of northern Japan. Fisheries Science 80:735–748.

Kurota, H., M. K. McAllister, G. L. Lawson, J. I. Nogueira, S. L. Teo, and B. A. Block. 2009. A sequential Bayesian methodology to estimate movement and exploitation rates using electronic and conventional tag data: application to Atlantic Bluefin Tuna (*Thunnus thynnus*). Canadian Journal of Fisheries and Aquatic Sciences 66:321–342.

Lester, S. E., B. S. Halpern, K. Grorud-Colvert, J. Lubchenco, B. I. Ruttenberg, S. D. Gaines, S. Airamé, and R. R. Warner. 2009. Biological effects within no-take marine reserves: a global synthesis. Marine Ecology Progress Series 384:33–46.

Munehara, H. 1996. Sperm transfer during copulation in the marine sculpin *Hemitripterus villosus* (Pisces: Scorpaeniformes) by means of a retractable genital duct and ovarian secretion in females. Copeia 1996:452–454.

Narimatsu, Y. 2006. Reproductive biology of Pacific Cod, a review especially referring to interannual variations of biomass, age and size at maturity and fecundity. Bulletin of the Fisheries Research Agency 4:137–146. (In Japanese.)

Narimatsu, Y., A. Yamanobe, and M. Takahashi. 2007. Reproductive cycle, age, and body size at maturity and fecundity of female Willowy Flounder *Tanakius kitaharai*. Fisheries Science 73:55–62.

Pincin, J. S., and M. J. Wilberg. 2012. Surplus production model accuracy in populations affected by a no-take marine protected area. Marine and Coastal Fisheries: Dynamics, Management, and Ecosystem Science [online serial] 4:511–525.

Pine, W. E., K. H. Pollock, J. E. Hightower, T. J. Kwak, and J. A. Rice. 2003. A review of tagging methods for estimating fish population size and components of mortality. Fisheries 28(10):10–23.

Punt, A. E., and R. D. Methot. 2004. Effects of marine protected areas on the assessment of marine fisheries. Pages 133–154 *in* S. J. Brooke, editors. Aquatic protected areas as fisheries management tools. American Fisheries Society, Symposium 42, Bethesda, Maryland.

R Development Core Team. 2014. R: a language and environment for statistical computing. R Foundation for Statistical Computing, Vienna.

Shimizu, Y., I. Yasuda, and S. Ito. 2001. Distribution and circulation of the coastal Oyashio intrusion. Journal of Physical Oceanography 31:1561–1578.

Sibert, J. R., J. Hampton, D. A. Fournier, and P. J. Bills. 1999. An advection–diffusion–reaction model for the estimation of fish movement parameters from tagging data, with application to Skipjack Tuna (*Katsuwonus pelamis*). Canadian Journal of Fisheries and Aquatic Sciences 56:925–938.

Spiegelhalter, D., A. Thomas, N. Best, and D. Lunn. 2003. WinBUGS users' manual. Medical Research Council, Biostatistics Unit, Cambridge, UK.

Sturtz, S., U. Ligges, and A. E. Gelman. 2005. R2WinBUGS: a package for running WinBUGS from R. Journal of Statistical Software [online serial] 12(3).

Wada, T., K. Kamiyama, S. Shimamura, T. Mizuno, and Y. Nemoto. 2012. Effectiveness of stock enhancement of a rare species, Spotted Halibut *Verasper variegatus*, in Fukushima, Japan. Aquaculture 364/365:230–239.

Wada, T., Y. Nemoto, S. Shimamura, T. Fujita, T. Mizuno, T. Sohtome, K. Kamiyama, T. Morita, and S. Igarashi. 2013. Effects of the nuclear disaster on marine products in Fukushima. Journal of Environmental Radioactivity 124:246–254.

Yamada, M., T. Sohtome, and T. Iwasaki. 2014. Comparisons of fishing efforts and fishery resource abundance off Fukushima before and after the Great East Japan Earthquake. Tohoku Demersal Fish Research 34:63–66. (In Japanese.)

Appendix: Additional Data

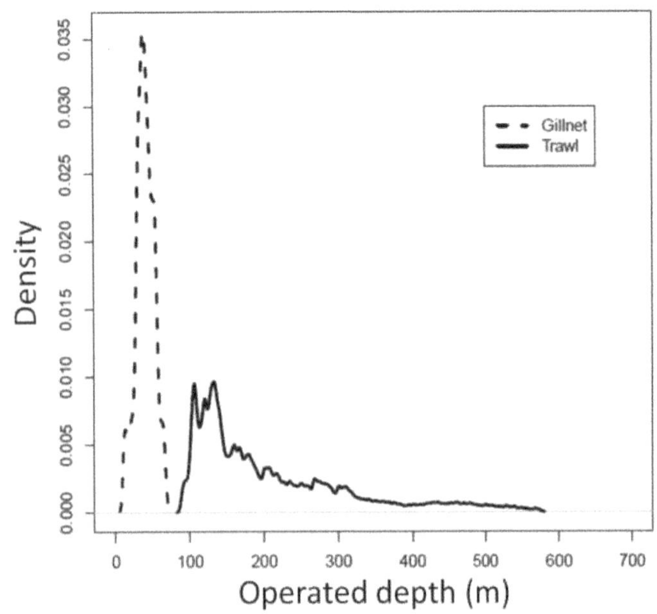

FIGURE A.1. Density distributions of gill-net operation depth (dotted line) and bottom trawl operation depth (solid line) in the waters off Fukushima, Japan, from 2000 to 2010 based on information from daily fishing logbooks (with outliers and typographical errors removed).

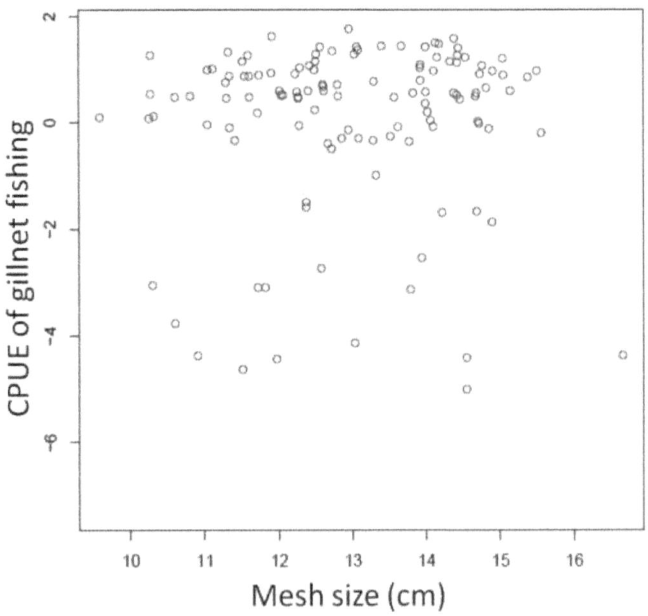

FIGURE A.2. Relationship between the monthly gill-net fishing CPUE values included in our analysis and the mean gill-net mesh size used during each month from 2000 to 2010 (correlation coefficient $r = 0.043$; $P = 0.636$).

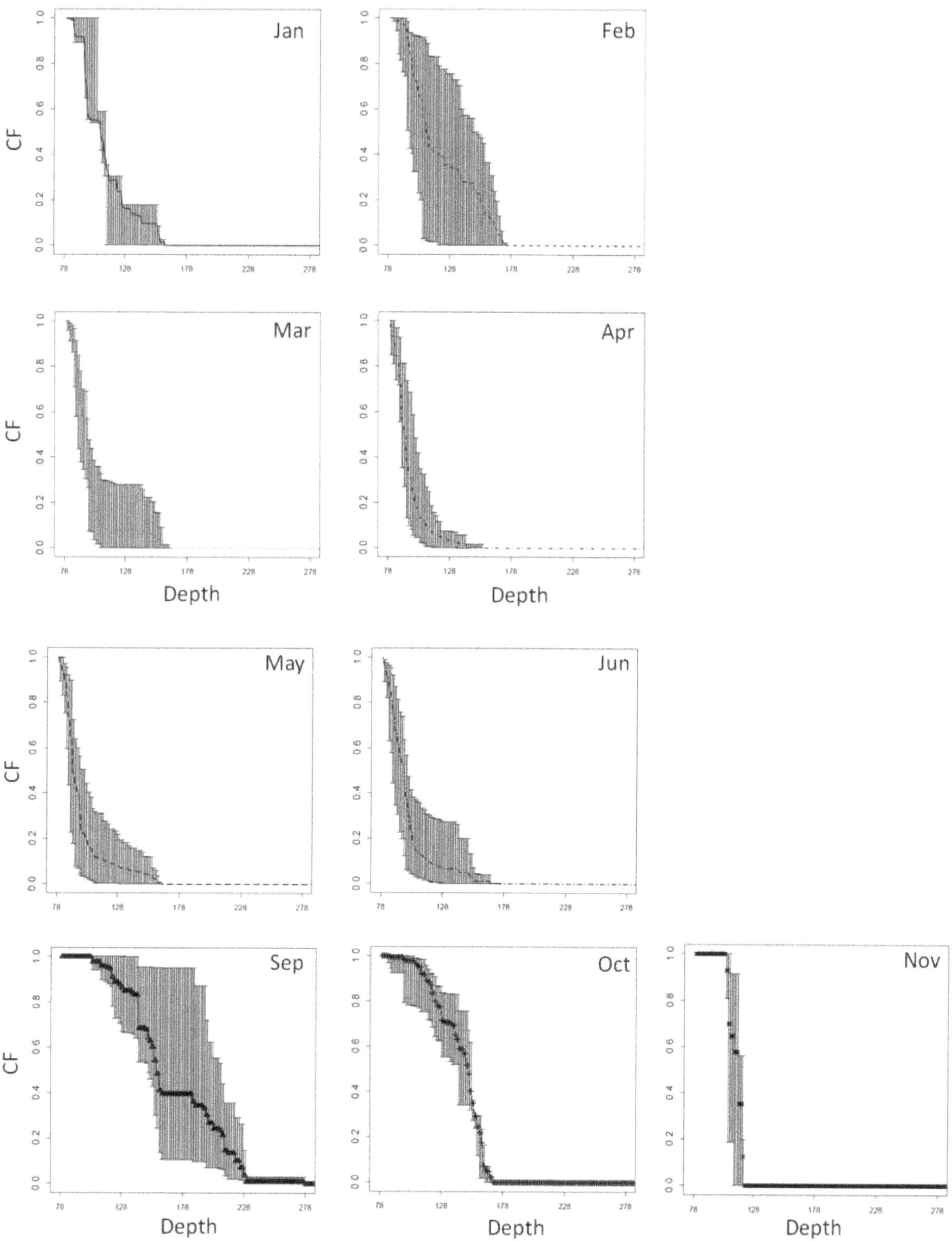

FIGURE A.3. Bootstrapped averages (with 95% confidence interval) of correction factors ($CF_{m,z}$) for each month. Note that there were no $CF_{m,z}$ values for July or August because bottom trawl fishing was prohibited during those months. There was no $CF_{m,z}$ in December because Sea Ravens were not recorded in the fishing logbooks for December of 2000–2010.

2

Changes in Red Snapper Diet and Trophic Ecology Following the Deepwater Horizon Oil Spill

Joseph H. Tarnecki* and William F. Patterson III

Dauphin Island Sea Lab, Department of Marine Sciences, University of South Alabama, 101 Bienville Boulevard, Dauphin Island, Alabama 36528, USA

Abstract

Red Snapper *Lutjanus campechanus* were sampled at 33 natural and 27 artificial reef sites in the northern Gulf of Mexico prior to (2009–2010) and after (2010–2011) to examine potential diet and trophic shifts following the Deepwater Horizon (DWH) oil spill. We dissected 708 stomachs for gut content analysis and processed 65 muscle tissue samples for stable isotope ratio-mass spectrometry analysis of $\delta^{13}C$, $\delta^{15}N$, and $\delta^{34}S$. Forty-eight percent of stomachs contained identifiable prey, which we grouped into seven categories: fish, decapods, cephalopods, stomatopods, gastropods, zooplankton, and other invertebrates. Based on these categories, Red Snapper diet was significantly different following the DWH oil spill, and was differentially affected by fish size. The interaction between habitat (natural versus artificial reefs) and DWH oil spill effects was also significant. Significant differences in diet among Red Snapper size-classes were due to low trophic position prey, such as pelagic zooplankton, being more abundant in the diet of larger (>500 mm) Red Snapper, while decapods and fish constituted a higher proportion of the diet of smaller individuals. Red Snapper consumed higher amounts of decapods at artificial (21.9% by mass) versus natural (14.8%) reef sites, but the habitat effect on diet was not significant. The habitat × DWH timing interaction was driven by a decrease in zooplankton consumed at both habitat types, increased benthic prey at natural reefs, and increased fish consumption at artificial reefs in post-DWH oil spill samples. Stable isotope data indicated a postspill increase in Red Snapper trophic position (^{15}N enrichment) and an increase in benthic versus pelagic prey (^{34}S depletion), both consistent with observed dietary shifts. Overall, results indicate shifts in Red Snapper diet and trophic position occurred following the DWH oil spill, thus the relative abundance of prey resources likely changed.

The Deepwater Horizon (DWH) oil spill was a catastrophic event for the Gulf of Mexico ecosystem resulting in a total volume discharged of approximately 780,000 m^3 between April 22 and July 15, 2010 (McNutt et al. 2012). Persistence of oil compounds in the water column (Allan et al. 2012; Reddy et al. 2012; Sammarco et al. 2013), impacts to plankton communities (Ortmann et al. 2012; Almeda et al. 2013), and flocculent particulate organic matter that transferred oil compounds to the benthos (Passow et al. 2012; Mason et al. 2014) have been well documented. Direct oiling and toxicological effects of the DWH oil spill on higher taxa, including fishes, sea turtles, birds, and marine mammals, also have been reported (Whitehead et al. 2012; Barron 2012; Schwacke et al. 2013; Murawski et al. 2014). However, little information exists to infer food web effects at higher trophic levels.

We examined changes in the diet and trophic position of Red Snapper *Lutjanus campechanus* following the DWH oil spill, which was only possible due to baseline data from a

Subject editor: Debra J. Murie, University of Florida, Gainesville

*Corresponding author: jtarnecki83@gmail.com

study that actually began 18 months prior to the spill but with very different objectives than to examine oil spill impacts. The original study was focused on examining differences in reef fish ecology on natural versus artificial reefs in the northern Gulf of Mexico. Understanding the relationship between artificial reefs and associated fish communities has important implications for whether artificial reefs are more likely to enhance reef fish production or merely aggregate fish, thus making them more vulnerable to fishing mortality (Bohnsack 1989; Lindberg 1997; Perkol-Finkel et al. 2006). Few fisheries issues in the northern Gulf are as controversial as the ecological versus fishery function of artificial reefs (Cowan et al. 2011), and this debate has mostly centered on Red Snapper.

Red Snapper was an ideal model species given their ecological and economic importance in the region, as well as their abundance on both artificial and natural reefs (Dance et al. 2011; Patterson et al. 2014). Previous studies have been conducted on Red Snapper feeding ecology, but nearly all of the research on this topic has been focused on their ecology on artificial reefs with little to no comparison to fish occupying natural reefs (e.g., Szedlmayer and Shipp 1994; McCawley and Cowan 2007; Dance et al. 2011). Others have clearly demonstrated that comparisons between artificial and natural habitats are critical for assessing the effects of artificial reefs on reef fish ecology (e.g., Carr and Hixon 1997; Perkol-Finkel et al. 2006). Therefore, the original focus of this study was to examine the feeding ecology of Red Snapper at artificial versus natural reefs in the northern Gulf. Analysis of potential habitat effects also required explicitly accounting for the effect of ontogeny given previous reports of ontogenetic shifts in Red Snapper feeding ecology (Ouzts and Szedlmayer 2003; McCawley and Cowan 2007; Wells et al. 2008).

In spring 2010, the occurrence of the DWH oil spill became an unplanned factor in this study. Food web impacts have been reported following previous large-scale spills (Teal and Howarth 1984; Jackson et al. 1989; Moreno et al. 2013), although chronic ecological impacts may not be apparent for many years following such events (Peterson et al. 2003). Environmental effects of the DWH oil spill have been reported across a variety of habitats and taxa, but effects of the spill on fishes have mostly been restricted to inshore or coastal species (e.g., Fodrie and Heck 2011; Whitehead et al. 2012; Pilcher et al. 2014). Our study occurred over a broad area (about 8,000 km^2) of the northcentral Gulf of Mexico continental shelf. The Red Snapper samples collected there prior to the DWH oil spill provided a unique opportunity to examine the effects of the spill on their diet and trophic ecology. This is significant, not only due to the ecological and economic importance of Red Snapper, but also because relatively few baseline data sets exist to examine the impacts of the DWH oil spill on upper trophic level taxa in the northern Gulf.

METHODS

Red Snapper were sampled between June of 2009 and August of 2011 from 33 natural and 27 artificial reefs south of Dauphin Island, Alabama, to southeast of Destin, Florida (Figure 1). Sampling was conducted onboard chartered fishing boats (home ports of Orange Beach, Alabama, and Pensacola and Destin, Florida) and also onboard the state of Florida's RV *Bellows*. Hook-and-line fishing was performed with two-hook rigs composed of 9/0, 12/0, or 15/0 Mustad circle hooks on 60-lb fluorocarbon leaders. Bait was kept uniform and consisted of cut squid, *Loligo* spp. or Mackerel Scad *Decapterus macarellus*, which allowed for easy identification and segregation from food during diet analysis.

Upon capture, Red Snapper fork length (FL) and total length (TL) were measured (mm) and sex was determined by examination of gonads. A sample ($\geq$50 g) of white muscle tissue was removed above the pectoral fin of each fish and placed in a sealed plastic bag on ice for transport to the laboratory, then stored in an ultralow freezer at $-80°C$ until processed. Fish stomachs were also removed at sea and immediately fixed in 10% buffered formalin. A two-factor ANOVA was

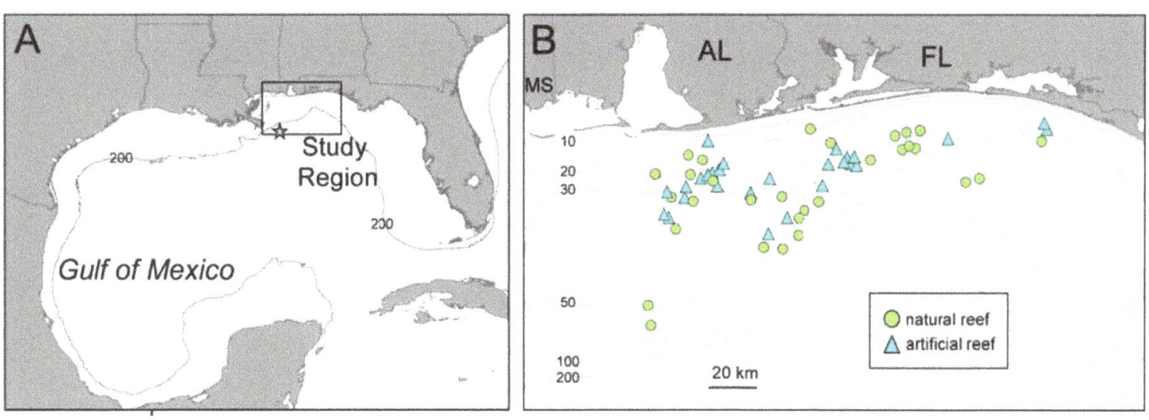

FIGURE 1. (**A**) Location of the study area in the northern Gulf of Mexico, showing the location of the Macondo wellhead (star), and (**B**) the natural (green circles [$n = 33$]) and artificial (blue triangles [$n = 27$]) reef sites where Red Snapper were sampled before and after the Deepwater Horizon oil spill; MS = Mississippi, AL = Alabama, FL = Florida. Isobath depths in m.

computed to test for significant differences in Red Snapper total length between habitat types and pre-DWH versus post-DWH. A priori, the experiment-wise error rate (α) was set to 0.05 for all statistical analyses.

Diet analysis.—Red Snapper stomachs were fixed in 10% formalin for at least 48 h and then were transferred to 70% isopropyl alcohol for preservation. Diet composition analysis was performed by identifying stomach contents to the lowest taxonomic level possible with the aid of a dissecting microscope for all stomach samples that had prey items present. Prey items were sorted by taxon and then dried for 48 h at 60°C to obtain dry mass.

Prey were sorted into one of seven categories for statistical analysis of diet: fish, decapods, cephalopods, stomatopods, gastropods, zooplankton, and other invertebrates. Mean percent dry mass was computed as the arithmetic mean among samples within a given factor, the factors being size-class (<400, 400–500, >500 mm), habitat type (natural versus artificial reefs), and DWH oil spill timing (before versus after April 22, 2010). A three-factor permutational multivariate ANOVA (PERMANOVA) was computed with the Primer statistical package (version 6; Anderson et al. 2008) to test for differences in Red Snapper diet by percent dry mass among size-classes, between habitat types, and pre versus post spill. Diet data were square-root-transformed, and a dummy variable with value = 1 was added to each sample to stabilize dispersion in the data prior to computing the Bray–Curtis similarity measure between each pair of samples (Clarke 2006). The PERMANOVA model then was computed in Primer with 10,000 permutations. Type-III sums of squares were utilized to determine significance.

Stable isotope analysis.—Stable isotope analysis was performed on a randomly selected subset of Red Snapper white muscle samples. Muscle tissue was dried at 60°C for at least 24 h, ground with a mechanical tissue grinder, and then pulverized into a fine powder with a glass or agate mortar and pestle. Between 0.2 and 0.5 g of pulverized dried tissue were placed in sterile plastic centrifuge tubes and shipped to a contract laboratory to perform analysis of $\delta^{15}N$, $\delta^{13}C$, and $\delta^{34}S$ with a Europa Scientific GSL/Geo 20-20 stable isotope ratio-mass spectrometer. The isotopic ratio of a sample relative to the ratio in an international standard is reported in the standard delta notation: $\delta X = [(R_{sample}/R_{standard}) - 1] \times 1{,}000$, where $X = {}^{15}N$, ${}^{13}C$, or ${}^{34}S$ and $R = {}^{15}N : {}^{14}N$, ${}^{13}C : {}^{12}C$, or ${}^{34}S : {}^{32}S$. Standards included nitrogen in air ($\delta^{15}N_{Air}$), Pee Dee Belemnite ($\delta^{13}C_{V\text{-}PBD}$), and Canyon Diablo Troilite ($\delta^{34}S_{V\text{-}CDT}$). International Atomic Energy Agency standard reference materials were run periodically as check samples to assess machine performance for each stable isotope ratio analyzed. Analytical precision was estimated from duplicate analysis of 20 randomly selected samples.

Trophic level was estimated for each Red Snapper sample following Post (2002): trophic level $= 1 + (\delta^{15}N_{fish} - \delta^{15}N_{prod})/\Delta_n$, where $\delta^{15}N_{fish}$ is the $\delta^{15}N$ of a given Red Snapper muscle sample, $\delta^{15}N_{prod}$ is the mean $\delta^{15}N_{prod}$ of primary producers in the system, and Δ_n is trophic fractionation per trophic level. Estimating mean $\delta^{15}N_{prod}$ of primary producers can be problematic if there is a diversity of primary producers present in the system (e.g., phytoplankton, benthic algae, seagrasses, marsh grasses) or if there is considerable temporal variability in $\delta^{15}N$. While seagrass and marsh habitats were present in the study region, they were not present on the shelf or within 30 km of the study reefs. Therefore, the predominant primary producers in the system are phytoplankton and benthic microalgae. There was no significant difference in mean $\delta^{15}N$ between phytoplankton and benthic microalgae samples ($n = 33$) collected in the northern Gulf shelf prior to the DWH oil spill (two-sample *t*-test, $P = 0.305$; W. Patterson, unreported data), which is similar to the pattern reported for $\delta^{15}N$ by Radabaugh et al. (2013) for the west Florida shelf, despite considerable variability in $\delta^{13}C$ between phytoplankton and benthic algae. Given the lack of difference in $\delta^{15}N$ among pre-DWH oil spill phytoplankton and benthic microalgae samples, the overall mean $\delta^{15}N$ of 5.82 (SE = 0.13) among combined sample types served as our estimate of $\delta^{15}N_{prod}$. Trophic fractionation (Δ_n) was assumed to equal 3.0‰, which is the mean measured for fishes feeding on high-protein diets (McCutchan et al. 2003; Vanderklift and Ponsard 2003; Rooker et al. 2006).

Values of $\delta^{13}C$ were corrected for percent lipid with the regression equation reported by Post et al. (2007) for aquatic animals: $CF = -3.32 + (0.99 \times C{:}N)$, where CF is the correction factor applied to $\delta^{13}C$ to account for percent lipid and C:N is a proxy for percent lipid. Hereafter, $\delta^{13}C$ always indicates lipid-corrected $\delta^{13}C$. To examine basal sources of C and S, the $\delta^{13}C$ and $\delta^{34}S$ values at the base of the food web were estimated following Post (2002):

$$\delta X_{base} = \delta X_{fish} - [(\text{trophic level} - 1) \times \Delta_x],$$

where δX_{base} = estimated $\delta^{13}C$ or $\delta^{34}S$ of the C or S source, respectively, at the base of the food web, $\delta X_{fish} = \delta^{13}C$ or $\delta^{34}S$ of a given Red Snapper sample, trophic level = estimated trophic level of a given Red Snapper sample derived from $\delta^{15}N$, and Δ_X = trophic fractionation (increase) in $\delta^{13}C$ or $\delta^{34}S$ per trophic level. A value of 1.5‰ was assumed for Δ_C, which is the mean reported by Sweeting et al. (2007) for fish muscle tissue. A value of 0.5‰ was assumed for Δ_S, which is the mean reported by McCutchan et al. (2003).

Linear regressions were computed between Red Snapper $\delta^{15}N$, $\delta^{13}C$, and $\delta^{34}S$ and total length to test for ontogenetic effects on those variables. Two-factor ANOVA models were computed to test the effects of habitat type and DWH oil spill timing on total length, estimated trophic level, $\delta^{13}C_{base}$, and $\delta^{34}S_{base}$.

RESULTS

Study reefs ranged in depth from 17 to 92 m. Red Snapper ranged from 238 to 890 mm TL, the means not differing (ANOVA) between habitat types ($F_{1,\,706} = 0.30$; $P = 0.586$) or between pre and post samples ($F_{1,\,706} = 2.45$; $P = 0.118$).

Stomach Content Analysis

A total of 708 Red Snapper stomachs were sampled. Of those, 20.5% ($n = 145$) were empty, 31.7% ($n = 220$) had only chyme present (i.e., no prey items could be identified), and 48.3% ($n = 343$) had identifiable prey items present. However, among the stomach samples with identifiable prey present, approximately 50% by dry mass (mean percent dry mass = 45–52% among factor levels) of their stomach contents had already been reduced to chyme, thus were unidentifiable. Among the identifiable prey items, 85 unique prey taxa were identified and then sorted into the seven prey categories (Table 1). Statistical analysis of Red Snapper diet was based on percent dry mass of these identifiable prey. Results of the PERMANOVA computed for Red Snapper diet were that size-class and DWH oil spill timing significantly affected diet, but habitat type did not (Table 2; Figure 2). However, the interaction between DWH oil spill timing and habitat type was significant.

Three of the seven prey categories constituted >80% of Red Snapper diet by dry mass. Overall, fish prey constituted nearly half (43.2%; Figure 2) of Red Snapper diet, but most (>90%) fish prey could not be identified even to family. Fish remains that could not be identified to order or below were classified as unknown fish. The smallest (<400 mm) Red Snapper size-class contained the highest percentage of fish prey (58.7%), while fish constituted 26.5% of Red Snapper diet in the 400–500 mm size-class, and 28.0% of the diet of Red Snapper >500 mm. Identified fish prey of Red Snapper sampled on natural reef sites consisted principally of nonreef benthic taxa, such as lizardfishes (Family Synodontidae), searobins, Red Porgy, and Southern Hake, while stomach samples of fish from artificial reefs contained a combination of nonreef benthic and pelagic fishes, such as harvestfish *Peprilus* sp. and herrings (Clupeidae). Reef fish taxa identified in Red Snapper stomach samples were rare and consisted of grunts (Haemulidae) and triggerfish (Balistidae).

Overall, zooplankton accounted for 22.6% of Red Snapper prey by dry mass and included 22 unique taxa. Zooplankton made up 14.1% of total prey mass within the <400 mm size-class, 25.3% for fish 400–500 mm, and 27.4% for individuals >500 mm. Although there was no significant difference in the amount of zooplankton consumed between natural reefs (22.6%) and artificial reefs (22.7%), there was a difference in zooplankton taxa consumed. On natural reefs, small pelagic gastropods within order Thecosomata constituted 52.8% of zooplankton prey, but they made a much smaller (5.5%) contribution to Red Snapper diet on artificial reefs. The greatest contributor to zooplankton prey on artificial reefs was the pelagic amphipod *Phrosina semiluna,* which constituted 84% of zooplankton prey on artificial reefs but only 11% on natural reefs.

Decapods contributed 17.6% of total prey mass, with 22 taxa identified among the diet samples. Decapod prey items tended to be more numerous in the stomachs of smaller Red Snapper but constituted a larger percentage of the diet of larger fish due to the consumption of larger decapods, such as box crabs and Florida lady crabs (Table 1). No difference in decapod diversity was observed in Red Snapper diets between habitat types, but Red Snapper did consume slightly higher amounts of decapods at artificial (21.9%) versus natural (14.8%) reefs.

Overall, cephalopods, stomatopods, gastropods, and other invertebrates combined made up less than 20% of Red Snapper diet by dry mass. The cephalopod category was composed of squid *Loligo* sp., and octopuses (Order Octopoda). Stomatopods (adult mantis shrimp) made up 4.7% of Red Snapper diet, stomatopod consumption on natural reefs being approximately 133% of that on artificial reefs. Benthic gastropods totaled <1% of total diet mass, and other invertebrates consisted of worms, echinoderms, isopods, sponges, barnacles, tunicates, and prey that could not be classified below Order Malacostraca. Other invertebrates constituted a fairly constant but low percentage (approximately 9%) of Red Snapper diet among all size- classes but were more predominant in stomach samples from natural reefs (Table 1; Figure 2).

The significant interaction between DWH oil spill and habitat type effects on Red Snapper diet precludes interpretation of the significant DWH oil spill main effect independently. Fish contribution to diet at artificial reefs more than doubled post-DWH; however, there was no change at natural reefs, and decapod consumption increased by 58.3% at natural reefs but declined by 19.8% at artificial reefs (Table 1; Figure 2). Red Snapper captured at both habitat types had lower zooplankton contribution to their diets after the spill, zooplankton contribution to Red Snapper diets decreasing 55.7% on natural and 64.7% on artificial reefs following the spill. Pteropods, mysid shrimp, and the hyperiid amphipod *Phorina semiluna* were the predominant zooplankton prey at natural reefs prior to the DWH oil spill but were entirely absent from post-DWH diet samples. Mysid shrimp and *Phorina semiluna* were the most abundant zooplankters in pre-DWH Red Snapper diet samples at artificial versus natural reefs, but were also entirely absent from Red Snapper diets at artificial reef sites after the DWH oil spill. Among the minor prey categories, the most substantial change following the DWH oil spill was the appearance of stomatopods in natural reef samples where they had been absent prior to the spill.

Stable Isotope Analysis

Red Snapper white muscle samples included tissue from 32 individuals sampled between June and November 2009 (pre-DWH), as well as 34 fish sampled between March and August

TABLE 1. Mean percent diet by dry mass computed from identifiable prey items within stomachs of 343 Red Snapper sampled at natural and artificial reefs in the northern Gulf of Mexico.

		Natural reefs		Artificial reefs	
Group	Taxon	Pre-DWH	Post-DWH	Pre-DWH	Post-DWH
Fish	Anguilliformes	0.00	0.49	0.77	0.00
	Balistidae	0.22	0.00	0.00	0.00
	Searobins *Bellator* spp.	0.00	0.66	0.00	0.00
	Clupeidae	0.00	0.00	0.00	1.15
	Atlantic Croaker *Micropogonias undulatus*	0.00	0.66	0.00	0.00
	Gulf Butterfish *Peprilus burti*	0.00	0.00	1.58	0.00
	Harvestfish *Peprilus* sp.	0.00	2.15	0.00	0.00
	Pleuronectiformes	0.00	0.00	0.00	1.15
	Red Porgy *Stenotomus caprinus*	0.00	1.73	0.00	0.00
	Synodontidae	0.00	0.95	0.00	0.18
	Unknown Fish	43.24	36.37	22.31	51.02
	Southern Hake *Urophycis floridana*	0.00	0.66	0.00	0.00
	Total Fish	43.46	43.65	24.65	52.50
Decapods	*Albunea* sp.	0.00	0.23	0.00	0.82
	Brachyura	0.00	3.95	0.26	0.97
	Yellow box crab *Calappa sulcata*	0.00	0.00	2.08	0.00
	Calappidae	0.00	0.66	0.00	0.00
	Caridea	0.00	0.66	0.00	0.00
	Decapoda	7.02	2.86	6.02	9.51
	Hippoidea	0.00	0.00	0.00	0.77
	White elbow crab *Leiolambrus nitidus*	0.00	0.00	0.00	1.15
	Fivespine purse crab *Myropsis quinguespinosa*	0.07	0.00	0.00	0.00
	Ogyrides sp.	0.01	0.00	0.00	0.00
	Florida lady crab *Ovalipes floridanus*	0.00	0.66	0.00	1.56
	Paguroidea	0.00	0.00	0.00	0.03
	Bladetooth elbow crab *Parthenope granulata*	1.27	0.00	2.08	0.00
	Penaeidae	1.01	3.59	2.74	1.15
	Brown shrimp *Penaeus aztecus*	0.00	0.04	0.00	0.00
	Portunidae	0.25	0.97	4.40	0.01
	Redhair swimming crab *Portunus gibbesii*	0.00	0.00	3.90	0.00
	Sargassum swimming crab *Portunus sayi*	0.00	0.29	0.00	0.00
	Longspine swimming crab *Portunus spinicarpus*	0.00	1.32	0.00	3.01
	Blotched swimming crab *Portunus spinimanus*	0.78	0.00	0.00	0.00
	Gulf frog crab *Raninoides louisianensis*	0.00	0.00	2.02	0.00
	Brown rock crab *Sicyonia brevirostris*	0.00	1.25	1.58	1.15
	Total Decapods	10.41	16.48	25.10	20.14
Cephaolopods	Cephalopoda	0.04	0.33	0.15	0.00
	Loligo sp.	0.00	2.23	0.21	1.73
	Octopoda	0.00	0.68	0.00	0.00
	Total Cephalopods	0.04	3.23	0.36	1.73
Stomatopods	Manits shrimp *Squilla* sp.	0.00	1.99	0.00	0.36
	Squillidae	0.00	0.00	0.10	1.15
	Stomatopoda	0.00	5.17	5.07	1.65
	Total Stomatopods	0.00	7.16	5.18	3.16
Gastropods	Gastropoda	0.10	0.57	0.00	1.21
	Total Gastropods	0.10	0.57	0.00	1.21

(*Continued on next page*)

TABLE 1. Continued.

Group	Taxon	Natural reefs		Artificial reefs	
		Pre-DWH	Post-DWH	Pre-DWH	Post-DWH
Other Inverts	*Aegathoa oculata*	0.00	0.02	0.00	0.00
	Ascidiacea	0.16	0.00	0.00	0.00
	Asellota	0.00	0.00	0.06	0.00
	Cirripedia	0.02	0.00	0.00	0.01
	Crustacea	1.62	0.00	3.77	0.00
	Echinodermata	0.00	0.49	0.00	0.09
	Invertebrata	4.31	0.69	1.92	0.88
	Isopoda	0.00	0.11	0.00	0.01
	Malacostraca	0.00	10.30	0.00	2.30
	Polychaeta	0.00	0.27	0.00	0.00
	Porifera	0.00	0.00	0.00	3.11
	Sessilia	0.02	0.00	0.00	0.00
	Sphaeroma quadridentata	1.68	0.00	0.07	0.00
	Tunicata	0.06	0.00	0.00	0.26
	Worm Tubes	0.00	0.00	0.00	0.89
	Total Other Inverts	7.85	11.89	5.82	7.55
Zooplankton	Amphipoda	0.02	0.68	1.05	0.70
	Aoridae	0.00	0.00	2.08	0.00
	Asteroidea larvae	0.00	0.00	0.06	0.00
	Brachyura megalopa	0.00	2.42	0.00	0.33
	Branchiopoda	19.39	0.00	0.21	0.00
	Carinarioidea	0.00	5.17	0.00	4.59
	Cladocera	1.68	2.42	2.08	0.22
	Decapod megalopa	0.00	0.02	0.00	0.33
	Decapod zoea	0.00	1.45	0.00	0.23
	Haemulidae larvae	0.01	0.00	0.00	0.00
	Hyperiidae	0.66	3.53	2.08	3.46
	Lophogastrida	0.00	0.01	0.00	0.00
	Mysidacea	0.00	0.00	4.12	0.00
	Mysidae	2.49	0.00	2.00	0.00
	Mysidopsis sp.	0.00	0.00	1.06	0.00
	Phronima sp.	0.00	0.00	2.32	0.00
	Phrosina semiluna	1.65	0.00	11.10	0.00
	Phyllosoma	0.00	0.13	0.00	0.03
	Scyllaridae larvae	0.00	0.00	2.08	0.00
	Sergestidae	0.00	0.81	0.00	0.00
	Squilla larvae	0.00	0.40	0.00	0.00
	Thecosomata	12.24	0.01	8.41	3.82
	Unknown Zooplankton	0.00	0.00	0.23	0.00
	Total Zooplankton	38.13	17.03	38.88	13.71

2011 (post-DWH). There was a significant linear relationship between $\delta^{15}N$ and total length (GLM procedure [SAS, Cary, North Carolina], $P = 0.035$), but regressions for $\delta^{13}C$ (GLM, $P = 0.981$) and $\delta^{34}S$ (GLM, $P = 0.130$) were not significant (Figure 3). Total length of fish sampled for stable isotope analysis was significantly different between pre-DWH (mean = 452 mm, SE = 17.8) and post-DWH (mean = 529 mm, SE = 29.7) samples (Table 3). Therefore, $\delta^{15}N$ data were length-detrended by subtracting the slope of the total length versus $\delta^{15}N$ regression times a given sample's total length from the sample's $\delta^{15}N$ value, and then trophic level was estimated as indicated above. The DWH oil spill timing effect

TABLE 2. Permutational multivariate ANOVA model testing the effect of fish size-class (< 400, 400–500, and >500 mm TL), habitat type (natural versus artificial reefs), and the timing of the Deepwater Horizon (DWH) oil spill (pre versus post) on Red Snapper diet.

Source	df	Type-III sums of squares	Mean square	Pseudo-*F*	*P*-value
Size	2	15,567	7,783	2.39	0.021
Habitat	1	5,503	5,503	1.69	0.150
DWH	1	21,225	21,225	6.52	0.001
Size × DWH	2	13,989	6,995	2.15	0.053
Size × habitat	2	9,454	4,727	1.45	0.193
DWH × habitat	1	8,903	8,903	2.73	0.049
Size × habitat × DWH	2	11,177	5,589	1.72	0.108
Residual	331	1,080,000	3,256		
Total	342	1,170,000			

was significant for Red Snapper trophic position, but there was no difference between habitat types; the interaction between habitat and DWH oil spill timing also was not significant (Table 3; Figure 4).

Variables $\delta^{13}C_{base}$ and $\delta^{34}S_{base}$ required a $\log_e$ transformation to meet the parametric assumption of normality (Table 3). The decrease in $\delta^{13}C_{base}$ following the DWH oil spill was statistically significant, and the DWH oil spill timing effect was

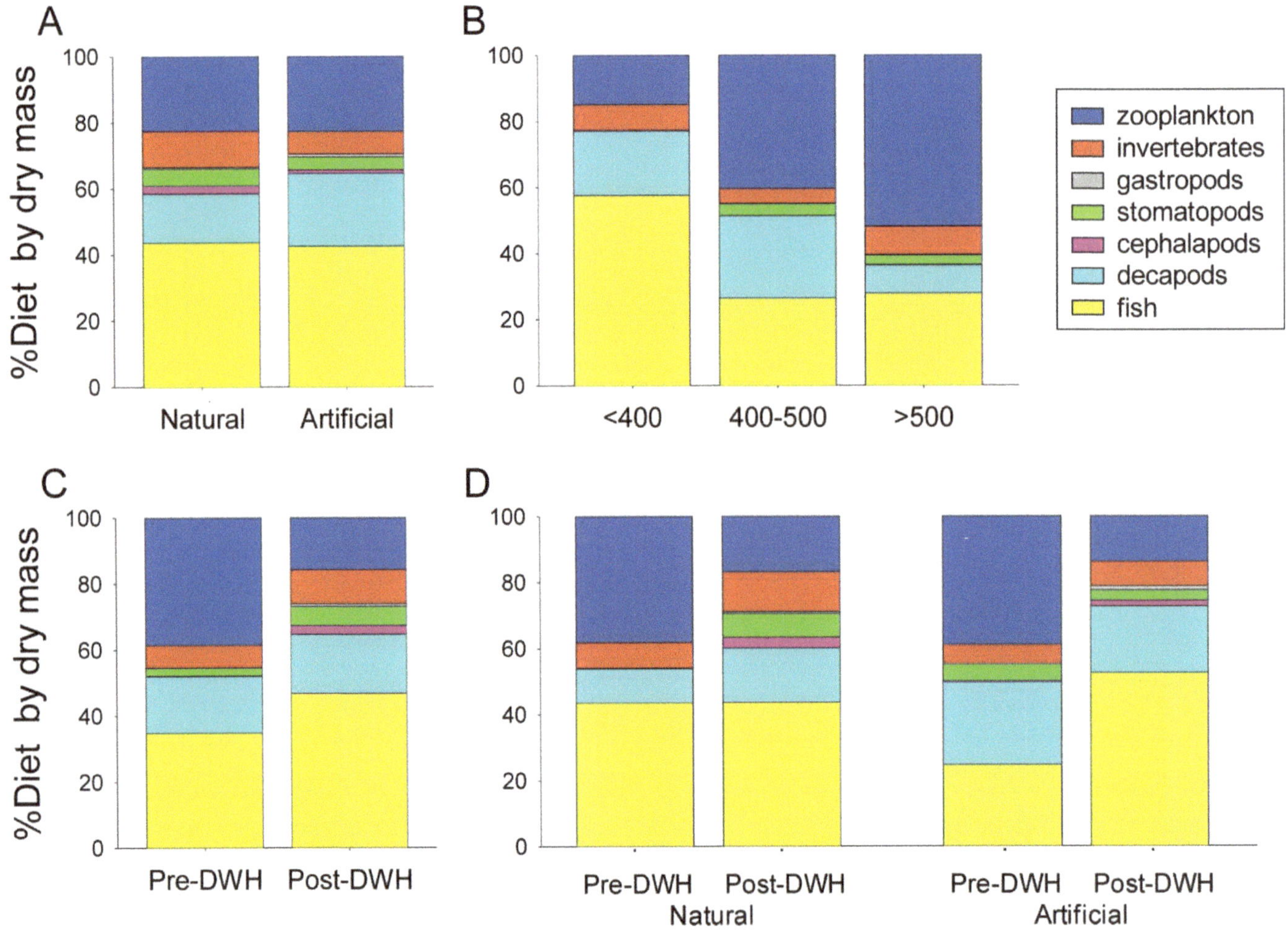

FIGURE 2. Mean percent Red Snapper diet by dry mass estimated among seven prey categories observed among stomach samples. Panels demonstrate the effect of **(A)** habitat (natural versus artificial reefs), **(B)** size-class (total length in mm), **(C)** the timing of the Deepwater Horizon (DWH) oil spill, and **(D)** the interaction between habitat type and DWH timing on Red Snapper diet among samples collected in the northern Gulf of Mexico from June 2009 through August 2011.

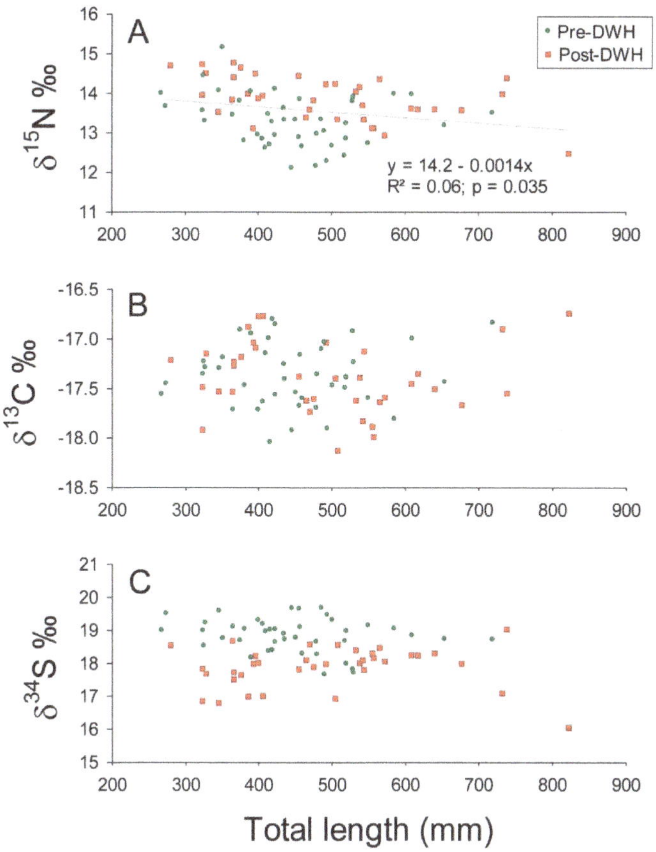

FIGURE 3. Scatterplots of white muscle **(A)** δ^{15}N, **(B)** lipid-corrected δ^{13}C, and **(C)** δ^{34}S versus total length for pre-Deepwater Horizon (DWH) and post-DWH oil spill Red Snapper samples. Regression fit to δ^{15}N data in panel A was significant but weak, and regressions for lipid-corrected δ^{13}C and δ^{34}S were nonsignificant.

also significant $\delta^{34}S_{base}$, both decreasing the year after spill (Table 3; Figure 4). Consistent with the lack of significant diet differences between natural and artificial reefs, none of the stable isotope variables displayed a significant habitat effect (Table 3).

DISCUSSION

Overall, data from this study demonstrate adult Red Snapper to be generalist mesopredators with broad diets. They are opportunistic foragers that consume prey along a size spectrum ranging from zooplankton to fish. However, nothing in their morphology would suggest they are able to strain or even pick zooplankton from the water column. Instead, they likely forage on swarms of plankton, such as pteropods or hyperiid amphipods (Fenwick 1978; Sanvicente-Añorve et al. 2013), when available. The phenomenon of Red Snapper subsidizing their diet with zooplankton was described previously by Ouzts and Szedlmayer (2003) and McCawley and Cowan (2007) for fish sampled at artificial reefs off Alabama in the northern Gulf of Mexico. McCawley and Cowan (2007) demonstrated that plankton consumption increased for larger Red Snapper, as is reported here, and that zooplankton consumption was highest in spring. They concluded that zooplankton consumption was probably either a response to artificial reefs being spaced too closely—thus, Red Snapper had to rely on unexpected prey resources to meet basal bioenergetics demands—or that adults subsidized their diet with zooplankton to facilitate maximum reproductive output during their protracted April to September spawning season (Jackson et al. 2007).

We found different zooplankton taxa were predominant in pre-DWH Red Snapper stomachs sampled at natural versus artificial reefs, but zooplankton percent diet by mass was nearly identical between habitat types. Therefore, it seems unlikely that artificial reef spacing explains zooplankton consumption, especially given the fact that Red Snapper density is an order of magnitude higher on artificial versus natural reefs in the system (Patterson et al. 2014). Instead, the presence of zooplankton highlights the extreme diversity of Red Snapper diet and the opportunism they display while foraging. It should also be noted that several zooplankton taxa identified in diet samples were relatively soft-bodied compared with decapods or fishes. Therefore, it is likely that zooplankton prey experienced higher digestion and evacuation rates than other prey types, thus may have constituted an even higher percentage of the Red Snapper diet than the overall 22.6% by mass observed among pre-DWH oil spill samples.

Habitat was not a significant factor with respect to Red Snapper diet or trophic position, and few reef-dependent taxa were present in Red Snapper stomach samples, regardless of habitat type. Densities of larger (>250 mm TL) reef fishes, including Red Snapper, on artificial reefs are typically 150–200% greater than their densities on natural reefs in the northern Gulf (Dance et al. 2011; Patterson et al. 2014), which may lead to greater competition for resources and decreased food web complexity (Bohnsack and Sutherland 1985). However, among our seven broad prey categories, only subtle differences in Red Snapper diet contribution were observed between habitat types, and those differences were probably related to local abundance rather than prey selectivity. The lack of reef-associated prey in Red Snapper stomachs indicates a reliance on nonreef foraging habitat (Bradley and Bryan 1975; Ouzts and Szedlmayer 2003; McCawley and Cowan 2007), regardless of whether fish were associated with natural or artificial reefs.

Many reef fishes, including snappers, have been shown to forage away from reef structure, thus translocate energy and nutrients horizontally back to reefs (Appledoorn et al. 2009; Luo et al. 2009; Berkström et al. 2012). Data presented here and by others suggest Red Snapper also translocate energy and nutrients from surrounding open substrates back to reefs, whether natural or artificial. Lack of habitat-specific differences in Red Snapper diet may at first seem to indicate no foraging advantage conveyed by artificial reef habitat. However, based on foraging distances (typically <100 m) estimated via

TABLE 3. ANOVA models testing for effects of habitat type (natural versus artificial reefs) and Deepwater Horizon (DWH) timing (pre versus post) on Red Snapper stable isotopes or stable isotope-derived variables.

Source	df	Type-III sum of squares	Mean square	Pseudo-*F*	*P*-value
		Total length			
Habitat	1	46,920	46,920	3.36	0.072
DWH	1	67,654	67,654	4.84	0.031
Habitat × DWH	1	8,792	8,792	0.63	0.431
Residual	62	866,056	13,969		
		Trophic level			
Habitat	1	0.045	0.045	1.46	0.232
DWH	1	0.287	0.287	9.25	0.003
Habitat × DWH	1	0.023	0.023	0.39	0.392
Residual	62	1.925	0.031		
		$\mathbf{Log}_e(-\delta^{13}\mathbf{C}_{\mathbf{base}})$			
Habitat	1	8.82×10^{-5}	8.82×10^{-5}	0.28	0.597
DWH	1	2.75×10^{-3}	2.75×10^{-3}	8.80	0.004
Habitat × DWH	1	2.01×10^{-4}	2.01×10^{-4}	0.65	0.425
Residual	62	1.93×10^{-2}	3.12×10^{-4}		
		$\mathbf{Log}_e(\delta^{34}\mathbf{S}_{\mathbf{base}})$			
Habitat	1	3.90×10^{-3}	3.90×10^{-3}	3.07	0.085
DWH	1	2.02×10^{-2}	2.02×10^{-2}	15.90	<0.001
Habitat × DWH	1	1.67×10^{-3}	1.67×10^{-3}	1.31	0.257
Residual	62	7.89×10^{-2}	1.27×10^{-3}		

acoustic telemetry data (Westmeyer et al. 2007; Topping and Szedlmayer 2011), deployment of artificial reefs appears to enable Red Snapper to exploit foraging habitats on the shelf they otherwise would be unlikely to visit, especially given that natural reef habitat in the northcentral Gulf of Mexico is concentrated in deeper (>40 m) waters than where the bulk of artificial reef zones are located (Parker et al. 1983; Schroeder et al. 1988; Patterson et al. 2014).

We observed ontogenetic shifts in Red Snapper diet and trophic position, but the decline in their estimated trophic level with increasing size was not expected a priori. Others have reported shifts to higher trophic position and to greater piscivory during development (Bradley and Bryan 1975; Wells et al. 2008). However, those studies described feeding and trophic shifts from settlement and early life, when a clear shift from planktonic to benthic feeding was observed. Here, we report a similar pattern among subadult and adult Red Snapper as that reported by McCawley and Cowan (2007) in which zooplankton constituted a substantial (>20%) percentage of the diet of larger, reproductively mature fish. Higher trophic-level prey, such as decapods and fish, were predominant in the diet of smaller (<400 mm TL) Red Snapper but became less predominant for larger individuals. Although decapod contribution to diet decreased as Red Snapper length increased, the mass of individual prey was notably greater in the diet of larger fish.

Analysis of δ^{15}N corroborated a decrease in Red Snapper trophic position with increasing size, which also indicates the presence of lower trophic-level prey in the diet of larger individuals was not ephemeral, given that turnover time of fish white muscle tissue is typically weeks to months (Buchheister and Latour 2010; Nelson et al. 2010). While δ^{15}N values and trophic level estimates derived from them confirm trends observed in diet data, δ^{13}C and δ^{34}S data provide information about Red Snapper trophic ecology that may not be apparent in diet data alone. For example, northern Gulf phytoplankton typically has a δ^{13}C signature of approximately −22‰ and a δ^{34}S signature of approximately 18‰, whereas benthic microalgae has a δ^{13}C signature of approximately −18‰ and a δ^{34}S signature of approximately10‰ (Moncrieff and Sullivan 2001; Fry 2006; Rooker et al. 2006). Therefore, estimates of $\delta^{13}C_{base}$ and $\delta^{34}S_{base}$ provide information with respect to the relative contribution of pelagic versus benthic production to Red Snapper biomass, and the lack of a significant relationship between δ^{13}C or δ^{34}S and length indicates neither the source of C nor S was related to fish size. The lack of a habitat effect on $\delta^{13}C_{base}$ and $\delta^{34}S_{base}$ indicates the relative contribution of pelagic versus benthic production was not significantly different between natural and artificial reefs, which further indicates that Red Snapper trophic ecology does not differ between these habitat types.

The factor that had the greatest impact on Red Snapper diet and trophic level was DWH oil spill timing. Pre-DWH diet

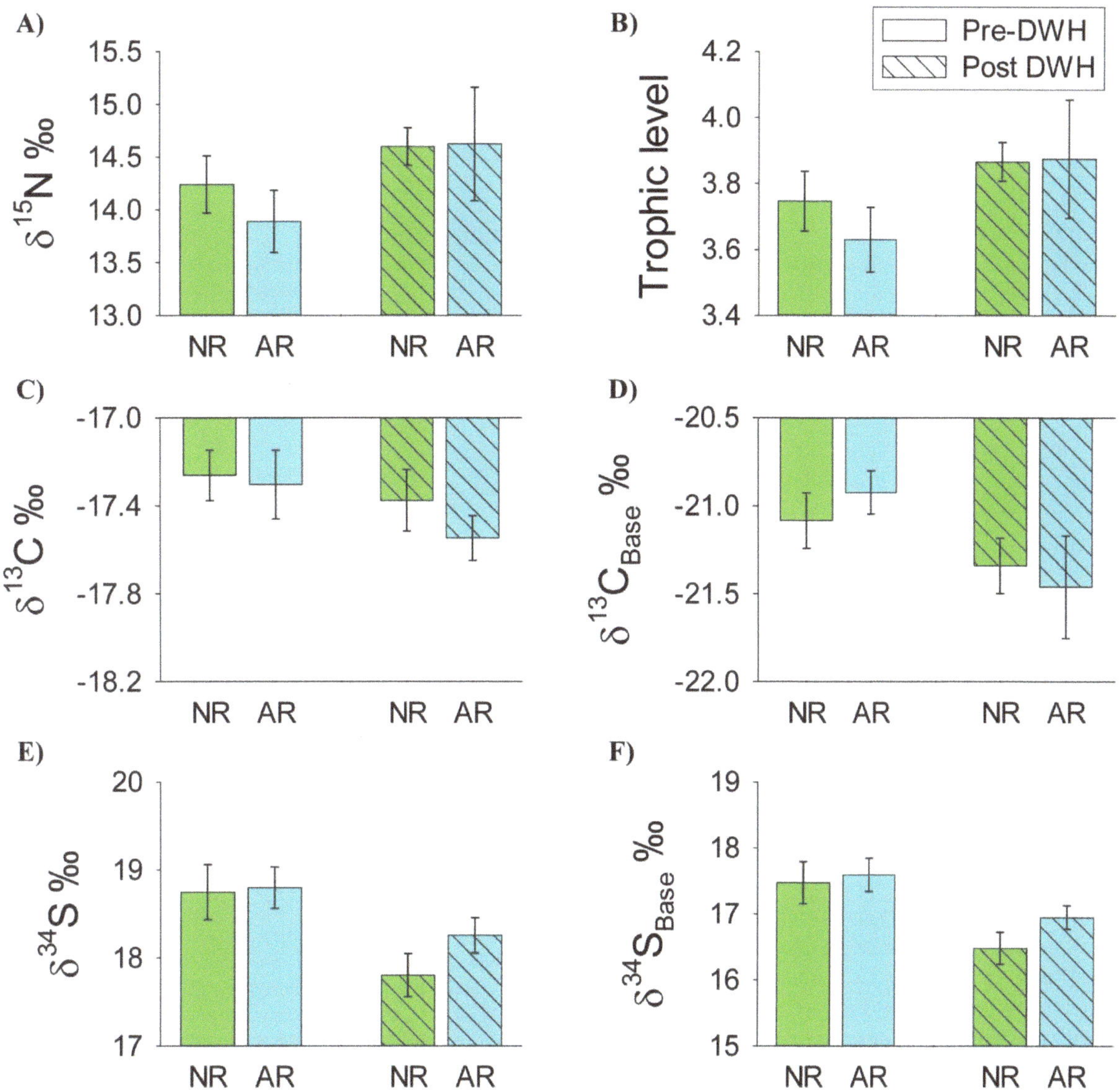

FIGURE 4. Mean (± 95% CI) habitat (NR = natural reef; AR = artificial reef) and Deepwater Horizon (DWH) oil spill timing-specific values for Red Snapper **(A)** white muscle δ^{15}N, **(B)** estimated trophic level, **(C)** lipid-corrected white muscle δ^{13}C, **(D)** estimated δ^{13}C at the base of the food web, **(E)** white muscle δ^{34}S, and **(F)** estimated δ^{34}S at the base of the food web for fish sampled between June 2009 and August 2011 in the northern Gulf of Mexico.

composition estimates presented here are consistent with earlier estimates reported by Ouzts and Szedlmayer (2003) and McCawley and Cowan (2007) (e.g., fish and crustaceans forming large percentages of Red Snapper diet, but zooplankton also contributing 15–20% of diet by mass), thus providing support that our pre-DWH diet data are generally representative of Red Snapper diet. Following the DWH oil spill, zooplankton consumption decreased markedly at both natural and artificial reefs, which was replaced with higher trophic level prey: decapods, cephalopods, and stomatopods at natural reefs and predominantly fish at artificial reefs. These shifts clearly were persistent for weeks to months as higher post-DWH δ^{15}N values and lower δ^{34}S values in Red Snapper muscle samples indicated a shift to higher trophic level and more benthic prey resources (i.e., pelagic zooplankton prey replaced by benthic decapods and fishes). Toxicological, immunological, and genetic effects on plankton communities were documented after the spill (Ortmann et al. 2012; Paul et al. 2013), and a decline in zooplankton consumption is consistent with high plankton mortality and a resultant blizzard of marine snow observed in the months following the spill (Passow et al. 2012). Not only did oil toxicity impact northern Gulf plankton communities following the spill, but the release of millions of gallons of Corexit dispersant likely magnified the toxic effects of the spill on plankton (Middaugh and Whiting 1995; Ortmann et al. 2012; Paul et al. 2013). Therefore, while increased higher trophic level prey following the DWH oil spill might suggest greater abundance of those taxa, a more plausible scenario is that zooplankton were less abundant in the months following the spill which resulted in much lower zooplankton consumption.

The one variable that appears inconsistent with DWH oil spill-related trends observed in diet and other stable isotope data is $\delta^{13}C$, specifically estimates of $\delta^{13}C_{base}$. Phytoplankton are depleted in ^{13}C relative to benthic algae (Moncrieff and Sullivan 2001; Fry 2006); thus, lower $\delta^{13}C_{base}$ values a year after the DWH oil spill should imply greater, not lower, plankton contribution to Red Snapper muscle C. However, hydrocarbons released during the DWH oil spill constitute another source of organic carbon on the shelf, and they were even more depleted in ^{13}C than phytoplankton (Chanton et al. 2012; Cherrier et al. 2013). Results from $\delta^{13}C$ mixing models computed by Cherrier et al. (2013) for northern Gulf particulate organic carbon (POC) sampled in summer 2011 and 2012 indicated 28–43% of POC was derived from fossil CH_4. Due to the sequestration of hydrocarbons beneath the seabed for millions of years, oil hydrocarbons are dead with respect to ^{14}C activity (Chanton et al. 2012). Therefore, extremely low POC $\Delta^{14}C$ values (e.g., less than −600‰) provide even stronger evidence that hydrocarbons from the DWH spill entered the food web (Cherrier et al. 2013). The fact that Red Snapper $\delta^{13}C_{base}$ estimates were significantly lower following the spill when other evidence suggests higher values would be expected may suggest that oil carbon reached higher levels of the food web, and did so as early as the year following the spill.

Red Snapper samples collected in 2009-2010 prior to the DWH oil spill provide a unique opportunity to examine potential impacts of the spill on their diet and trophic position, as well as future examination of ecosystem resiliency. These relatively large generalist mesopredators may be an ideal species to examine for such impacts because they have such a broad diet, ranging from zooplankton to fish, thus can easily shift to foraging on locally abundant prey taxa. The corroboration of shifts observed in diet data with trophic shifts inferred from muscle stable isotope ratios provides clear evidence of DWH oil spill impacts to Red Snapper feeding ecology, and likely the northern Gulf food web in general. Where pre-DWH muscle samples or stable isotope data exist, stable isotope analysis also can be employed to examine DWH oil spill effects for other northern Gulf reef fishes for which pre-DWH oil spill diet data are unlikely to be as extensive as they are for Red Snapper.

ACKNOWLEDGMENTS

This research was made possible, in part, by a grant from BP/The Gulf of Mexico Research Initiative (data available at https://data.gulfresearchinitiative.org), and in part, by funding from the Florida Fish Wildlife Research Institute and National Oceanic and Atmospheric Administration Fisheries' Cooperative Research Program. We thank Captains S. Wilson, J. Greene, S. Kelley, and G. Jarvis, along with their crews, for providing access to their fishing vessels and considerable help in the field. We also acknowledge J. Neese, H. Moncrief, R. Scharer, and S. Singleton, who aided sampling and sample processing. Lastly, we acknowledge the reviews of three anonymous reviewers, which greatly improved an earlier draft of this paper.

REFERENCES

Allan, S. E., B. W. Smith, and K. A. Anderson. 2012. Impact of the Deepwater Horizon oil spill on bioavailable polycyclic aromatic hydrocarbons in Gulf of Mexico coastal waters. Environmental Science and Technology 46:2033–2039.

Almeda, R., Z. Wambaugh, Z. Wang, C. Hyatt, Z. Liu, and E. J. Buskey. 2013. Interactions between zooplankton and crude oil: toxic effects and bioaccumulation of polycyclic aromatic hydrocarbons. PLoS (Public Library of Science) ONE [online serial] 8(6):e67212.

Anderson, M. J., R. N. Gorley, and K. R. Clarke. 2008. PERMANOVA+ for PRIMER: guide to software and statistical methods. PRIMER-E, Plymouth, UK.

Appeldoorn, R. S., A. Aguilar-Perera, B. L. K. Bouwmeester, G. D. Dennis, R. L. Hill, W. Merten, C. W. Recksiek, and S. J. Williams. 2009. Movement of fishes (grunts: Haemulidae) across the coral reef seascape: a review of scales, patterns and processes. Caribbean Journal of Science 45:304–316.

Barron, M. G. 2012. Ecological impacts of the Deepwater Horizon oil spill: implications for immunotoxicity. Toxicological Pathology 40:315–320.

Berkström, C., M. Gullström, R. Lindborg, A. W. Mwandya, S. A. S. Yahya, N. Kautsky, and M. Nyström. 2012. Exploring knowns and unknowns in tropical seascape connectivity with insights from East African coral reefs. Estuarine Coastal and Shelf Science 107:1–21.

Bohnsack, J. A. 1989. Are high densities of fishes at artificial reefs the result of habitat limitation or behavioral preference? Bulletin of Marine Science 44:631–645.

Bohnsack, J. A., and D. L. Sutherland. 1985. Artificial reef research: a review with recommendations for future priorities. Bulletin of Marine Science 37:11–39.

Bradley, E., and C. E. Bryan. 1975. Life history and fishery of the Red Snapper (*Lutjanus campechanus*) in the northwestern Gulf of Mexico: 1970–1974. Proceedings of the Gulf and Caribbean Fisheries Institute 27:77–106.

Buchheister, A., and R. J. Latour. 2010. Turnover and fractionation of carbon and nitrogen stable isotopes in tissues of a migratory coastal predator, Summer Flounder (*Paralichthys dentatus*). Canadian Journal of Fisheries and Aquatic Sciences 67:445–461.

Carr, M. H., and M. A. Hixon. 1997. Artificial reefs: the importance of comparisons with natural reefs. Fisheries 22(4):28–33.

Chanton, J. P., J. Cherrier, R. M. Wilson, J. Sarkodee-Adoo, S. Bosman, A. Mickle, and W. M. Graham. 2012. Radiocarbon evidence that carbon from the Deepwater Horizon spill entered the planktonic food web of the Gulf of Mexico. Environmental Research Letters [online serial] 7:045303.

Cherrier, J., J. Sarkodee-Adoo, T. P. Guilderson, and J. P. Chanton. 2013. Fossil carbon in particulate organic matter in the Gulf of Mexico following the Deepwater Horizon event. Environmental and Science Technical Letters 2014:108–112.

Clarke, K. R., P. J. Somerfield, and M. G. Chapman. 2006. On resemblance measures for ecological studies, including taxonomic dissimilarities and a zero-adjusted Bray-Curtis coefficient for denuded assemblages. Journal of Experimental Marine Biology and Ecology 330:55–80.

Cowan, J. H. Jr., C. B. Grimes, W. F. Patterson III, C. J. Walters, A. C. Jones, W. J. Lindberg, D. J. Sheehy, W. E. Pine III, J. E. Powers, M. D. Campbell, K. C. Lindeman, S. L. Diamond, R. Hilborn, H. T. Gibson, and K. A. Rose. 2011. Red Snapper management in the Gulf of Mexico: science- or faith-based? Reviews in Fish Biology and Fisheries 21:187–204.

Dance, M. A., W. F. Patterson III, and D. T. Addis. 2011. Fish community and trophic structure at artificial reef sites in the northeastern Gulf of Mexico. Bulletin of Marine Science 87:301–324.

Fenwick, G. D. 1978. Plankton swarms and their predators at the Snares Islands. New Zealand Journal of Marine Freshwater Research 12:223–224.

Fodrie, F. J., and K. L. Heck Jr. 2011. Response of coastal fishes to the Gulf of Mexico oil disaster. PLoS (Public Library of Science) ONE [online serial] 6(7):e21609.

Fry, B. 2006. Stable isotope ecology. Springer, New York.

Jackson, J. B. C., J. D. Cubit, B. D. Keller, V. Batista, K. Burns, H. M. Caffey, R. L. Caldwall, S. D. Garrity, C. D. Getter, C. Gonzalez, H. M. Guzman, K. W. Kaufmann, A. H. Knap, S. C. Levings, M. J. Marshall, R. Steger, R. C. Thompson, and E. Weil. 1989. Ecological effects of a major oil spill on Panamanian coastal marine communities. Science 243:37–44.

Jackson, M. W., J. H. Cowan Jr., and D. H. Nieland. 2007. Demographic differences in northern Gulf of Mexico Red Snapper reproductive maturation: implications for the unit stock hypothesis. Pages 217–228 *in* W. F. Patterson III, J. H. Cowan Jr., G. R. Fitzhugh, and D. L. Nieland, editors. Red Snapper ecology and fisheries in the U.S. Gulf of Mexico. American Fisheries Society, Symposium 60, Bethesda, Maryland.

Lindberg, W. J. 1997. Can science solve the attraction versus production debate? Fisheries 22(4):10–13.

Luo, J., J. E. Serafy, S. Sponaugle, P. B. Teare, and D. Kieckbusch. 2009. Movement of Gray Snapper, *Lutjanus griseus*, among subtropical seagrass, mangrove, and coral reef habitats. Marine Ecological Progress Series 380:255–269.

Mason, O. U., N. M. Scott, A. Gonzalez, A. Robbins-Pianka, J. Bælum, J. Kimbrel, N. J. Bouskill, E. Prestat, S. Borglin, D. C. Joyner, J. L. Fortney, D. Jurelevicius, W. T. Stringfellow, L. Alvarez-Cohen, T. C. Hazen, R. Knight, J. A. Gilbert, and J. K Jansson. 2014. Metagenomics reveals sediment microbial community response to Deepwater Horizon oil spill. ISME Journal 8:1464–1475.

McCawley, J., and J. H. Cowan Jr. 2007. Seasonal and size specific diet and prey demand of Red Snapper on Alabama artificial reefs: implications for management. Pages 77–104 *in* W. F. Patterson III, J. H. Cowan Jr., G. R. Fitzhugh, and D. L. Nieland, editors. Red Snapper ecology and fisheries in the U.S. Gulf of Mexico. American Fisheries Society, Symposium 60, Bethesda, Maryland.

McCutchan, J. H. Jr., W. M. Lewis Jr., C. Kendall, and C. C. McGrath. 2003. Variation in trophic shift for stable isotope ratios of carbon, nitrogen, and sulfur. Oikos 102:378–390.

McNutt, M. K., R. Camilli, T. J. Crone, G. D. Guthrie, P. A. Hsieh, T. B. Ryerson, O. Savas, and F. Shaffer. 2012. Review of flow rate estimates of the Deepwater Horizon oil spill. Proceedings of the National Academy of Sciences of the USA 109:20260–20267.

Middaugh, D. P., and D. D. Whiting. 1995. Responses of embryonic and larval silversides, *Menidia beryllina*, to no. 2 fuel oil and oil dispersants in seawater. Archives of Environmental Contamination and Toxicology 29:535–539.

Moncrieff, C. A., and M. J. Sullivan. 2001. Trophic importance of epiphytic algae in subtropical seagrass beds: evidence from multiple stable isotopes. Marine Ecological Progress Series 215:93–106.

Moreno, R., L. Jover, C. Diez, F. Sardà, and C. Sanpera. 2013. Ten years after the *Prestige* oil spill: seabird trophic ecology as indicator of long-term effects on the coastal marine ecosystem. PLoS (Public Library of Science) ONE [online serial] 8(10):e77360.

Murawski, S. A., W. T. Hogarth, E. B. Peebles, and L. Barbeiri. 2014. Prevalence of fish diseases in the Gulf of Mexico post Deepwater Horizon. Transactions of the American Fisheries Society 143:1084–1097.

Nelson, J., J. Chanton, F. Coleman, and C. Koenig. 2010. Patterns of stable carbon isotope turnover in Gag *Mycteroperca microlepis,* an economically important marine piscivore determined with a non-lethal surgical biopsy procedure. Environmental Biology of Fishes 90:243–252.

Ortmann, A. C., J. Anders, N. Shelton, L. Gong, A. G. Moss, and R. H. Condon. 2012. Dispersed oil disrupts microbial pathways in pelagic food webs. PLoS (Public Library of Science) ONE [online serial] 7(7):e42548.

Ouzts, A. C., and S. T. Szedlmayer. 2003. Diel feeding patterns of Red Snapper on artificial reefs in the north-central Gulf of Mexico. Transaction of the American Fisheries Society 132:1186–1193.

Parker, R. O. Jr., D. R. Colby, and T. D. Willis. 1983. Estimated amount of reef habitat on a portion of the U.S. South Atlantic and Gulf of Mexico continental shelf. Bulletin of Marine Science 33:935–940.

Paul, J. H., D. Hollander, P. Coble, K. L. Daly, S. Murasko, D. English, J. Basso, J. Delaney, L. McDaniel, and C. W. Kovach. 2013. Toxicity and mutagenicity of Gulf of Mexico waters during and after the Deepwater Horizon oil spill. Environmental Science and Technology 4:9651–9659.

Passow, U., K. Ziervogel, V. Asper, and A. Diercks. 2012. Marine snow formation in the aftermath of the Deepwater Horizon oil spill in the Gulf of Mexico. Environmental Research Letters [online serial] 7:035301.

Patterson, W. F. III, J. H. Tarnecki, D. T. Addis, and L. R. Barbieri. 2014. Reef fish community structure at natural versus artificial reefs in the northern Gulf of Mexico. Proceedings of the Gulf Caribbean Fisheries Institute 66:4–8.

Perkol-Finkel, S., N. Shashar, and Y. Benayahu. 2006. Can artificial reefs mimic natural reef communities? The roles of structural features and age. Marine Environmental Research 61:121–135.

Peterson, C. H., S. D. Rice, J. W. Short, D. Esler, J. L. Bodkin, B. E. Ballachey, and D. B Irons. 2003. Long-term ecosystem response to the *Exxon Valdez* oil spill. Science 302:2082–2086.

Pilcher, W., S. Miles, S. Tang, G. Mayer, and A. Whitehead. 2014. Genomic and genotoxic responses to controlled weathered-oil exposures confirm and extend field studies on impacts of the Deepwater Horizon oil spill on native killifish. PLoS (Public Library of Science) ONE [online serial] 9:e106351.

Post, D. M. 2002. Using stable isotopes to estimate trophic position: models, methods, and assumptions. Ecology 83:703–718.

Post, D. M., C. A. Layman, D. A. Arrington, G. Takimoto, J. Quattrochi, and C. G. Montaña. 2007. Getting to the fat of the matter: models, methods and assumptions for dealing with lipids in stable isotope analyses. Oecologia 152:179–189.

Radabaugh, K. R., D. J. Hollander, and E. B. Peebles. 2013. Seasonal $\delta^{13}C$ and $\delta^{15}N$ isoscapes of fish populations along a continental shelf trophic gradient. Continental Shelf Research 68:112–122.

Reddy, C. M., J. S. Arey, J. S. Seewald, S. P. Sylva, K. L. Lemkau, R. K. Nelson, C. A. Carmichael, C. P. McIntyre, J. Fenwick, G. T. Ventura, B. A. S. Van Moody, and R. Camilli. 2012. Composition and fate of gas and oil released to the water column during the Deepwater Horizon oil spill. Proceedings of the National Academy of Sciences of the USA 50:20229–20234.

Rooker, J. R., J. P. Turner, and S. A. Holt. 2006. Trophic ecology of Sargassum-associated fishes in the Gulf of Mexico determined from stable isotopes and fatty acids. Marine Ecology Progress Series 313:249–259.

Sammarco, P. W., S. R. Kolian, R. A. F. Warby, J. L. Bouldin, W. A. Subra, and S. A. Porter. 2013. Distribution and concentrations of petroleum hydrocarbons associated with the BP/Deepwater Horizon oil spill, Gulf of Mexico. Marine Pollution Bulletin 73:129–143.

Sanvicente-Añorve, L., E. Lemus-Santana, C. Flores-Coto, and M. Alatorre-Mendieta. 2013. Vertical segregation of holoplanktonic molluscs in the epipelagic layer, southern Gulf of Mexico. Helgoland Marine Research 67:397–405.

Schroeder, W. W., A. W. Schultz, and J. J. Dindo. 1988. Inner-shelf hardbottom areas, northeastern Gulf of Mexico. Gulf Coast Association of Geological Societies Transactions 38:535–541.

Schwacke, L. H., C. R. Smith, F. I. Townsend, R. S. Wells, L. B. Hart, B. C. Balmer, T. K. Collier, S. DeGuise, M. M. Fry, L. J. Guillette Jr., S. V. Lamb, S. M. Lane, W. E. McFree, N. J. Place, M. C. Tumlin, G. M. Ylitalo, E. S. Zolman, and T. K. Rowles. 2013. Health of common bottlenose dolphins (*Tursiops truncatus*) in Barataria Bay, Louisiana, following the Deepwater Horizon oil spill. Environmental Science and Technology 48:93–103.

Sweeting, C. J., J. T. Barry, N. V. C. Polunin, and S. Jennings. 2007. Effects of body size and environment on diet-tissue $\delta^{13}C$ fractionation in fishes. Journal of Experimental Marine Biology and Ecology 352:165–176.

Szedlmayer, S. T., and R. L. Shipp. 1994. Movement and growth of Red Snapper, *Lutjanus campechanus*, from an artificial reef area in the northeastern Gulf of Mexico. Bulletin of Marine Science 55:887–896.

Teal, J. M., and R. W. Howarth. 1984. Oil spill studies: a review of ecological impacts. Environmental Management 8:27–43.

Topping, D. T., and S. T. Szedlmayer. 2011. Home range and movement patterns of Red Snapper (*Lutjanus campechanus*) on artificial reefs. Fisheries Research 112:77–84.

Vanderklift, M. A., and S. Ponsard. 2003. Sources of variation in consumer-diet delta N-15 enrichment: a meta-analysis. Oecologia 136: 169–182.

Wells, D. R. J., J. H. Cowan Jr., and B. Fry B. 2008. Feeding ecology of Red Snapper *Lutjanus campechanus* in the northern Gulf of Mexico. Marine Ecology Progress Series 61:213–225.

Westmeyer, M. P., C. A. Wilson, and D. L. Nieland. 2007. Fidelity of Red Snapper to petroleum platforms in the northern Gulf of Mexico. Pages 105–122 *in* W. F. Patterson III, J. H. Cowan Jr., G. R. Fitzhugh, and D. L. Nieland, editors. Red Snapper ecology and fisheries in the U.S. Gulf of Mexico. American Fisheries Society, Symposium 60, Bethesda, Maryland.

Whitehead, A., B. Dubansky, C. Bodinier, T. I. Garcia, S. Miles, C. Pilley, V. Raghunathan, J. L. Roach, N. Walker, R. B. Walter, C. D. Rice, and F. Galvez. 2012. Genomic and physiological footprint of the Deepwater Horizon oil spill on resident marsh fishes. Proceedings of the National Academy of Sciences of the USA 109:20298–20302.

3

Estimating the Sex Composition of the Summer Flounder Catch using Fishery-Independent Data

Jason M. Morson* and Eleanor A. Bochenek
Haskin Shellfish Research Laboratory–Rutgers, The State University of New Jersey, 6959 Miller Avenue, Port Norris, New Jersey 08349, USA

Eric N. Powell
Gulf Coast Research Laboratory, University of Southern Mississippi, 703 East Beach Drive, Ocean Springs, Mississippi 39564, USA

Emerson C. Hasbrouck
Cornell University Cooperative Extension, Marine Program, 423 Griffing Avenue, Riverhead, New York 11901, USA

Jennifer E. Gius
Haskin Shellfish Research Laboratory–Rutgers, The State University of New Jersey, 6959 Miller Avenue, Port Norris, New Jersey 08349, USA

Charles F. Cotton[1]
Virginia Institute of Marine Science, 1208 Greate Road, Gloucester Point, Virginia 23062, USA

Kristin Gerbino and Tara Froehlich
Cornell University Cooperative Extension, Marine Program, 423 Griffing Avenue, Riverhead, New York 11901, USA

Abstract

Models that account for sex-specific behavior and population dynamics are becoming more common in the stock assessment of sexually dimorphic fishes. However, such models can be data intensive and require some knowledge or assumptions about the sex ratio of fishery landings. A recent stock assessment review of Summer Flounder *Paralichthys dentatus* identified the need to account for sex-specific fishing mortality in the assessment model; however, no data on the sex composition of the catch were available. Fishery-independent, sex-specific information for this species is collected annually by the National Marine Fisheries Service's Northeast Fisheries Science Center during their bottom trawl survey. Sex at age from the survey could be applied to the fishery landings if the probability of landing a given sex at a given age is equivalent for fish collected by the survey and those in the landings. To generate the first regionally comprehensive database on the sex ratio of Summer Flounder landings and to determine the efficacy of using survey sex-at-age keys to estimate the sex of landed fish, we recorded the sex

Subject editor: Carl Walters, University of British Columbia, Canada

*Corresponding author: jmorson@hsrl.rutgers.edu

[1]Present address: Coastal and Marine Laboratory, Florida State University, 3618 Coastal Highway 98, St. Teresa, Florida 32358, USA.

composition of the commercial and recreational catches of Summer Flounder ($n = 31{,}912$) in 2010 and 2011. When (1) trawl survey length data were left-truncated to simulate the minimum retention sizes in the fisheries and (2) age–length keys generated from fishery-dependent data were applied to length frequency distributions from the survey to simulate the growth rates of landed fish, the sex-at-age pattern in the survey-derived data closely resembled the patterns in the catch. However, statistically significant differences in sex at age remained between the catch and the survey-derived data. We hypothesize that these differences are attributable to differences in the spatiotemporal distributions of the sexes and of the survey and fishing effort.

Recent work suggests that the inclusion of sex structure in stock assessment models is important for sexually dimorphic fishes. For example, if growth rates or natural mortality differ between male and female fish (Wang et al. 2005; Su et al. 2013), if sex-specific migratory behavior and fishing effort are not evenly distributed in space and time (Okamura et al. 2014), or if fishery selectivity is sex specific (Myers et al. 2014), then the absence of sex structure in the assessment model can bias biological reference points. In light of these recent findings, sex-structured assessment models are becoming more common in describing the population dynamics of fishes that have sexually dimorphic characteristics and/or behavior (Clark and Hare 2006; Wang et al. 2007; Fenske et al. 2011; Su et al. 2011; McGilliard et al. 2013).

Female Summer Flounder *Paralichthys dentatus* grow faster (Poole 1961; King et al. 2001) and mature at a larger size (Able and Kaiser 1994; Packer et al. 1999) than males. In addition, the sex ratio of young fish is skewed in favor of males, probably due to a complex interaction between temperature and the biochemistry of sex determination (Luckenbach et al. 2009). However, as the cohort ages, the balance in sex ratio shifts toward females (Smith and Daiber 1977; Bonzek et al. 2009), a pattern that indicates a higher natural mortality rate among males (Maunder and Wong 2011). Sexually dimorphic life history characteristics of Summer Flounder suggest that a sex-structured assessment model would be appropriate for this species (Jones et al. 2013; NEFSC 2013), yet the most recent assessment applied a statistical catch-at-age model in which the sexes were combined (NEFSC 2013).

Sex-structured assessment models can be difficult to implement because they are data intensive, requiring either some prior knowledge about the sex composition of the catch or that assumptions be made about the sex-specific selectivity curve and its spatial and temporal stability (Methot and Wetzel 2013; Myers et al. 2014). Sex-specific information is not collected from Summer Flounder landings as part of the annual assessment process because (1) like other flatfishes, Summer Flounder display no external characteristics that can be used to identify sex; and (2) Summer Flounder cannot be dissected by port agents and still retain their market value. However, dissection is used to collect sex-specific data on Summer Flounder that are captured by the National Marine Fisheries Service's (NMFS) Northeast Fisheries Science Center (NEFSC) during its bottom trawl survey, which is conducted twice annually in the spring and fall (NEFSC 2013).

If fishery-independent survey data are sex specific and if the probability that a fish of a given length or age is female (or male) is similar between the survey and the fishery catch, then a sex-at-length or sex-at-age key developed from survey data might be applicable directly to length or age data from the catch (Clark 2004). However, the sex composition of the fishery catch is influenced by a combination of the following factors: (1) differences in the size-, age-, and sex-specific retention selectivity imposed by the minimum retention size (MRS) in each fishery (Kendall and Quinn 2013); (2) the size-, age-, and sex-specific vulnerability of fish resulting from their interaction with the fishing gears (Ryer 2008; Somerton et al. 2011) and/or their availability to the fishing gears (Okamura et al. 2014; Sampson 2014); and (3) the spatial-, temporal-, and sex-specific growth rates of the fish that are retained by the different sampling gears (Poole 1961; Planes et al. 1999; Wilderbuer and Turnock 2009). For instance, different MRS limits are implemented in the recreational and commercial Summer Flounder fisheries (NEFSC 2013), and some limited evidence suggests that at a given length, males' and females' availability or vulnerability to fishing effort may vary seasonally and between the recreational fishery and the NMFS-NEFSC survey (Morson et al. 2012). These fishery-specific processes can generate differences in the sex composition within the fishery catch and between the catch and the survey; thus, it may be incorrect to assume that the probability of a fish being female (or male) at a given length or age is equivalent between catch data and survey data.

The goal of this work was to test the efficacy of applying NMFS-NEFSC survey-derived sex-at-age keys to fishery catch data, thus allowing the potential implementation of a sex-specific model in the Summer Flounder stock assessment to be examined. We collected sex-specific data from the commercial and recreational fisheries and compared the probability distribution for sex at age from the catch data to probability distributions generated by simulating fishery-dependent processes using the survey data.

METHODS

Data collection.—Summer Flounder along the northeastern USA and the U.S. mid-Atlantic continental shelf were collected in 2010 and 2011 by the NMFS-NEFSC fishery-independent survey (NFS data) and by port sampling of the commercial fishery landings (CFS data) and recreational hook-and-line fishery landings (RFS data; Table 1). For all fish, TL, sex, and catch location were recorded. For a subset of the sampled fish, otoliths and/or scales were removed for use in age determinations (Table 1). Otoliths were removed from carcasses collected from the RFS; for the CFS and the NFS samples, scales were collected from specimens up to 60 cm TL, and otoliths were obtained from fish larger than 60 cm TL. The right-side otolith was selected from each fish and was processed by center alignment in a merchandise tag, embedded in black polyester resin, and then cross-sectioned through the otolith core with a low-speed sectioning saw. After processing, otoliths were examined by using a Leica MZ-6 dissecting microscope with transmitted light at 8–60× magnification. Annuli from otoliths were identified as narrow opaque bands and were counted along the ventral side of the sulcal groove. Scales were processed by selecting five or six samples from each fish and creating impressions on acetate sheets. The scale impressions were examined using a microfiche reader with 20- and 29-mm lenses. Annuli in scales were identified by using established protocols (Pentilla and Dery 1988).

The NMFS-NEFSC provided NFS data for Summer Flounder sampled from the spring and fall bottom trawl survey cruises at predetermined sampling strata and at depths of 27–365 m from Cape Hatteras, North Carolina, to Nova Scotia, Canada. Detailed descriptions of the bottom trawl survey design are provided by Azarovitz (1981) and NEFSC (1988). To summarize, between 350 and 400 stations were sampled each year with a bottom trawl that had a 1.5-cm stretch-mesh liner and rollers for sampling hard substrate. Tows were conducted for 30 min per station at a tow speed of 6.48 km/h (3.5 knots). The CFS and RFS data were collected by sampling Summer Flounder that were landed at marinas, packing houses, and fishing docks from North Carolina to Massachusetts (Table 2).

TABLE 1. Total number of Summer Flounder that were sampled for sex and TL, and the subset of those fish that were also aged via examination of annuli in otoliths and scales (see Methods; NMFS = National Marine Fisheries Service; NEFSC = Northeast Fisheries Science Center).

Data source	Sex and length data	Age data
NMFS-NEFSC bottom trawl survey	4,828	3,935
Commercial fishery port sampling	18,685	11,809
Recreational fishery port sampling	13,232	7,276
Total	36,745	23,020

TABLE 2. Port locations where samples were collected from the recreational (RFS) and commercial (CFS) fisheries for Summer Flounder.

State	RFS port	CFS port
North Carolina		Wanchese
Virginia	Virginia Beach	Hampton
	Hampton	Newport News
	Buckroe	
	James River	
	Capeville	
	Wachapreague	
	Yorktown	
	Cape Charles	
	Mathews	
New Jersey	Cape May	Cape May
	Fortescue	Barnegat Light
	Barnegat Light	Point Pleasant
	Point Pleasant	
	Atlantic Highlands	
New York	Freeport	Point Lookout
	Huntington	Hampton Bays
	Captree	Mattituck
	Port Jefferson	East Hampton
	Moriches	Montauk
	Riverhead	
	Hampton Bays	
	Mattituck	
	Greenport	
	East Hampton	
	Montauk	
Maryland	Ocean City	
Delaware	Lewes	
Connecticut	Niantic	Stonington
Rhode Island	Point Judith	Point Judith
Massachusetts	New Bedford	Westport
	Hyannis	New Bedford
		Hyannis

Sampling of Summer Flounder landings in the commercial fishery occurred year-round wherever the fishery was operational. The minimum mesh size in the commercial fishery was a 14-cm diamond mesh or 15-cm square mesh applied throughout the net body, extension, and cod end. For each visit to a commercial dock, samples of up to 100 fish were collected haphazardly from each market category (medium, large, and jumbo sizes) available from a given fishing trip. The market category was determined by each dock or packing house; generally, small fish ranged from 35 to 45 cm, medium fish ranged from 45 to 55 cm, and jumbo fish were greater than 55 cm. At these sizes, all fish sampled were sexually mature.

Summer Flounder cannot be sexed based on external characteristics; therefore, to avoid reducing the market value of landed fish, we developed a minimally invasive technique for determining sex (Figure 1). On the pigmented side of the fish, an approximately 2.5-cm incision was made in an area halfway between the anterior end of the anal fin and the center of the pectoral fin. Forceps were then used to pull the gonads out through the incision. Orange eggs or empty ovaries were indicative of females; males were distinguished by the presence of white testis tissue. Both males and females were easily identifiable by this method, even for sampling that took place outside of the spawning season.

The RFS ports were sampled once per week during the recreational fishing season (May–September) to collect racks (filleted carcasses) of all Summer Flounder that were caught on the sampling day by all participating boats. Once the Summer Flounder were filleted, the sex organs remained intact and visible on the rack. Boat captains and crews saved fish racks in a bin, and those racks were collected upon our arrival at a participating dock. To increase the number of fish available for collection, freezers were placed at each dock. Bags and waterproof tags were provided to the boat captains and crews and were made available near the freezers so that samples could be accurately labeled with the date and location of the catch. Participating boat captains were asked to deposit all fish racks from the day's catch into these tagged bags and to place the bags in the freezers. We emptied the freezers upon our arrival at a dock to collect fresh racks. Morson et al. (2012) concluded that the change in Summer Flounder length from freezing was negligible. To ensure that the samples represented the sex, length, and age of harvested Summer Flounder, all fish that were caught during a recreational fishing trip were sampled without regard to size. Sex was determined by macroscopic investigation of exposed gonad on the filleted fish carcass. Over 99% of the fish in the RFS collections had reproductive organs that were intact and readily visible to the naked eye.

A

B

FIGURE 1. Photographs showing a technique for determining the sex of a Summer Flounder without dissecting the entire fish: **(A)** female and **(B)** male. A small incision was made along the gut cavity, and the gonads were pulled through the incision with a pair of forceps.

For all commercial and recreational samples, records were kept of the NMFS-NEFSC statistical area where the fish were caught (Figure 2; Table 3). The commercial and recreational fisheries operate in different statistical areas, so their fishing locations did not completely overlap. For the NFS, data were collected along transects in a stratified sampling design. To permit spatially similar comparisons between the NFS data and the RFS or CFS data, the latitude and longitude of each NFS tow were used to assign NFS samples to NMFS-NEFSC statistical areas.

Data analysis.—Age was determined for 23,020 of the 36,745 Summer Flounder sampled from the RFS, CFS, and NFS (Table 1). To utilize the full data set for determining how well the NFS sex-at-age data described the RFS or CFS sex at age, it was necessary to generate appropriately partitioned age–length keys (ALKs) for application to fish that lacked empirically determined age data. Age–length data were partitioned by sex, region, and data source (NFS, CFS, or RFS). Region was defined as either "north" (including all fish that were caught or collected north of 40°N) or "south" (including all fish that were caught or collected south of 40°N). To compare ALKs, the age distribution at a given length was modeled using multinomial logistic regression (Gerritsen et al. 2006). The multinomial log-linear models were fitted via neural networks by using the multinom() function in the R package "nnet" (Venables and Ripley 2002). This approach provided a simple, robust method for identifying significant differences between ALKs generated from data partitioned by sex, region, and data source. The best-fitting model identified the most appropriate stratification of the data, and ALKs were generated based on that stratification. For all fish that were not aged empirically, the ALKs were then applied based on the probability of each age given the length of each fish; this was done by using the ageKey function in the R package FSA (Isermann and Knight 2005; Ogle 2014).

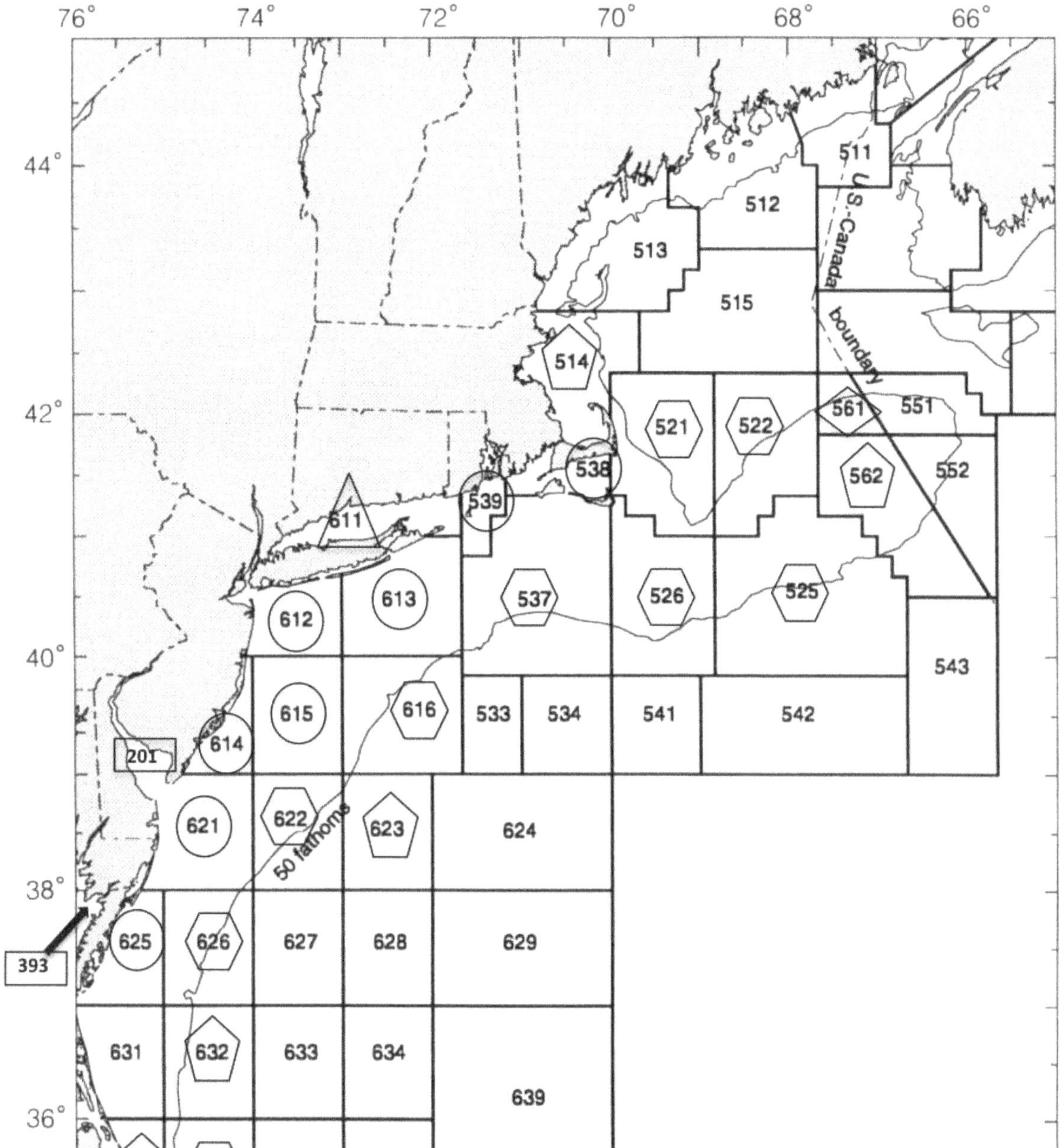

FIGURE 2. Map showing the National Marine Fisheries Service (NMFS) Northeast Fisheries Science Center's (NEFSC) statistical area units along the northeastern USA and the western mid-Atlantic continental shelf (NEFSC 1997). The shape enclosing each statistical area's number indicates the data source as the NMFS-NEFSC bottom trawl survey (NFS), commercial fishery port sampling (CFS), recreational fishery port sampling (RFS), or some combination thereof (rectangles = RFS only; diamonds = CFS only; pentagons = NFS only; triangles = RFS and CFS; hexagons = NFS and CFS; circles = RFS, CFS, and NFS).

We used logistic regression to evaluate the efficacy of applying a sex-at-age key developed from NFS data to the CFS and RFS age data. The probability of landing a female Summer Flounder at a given age was modeled as the dependent variable, and data source (NFS, CFS, or RFS) was modeled as the independent variable (Wilson and Hardy 2002; Morson et al. 2012). Significant differences between the NFS data source and both fishery data sources suggested that the sex-at-age key developed from raw NFS data would not be appropriate for describing the sex composition from either fishery. We hypothesized that the variability in sex at age among the data sets could be explained by (1) the size-, age-,

TABLE 3. Total number of Summer Flounder sampled from each National Marine Fisheries Service (NMFS) Northeast Fisheries Science Center (NEFSC) statistical area (see Figure 2). Samples were obtained by the NMFS-NEFSC bottom trawl survey (NFS), commercial fishery port sampling (CFS), and recreational fishery port sampling (RFS).

Statistical area	NFS	CFS	RFS
201			930
393			419
514	12		
521	3	40	
522	14	120	
525	71	401	
526	230	157	
537	741	1,596	
538	5	454	1,278
539	270	2,558	1,065
561		39	
562	27		
611		2,028	1,078
612	7	842	3,454
613	658	3,985	839
614	112	687	45
615	262	3,343	40
616	441	822	
621	421	434	4,080
622	289	508	
623	18		
625	178	176	22
626	493	488	
631	141		15
632	114		
635	293		
636	44	15	

and sex-specific retention selectivity imposed by the MRS of 35 cm for the commercial fishery and the MRSs of 46–53 cm for the recreational fishery; and (2) the differences in growth rates among fish collected from the three data sources. Therefore, the effects of these fishery-specific processes on the fishery-independent survey data were simulated separately for the CFS and RFS, and the resultant sex ratios at age were reanalyzed.

First, to simulate the length-specific retention selectivity of the recreational and commercial fisheries, the NFS data were left-truncated at the MRS for each fishery. This approach provided an NFS-derived measure of Summer Flounder sex at age that was comparable to the left-truncated sex composition in each fishery-dependent data set. The 35-cm MRS was implemented coastwide in the commercial fishery during the sample collection period, so NFS and CFS data were left-truncated at 35 cm for the CFS comparison. The MRS for the recreational fishery varied from 46 to 53 cm depending on the landing year and the state. To avoid too many pairwise comparisons based on too few data, we left-truncated the NFS and RFS data at 46 cm and at 53 cm in two separate comparisons that together encompassed the full range of potential comparisons.

To simulate the observed growth rates in the commercial and recreational fisheries, sex- and region-specific ALKs generated from the CFS and RFS data were applied to the left-truncated sex- and region-specific NFS length data. This allowed us to assign age to the left-truncated distribution of lengths in the NFS data as if the NFS Summer Flounder had grown at the same rates as the fish in each respective fishery.

Finally, as a proxy for evaluating the length-specific vulnerability of the sexes to the commercial and recreational fisheries, the left-truncated, NFS-derived sex ratio at length was compared with the measured sex at length from the CFS and RFS data. The NFS utilized a smaller mesh size than the regulation size for the commercial fishery, and the MRS for the recreational fishery is set higher than that for the commercial fishery; therefore, all left-truncated NFS data in these comparisons originated from fish that were fully recruited to the survey. For each pairwise comparison, a logistic regression model was fitted to estimate the probability that a landed fish of a given length was female.

All statistical analyses were conducted in R version 3.0.3 (R Development Core Team 2014). Model selection was based on Akaike's information criterion (AIC). In evaluating two competing models, a model with an AIC score that was at least 2 units lower than the other model's AIC was accepted as the better-fitting model (Arnold 2010).

RESULTS

Summer Flounder from 27 different NMFS-NEFSC statistical areas were sampled during the NFS, CFS, and RFS programs (Table 3). The CFS samples were distributed inshore and offshore across similar statistical areas as the NFS samples, whereas the RFS samples were limited to statistical areas located inshore (Figure 2; Table 3).

The probability that a Summer Flounder of a given length was a given age depended on the sex, region of capture, and data source (Table 4). Sex was the most important effect in the model, followed by data source and region.

The sex ratio at age in the NFS data was significantly different than those in the CFS data ($P < 0.0001$) and the RFS data ($P < 0.0001$) when comparisons included all NFS data (Figures 3a, 4a). In both fisheries, a higher proportion of the catch at a given age was female (Figures 3a, 4a, 5a). However, fish that were landed under an MRS had a higher mean length at age than fish that were collected by the NFS. The higher the MRS imposed on the landed fish, the higher was the mean length at age (Figure 6). This effect was apparent for all ages that contained the MRS within their length frequency. To account for this,

TABLE 4. Values of Akaike's information criterion (AIC) for eight potential models describing the probability that a Summer Flounder was a given age based on fish length (null model) and fish sex, region of capture (north or south of 40°N), and data source (recreational fishery port sampling, commercial fishery port sampling, or fishery-independent bottom trawl survey). Delta AIC (ΔAIC) is the difference between the specified model and the best-performing model (i.e., the model with the lowest AIC value).

Model	df	AIC	ΔAIC	Residual deviance
Sex, region, data source	65	59,836	0	59,668
Sex, region	56	60,226	390	60,114
Sex, data source	70	60,268	432	60,128
Sex	42	60,605	769	60,521
Region, data source	70	65,590	5,754	65,450
Data source	56	65,949	6,113	65,837
Region	42	66,242	6,406	66,158
Null	28	66,550	6,714	66,494

the NFS data were left-truncated to simulate the MRS values used in the commercial and recreational fisheries; the resulting sex-ratio-at-age pattern more closely approximated the pattern in the data collected directly from the fisheries (Figures 3b, 4b, 5b). The proportions at age in the NFS data shifted dramatically toward females at younger ages (Figures 3b, 4b, 5b). Separate application of sex- and region-specific ALKs generated from CFS and RFS data to the left-truncated, sex- and region-specific NFS length data resulted in a sex-ratio-at-age pattern that further resembled the patterns shown in the CFS and RFS data (Figures 3c, 4c, 5c). The shift in sex ratio at age suggests that the growth differences between Summer Flounder that are landed in the fisheries and those that are collected during the NFS are sex specific. Females that were landed in the commercial and recreational fisheries had grown faster than females that were collected by the NFS, and males that were landed in the fisheries had grown more slowly than NFS-collected males. Thus, the application of fishery-specific ALKs produced a shift in the sex ratio of younger fish further toward females and a shift in the sex ratio of older fish toward males (Figures 3b, c; 4b, c; 5b, c). For example, in the comparison of NFS and CFS data (Figure 3b, c), the survey-derived sex at age 1 increased from approximately 40% female (Figure 3b) to approximately 65% female (Figure 3c), whereas the survey-derived sex at age 6 decreased from approximately 75% female (Figure 3b) to approximately 50% female (Figure 3c). Nevertheless, statistically significant differences remained between the CFS data ($P < 0.0001$) or RFS data ($P < 0.0001$) and the NFS-derived data that were manipulated to simulate the fishery-specific MRSs and growth processes within catches from the two fisheries.

Because accounting for the length-specific retention selectivity and differences in growth rate between Summer Flounder that were landed in the fisheries and those collected during the survey did not remove all of the variability in sex at age between the NFS data and the catch data, we attempted to estimate the difference in availability of the sexes to the survey and the fisheries. The length-specific vulnerability of the sexes to the gear (modeled as the probability that a landed fish of a given length was female) was similar but statistically different ($P = 0.0037$) between the left-truncated NFS data and the CFS data (Figure 7a). The probability that a landed fish was female was slightly higher for the CFS data than for the left-truncated NFS data (Figure 7a). In addition, the proportion female at length increased at similar rates in both the CFS data set and the left-truncated NFS data set until each reached approximately 45 cm—similar to the smallest MRS for the recreational fishery. At that point, the proportion female at length remained unchanged or increased only slightly from each 1-cm increment to the next until a length of approximately 50 cm was reached, when the rate for both data sets returned to a logistic form (Figure 7). The length-specific probability that a landed fish was female was much higher for the RFS data than for the left-truncated NFS data ($P < 0.0001$; Figure 7b). In fact, for a Summer Flounder landed in the recreational fishery, the probability that the fish was female was 80% at the smallest MRS (46 cm) and rapidly approached 100% with increasing fish size (Figure 7b).

DISCUSSION

To determine whether the sex composition of Summer Flounder from a fishery-independent trawl survey could be used to estimate the sex composition of the commercial and recreational catch without collecting sex-specific data directly from the fisheries, we simulated fishery-specific processes on the survey data, including the MRS in each fishery and the growth rates of fish landed in both fisheries, and we compared the survey-derived estimates to the actual sex composition in the fishery catch.

Summer Flounder that were landed in the commercial and recreational fisheries had grown at different rates than survey-collected fish and at different rates from one another, suggesting that each gear is sampling a different portion of the population. Furthermore, this effect was sex specific: females that were landed in the fisheries had grown faster than those collected during the survey, whereas males that were collected by the survey had grown faster than males landed in the fisheries. This suggests that fisheries are selectively targeting the faster-growing females and slower-growing males. Stari et al. (2010) proposed that sampling location or gear selectivity could explain differences in ALKs for North Sea Haddock *Melanogrammus aeglefinus* collected with different gear types. We propose that the differences among ALKs in the present study could have resulted from (1) the sex-specific spatial segregation of Summer Flounder, as previously suggested by Morson et al. (2012); (2) the influence of

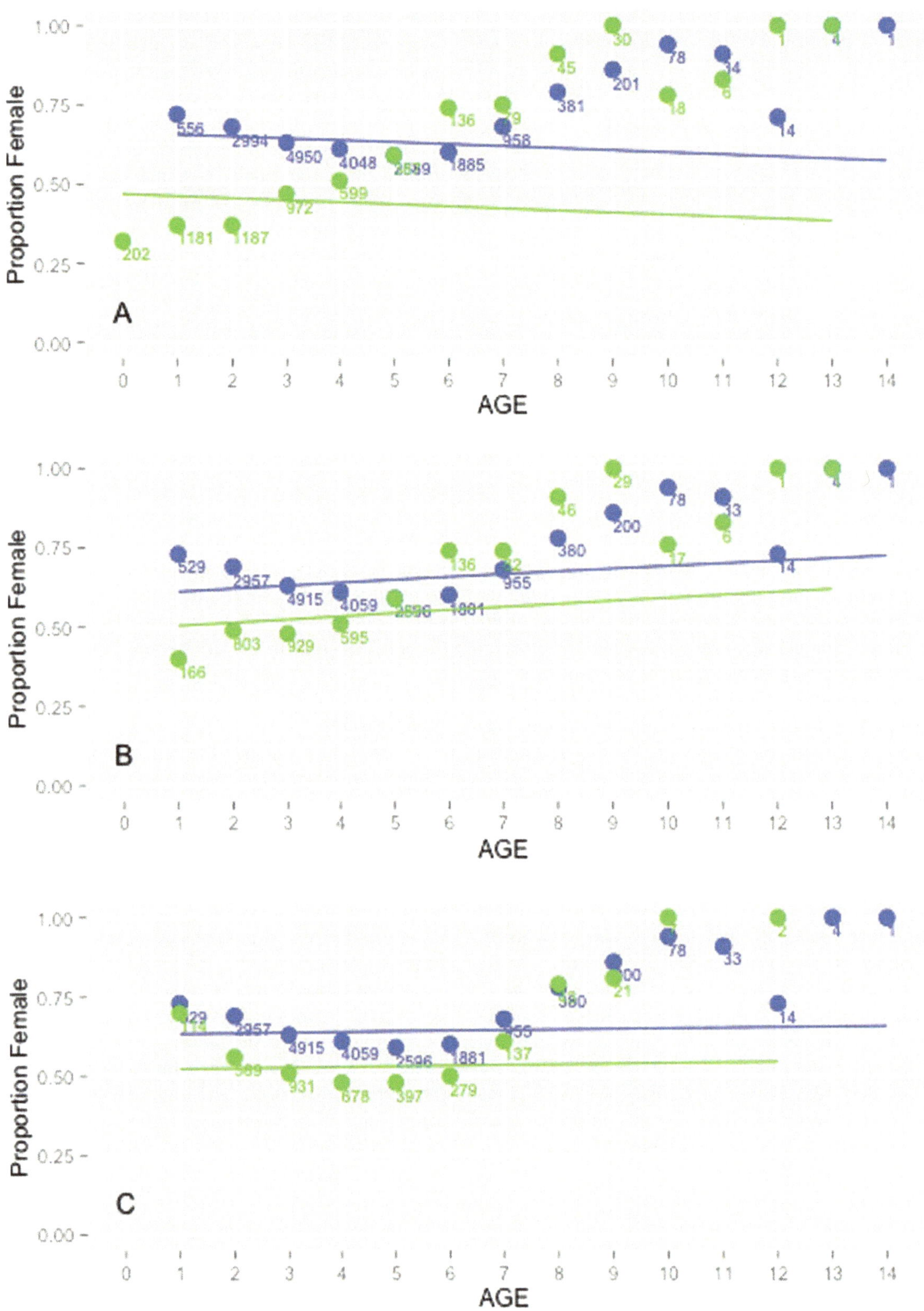

FIGURE 3. Proportion female at age for Summer Flounder sampled in the commercial fishery (blue dots) and the National Marine Fisheries Service Northeast Fisheries Science Center's bottom trawl survey (NFS data; green dots). Lines represent the probability that a landed fish was female, as estimated by logistic regression (see text). Commercial fishery data are compared with the NFS data in the following forms: **(A)** raw NFS data, **(B)** NFS data that were left-truncated at 35 cm to simulate the minimum retention size in the commercial fishery, and **(C)** NFS data that were left-truncated at 35 cm and assigned ages based on age–length keys generated from commercial fishery data. The number of sampled fish is shown next to each data point.

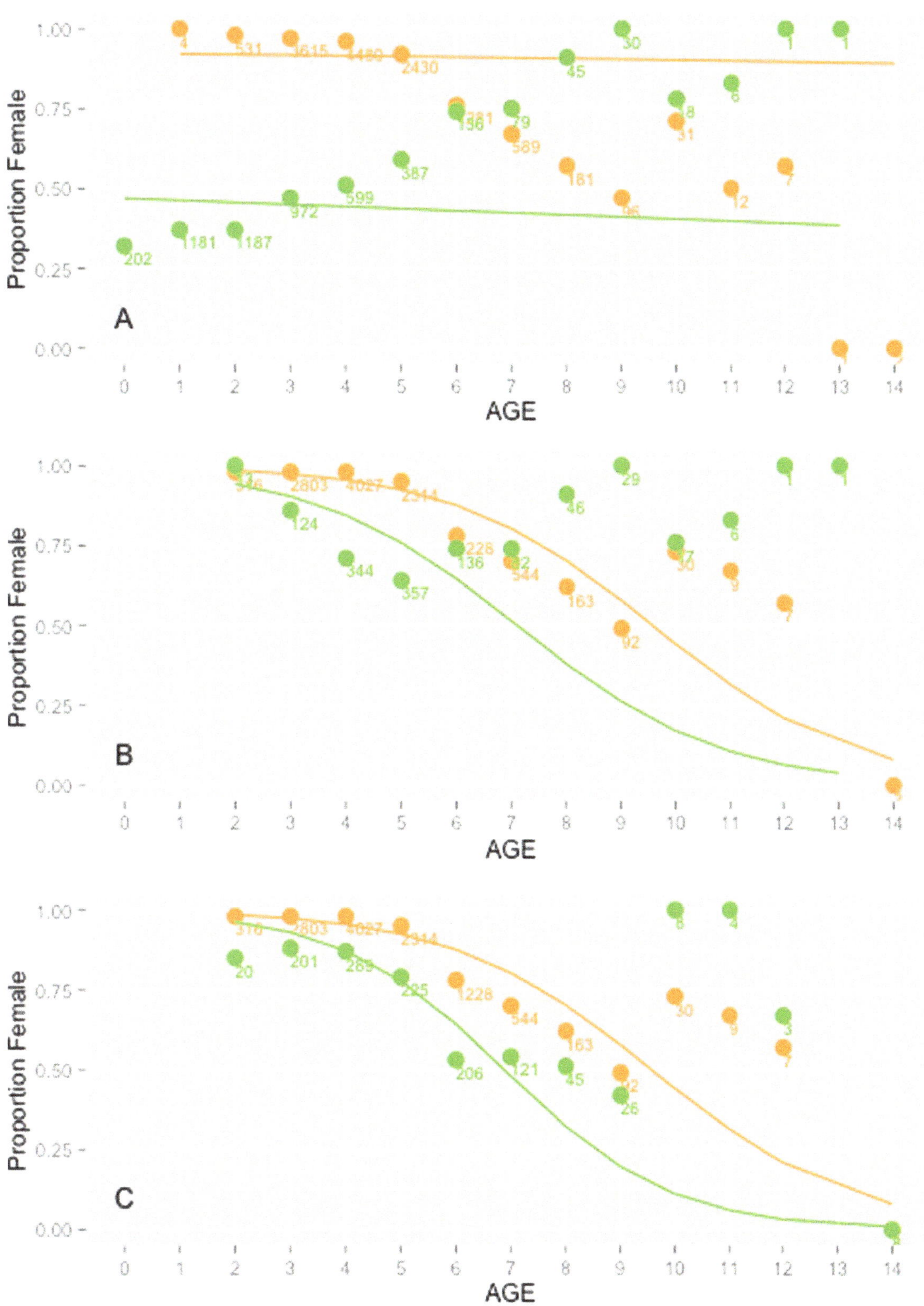

FIGURE 4. Proportion female at age for Summer Flounder sampled in the recreational fishery (orange dots) and the National Marine Fisheries Service Northeast Fisheries Science Center's bottom trawl survey (NFS data; green dots). Lines represent the probability that a landed fish was female, as estimated by logistic regression (see text). Recreational fishery data are compared with NFS data in the following forms: **(A)** raw NFS data, **(B)** NFS data that were left-truncated at 46 cm to simulate the smallest minimum retention size in the recreational fishery, and **(C)** NFS data that were left-truncated at 46 cm and assigned ages based on age–length keys generated from recreational fishery data. The number of sampled fish is shown next to each data point.

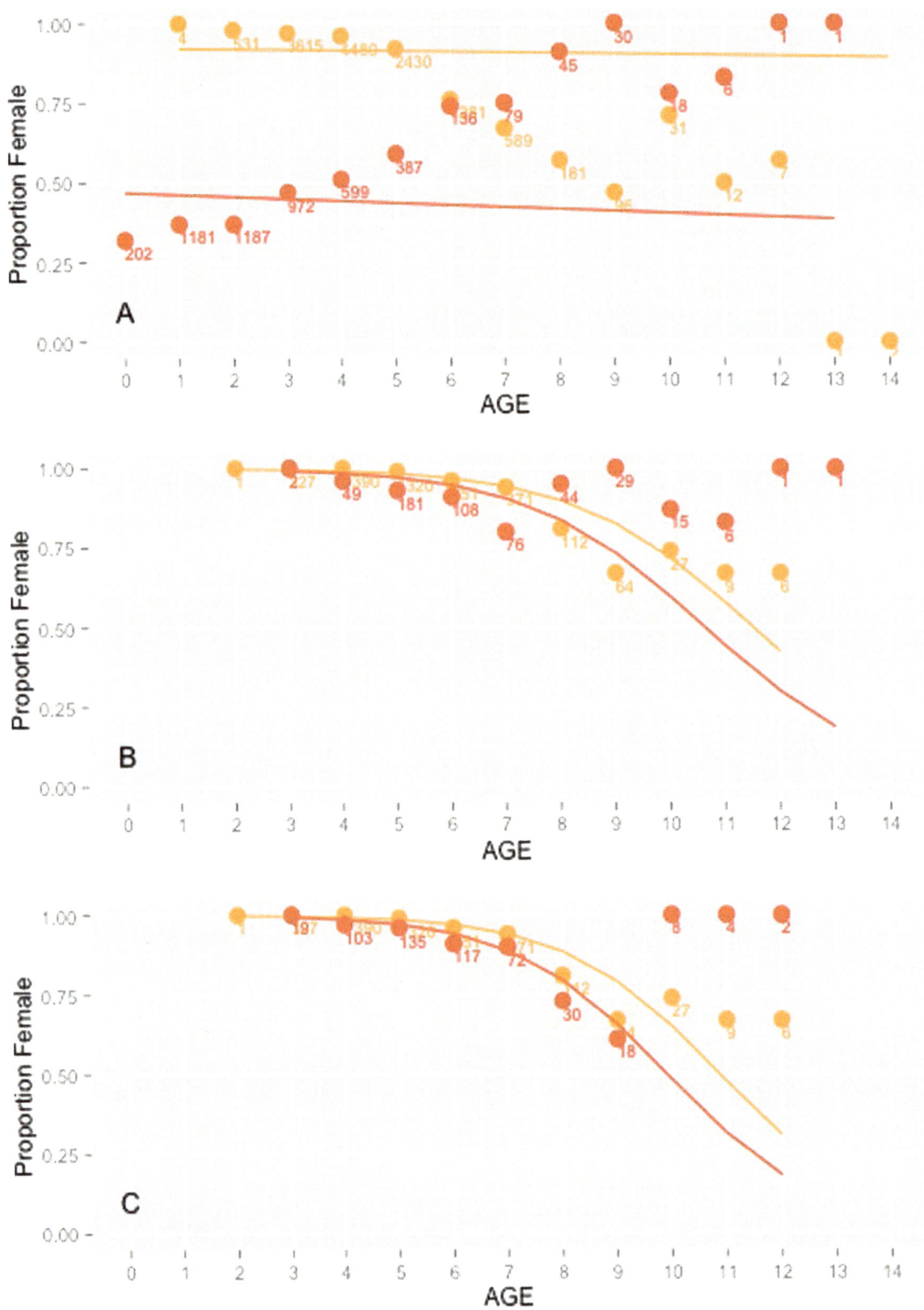

FIGURE 5. Proportion female at age for Summer Flounder sampled in the recreational fishery (orange dots) and the National Marine Fisheries Service Northeast Fisheries Science Center's bottom trawl survey (NFS data; red dots). Lines represent the probability that a landed fish was female, as estimated by logistic regression (see text). Recreational fishery data are compared with NFS data in the following forms: **(A)** raw NFS data, **(B)** NFS data that were left-truncated at 53 cm to simulate the largest minimum retention size in the recreational fishery, and **(C)** NFS data that were left-truncated at 53 cm and assigned ages based on age–length keys generated from recreational fishery data. The number of sampled fish is shown next to each data point.

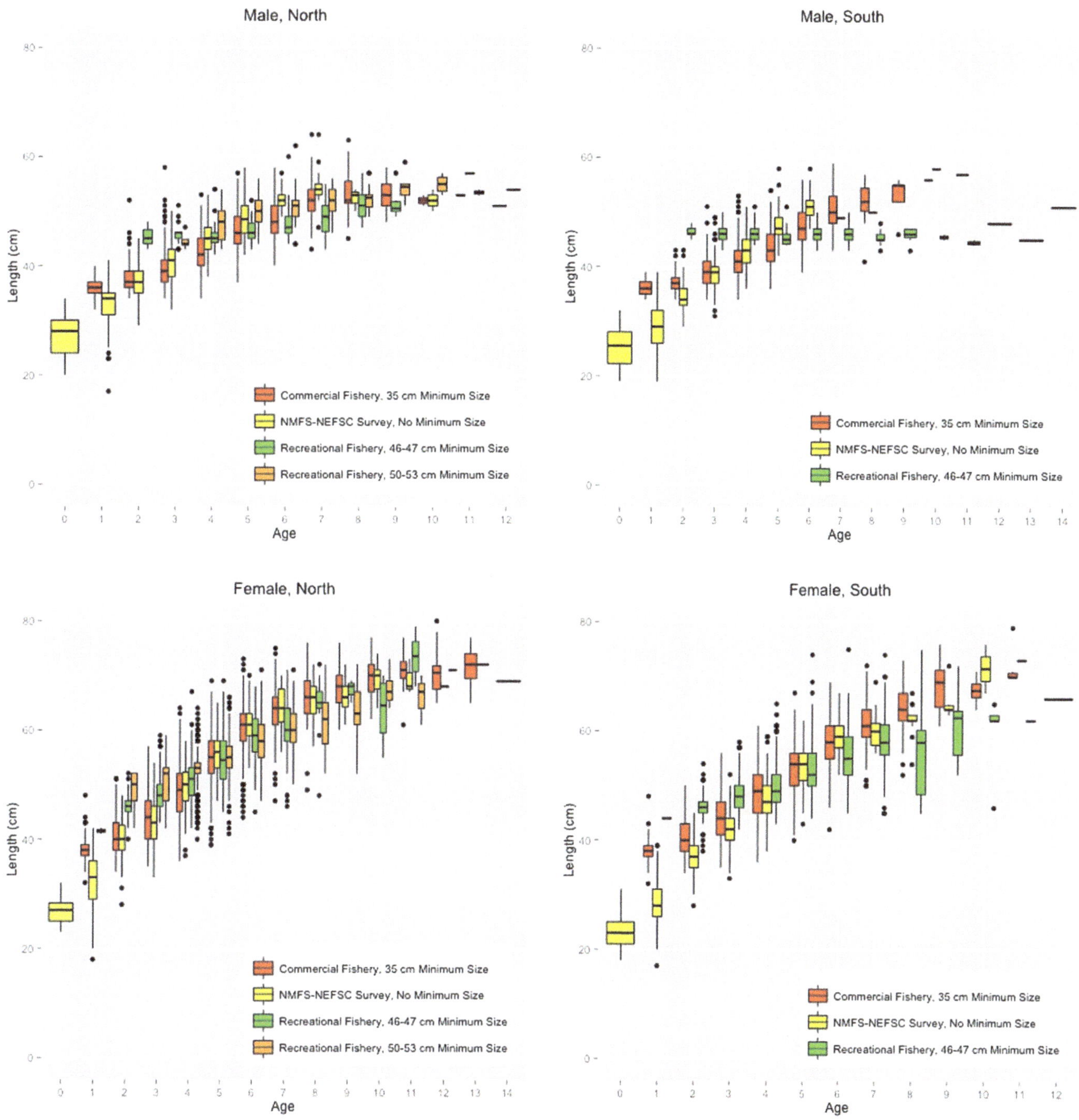

FIGURE 6. Box-and-whisker plots of the mean TL at age for Summer Flounder under different minimum retention size regulations in the commercial and recreational fisheries (NMFS–NEFSC survey = National Marine Fisheries Service [NMFS] Northeast Fisheries Science Center [NEFSC] bottom trawl survey). Plots are organized by region (north and south of 40°N) and sex. The black horizontal line within each box is the mean, the ends of the box represent the interquartile range, the ends of whiskers denote the 5th and 95th percentiles, and the dots are outliers.

region on growth rates, which was previously identified by Kraus and Musick (2001) and further confirmed with this work; and (3) the spatial and temporal distribution of fishing effort, although the influence of age-specific gear selectivity cannot be ruled out. We discuss the spatiotemporal distributions of the fish and of the fishing effort in more detail below.

The MRS in each fishery, operating on the sex-specific growth rates (Poole 1961; King et al. 2001), forces a higher fraction of females at age in the catch than would be expected

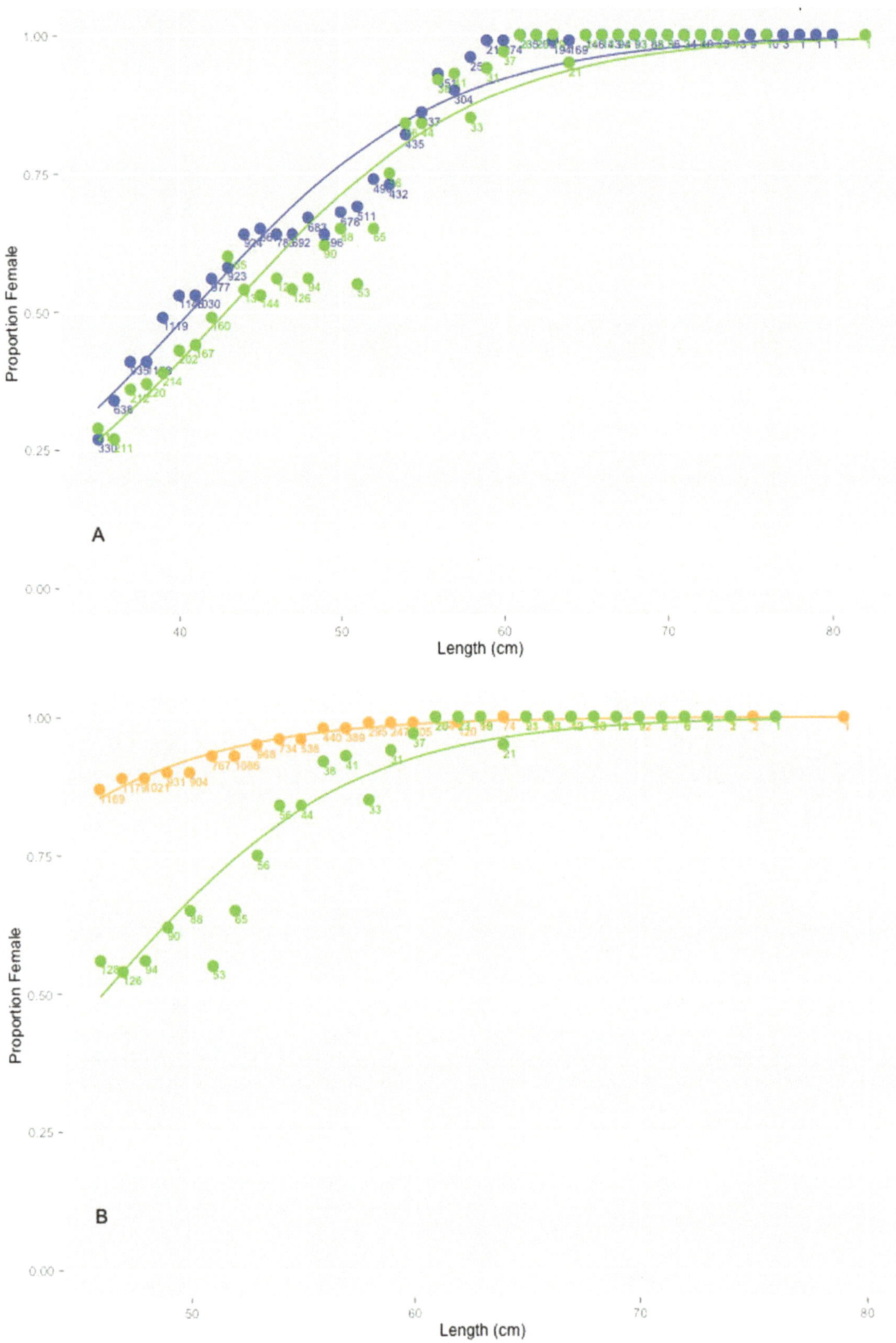

FIGURE 7. Proportion female at length (TL) for Summer Flounder sampled in the commercial fishery (blue dots), the recreational fishery (orange dots), and the National Marine Fisheries Service Northeast Fisheries Science Center's bottom trawl survey (NFS data; green dots). Lines represent the probability that a landed fish was female, as estimated by logistic regression (see text). The NFS data are truncated according to the minimum retention size in **(A)** the commercial fishery (35 cm) and **(B)** the recreational fishery (46 cm). The number of sampled fish is shown next to each data point.

in a non-culled, fishery-independent sample. We were able to largely correct for this by left-truncating the NFS data to mimic the MRS in each fishery; the transformation of the survey data recovered a sex-at-age pattern that more closely resembled the pattern measured directly from the fishery catch. Not surprisingly, when we applied ALKs developed from fishery-dependent data to the left-truncated NFS data, the result was a sex-at-age pattern in the survey-derived data that even more closely resembled the pattern measured directly from the catch.

This stepwise approach to simulating fishery-specific effects on fishery-independent data recovered the sex-at-age pattern observed in the fisheries; such an approach may therefore be applicable to other fisheries for which biological data from the catch are limited and for which the estimation of sex-specific parameters in an assessment is desirable. However, these manipulations did not yield the exact proportions at age measured directly from the catch. We hypothesize that this was due to some difference in the spatiotemporal distributions of stock structure and fishing effort such that the males' and females' availability to the NMFS-NEFSC survey and the fishing fleets was not equivalent. As evidence, we showed that sex at length (used as a proxy for fish availability to the fishing gears) varied slightly between the commercial fishery and the NFS and differed dramatically between the recreational fishery and the survey. The recreational fishery occurs inshore from late spring to early fall and the commercial fishery occurs year-round inshore and offshore, whereas the NFS takes place offshore, once in March and once in September. Because there was no spatial or temporal overlap between the recreational fishery and the survey, we would expect that if there is spatial and temporal structure in the population's sex composition, then differences in the spatiotemporal availability of the sexes to the survey and the recreational fishery will be reflected in the catch. Indeed, this was manifested by the large difference in sex ratio at length between the RFS and NFS data.

The commercial fishery overlapped with the survey both in time and in space, so the expectation is that the sex-specific availability of fish to the commercial fishery and the survey will be similar. Such a finding was also observed as the fairly close match in sex at length between the CFS and NFS data. To our knowledge, the present study provides the first evidence that Summer Flounder exhibit sex-specific segregation spatially and temporally.

Spatial and temporal segregation of fish by sex is a common phenomenon and occurs in other sexually dimorphic flatfishes (Morgan and Trippel 1996; Swain 1997; Swain and Morin 1997; Gorchinsky 1998; Richards et al. 2008; Gerritsen et al. 2010; Sahin and Gunes 2010; Loher and Hobden 2012; Loher et al., in press). Explanations for sex-specific segregation include temperature selection (Swain 1997) and/or depth selection (Swain and Morin 1997), sex-specific life history strategies that promote sexually dimorphic characteristics (Gerritsen et al. 2010), and sex-specific foraging intensity and the spatial and temporal distribution of the forage base (Swain and Morin 1996). Any of these postulations could be advanced to explain the sex-specific distribution of Summer Flounder in space and time. The importance, however, of estimating and understanding the sex-specific spatial and temporal distribution of Summer Flounder goes beyond these interesting ecological implications. Okamura et al. (2014) showed that biological reference points were highly sensitive to assumptions about effort allocation in space and time when seasonal migrations of offshore Lingcod *Ophiodon elongatus* were sex specific. A similar analysis for Summer Flounder is beyond the scope of this paper, but minimally the application of sex-aggregated ALKs to landings data should result in highly biased estimates of catch at age given that (1) female Summer Flounder grow much faster than males and (2) the recreational fishery lands females almost exclusively. Accounting for the complex spatiotemporal dynamics in the distribution of sex composition and fishing effort will undoubtedly complicate future stock assessments of Summer Flounder.

Given the importance of having an accurate estimate for the sex composition of the catch and given the failure of our method to recover an exact match by applying sex-specific processes to survey data due to what appear to be complex, time- and space-varying sex-specific fishery selectivities, it makes sense to explore other options for collecting sex composition data directly from fishery landings. For the recreational fishery, this would be easy to incorporate into any port sampling program that already collects length frequency data from the catch. We demonstrated that even when Summer Flounder are filleted at sea, the gonad remains intact and readily visible to the naked eye. An observer measuring the length of a Summer Flounder can thus determine its sex macroscopically, with no additional time or cost. Collecting sex-specific information from the commercial fishery may prove more difficult because fish are not filleted at sea to expose the gonads, and the fish are typically sold whole to market, which precludes full dissection of the fish to determine sex. We introduced a method for determining sex by using a small incision in the gut cavity. Although this may still be too invasive for some fishing ports or fish sellers or buyers, we encountered no resistance on the part of the many commercial docks that we sampled. The International Pacific Halibut Commission used microsatellite-based genetic techniques to measure the sex composition of the catch for Pacific Halibut *Hippoglossus stenolepis* (Loher et al., in press). This reduced the influence of spatial and temporal variability in survey and fishing effort when deriving sex composition of the catch using survey data, as it provided a direct measure of sex composition of the catch. A cost–benefit analysis comparing the purchase and dissection of fish versus the development of microsatellite techniques may reveal the most cost- and time-effective method for determining the sex composition of the commercial catch. No matter how the data are obtained, if a program is instituted to collect sex composition data directly from the Summer Flounder catch, that program would have to mirror the spatial and temporal distribution of the fishing effort since it appears that males and females are disaggregated in time and space.

The trends we observed in the sex, length, and age of the catch may interact with fishing regulations and induce changes in population structure. The mean length at age of individuals in the Summer Flounder population is declining over time, with none of the accompanying changes in length–weight relationship or condition index (NEFSC 2013) that one might expect from density-dependent influences on individual fish growth (Rose et al. 2001). Given the dynamic nature of climate change in the northwestern Atlantic Ocean (Scavia et al. 2002) and its influence on the fish stocks of this region (Rothschild and Jiao 2012; Sparrevohn et al. 2013), a simple, environmentally determined change in growth rate for the Summer Flounder stock cannot be discounted. Nonetheless, we presented a trend of increasing mean length at age with increasing MRS in the fishery landings. The MRS implemented for a fishery is selecting the fastest-growing fish at a given age while leaving behind the slower-growing fish to reproduce another year. A growing body of literature addresses the potential evolutionary effects of size-selective fishing (e.g., Conover and Munch 2002; Law 2007). Varying opinions exist concerning the time scale and fishing pressure that are needed to induce evolutionary change as opposed to simple phenotypic change in a fish population (e.g., Anderson and Brander 2009; Enberg et al. 2009). However, we cannot rule out the possibility that size-selective fishing pressure on the Summer Flounder population is causing a shift in the genetic structure of the stock, favoring slower growth and resulting in a decrease in the mean length at age over time.

Additionally, the proportion female at age is declining in the Summer Flounder population (NEFSC 2013). If an MRS selects the faster-growing fish at a given age in the population, those fish must be disproportionately female because of the difference in growth rates between the sexes. Our sampling found that the sex-at-age relationship from the recreational catch—and, to a lesser extent, from the commercial catch—was skewed toward females in comparison with the sex at age that would be expected from survey data. Simulation of MRS effects showed that this result accrued in part from the implementation of the MRS in each fishery. For example, the proportion of fish that were female increased with increasing TL in the survey as well as in the fisheries. Interestingly, in both the CFS data and the NFS data, the sex ratio remained relatively unchanged at around the 5-cm increment that encompassed the range of MRS values used for the recreational fishery (45–50 cm), despite the expectation that the sex ratio should increase as it did at each length increment before and after the 45–50-cm increment. This observation suggests that the MRSs implemented in the recreational fishery have altered the population sex ratio within this range of sizes. Therefore, one potential explanation for the trend of a decreasing proportion female at age in the population over the last decade (NEFSC 2013) is the increase in MRSs used to rebuild the stock over that same time period and the associated increases in female-specific fishing mortality. Sex-specific differences in fishery selectivity have been reported for other sexually dimorphic flatfishes that are regulated by MRSs (Briggs 1965; Loher and Hobden 2012). In fact, the sex composition estimates for commercial landings of Pacific Halibut indicate that males may never reach the mean sizes associated with full fishery selectivity at age (Clark and Hare 2006).

If current management strategies are driving changes in the structure of the Summer Flounder stock, it is important for managers to consider the potential adverse effects of those changes. Size-selective fishing can alter the sex ratio of a population (Kendall and Quinn 2013) and can influence the reproductive potential of the stock (Rowe and Hutchings 2003; Claereboudt et al. 2004; Rijnsdorp et al. 2010), and genetic traits that are selected against by implementing an MRS as a management strategy may be difficult to recover (Enberg et al. 2009).

ACKNOWLEDGMENTS

We would like to acknowledge all of the participating fishing docks, for-hire vessel captains, and mates that participated in this project. We thank the NMFS-NEFSC Aging Center, especially Eric Robillard; Integrated Statistics; Mark Luckenbach, Erin Reilly, Patti McGrath, and Roger Mann (Virginia Institute of Marine Science); Joe Cimino and Joe Grist (Virginia Marine Resources Commission), Chris Batsavage (North Carolina Division of Marine Fisheries); Jacqueline Wilson, Joe Costanzo, and Emerson G. Hasbrouck (Cornell Cooperative Marine Program); and the many technicians at the Haskin Shellfish Research Laboratory, Rutgers University. We are especially grateful to the Partnership for Marine Fisheries Science for providing 2 years of funding for this project through National Oceanic and Atmospheric Administration NMFS awards NA10NMF4720017 and NA10NMF4720402 and to the New Jersey Sea Grant College Program (Award NJSG-15-892).

REFERENCES

Able, K. W., and S. C. Kaiser. 1994. Synthesis of Summer Flounder habitat parameters. National Oceanic and Atmospheric Administration, Coastal Ocean Office, Decision Analysis Series 1, Silver Spring, Maryland.

Anderson, K. H., and K. Brander. 2009. Expected rate of fisheries-induced evolution is slow. Proceedings of the National Academy of Sciences of the USA 106:11657–11660.

Arnold, T. W. 2010. Uninformative parameters and model selection using Akaike's information criterion. Journal of Wildlife Management 74:1175–1178.

Azarovitz, T. R. 1981. A brief historical review of the Woods Hole Laboratory trawl survey time series. Canadian Special Publication of Fisheries and Aquatic Sciences 58:62–67.

Bonzek, C. F., J. Gartland, J. D. Lange Jr., and R. J. Latour. 2009. Northeast area monitoring and assessment program (NEAMAP) near shore trawl survey final report. Atlantic States Marine Fisheries Commission, Washington, D.C.

Briggs, P. T. 1965. The sport fisheries for Winter Flounder in several bays of Long Island. New York Fish and Game Journal 12:48–70.

Claereboudt, M. R., J. L. McIlwain, H. S. Al-Oufi, and A. A. Ambu-Ali. 2004. Relationships between fishing gear, size frequency and reproductive patterns for the Kingfish (*Scomberomorus commerson*, Lacepede) fishery in the Gulf of Oman. Pages 56–67 *in* A. I. L. Payne, C. M. O'Brien, and S. I. Rogers, editors. Management of shared fish stocks. Blackwell Scientific Publications, Oxford, UK.

Clark, W. G. 2004. A method of estimating the sex composition of the commercial landings from setline survey data. Pages 111–162 *in* L. Sadorus, editor. Report of assessment and research activities 2003. International Pacific Halibut Commission, Seattle.

Clark, W. G., and S. R. Hare. 2006. Assessment and management of Pacific Halibut: data, methods, and policy. International Pacific Halibut Commission, Scientific Report 83, Seattle.

Conover, D. O., and S. B. Munch. 2002. Sustaining fishery yields over evolutionary time scales. Science 297:94–96.

Enberg, K., C. Jorgensen, E. S. Dunlop, M. Heino, and U. Dieckmann. 2009. Implications of fisheries-induced evolution for stock rebuilding and recovery. Evolutionary Applications 2:394–414.

Fenske, K. H., M. J. Wilberg, D. H. Secor, and M. C. Fabrizio. 2011. An age- and sex- structured assessment model for American Eels (*Anguilla rostrata*) in the Potomac River, MD. Canadian Journal of Fisheries and Aquatic Sciences 68:1024–1037.

Gerritsen, H. D., D. McGrath, and C. Lordan. 2006. A simple method for comparing age–length keys reveals significant regional differences within a single stock of Haddock (*Melanogrammus aeglefinus*). ICES Journal of Marine Science 63:1096–1100.

Gerritsen, H. D., D. McGrath, C. Lordan, and X. Harlay. 2010. Differences in habitat selection of male and female Megrim (*Lepidorhombus whiffiagonis*, Walbaum) to the west of Ireland: a result of differences in life-history strategies between the sexes? Journal of Sea Research 64:487–493.

Gorchinsky, K. V. 1998. Spatial and temporal variations in Greenland Halibut (*Reinhardtius hippoglossoides*) sex ratio-at-age in NAFO divisions 0B, 2GH, 3K and 3LM. North Atlantic Fisheries Organization Scientific Council Studies 31:73–78.

Isermann, D. A., and C. T. Knight. 2005. A computer program for age length keys incorporating age assignment to individual fish. North American Journal of Fisheries Management 25:1153–1160.

Jones, C. M., R. Cook, J. Simmonds, and H. Sparholt. 2013. Summary report of the 57th northeast regional stock assessment review committee (SARC 57). National Marine Fisheries Service, Northeast Fisheries Science Center, Woods Hole, Massachusetts.

Kendall, N. W., and T. P. Quinn. 2013. Size-selective fishing affects sex ratios and the opportunity for sexual selection in Alaskan Sockeye Salmon *Oncorhynchus nerka*. Oikos 122:411–420.

King, N. J., G. C. Nardi, and C. J. Jones. 2001. Sex-linked growth divergence of Summer Flounder from a commercial farm: are males worth the effort? Journal of Applied Aquaculture 11:77–78.

Kraus, R. T., and J. A. Musick. 2001. A brief interpretation of Summer Flounder, *Paralichthys dentatus*, movements and stock structure with tagging data on juveniles. Marine Fisheries Review 63:1–6.

Law, R. 2007. Fisheries-induced evolution: present status and future directions. Marine Ecology Progress Series 335:271–277.

Loher, T., and J. Hobden. 2012. Length and sex effects on the spatial structure of catches of Pacific Halibut (*Hippoglossus stenolepis*) on longline gear. U.S. National Marine Fisheries Service Fishery Bulletin 110:46–51.

Loher, T., M. Woods, I. Jimenez-Hidalgo, and L. Hauser. In press. Variance in age-specific sex composition of Pacific Halibut catches, and comparison of statistical and genetic methods for reconstructing sex ratios. Journal of Sea Research. DOI: 10.1016/j.seares.2015.06.004.

Luckenbach, J. A., R. J. Borski, H. V. Daniels, and J. Godwin. 2009. Sex determination in flatfishes: mechanisms and environmental influences. Seminars in Cell and Developmental Biology 20:256–263.

Maunder, M. N., and R. A. Wong. 2011. Approaches for estimating natural mortality: application to Summer Flounder (*Paralichthys dentatus*) in the U.S. mid-Atlantic. Fisheries Research 111:92–99.

McGilliard, C. R., W. Palsson, W. Stockhausen, and J. Ianelli. 2013. Assessment of the deepwater flatfish stock in the Gulf of Alaska. Pages 403–536 *in* Stock Assessment and Fishery Evaluation Report for the Groundfish Resources of the Gulf of Alaska. North Pacific Fishery Management Council, Anchorage, Alaska.

Methot, R. D. Jr., and C. R. Wetzel. 2013. Stock synthesis: a biological and statistical framework for fish stock assessment and fishery management. Fisheries Research 142:86–99.

Morgan, M. J., and E. A. Trippel. 1996. Skewed sex ratios in spawning shoals of Atlantic Cod (*Gadus morhua*). ICES Journal of Marine Science 53:820–826.

Morson, J. M., E. A. Bochenek, E. N. Powell, and J. E. Gius. 2012. Sex-at-length of Summer Flounder landed in the New Jersey recreational party boat fishery. North American Journal of Fisheries Management 32:1201–1210.

Myers, R. A., M. W. Smith, J. M. Hoenig, N. Kmeicik, M. A. Luehring, M. T. Drake, P. J. Schmalz, and G. G. Sass. 2014. Size- and sex-specific capture and harvest selectivity of Walleyes from tagging studies. Transactions of the American Fisheries Society 143:438–450.

NEFSC (Northeast Fisheries Science Center). 1988. An evaluation of the bottom trawl survey program of the Northeast Fisheries Science Center. NOAA Technical Memorandum NMFS-F/NEC-52.

NEFSC (Northeast Fisheries Science Center). 1997. Report to the 24th northeast regional stock assessment workshop (24th SAW): stock assessment review committee (SARC) consensus summary of assessments. NEFSC, Reference Document 97-12, Woods Hole, Massachusetts.

NEFSC (Northeast Fisheries Science Center). 2013. 57th Northeast regional stock assessment workshop (57th SAW) assessment report. NEFSC, Reference Document 13-16, Woods Hole, Massachusetts.

Ogle, D. H. 2014. FSA: fisheries stock analysis. R package version 0.4.16. Available: https://fishr.wordpress.com/fsa/. (September 2015).

Okamura, H., M. K. McAllister, M. Ichinokawa, L. Yamanaka, and K. Holt. 2014. Evaluation of the sensitivity of biological reference points to the spatio-temporal distribution of fishing effort when seasonal migrations are sex-specific. Fisheries Research 158:116–123.

Packer, D. B., S. J. Griesbach, P. L. Berrien, C. A. Zetlin, D. L. Johnson, and W. W. Morse. 1999. Essential fish habitat source document: Summer Flounder, *Paralichthys dentatus*, life history and habitat characteristics. NOAA Technical Memorandum NMFS-NE-151.

Pentilla, J., and L. M. Dery. 1988. Age determination methods for northwest Atlantic species. NOAA Technical Report NMFS-72.

Planes, S., E. Macpherson, F. Biagi, A. Garcia-Rubies, J. Harmelin, M. Harmelin-Vivien, J.-Y. Jouvenel, L. Tunesi, L. Vigliola, and R. Galzin. 1999. Spatio-temporal variability in growth of juvenile sparid fishes from the Mediterranean littoral zone. Journal of the Marine Biological Association of the UK 79:137–143.

Poole, J. C. 1961. Age and growth of the fluke in Great South Bay and their significance to the sport fishery. New York Fish and Game Journal 8:1–18.

R Development Core Team. 2014. R: a language and environment for statistical computing. R Foundation for Statistical Computing, Vienna.

Richards, R. A., P. C. Nitschke, and K. A. Sosebee. 2008. Population biology of monkfish (*Lophius americanus*). ICES Journal of Marine Science 65:1291–1305.

Rijnsdorp, A. D., D. J. G. van Damme, and P. R. Witthames. 2010. Implications of fisheries-induced changes in stock structure and reproductive potential for stock recovery of a sex-dimorphic species, North Sea Plaice. ICES Journal of Marine Science 67:1931–1938.

Rose, K. A., J. H. Cowan Jr., K. O. Winemiller, R. A. Myers, and R. Hilborn. 2001. Compensatory density dependence in fish populations: importance, controversy, understanding and prognosis. Fish and Fisheries 2:293–327.

Rothschild, B. J., and Y. Jiao. 2012. Characterizing variation in northwest Atlantic fish-stock abundance. ICES Journal of Marine Science 69:913–922

Rowe, S., and J. A. Hutchings. 2003. Mating systems and the conservation of commercially exploited marine fish. Trends in Ecology and Evolution 18:567–572.

Ryer, C. H. 2008. A review of flatfish behavior relative to trawls. Fisheries Research 90:138–146.

Sahin, T., and E. Gunes. 2010. Seasonal variation in length, weight, and sex distribution of flounder (*Platichthys flesus luscus*) in the southeastern Black Sea. Journal of Fisheries Sciences 4:238–245.

Sampson, D. B. 2014. Fishery selection and its relevance to stock assessment and fishery management. Fisheries Research 158:5–14.

Scavia, D., J. C. Field, D. F. Boesch, R. W. Buddemeier, V. Burkett, D. R. Cayan, M. Fogarty, M. A. Harwell, R. W. Howarth, C. Mason, D. J. Reed, T. C. Royer, A. H. Sallenger, and J. G. Titus. 2002. Climate change impacts on U.S. coastal and marine ecosystems. Estuaries 25:149–164.

Smith, R. W., and F. C. Daiber. 1977. Biology of Summer Flounder, *Paralichthys dentatus*, in Delaware Bay. U.S. National Marine Fisheries Service Fishery Bulletin 75:823–830.

Somerton, D. A., K. Williams, P. G. von Szalay, and C. S. Rose. 2011. Using acoustics to estimate the fish-length selectivity of trawl mesh. ICES Journal of Marine Science 68:1558–1565.

Sparrevohn, C. R., M. Lendegren, and B. R. Mackenizie. 2013. Climate-induced responses of commercially important flatfish species during the 20th century. Fisheries Oceanography 22:400–408.

Stari, T., K. F. Preedy, E. McKenzie, W. S. C. Gurney, M. R. Heath, P. A. Kunzlik, and D. C. Speirs. 2010. Smooth age length keys: observations and implications for data collection on North Sea Haddock. Fisheries Research 105:2–12.

Swain, D. P. 1997. Sex-specific temperature distribution of American Plaice (*Hippoglossoides platessoides*) and its relation to age and abundance. Canadian Journal of Fisheries and Aquatic Sciences 54:1077–1087.

Swain, D. P., and R. Morin. 1996. Relationship between geographic distribution and abundance of the American Plaice (*Hippoglossoides platessoides*) in the southern Gulf of St. Lawrence. Canadian Journal of Fisheries and Aquatic Sciences 53:106–119.

Swain, D. P., and R. Morin. 1997. Effects of age, sex and abundance on the bathymetric pattern of American Plaice in the southern Gulf of St. Lawrence. Journal of Fish Biology 50:181–200.

Su, N.-J., C.-L. Sun, A. E. Punt, S.-Z. Yeh, W.-C. Chaing, Y.-J. Chang, and H.-Y. Chang. 2013. Effects of sexual dimorphism on population parameters and exploitation ratios of Blue Marlin (*Makaira nigricans*) in the northwest Pacific Ocean. Aquatic Living Resources 26:19–24.

Su, N.-J., C.-L. Sun, A. E. Punt, S.-Z. Yeh, and G. DiNardo. 2011. Evaluation of a spatially sex-specific assessment method incorporating a habitat preference model for Blue Marlin (*Makaira nigricans*) in the Pacific Ocean. Fisheries Oceanography 20:415–433.

Venables, W. N., and B. D. Ripley. 2002. Modern applied statistics with S, 4th edition. Springer, New York.

Wang, S.-P., C.-L. Sun, A. E. Punt, and S.-Z. Yeh. 2005. Evaluation of a sex-specific age-structured assessment method for the Swordfish, *Xiphias gladius*, in the North Pacific Ocean. Fisheries Research 73:79–97.

Wang, S.-P., C.-L. Sun, A. E. Punt, and S.-Z. Yeh. 2007. Application of a sex-specific age-structured assessment method for the Swordfish, *Xiphias gladius*, in the North Pacific Ocean. Fisheries Research 84:282–300.

Wilderbuer, T. K., and B. J. Turnock. 2009. Sex-specific natural mortality of Arrowtooth Flounder in Alaska: implications of a skewed sex ratio on exploitation and management. North American Journal of Fisheries Management 29:306–322

Wilson, K., and I. Hardy. 2002. Statistical analysis of sex ratios: an introduction. Pages 48–92 *in* C. W. Ian, editor. Sex ratios: concepts and research methods. Cambridge University Press, Cambridge, UK.

4

Calico Scallop *Argopecten gibbus* Abundance on the Cape Canaveral Bed and on Florida's Gulf of Mexico Shelf

Stephen P. Geiger*
Florida Fish and Wildlife Conservation Commission, Fish and Wildlife Research Institute, 100 8th Avenue Southeast, St. Petersburg, Florida 33701, USA

William S. Arnold
National Oceanic and Atmospheric Administration, Southeast Regional Office, 263 13th Avenue South, St. Petersburg, Florida 33701, USA

Sarah Stephenson and Keith Fischer
Florida Fish and Wildlife Conservation Commission, Fish and Wildlife Research Institute, 100 8th Avenue Southeast, St. Petersburg, Florida 33701, USA

Abstract

Annual landings of the calico scallop *Argopecten gibbus* fishery in the southeastern United States increased from less than 3,000 kg in 1959 to 19.5 million kg of adductor muscle meat in 1984. The fishery began to collapse in early 1986 in Florida and fell below 1 million kg/year in the mid-1990s. From 2002 through 2011, the only reported landings were 550 kg in 2009, but annual landings of less than 200,000 kg beginning in 2012 indicate the resource rebounded enough in that year to create renewed interest in the fishery. When the fishery was developing, the National Marine Fisheries Service (NMFS) sponsored 59 cruises (1956 through the mid-1970s) to map the distribution of the stock and found the calico scallop stocks on Florida's east coast principally occurred around the 40-m depth contour near the shelf break. By the early 1980s, surveys were no longer routinely conducted. Two surveys in the mid-1990s were dedicated to investigating the collapse of the stock. The present study consisted of eight cruises from 2004 to 2006: four each to the beds off Cape Canaveral on the eastern coast and the poorly documented beds off southwestern Florida. Cruises studying baitfish and cruises in the Gulf of Mexico by the Southeast Area Monitoring and Assessment Program (SEAMAP) yielded data that provided additional calico scallop abundance estimates. Abundant, albeit small, scallops were found on both coasts during three of the four cruises to each coast though none were detected in fall 2004 on either coast. Our results concur with those historic NMFS surveys conducted prior to the period of heaviest exploitation of the resource (1974–1999); the center of distribution for calico scallops on the Cape Canaveral beds still occurs near the 40-m depth contour. The essential habitat for calico scallops, shell base, which was suspected to have declined during the most active fishery periods, was found to have persisted consistently at that depth. There were few scallops with a shell height greater than 40 mm, the minimal size considered acceptable for commercial exploitation.

Subject editor: Patrick Sullivan, Cornell University, Ithaca, New York

*Corresponding author: steve.geiger@myfwc.com

Fisheries are prone to both long-term trends and very short-term fluctuations in abundance. The changes may be cyclic on both short (Beukema et al. 2001) and long terms (Kimmel et al. 2013), episodic (Powell et al. 2008), or reflect relatively permanent shifts related to changes in the ecosystem (Peterson et al. 2008). In the long-lived Atlantic sea scallop *Placopecten magellanicus* the increase in abundance observed from 1999 through 2005 occurred after new management measures were implemented in 1994 (Hart and Rago 2006). Many scallop species have life spans as short as 1–2 years (Marshall 1963; Wolff 1988), and because many adults have only one chance to spawn, the populations are prone to large and rapid swings in abundance (Maeda-Martínez et al. 1993; Wolff and Mendo 2000). An adaptive management plan that considers such population swings might be more effective than a fixed harvest rate.

The calico scallop *Argopecten gibbus* of the southeastern United States, which can reach maturity at 3 months and typically lives 18–24 months (Blake and Shumway 2006), has also exhibited highly variable abundance. Federal port samplers conducted 11 harvest surveys of seafood dealers from 1927 through 1945, and, of those, three indicated a species of scallop, distinctly separate from the bay scallop *Argopecten irradians*, had been landed in Florida's southwestern counties, Pinellas and Sarasota (Fiedler 1938). The earliest landings of what were thought to be calico scallops were made in Sarasota County in 1927, when 5,800 kg of meat (only the adductor muscle meat is consumed in the United States) was recorded. This other species was recorded as sea scallop, but the Atlantic sea scallop is not found in Florida, while calico scallops are known to be present in abundance immediately offshore of Sarasota, so *Argopecten gibbus* was most likely the identity of the species, though we can never be certain.

The National Marine Fisheries Service (NMFS) began to conduct focused exploratory dredge and trawl surveys for calico scallops in the late 1950s (Bullis and Ingle 1958). Data from numerous NMFS survey vessels were prepared as unpublished cruise reports and were presented to the commercial fishing industry mainly through summary reports (L. L. May, R. O. Maurer Jr., and J. B. Rivers, NMFS, unpublished data; L. L. May, D. L. Sutherland, S. B. Drummond, and J. B. Rivers, NMFS, unpublished data) (Table 1). The fishery had become firmly established by 1961, the most consistently productive stocks located in the Cape Canaveral beds off east-central Florida. The calico scallop fishery was active from 1961 through 2001 (Figure 1; data compiled from NMFS annual landings query for calico scallops [http://www.st.nmfs.noaa.gov/st1/commercial/landings/annual_landings.html] and prior Reports of the Commissioner, U.S. Commission of Fish and Fisheries), having first developed in the northeastern Gulf of Mexico (Bullis and Ingle 1958) before stocks were also found in North Carolina (Cummins 1971), South Carolina (SAFMC 1998), and eventually east-central Florida (Taylor 1967; Cummins 1971). Small numbers of calico scallops have occasionally been landed in Georgia as bycatch in commercial shrimp trawls and have probably not been directly targeted for there. The NMFS conducted surveys through 1968 using dredges, or to a lesser extent various trawls, primarily to identify commercially fishable calico scallop stocks. These surveys targeted known resource areas and minimized random exploration for new fishing grounds. Data were recorded as presence or absence rather than quantitatively when samples were below a threshold that the investigators deemed too low for commercial production, i.e., about one bushel (~35 L) per 1-h tow. Surveys from 1969 through 1976 relied principally on video transects with occasional dredge samples for ground truthing (May, Maurer, and Rivers, unpublished; May, Sutherland, Drummond, and Rivers, unpublished). No directed calico scallop surveys were made from 1976 until the mid-1980s.

The calico scallop fishery achieved a record level of production in 1984 (20 million kg of meat) and employed more than 2,600 people in Florida. The fishery collapse on Florida's east coast began in early 1986 to the point that no scallops were available for harvest on the Cape Canaveral beds (Rockwood and Pompe 1988); it rebounded in 1987. Average annual direct spending by the calico scallop fishing industry during its peak (1984–1987) was around US$23 million, and total indirect economic impact in the region averaged more than $86 million/year (Rockwood and Pompe 1988). Statewide landings in Florida went through another cycle of decline and rebound in the 1990s but still averaged 0.8 million kg of meat. Calico scallop harvests were essentially nonexistent from 2002 through 2012. Modest harvests were achieved in 2013, and 5,100 kg of calico scallop meat was landed in Florida. In 2014 harvests were reported in federal waters off Florida, but catches were landed in other states (federal landings data were not available as of publication date). North Carolina has had no commercial landings of calico scallops in more than a decade.

Many factors may influence the interannual success of the Cape Canaveral calico scallop fishery including disease (Moyer et al. 1993), predation (Schwartz and Porter 1977), senility (Carpenter 1967; Roe et al. 1971), and commercial harvest. Settling scallop larvae requires a hard and stable substrate for attachment and typically use the valves of living and dead mollusk shells (Schwartz and Porter 1977). Traditionally, most calico scallop harvests were made with modified otter trawls using 7.6-cm (3 in) mesh. The process swept the bottom of all living and dead matter, and the entire catch was taken dockside for sorting both calico scallops and nontarget fauna, flora, and substrate. This activity would remove essential habitat for juvenile calico scallops (Nelson et al. 1992), though at least one harvester processed the catch at sea and returned the bycatch to the scallop grounds (SAFMC 1998). This same fisher, one of the last to harvest calico scallops in Florida in the 1990s, indicated he would travel the entire state coastline and would harvest from many nontraditional beds, including the nearshore zone off southwestern Florida around Sanibel–Captiva and Marco islands. Increasingly, the loss of shell is a worldwide concern for shellfisheries (Kraeuter et al. 2003; Guay and Himmelman

TABLE 1. The listed NOAA survey cruises with historical data were conducted from several research vessels: RV *Silver Bay* (SB), RV *Oregon* (OR), RV *Bowers* (BO), RV *Oregon II* (O2), RV *Delaware II* (D2), RV *Bellows* (BW), RV *GeoQuest* (GQ), and RV *Bonnie E* (BE). For most cruises, a general description and hand-drawn map were provided in the cruise report. In the column, Data description, latitude–longitude (LL) indicates that the cruise log provided station location and catch rates (of calico scallops). Common commercial species (CCS) indicates that calico scallops were not necessarily the primary focus of the sampling but that scallops were recorded. For RV *Bowers* cruises, spatial data are generally provided in terms of LORAN lines accompanied with indications of the depth along the LORAN line at which a catch of scallops was made. When the data description indicates text and map (TM) only, we have been unable to locate the detailed cruise logs with station and catch data. LL indicates that the latitude and longitude of the samples were recorded; P–A indicates that presence and absence of scallops was noted, but not abundance; *N* indicates the number of useful station samples. For some cruises (*), only those catches exceeding 1 bushel (~35 L) were recorded, so the other stations cannot reliably be considered to have had no scallops in their catch; in those cases (OR), *N* is equal to the number of catches exceeding 1 bushel, or when both values are reported, the total number of scallop dredges is included as well (*N*:*N*). GoM: Gulf of Mexico; Atl: Atlantic; FL: Florida; GA: Georgia; NC: North Carolina; SC: South Carolina; Cal. scal.: calico scallops.

Vessel	Cruise	Dates	Location	Data description	*N*
SB	2	Jul 12–29, 1957	GoM, FL	TM, LL, P–A	
	10	Jul 17–Aug 4, 1958	GoM, FL	TM, LL, P–A	
	18	Sep 2–29, 1959	Atl, NC	TM, LL, CCS, P–A	
	20	Nov 21–Dec 13, 1959	Atl, NC	TM, LL, CCS	3
	21	Jan13–29, 1960	Atl, GA–FL	TM, LL, CCS, P–A	
	22	Feb 16–Mar 18, 1960	Atl, NC–SC	TM, LL, CCS.	50
	23	Apr 13–May 6, 1960	Atl, FL	TM, LL	199
	24	May 26–Jun 14, 1960	Atl, FL	TM, LL	41
	25	Jul 13–30, 1960	Atl, SC–NC	TM, LL, CCS	
	26	Oct 11–Nov 15, 1960	Atl, FL	TM, P–A	
	27	Dec 1–16, 1960	Atl, SC	TM	
	28	Jan 18–Feb 10, 1961	Atl, FL	TM	
	29	Feb 27–Mar 11, 1961	Atl, NC	TM, LL, CCS	61
	30	Apr 17–May 11, 1961	Atl, FL	TM, LL, CCS	132
	31	Jul 5–21, 1961	Atl, FL	TM, LL, CCS	118
	32	Aug 18–23, 1961	Atl, NC	TM, LL, CCS	
	33	Sep 23–Oct 6, 1961	Atl, FL	TM, LL, CCS	84
	35	Nov 28–Dec 15, 1961	Atl, FL	TM	
	36	Jan 15–Feb 6, 1961	Atl, FL	TM, LL	22
	39	May 24–Jun 12, 1962	Atl, NC	TM, LL, CCS	1
	41	Aug 21–Sep 7, 1962	Atl, FL	TM, LL	103
	42	Sep 24–Oct 2, 1962	Atl, FL	TM	
	47	Mar 25–Apr 1, 1963	Atl, FL	TM	
	51	Nov 6–19, 1963	Atl, FL	Text, LL	196
	54	Feb 4–16, 1964	Atl, FL	TM	
	55	Feb 25–Mar 4, 1964	Atl, FL	TM, LL, CCS	20
OR	123	Nov 6–20, 1967	Atl, FL–GA	TM, LL, CCS*	27:110
	124	Dec 4–19, 1967	Atl, FL	TM	
	126	Feb 12–29, 1968	Atl, FL	TM, LL, CCS*	120:225
	128	Apr 12–27, 1968	Atl, FL	TM, LL, CCS*	76:130
	130	Jun 10–26, 1968	Atl, FL	TM, LL, CCS*	83:181
	132	Aug 22–26, 1968	Atl, FL	TM, LL, CCS*	37:99
	134	Oct 8–25, 1968	Atl, FL	TM, CCS*	
	136	Dec 9–20, 1968	Atl, FL	TM, LL, CCS*	87:195
BO	85	Jul 21–Aug 5, 1969	Atl, FL	TM (LORAN)	
	93	Jun 17–25, 1970	Atl, FL	TM (LORAN)	
	98	Oct 21–Nov 2, 1970	GoM, FL	TM (LORAN)	
	99	Jan 21–27, 1971	Atl, FL	Cal. scal., no spatial data	
	100	Mar 10–17, 1971	Atl, FL	Cal. scal., no spatial data	
	101	Apr 15–21, 1971	Atl, FL	Cal. scal., no spatial data	

TABLE 1. Continued.

Vessel	Cruise	Dates	Location	Data description	*N*
	102	Apr 28–May 8, 1971	Atl, GA–SC	TM	
	104	Jun 3–23, 1971	Atl, FL	TMs, LORAN lines	
	106	Jul 17–Aug 17, 1971	GoM, FL	TM	
	109	May 1–Jun 2, 1972	Atl, FL	TM, LORAN lines	
	110	Jul 12–17, 1972	Atl, FL	TM, LORAN lines	
	112	Sep 13–Oct 10, 1972	Atl, NC	TM, LORAN	
	114	Jan 9–26, 1973	Atl, FL	TM	
	120	Oct 25–Nov 18, 1973	Atl, FL	TM, LORAN	
	122	Apr 22–May 16, 1974	Atl, NC	TM	
	124	Aug 13–Oct 23, 1974	Atl, FL	TM	
	128	Apr 26–May 13, 1975	GoM	Text	
	138	Mar 9–26, 1976	Atl, FL	Text	
O2	59	Jun 10–27, 1975	Atl, FL	TM, LORAN	
	65	Mar 19–Apr 8, 1976	Atl, GoM	Text	
	76	Mar 27–Apr 11, 1977	Atl, FL	Text	
	122	Oct 14–Nov 20, 1981	GoM, Atl	Some text	
	165	Mar 16–22, 1987	Atl	LL, CCS	40
D2		Jun 1982	Atl, FL	Text, LL	17
BW		Aug 2–8, 1989	Atl, FL	TM, LL	40
GQ		Nov 20–27, 2004	Atl, FL	TM, LL	59
		Apr 29–May 13, 2005	Atl, FL	TM, LL	59
		Oct 11–18, 2005	Atl, FL	TM, LL	58
		Jul 18–21, 2006	Atl, FL	TM, LL	54
BE		Oct 27–31, 2004	GoM FL	TM, LL	49
		Sep 15–Nov 7, 2005	GoM FL	TM, LL	63
		Apr 17–21, 2006	GoM FL	TM, LL	58
		Aug 21–Sep 22, 2006	GoM FL	TM, LL	60

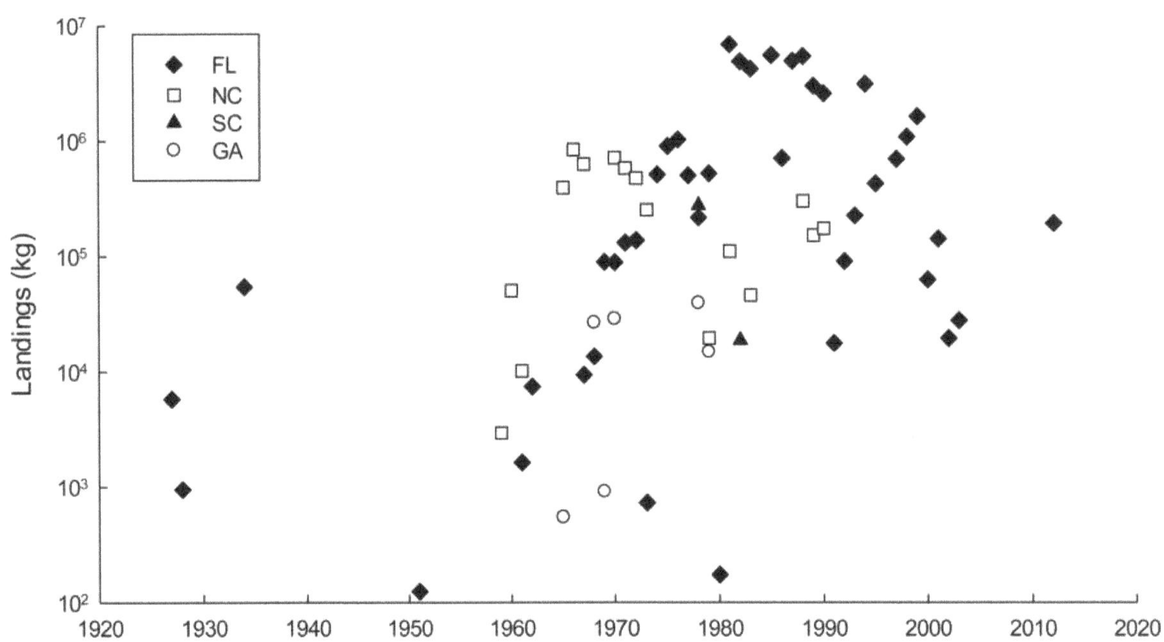

FIGURE 1. Recorded landings of calico scallop meat from southeastern U.S. states: Florida (FL), North Carolina (NC), South Carolina (SC), and Georgia (GA). See text for landings data sources.

2004; Bourgeois et al. 2006). The hypothesis that the removal of calico scallop shell limits the recruitment of scallop larvae is controversial within the industry (SAFMC 1998), and the South Atlantic Fishery Management Council has an ongoing need for habitat surveys to assess the impacts of trawling (SAFMC 1998). According to NMFS landings data, 73 million kg of calico scallop meat was harvested from Florida waters from 1976 through 2000, which represents roughly 10% by weight of the scallops harvested (shell and viscera are not used or reported) and does not measure bycatch, which is highly variable and for which it is difficult to find precise data. The total removal of substrate that serves as habitat would have been colossal given almost four decades of fishing.

Recent landings suggest that stocks of calico scallops may be large enough again to support profitable commercial harvests, and information on their status could be used to guide the management of the fishery. In this study we present information concerning the distribution, abundance, and size distribution of calico scallops and their shell during 2005–2013. Although a comparison with historic studies would be ideal, direct quantitative comparisons of contemporary work with historic studies were not feasible. Historic surveys focused on reporting commercial quantities of catch only and thus excluded zero- and low-catch data. Nevertheless, gross differences in distribution and abundance of calico scallops are compared between the two periods as much as possible.

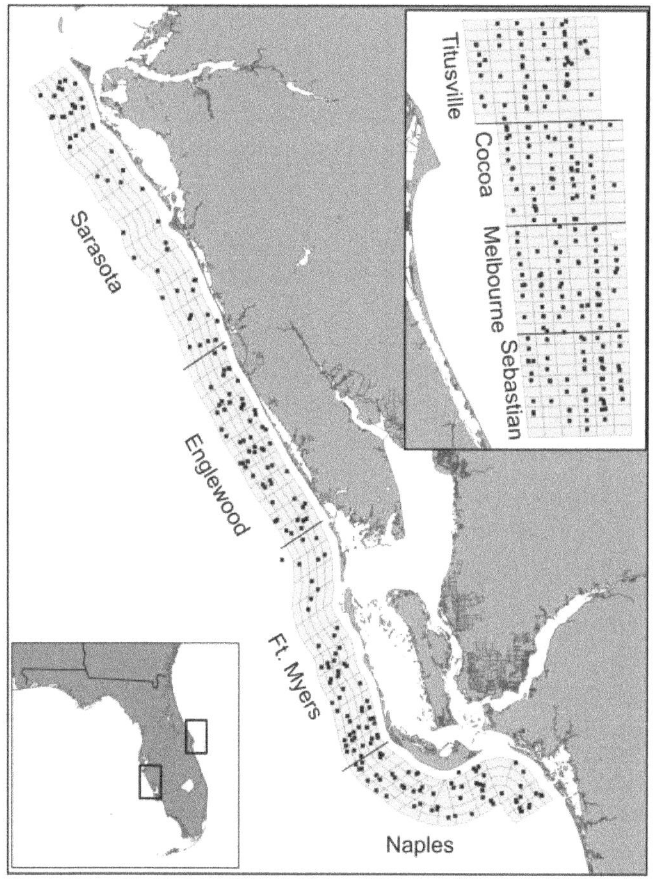

FIGURE 2. Florida east-central and southwestern study sites, 2004–2006. Light gray shaded areas represent the gridded sampling zones; black squares represent stations at which tows were conducted.

METHODS

2004–2006 study.—Dredge surveys were conducted in four sampling periods on each coast. The Cape Canaveral fishing zone located off east-central Florida was sampled during November 2004, April–May 2005, October 2005, and July 2006 from the RV *Geoquest*. The fished zone of the nearshore southwestern Florida shelf was sampled during October 2004, September–November 2005, April 2006 (RV *Bonnie E*), and August–September 2006 (RV *Veliger*, a 6.1-m [20 ft] runabout, and RV *Callinectes*, a 7.9-m [26 ft] runabout).

During each sampling period we conducted ~15 scallop dredges (Table 1) in each of four latitudinal zones on each coast: east (Sebastian, 27.797–28.045°N; Melbourne, 28.045–28.305°N; Cocoa Beach, 28.305–28.551°N; Titusville, 28.551–28.881°N) and west (Naples, 26.34–26.45°N; Fort Myers, 26.45–26.8°N; Englewood, 26.8–27.1°N; Sarasota, 27.1–27.5°N) (Figure 2). Each zone was divided into a grid of 60 stations resulting in a total of 240 stations for each coast. Stations sampled were randomly chosen in the depth ranges of roughly 20–80 m on the east coast and 5–15 m on the west coast. These depth zones encompassed a large majority of the locations for which we were able to document that historical calico scallop harvests had occurred. On each sampling day we attempted to sample the randomized stations, although we occasionally sampled a different station (typically one closer to shore or that would reduce total daily transit) when weather was inclement or it was getting dark (for safety reasons we sampled only during daylight hours).

Dredges were towed at 1–2 knots for 5 min per sample. The dredge mouth was 0.69 m wide and 0.50 m high, and had a bag length of 1.2 m when a mechanical winch was available (RV *GeoQuest* and RV *Bonnie E*) or 0.42 m wide and 0.25 m high, with a bag length 0.7 m (for the final west coast cruise only). The 0.42-m dredge was deployed from the stern cleat of a standard runabout and retrieved by hand for our final cruise. Both frames were equipped with a net that had a 3-cm (1.25 in)-stretch mesh and a 1-m bridle. Data were recorded as the weight and number of calico scallops caught ($kg{\cdot}min^{-1}{\cdot}m^{-1}$ and $n{\cdot}min^{-1}{\cdot}m^{-1}$, respectively) standardized to the width of the mouth of the net (m) and time towed (min).

During each tow we recorded initial and final GPS coordinates as an estimate of the length of the tow. We also recorded start and stop times, water column depth, and the length of the wire deployed. The appropriate wire angle had been determined on preliminary research cruises (data not included), during which we attached a downward-facing, real-time video system to the upper edge of the mouth of the dredge. The length of the cable deployed and tow speed were modified until the lower edge of the mouth of the frame consistently remained in contact with the sediment surface but did not dig deeply into it. The optimal wire-to-depth ratio was about 3.5:1 at 2 knots and increased with depth and vessel speed. The optimal attachment point for the bridle was at roughly 16.5 cm from the bottom of the frame. In practice, the wire was deployed until approximately a 3:1 ratio (west coast) to 4:1 ratio (east coast) was reached, and then the boat speed was decreased until tension occurred on the wire, indicating that the net was towing on the sediment surface. At the east coast stations we were limited to ~65 m maximal depth because the capacity of the available winch was ~180 m of wire.

When processing a dredge sample, we weighed the entire catch. In most cases, we sorted the entire catch, but we split some very large samples (>20 kg) into two to four equal subsamples by weight and sorted only one of those subsamples. All scallops from the portion of the sample that was sorted were removed and weighed. Next, the scallop shells and other mollusk shells were each separated and weighed. Shell heights were measured for a maximum of 30 live scallops from each dredge and the first 15 scallops collected in each zone during each season were preserved for histological analysis of reproduction and disease (authors' unpublished data). Many of the spat detected were originally recorded on data sheets to be adults but were later classified as loose spat because their shell heights were less than 23 mm, the size at which sexual maturity has been reported based on gonad color changes (Miller et al. 1979). Scallops as small as 22 mm had ripe gonads and contained both male and female gametes. These were observed in histological sections of gonad examined for *Marteilia*, a protozoan parasite. Any observations of spat (juveniles < 23 mm shell height) were recorded as either loose or attached to substrate. In an attempt to identify settlement substrate and the proportion of total calico shell, for any tow in which spat were observed we calculated the amount of other shell and all other catch. Assuming that the spat had no substrate preference, we calculated the expected frequency distribution of settled spat as the proportion of each substrate type multiplied by the total number of spat observed in the sample. The observed frequency of attached spat and their settlement substrate were compared with this theoretical expected frequency using a simple chi-square test. Loose scallops < 23 mm shell height were considered to be spat as were any attached scallops. Catch data were standardized for time, dredge width, and any subsampling. Graphic representations of the data were created using ArcGIS.

The hypothesis that the density of scallops, scallop spat, scallop shells, and other mollusk shells were equal among stations was tested by ANOVA, with the four latitudinal zones as a fixed factor. Season and coast also served as fixed factors. Because heterogeneity of variances was a common character of the data, Kruskal–Wallis or Mann–Whitney tests were applied to individual hypotheses when appropriate. The null hypothesis that we tested was that all stations had equal density, and that no seasonal variation occurred. The alternative hypothesis anticipated many stations (or seasons) with densities of zero, interspersed with fewer stations (or seasons) supporting relatively high densities.

Supplemental and historical data sets.—Additional Gulf of Mexico data collections from two bottom trawl surveys were provided by the Fisheries Independent Monitoring (FIM) group of the Florida Fish and Wildlife Conservation Commission's (FWC) Fish and Wildlife Research Institute (FWRI). Data from FWRI's annual spring baitfish survey were summarized from 2009 to 2013. Data were also obtained from summer (June and July) and fall (October and November) groundfish trawl surveys conducted by SEAMAP and were summarized for the northern Gulf of Mexico (Gulf) east of the Mississippi River, including Florida's pilot trawl program in 2008 and 2009 in the northeastern Gulf, which was expanded to full-scale sampling in 2010 to the entire eastern Gulf. The focus of these two surveys was fish stock assessment, but both contained data useful to the present calico scallop study.

The study area for the spring baitfish survey encompassed the nearshore waters of Tampa Bay and Charlotte Harbor from 6 to 28 m depth. The study area was divided into two latitude strata based on degrees of latitude. The central stratum spanned from 27°N to 28°N and the southern stratum spanned from 26°N to 27°N. These strata were further divided into two depth strata. The inshore stratum covered depths from 6 to 11 m; the offshore stratum covered depths from 12 to 28 m. For every survey, three transects (lines of latitude) were randomly chosen in each latitude stratum. These transects started at 6 m and ended at 28 m depth, so transect length depended on the slope. Six trawl stations, three in the inshore stratum and three in the offshore stratum, were then randomly chosen for each transect. A balloon trawl (nominal opening, 9.9 m wide and 4.5 m high; 36.5-m trawl bridle lines; cod end stretch mesh, 43.75 mm; 2.2-m^2 china V-doors with nominal weights of 408.2 kg) with a 19.8-m headrope was towed at 3 knots for 30 min at each station. Stations that could not be trawled due to obstructions or hard bottom were skipped and an alternate station selected and sampled. The entire catch for most trawls was processed, though for a few large catches, a subsample was randomly selected for processing and the total value proportionately extrapolated. As many as 50 scallops were randomly selected and then measured (hinge to outer margin of the shell) for determination of length distribution. We present only an overview of the calico scallop data extracted from that larger data set.

The Gulf of Mexico SEAMAP trawl survey used a standard 12.8-m shrimp trawl and a stratified-random sampling design, with stratification based on a combination of depth (23 depth strata from 9 to 110 m), time of day (day or night), and NMFS

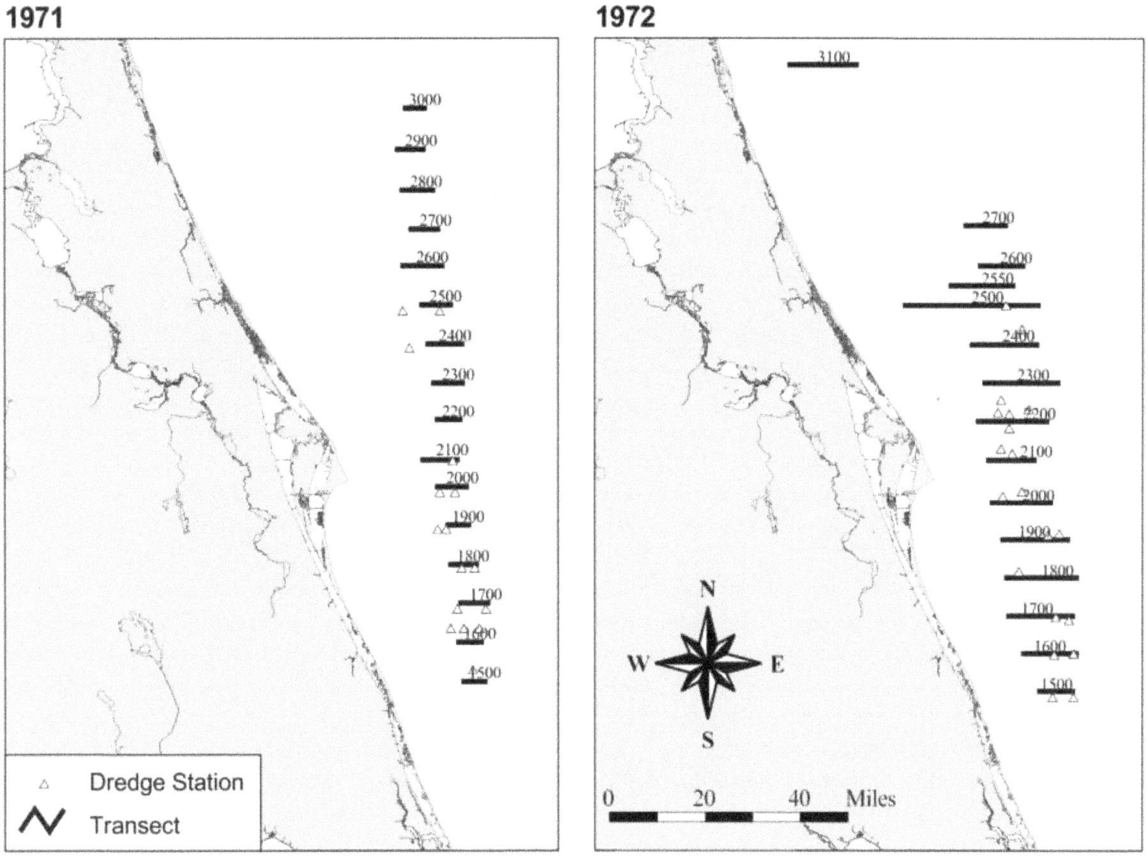

FIGURE 3. NOAA survey locations sampled from RV *George M. Bowers*, 1971–1972. Approximate latitude of each transect line is indicated by the LORAN line (*xx*00), which was the technology used at the time of the survey.

statistical reporting zones. These surveys began in 1981 with summer and fall surveys of coastal waters (9–110 m) from Texas to Alabama (see Switzer et al. 2009 for a full description of survey design). A Florida bottom trawl component was added in 2008 in which a 2-year pilot study in the northeastern Gulf of Mexico evaluated the feasibility of trawling along the west Florida shelf. Full-scale sampling in Florida began in 2010 along its entire Gulf coast. The same year, a new survey design was implemented for the Gulf-wide SEAMAP trawl survey. Overall sampling effort was allocated proportionally among NMFS statistical reporting zones based on proportional availability of sampling habitat (each zone had a different extent of bottom with depths of 9–110 m). In each zone sampling sites were chosen following a simple random survey design. Trawling effort was also standardized to 30 min at each station. The entire catch for most trawls was processed, though for a few large catches a subsample was randomly selected for processing and the total value proportionately extrapolated. As many as 20 scallops were selected and measured (hinge to bill) to determine length distribution.

For a comparison of long-term changes, we categorized historical collections into four data sets: MV *Silver Bay*, 1960–62; RV *Oregon*, 1967–68; RV *Oregon II* and RV *Bellows*, 1980s; and RV *GeoQuest*, 2004–06. Additional information from the 1950s (RV *Oregon*, RV *Delaware*, RV *Bowers*, and RV *Pelican*) and 1970s (RV *Bowers* and RV *Oregon II*) (Figure 3) largely included text-based descriptive reports and exploratory work over a wide region, which were unsuitable for analysis. When possible, we calculated mean catch rates and then standardized individual catches to those mean rates and to tow times and net size. The minimum criteria for inclusion in a data set for a tow included data for latitude and longitude and an indication about the presence or absence of calico scallops.

RESULTS

2004–2006 Study

Eight cruises and 491 tows were attempted, 27 of which were eliminated from the analyses due to lost or malfunctioning gear or when the tow lasted <1 min due to snags. On the east coast 238 of 245 tows were valid, and on the west coast 226 of 246 tows were valid. Over the course of the study 167 of the 240 east coast grid cells and 166 of the 240 west coast grid cells were sampled.

The total catch did not differ between east and west coast samples (Mann–Whitney U-test: $Z = 1.13$, $P = 0.259$). The overall average total catch was 2.10 kg·min^{-1}·m^{-1} (6.44 kg/5-min tow, standardized to the width of the mouth of the net). On the east coast the average catch was 2.30 kg·min^{-1}·m^{-1}, the median sample yielded 0.67 kg·min^{-1}·m^{-1} with a range of 0–51.7 kg·min^{-1}·m^{-1}. On the west coast the average total catch was 1.90 kg·min^{-1}·m^{-1}; the median sample yielded 0.64 kg·min^{-1}·m^{-1} with a range of 0–36.2 kg·min^{-1}·m^{-1}. High within-coast variability precluded detection of differences between the coasts. Catch rates varied seasonally (Kruskal–Wallis test: $H_{3,\ 465} = 21.93$, $P < 0.001$). The initial cruises, in fall 2004, had roughly twice as much total catch as the fall 2005 cruises. The spring and summer cruises were intermediate.

On the east coast there was no statistical difference in total catch between zones; the Sebastian site yielded the highest mean catch but it was not statistically higher than at other sites. There was seasonal variability ($H = 23.69$, $P < 0.001$), with the fall of 2004 yielding the highest catch rates. On the west coast there were differences in total catch between zones ($H = 49.03$, $P < 0.001$). Catch was highest in the Naples zone and lowest in the Englewood zone. There was also seasonal variability ($H = 13.62$, $P = 0.004$), and the fall 2005 cruise yielded the lowest catch rates and other seasons yielded higher catch rates and were not significantly different from one another. There was a very high catch rate in the spring of 2006 in the Naples zone. In other seasons, the difference between west coast zones was reduced.

A total of 3,314 scallops, 1,911 of which were adults, were collected from 154 samples; average shell height of those scallops was 31.4 mm for adults and 10.8 mm for juveniles. On the eastern coast, the greatest number of calico scallops collected in any sample was 205, from a station east of Sebastian Inlet (49.5 scallops·min^{-1}·m^{-1}). Calico scallops were collected in 83 of the 167 grid cells sampled off the eastern coast on at least one cruise (Figure 4). The greatest number of calico scallops collected from any single west coast sample was 213 (253.6 scallops·min^{-1}·m^{-1}), from the region southwest of Sanibel Island. Calico scallops were collected in 45 of the 166 west coast grid cells sampled on at least one cruise (Figure 5).

There was strong seasonal variability in abundance of adult calico scallops ($H = 43.71$, $P < 0.001$). Adults were most abundant on summer cruises, less abundant on fall cruises, and essentially absent in the spring. This pattern was observed on both coasts. Adult scallops were more abundant in east coast samples than in west coast samples ($Z = 3.92$, $P < 0.001$). The median east coast sample yielded 0 scallops (half of the tows had no adult scallops) and an average of 2.11 and a range of 0–49.52 scallops·min^{-1}·m^{-1}. The median west coast sample also yielded 0 scallops, but had an average of 2.35 and a range of 0–253.6 scallops·min^{-1}·m^{-1}. There was no statistical difference in scallop abundance between the east coast zones ($H = 5.87$, $P = 0.118$), but there was a strong seasonal difference ($H = 44.3$, $P = 0.001$) (Figure 4). The yield during the summer cruise exceeded all others; yields were similar for the fall cruises, and the spring cruise had the lowest yield. In the fall 2004 cruise, there were three distinct centers of abundance, all centered shoreward of the 40-m depth contour. In the spring 2005 cruise, adult scallops were not present on the Cape Canaveral fishing grounds. On the west coast there was a strong difference between zones ($H = 46.34$, $P < 0.001$). A consistent trend was for higher scallop abundance southwest of Sanibel Island near the border of the Naples and Fort Myers zones (Figure 5). While the abundance varied seasonally ($H = 15.75$, $P = 0.001$), there were always more scallops in the Naples zone and few north of Boca Grande pass.

Shell height was measured for 1,726 scallops (spat and adults), 1,412 from the east coast and 314 from the west coast (Figure 6). Over the course of eight cruises only 7% of the scallops measured were 40 mm or larger.

The catch rate of spat was highest in samples collected during the summer cruises, a result observed on each coast. Spat were more abundant on east coast scallop beds than on the west coast ($Z = 4.39$, $P < 0.001$). There were 0.96 spat·min^{-1}·m^{-1} (range, 0–26.45) collected in east coast samples and 0.15 spat·min^{-1}·m^{-1} (range, 0–8.21) collected in west coast samples. A high proportion of samples contained no spat. On the east coast, spat were more abundant in samples from the summer cruise than from the other cruises ($H = 93.73$, $P < 0.001$). There was no significant difference between zones ($H = 4.11$, $P = 0.25$) (Figure 4). On the west coast, there were more spat in samples from the Naples zone than from other zones ($H = 16.60$, $P = 0.001$). There were also more spat in samples collected during the spring and summer than during fall cruises ($H = 8.76$, $P = 0.033$). Each of the four west coast zones had at least some spat during one or more cruises. Adult calico scallop abundance in the Englewood and Sarasota zones increased in fall 2006 following the peak in spat observed in spring 2006 samples from those zones (Figure 5).

Spat were more abundant than expected on scallop shell and less abundant than expected on other mollusk shell or on other substrates. This pattern held for analyses of all spat (both coasts combined) ($\chi^2 = 35.83$, $P < 0.001$) and total east coast samples ($\chi^2 = 35.83$, $P < 0.001$) but not total west coast samples ($\chi^2 = 3.11$, $P > 0.05$), although the trend was similar on the west coast where small sample size may have precluded detection of true differences. In each east coast zone during the summer 2006 cruise, the only cruise with a large enough sample to allow examination of spatial variability within a single season, more spat than expected were found on calico scallop shell (Titusville: $\chi^2 = 8.55$, $P < 0.05$; Cocoa Beach: $\chi^2 = 6.93$, $P < 0.05$; Melbourne: $\chi^2 = 13.36$, $P < 0.01$; Sebastian: $\chi^2 = 18.48$, $P > 0.001$) and fewer spat than expected were found on other substrates. In west coast samples there was only one scallop larger than 23 mm that was still byssally attached (2.3% of

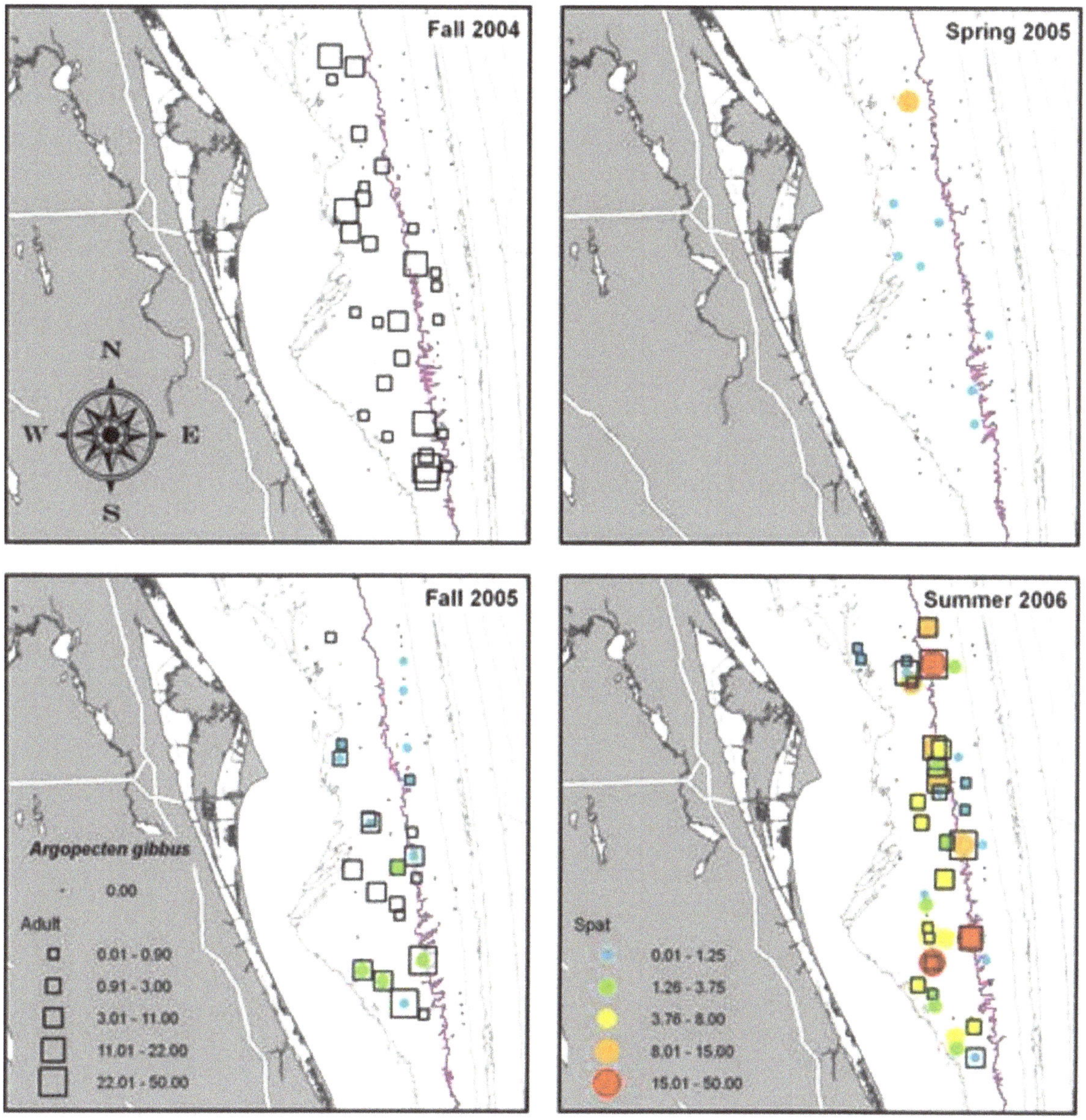

FIGURE 4. Cape Canaveral calico scallop catches (scallops·min^{-1}·m^{-1}), 2004–2006. The 40-m contour line is accentuated in purple. The inshore line is 20 m, successive offshore lines are at 60, 80, 100, 200, 400, and 500 m.

attached scallops were ≥23 mm). In east coast samples there were seven scallops 23 mm or larger still byssally attached (7.4% of attached scallops were ≥23 mm).

There were no seasonal differences in calico scallop shell when the factor "coast" was ignored, but the total weight of shell tended to be most abundant in the two summer cruises, similarly abundant in the four fall cruises, and least abundant in the spring. The prevalence of other shell, however, did differ significantly between seasons, being most abundant in summer, similarly lower in the fall cruises, and least in the spring ($H = 10.71$, $P = 0.013$). When data from each season were pooled to examine differences between coasts, the east coast had much more scallop shell than did the west coast ($Z = 12.36$, $P < 0.001$); there was roughly 10 times more calico scallop shell on the east coast than there was on the west coast. However, the abundance of other shell substrate did not vary by coast ($Z = -1.93$, $P = 0.055$).

On the east coast, calico scallop shell was more abundant in the Cocoa Beach and Sebastian zones than in the Melbourne and Titusville zones ($H = 10.68$, $P = 0.014$). Scallop shell was most abundant in the summer, less abundant during the two fall cruises, and least abundant during spring ($H = 15.72$, $P = 0.001$) (Figure 7). Scallop shell was present in 92% of the valid tows; other shell was present in 95% of the valid tows, albeit at lower density. The prevalence of other shell was more abundant in the Cocoa and Sebastian zones than in the

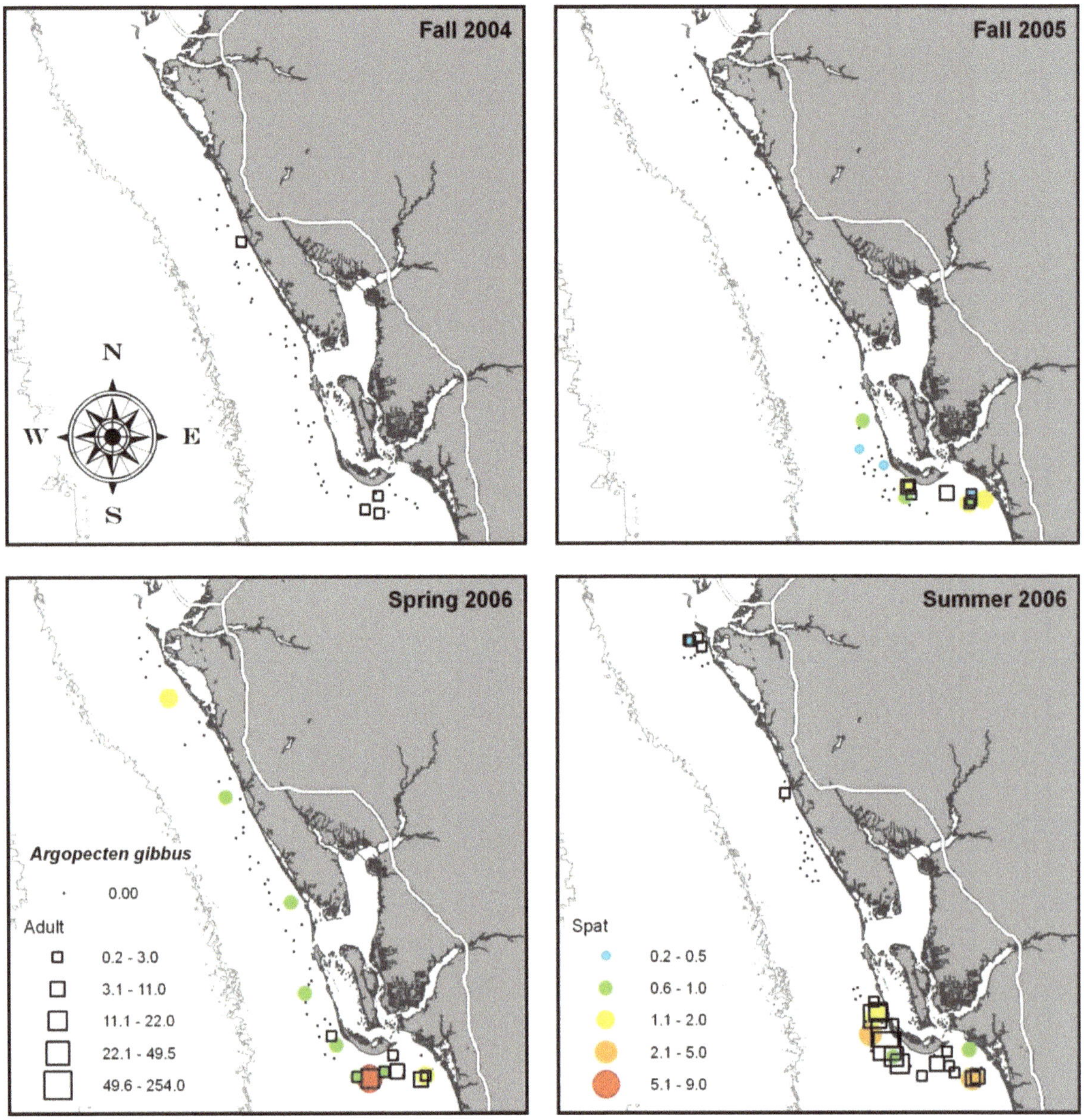

FIGURE 5. Gulf of Mexico calico scallop catches (scallops·min^{-1}·m^{-1}), 2004–2006. The main contour line represents the 20-m isobath.

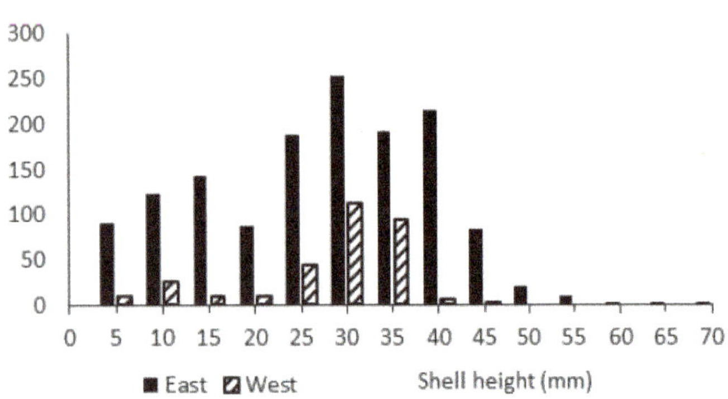

FIGURE 6. Calico scallop size-frequency distribution for all 2004–2006 east coast and west coast collections combined.

Melbourne and Titusville zones ($H = 9.56$, $P = 0.023$). Other shell was more abundant in shallower waters, especially where the 20-m contour extended seaward into the sampling grid in the southwestern corner of the Sebastian zone and northwestern corner of the Cocoa Beach zone. Venerid bivalves were especially abundant in these shoal waters. The abundance of other shell also varied seasonally ($H = 15.38$, $P = 0.002$).

On the west coast, calico scallop shell abundance was highest in the south and less in the north ($H = 20.01$, $P < 0.001$) (Figure 8). The abundance of scallop shell on the west coast varied seasonally ($H = 3.99$, $P = 0.262$) and interactive effects were observed. There was a high abundance of shell in the Fort Myers zone during fall 2004, but this was not detected in later cruises. Scallop shell was more widely distributed (68% of

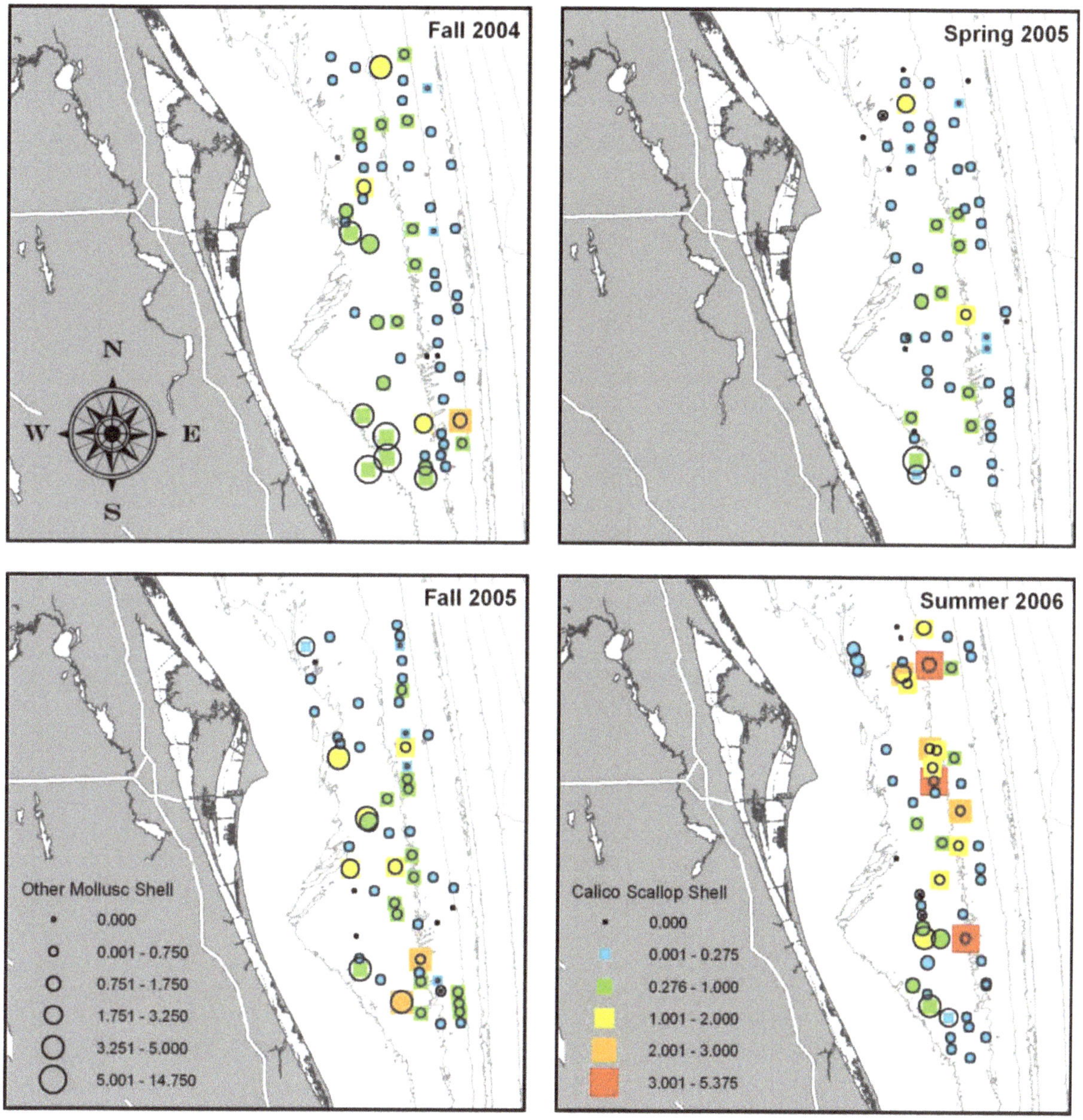

FIGURE 7. Mollusk shells collected (kg·min^{-1}·m^{-1}) in Cape Canaveral samples, 2004–2006.

valid tows) than either adult (17% of valid tows) or juvenile scallops (12% of valid tows). Other mollusk shell was most abundant in the Naples zone but was widespread ($H = 52.11$, $P < 0.001$) (other shell was collected in 84% valid tows) and did not vary seasonally ($H = 1.35$, $P = 0.716$). The Naples zone typically had twice as much other shell substrate than any other zone.

Supplemental and Historical Data Sets

Calico scallops were observed in of 67 of 1,497 SEAMAP tows (Figure 9). The average number of scallops collected in tows that collected scallops was 101.2 per tow (0.4 scallops·min^{-1}·m^{-1}) and ranged from 1 to 1,629 (10.9 scallops·min^{-1}·m^{-1}). The average starting depth for the tows in which scallops were caught was 38 m (19 fathoms), and the average depth of the top 16 tows, i.e., those with 20 or more scallops, was also 38 m (19 fathoms). Calico scallops were observed in two main areas: on the continental shelf from south of Cape San Blas westward to the waters off Alabama and in nearshore areas between Sanibel Island and Tampa Bay. A scattered distribution of scallops also was observed across the west Florida shelf, with the exception of the Big Bend region shoreward of ~20 m depth.

A list of historical data sets for the east coast is provided and includes 1,297 tows of nets or dredges (Table 1). The center of

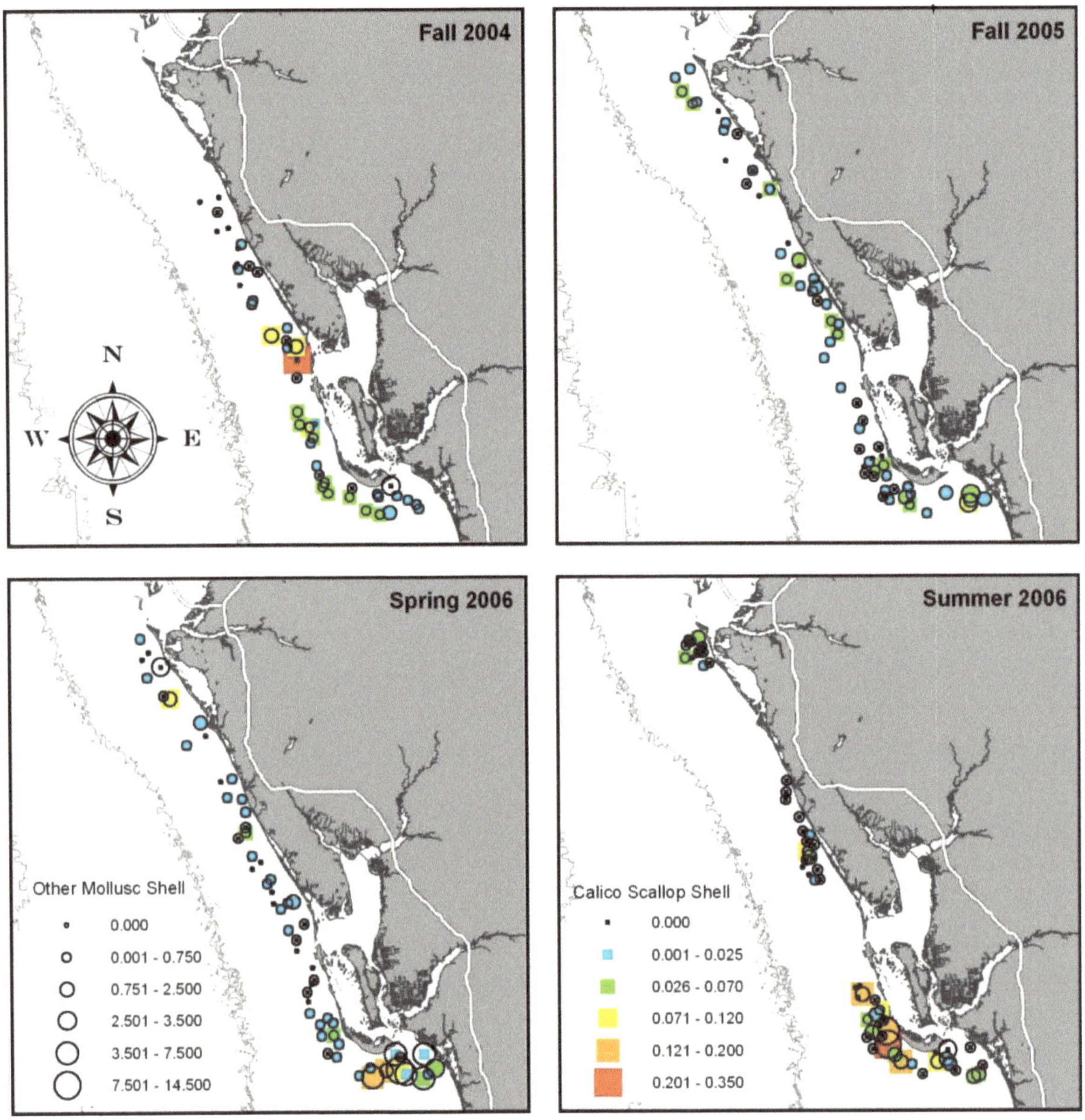

FIGURE 8. Mollusk shells collected (kg·min^{-1}·m^{-1}) in Gulf of Mexico samples, 2004–2006.

the geographic distribution of calico scallops on the east coast was the region of 37–48 m (20–26 fathoms) off Cape Canaveral (Figure 10), which is nearly identical to that observed in the present study. Calico scallops were, and remain, widely distributed on the coastal shelf off Florida, both in the South Atlantic Ocean and Gulf of Mexico. The abundance and distribution of calico scallops observed in the present 2004–2006 study (38% of Atlantic stations, range, 14–56%; 23% of Gulf stations, range, 12–35%) was lower than that reported in previous surveys (Table 2). The Silver Bay cruises had calico scallops present in 71% of the dredges and in 40% of the trawls in Atlantic waters. In trawls conducted in the Gulf for SEAMAP, scallops were found in 4.5% of the samples.

DISCUSSION

Calico scallop beds in this study appeared to cover similar geographic areas as in previous surveys, but the abundance of both scallops and scallop shell in any area was found to change dramatically over several months. Variations in the abundance of calico scallops continue to have a strong temporal component, and large beds can appear and disappear over a short period, which is similar to such findings as those of Roe et al. (1971) and those summarized by Blake and Shumway (2006). There is clearly a widespread distribution on both coasts, which suggests that dense patches still form. Within the Cape Canaveral bed, calico scallops remain the dominant macrobenthic animal, and scallop shell is the dominant substrate in a

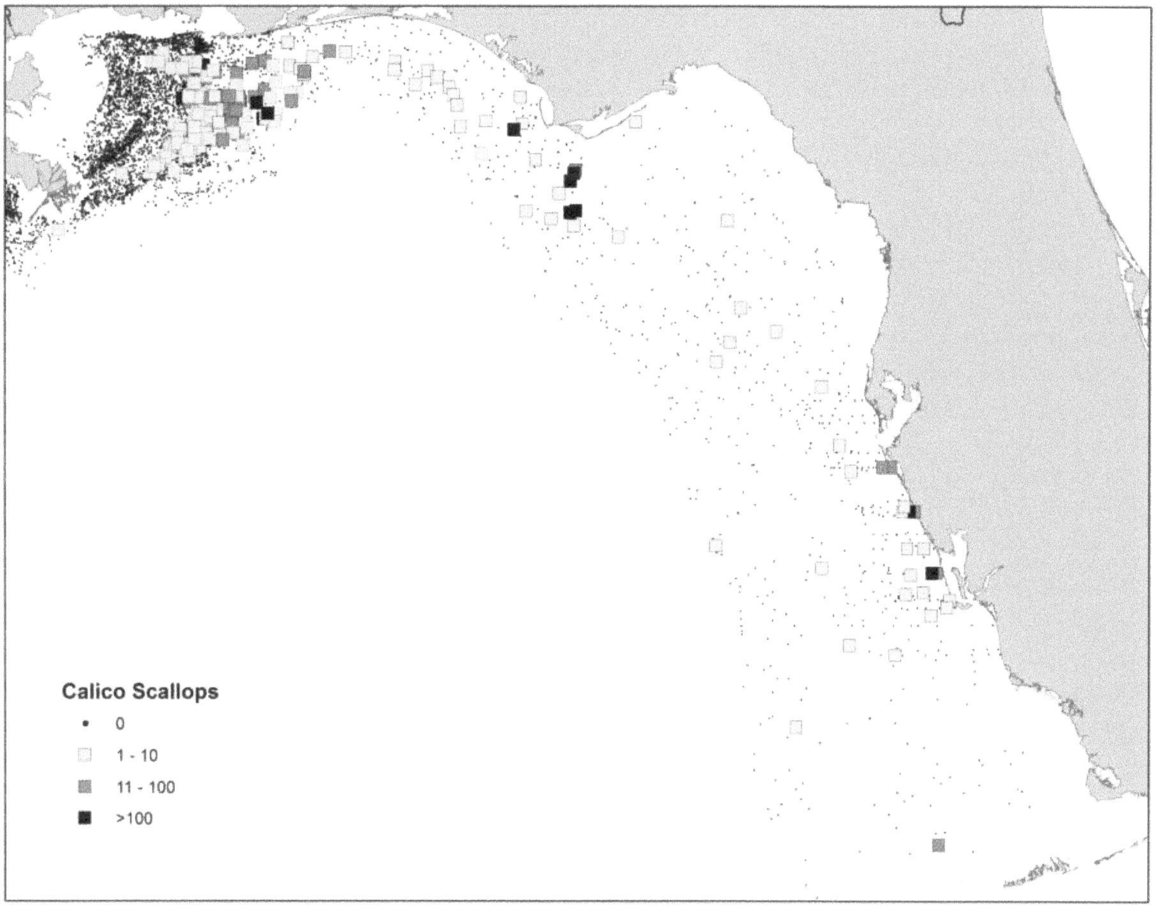

FIGURE 9. Number of calico scallops observed at stations sampled via towed dredges during SEAMAP and baitfish cruises, 2008–2014.

majority of the samples. This is consistent with historical studies, where at times scallops were the only species reported, comprising the total catch. It would appear that in the present study fewer scallops were found on the Cape Canaveral fishing grounds than in previous surveys, which resulted in the reduced total catch weights observed, but abundant shell was available as settlement substrate. During summer 2006, abundance in west coast scallop beds in some tows equaled and exceeded those on the east coast, but these patches were smaller and did not cover large portions of any given zone in any cruise.

Potential problems exist in attempting to compare current work with past studies. One problem is the exclusion of samples for which the intended target was calico scallops, but no catch data were recorded. Some historical reports indicate that catches of less than 1 bushel (~35 L) of any commercial species were not reported. For those studies, calculation of mean catch or even presence-to-absence information is impossible, and the only reliable data are the locations where commercial quantities were reported. It is apparent from past studies and those reported in this study that at a given moment much of the distribution of the species may be outside of the areas of highest concentration. Also, the RV *Oregon* and RV *Bowers* cruises (Table 1) did not use a random sample design; instead the surveys focused on commercially fishable areas, and the objective was to outline the limits of the commercial beds to maximize harvest rates and efficiency.

Another source of bias is the use of a very small dredge in the present study, comparable with the use of a try net in shrimp fisheries, whereas historically commercial gear was used in surveys. Differences in efficiency between gear types also probably contributed to the quantitative differences we observed. Dredges are known to have widely variable efficiencies (Fifas and Berthou 1999) and scallops can avoid them (Caddy 1968; Gedamke et al. 2004). Although in shrimping the efficiency of try nets may be fairly reliable and comparable to that of working gear (Cody and Fuls 1985), similar data for dredges are not available. The intent of this study was not to determine gear efficiency but to explore whether stocks of calico scallops were present at densities that may warrant commercial exploration. Future studies using commercial-scale gear should be able to conduct meaningful quantitative

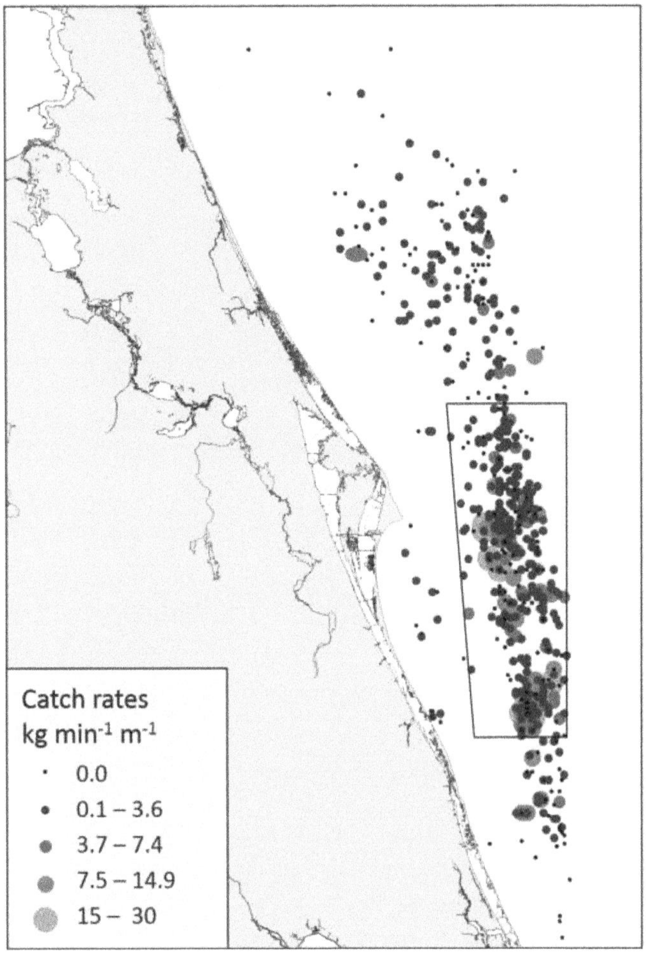

FIGURE 10. Calico scallops catches (kg·min^{-1}·m^{-1}) by MV *Silver Bay*, 1959–1964. See text for a description of the archival data discovery.

historical comparisons using the data set outlined in this study, which will aid the industry in evaluating the stock's commercial potential.

One objective of our study was to determine whether there was any remaining shell bed, considered to be the principal settlement substrate for spat and thus to be essential fisheries habitat. Our focus on shell abundance differed from that of historical studies, which indicated shell presence only when commercial species were not abundant, so direct comparisons do not seem justified. Shell was the principal catch in many tows and could fill the nets quickly. We could not conclude that the shell bed had been degraded to the point where essential fisheries habitat was limiting due to the high catch rate of scallop shell in our samples. In fact, in the present study there were almost no dredges on either coast in which no shell was captured. The condition of the calico scallop bed appeared to be close to its natural, prefishery state at the time of sampling, 2004–2006, when the commercial harvest had been minimal or nonexistent for roughly a decade; the shell bed has remained largely undisturbed for almost another decade since the present study was completed.

The ratio of scallop shell to other shell was much higher on the east coast than on the west coast and was usually the dominant shell type and more abundant than all other species' shells combined. On the west coast, the majority of the shell present was from other mollusks. While scallop spat were shown to prefer scallop shell over either other mollusk shell or other nonshell substrates, they nevertheless used the most abundant substrate available. These alternative substrates were most commonly shell from other mollusks but also included biofouled rock, algae, living organisms (including an unidentified oyster drill), and even moderately fouled trash. Bremec et al. (2008), in a study of the Patagonian scallop *Zygochlamys patagonica*, showed that it might not be the shell that the spat prefer as their substrate, but the fouling organisms for which the shell provides structure. In samples we collected, the fouling organisms on scallop shells were very diverse, as described by Schwartz and Porter (1977), so without directed study, we cannot say what specific substrate the scallops are choosing, the shell or fouling organisms.

The past practice of culling on shore rather than at sea should be avoided. Landings of the entire catch remove both substrate and the associated community, resulting in degradation of valuable low-relief habitat and adding the expense of disposal onshore. Adjoining habitats include inshore low-relief shrimp grounds as well as offshore high-relief habitat (Koenig et al. 2000). This high-relief area is known to harbor commercially valuable fishes such as snappers and groupers, which use the calico scallop beds as forage areas (Schwartz and Porter1977) and a pathway to estuarine juvenile habitat.

The amount of variability in the abundance of scallops between zones was not consistent between coasts. There was no significant variation between east coast zones, which was not entirely unexpected since the sample design was limited to a geographic area known to harbor calico scallops and covered the majority of the depth zone in which they are known to occur. True differences may be masked by high seasonal variation and very large catches in the southwestern corner of the Sebastian zone, where some random sampling stations were located shoreward of the 20-m contour in a region of very abundant marine life and large catches of sediment. The southern end of the sampling grid is also a region where the shelf is relatively narrow, resulting in a rapid increase in water depth and outcroppings of diverse hard-bottom communities along the slope throughout the depth range of the Sebastian zone. The high diversity and productivity may also reflect the proximity of both the Gulf Stream and upwelled bottom waters (Lee et al. 1991; Fiechter and Mooers 2003).

There was significant variation in the abundance of calico scallops between west coast zones. The area southwest of Sanibel Island was identified as a potential location for harvestable stocks in interviews with former calico scallop harvesters. It is a region that receives the ebb-tidal flow from San Carlos Bay

TABLE 2. Summary statistics from six east coast data sets obtained from RV *Silver Bay* (SB) 1959–1964, RV *Oregon* (OR) 1967–1968, RV *Delaware II* (D2) 1982, RV *Oregon II* (O2) 1987, RV *Bellows* (BW) 1989, and RV *GeoQuest* (GQ) 2004–2006 and from SEAMAP (SM) 2008–2014. Data are mean ± SD and maximum $kg \cdot min^{-1} \cdot m^{-1}$ of net width, if available. %P = the percentage of stations at which calico scallops were collected; %CQ = the percentage of stations at which scallops were collected in commercially viable quantities, ~2.5 $kg \cdot min^{-1} \cdot m^{-1}$ for dredges or 0.5 $kg \cdot min^{-1} \cdot m^{-1}$ for nets. *In the OR and D2 the values for dredge, estimates are made from only those stations at which commercial quantities were present, since values of <1 bushel were not reported.

	Dredge			Trawl		
Vessel	Mean $kg \cdot min^{-1} \cdot m^{-1}$	%P	%CQ	Mean $kg \cdot min^{-1} \cdot m^{-1}$	%P	%CQ
SB	10.4 ± 16.8 120.3	71	37	0.67 ± 1.52 8.47	40	10
OR*	14.2 ± 13.1 69.8		46	1.8 ± 2.3 7.50	50	40
O2		43	2		5	0
BW				0.13 ± 0.25 1.25	85	12
D2*	21.0 ± 17.5 75.2		16			
GQ	0.02 ± 0.07 0.55	38	2			
SM					4.5	0

and the Caloosahatchee River, known to be rich in nutrients (Doering et al. 2006), thus increasing primary productivity. Similar increases in benthic productivity appear to occur in the waters off the mouth of Tampa Bay (Vargo et al. 2008). The outlets of both these estuaries have patches of abundant scallops. The west Florida shelf is also susceptible to prolonged periods of northeast winds, which can drive surface waters offshore and induce upwelling in nearshore regions (Yang and Weisberg 1999). The two processes are not mutually exclusive and both increase primary productivity, creating hot spots of diversity and productivity similar to those in more thoroughly described east coast habitats. The data collected during Gulf of Mexico SEAMAP cruises support the hypothesis that calico scallops are more abundant along the shelf edge at 32–40 m and may connect the newly described region near Sanibel Island (Fort Myers and Naples zones) with previously recognized commercial beds off Cape San Blas (Bullis and Ingle 1958) and near the Tortugas shrimp grounds, as well as farther west toward the Mississippi River. At times calico scallops may be much more abundant and widespread on the west Florida shelf than what we observed. Additional anecdotal evidence, such as wracks of dying calico scallops along the Gulf beaches and after strong west-wind events and occasional reports of calico scallop aggregations on and around artificial reefs, underscore the dynamic nature of the coastal zone of this supposedly low-energy and relatively low-productivity region.

Parasitic infection by a protozoan presumed to be *Marteilia* sp. was seen in every calico scallop inspected from both Atlantic and Gulf waters, and late-stage disease, for which pathologies are observed, was common. *Marteilia* or a *Marteilia*-like disease coincided with massive mortalities in the calico scallop fishery in 1986 (Moyer et al. 1993) and, if it is limiting growth to sub-market size, is still detracting from the commercial viability of this resource. Few scallops seen in our samples were of commercial size, i.e., 40 mm or larger, a size indicated by Blake and Moyer (1991) to be most profitable (200 scallops per pint). Counts of 60–110 scallops per pint and shell heights exceeding 50 mm were observed in the 1960s (Roe et al. 1971). Sea scallops command higher prices and can outcompete the bay scallop and presumably calico scallop product (Repetto 2002; Blake and Shumway 2006; Naidu and Robert 2006). Seasonal and annual variability in the calico scallop fishery appears to discourage harvesters from reinvesting in this fishery. The lack of scallops of a commercially valuable size may reflect the season during which samples were collected, though commercially viable sizes have been collected in all seasons, and the spatial and temporal patchiness of the species' distribution suggests that at least some samples of predominantly market-size scallops should have been found.

Follow-up studies to evaluate this resource and its rebound to fishable levels in the Atlantic Ocean and Gulf of Mexico are warranted. Future studies should focus on standardized gear, random sampling of adults, juveniles, and habitat (shell substrate), and disease prevalence during all seasons. Surveys in the Gulf of Mexico should focus on nearshore resources, which may exist at the outflows to Tampa Bay, Charlotte Harbor, and the Caloosahatchee River and along the deeper margin of the shelf along the Florida Panhandle and westward.

ACKNOWLEDGMENTS

Capt. William Burkhardt was an inspiration for portions of this research. He provided valuable comments to the South Atlantic Fishery Management Council during the 1998 attempt to write a fishery management plan and met with the authors to guide the sample design. C. Beals, J. Cobb, A. Granholm, M. Parker, and B. Pittinger assisted with field sampling. Copies of the unpublished cruise reports were obtained from the NOAA library in Pascagoula, Mississippi, through the assistance of L. Fontroy. W. Stringer and J. Balsille of the Florida Geological Survey served as captain and crew of the RV *GeoQuest* from which all east coast samples were gathered. Funding for this study was provided by NOAA grant NA04NMF4330078 through the MARFIN program.

REFERENCES

Beukema, J. J., R. Dekker, K. Essink, and H. Michaelis. 2001. Synchronized reproductive success of the main bivalve species in the Wadden Sea: causes and consequences. Marine Ecology Progress Series 211:143–155.

Blake, N. J., and M. A. Moyer. 1991. The calico scallop, *Argopecten gibbus*, fishery of Cape Canaveral, Florida. Pages 899–909 *in* S. Shumway, editor. Scallops: biology, ecology, and aquaculture. Elsevier, New York.

Blake, N. J., and S. E. Shumway. 2006. Bay scallop and calico scallop fisheries, culture and enhancement in eastern North America. Pages 945–964 *in* S. E. Shumway and G. J. Parsons, editors, Scallops: biology, ecology and aquaculture. Elsevier, Amsterdam.

Bourgeois, M., J.-C. Brêthes, and M. Nadeau. 2006. Substrate effects on survival, growth and dispersal of juvenile sea scallop, *Placopecten magellanicus* (Gmelin 1791). Journal of Shellfish Research 25:43–49.

Bremec, C., M. Escolar, L. Schejter, and G. Genzano. 2008. Primary settlement substrate of scallop, *Zygochlamys patagonica* (King and Broderip, 1832) (Mollusca: Pectinidae) in fishing grounds in the Argentine Sea. Journal of Shellfish Research 27:273–280.

Bullis, H. R. Jr., and R. M. Ingle. 1958. A new fishery for scallops in western Florida. Gulf and Caribbean Fisheries Institute 11:75–78.

Caddy, J. F. 1968. Underwater observations on scallop (*Placopecten magellanicus*) behavior and drag efficiency. Journal of the Fisheries Research Board of Canada 25:2123–2141.

Carpenter, J. S. 1967. History of scallop and clam explorations in the Gulf of Mexico. Commercial Fisheries Review 29:47–53.

Cody, T. J., and B. E. Fuls. 1985. Comparison of the catch rates of three trawls in offshore Texas waters. Pages 19–29 *in* J. W. Watson and N. Bane, editors. Proceedings of the SEAMAP sampling gear workshop. Gulf States Marine Fisheries Commission 12, Ocean Springs, Mississippi.

Cummins, R. Jr. 1971. Calico scallops of the southeastern United States, 1959–1969. NOAA Technical Report SSRF 627:1–22.

Doering, P. H., R. H. Chamberlain, and K. M. Hauner. 2006. Chlorophyll *a* and its use as an indicator of eutrophication in the Caloosahatchee estuary. Florida Scientist 69(Supplement 2):51–72.

Fiechter, J., and C. N. K. Mooers. 2003. Simulation of frontal eddies on the east Florida shelf. Geophysical Research Letters [online serial] 30:2151.

Fiedler, R. H. 1938. Fishery industries of the United States, 1936. *In* Report of the Commissioner for 1937, appendix II. U.S. Commission of Fish and Fisheries, Washington, D.C.

Fifas, S, and P. Berthou. 1999. An efficiency model of a scallop (*Pecten maximus*, L.) experimental dredge: sensitivity study. ICES Journal of Marine Science 56:489–499.

Gedamke, T., W. D. Dupaul, and J. M. Hoenig. 2004. A spatially explicit open-ocean Delury analysis to estimate gear efficiency in the dredge fishery for sea scallop *Placopecten magellanicus*. North American Journal of Fisheries Management 24:335–351.

Guay, M., and J. H. Himmelman. 2004. Would adding scallop shells (*Chlamys islandica*) to the sea bottom enhance recruitment of commercial species? Journal of Experimental Marine Biology and Ecology 312:299–317.

Hart, D. R., and P. J. Rago. 2006. Long-term dynamics of U.S. Atlantic sea scallop *Placopecten magellanicus* populations. North American Journal of Fisheries Management 26:490–501.

Kimmel, D. G., M. Tarnowski, and R. I. E. Newell. 2013. Long-term (1939–2008) spatial patterns in juvenile eastern oyster (*Crassostrea virginica*, Gmelin 1791) abundance in the Maryland portion of the Chesapeake Bay. Journal of Shellfish Research 31:1023–1031.

Koenig, C. C., F. C. Coleman, C. B. Grimes, G. R. Fitzhugh, K. M. Scanlon, C. T. Gledhill, and M. Grace. 2000. Protection of fish spawning habitat for the conservation of warm-temperate reef-fish fisheries of shelf-edge reefs of Florida. Bulletin of Marine Science 66:593–616.

Kraeuter, J. N., M. J. Kennish, J. Dobarro, S. R. Fegley, and G. E. Flimlin Jr. 2003. Rehabilitation of the northern quahog (hard clam) (*Mercenaria mercenaria*) habitats by shelling: 11 years in Barnegat Bay New Jersey. Journal of Shellfish Research 22:61–67.

Lee, T. N., J. A. Yoder, and L. P. Atkinson 1991. Gulf Stream frontal eddy influence on productivity of the southeast U.S. continental shelf. Journal of Geophysical Research Oceans 96:22191–22205.

Maeda-Martínez, A. N., T. Reynoso-Granados, F. Solís-Marin, A. Leija-Tristán, D. Aurioles-Gamboa, C. Salinas-Zavala, D. Lluch-Cota, P. Ormart-Castro, and E. Felix-Pico. 1993. A model to explain the formation of catarina scallop, *Argopecten circularis* (Sowerby, 1835), beds in Magdalena Bay, Mexico. Aquaculture and Fisheries Management 24:323–339.

Marshall, N. 1963. Mortality rates and the life span of the bay scallop, *Aequipecten irradians*. Proceedings of the National Shellfish Association 54: 87–92.

Miller, G. C., D. M. Allen, T. J. Costello, and J. H. Hudson. 1979. Maturation of the calico scallop, *Argopecten gibbus*, determined by ovarian color changes. Northeast Gulf Science 3:93–103.

Moyer, M. A., N. J. Blake, and W. S. Arnold. 1993. An ascetosporan disease causing mass mortality in the Atlantic calico scallop, *Argopecten gibbus* (Linnaeus, 1758). Journal of Shellfish Research 12:305–310.

Naidu, K. S., and G. Robert. 2006. Fisheries sea scallop, *Placopecten magellanicus*. Pages 869–906 *in* S. E. Shumway and G. J. Parsons, editors. Scallops: biology, ecology and aquaculture. Elsevier, Amsterdam.

Nelson, W. G., J. C. Sainsbury, and M. A. Sager 1992. Biological effects of scallop trawling: estimation of by-catch composition, mortality and effects on bottom communities. Final Report to Florida Department of Natural Resources, Marine Research Institute, St. Petersburg.

Peterson, J. K., J. W. Hansen, M. B. Laursen, P. Clausen, J. Carstensen, and D. J. Conley. 2008. Regime shift in a coastal marine ecosystem. Ecological Applications 18:497–510.

Powell, E. N., K. A. Ashton-Alcox, J. N. Kraeuter, S. E. Ford, and D. Bushek. 2008. Long-term trends in oyster population dynamics in Delaware Bay: regime shifts and response to disease. Journal of Shellfish Research 27: 729–755.

Repetto, R. 2002. Creating asset accounts for a commercial fishery out of equilibrium: a case study of the Atlantic sea scallop fishery. Review of Income and Wealth 48:245–259.

Rockwood, C. E., and J. Pompe. 1988. Economic impact of the Brevard County calico scallop fishery. Report to the Calico Scallop Preservation Association, Tallahassee, Florida.

Roe, R. B., R. Cummins Jr., and H. R. Bullis Jr. 1971. Calico scallop distribution, abundance, and yield off eastern Florida. U.S. National Marine Fisheries Service Fishery Bulletin 69:399–409.

SAFMC (South Atlantic Fishery Management Council). 1998. Final fishery management plan for the calico scallop fishery in the South Atlantic region. South Atlantic Fishery Management Council, Charleston, South Carolina.

Schwartz, F. J., and H. J. Porter 1977. Fishes, macroinvertebrates, and their ecological inter-relationships with a calico scallop bed off North Carolina. U.S. National Marine Fisheries Service Fishery Bulletin 75:427–446.

Switzer, T. S., E. J. Chesney, and D. M. Baltz. 2009. Habitat selection by flatfishes in the northern Gulf of Mexico: implications for susceptibility to hypoxia. Journal of Experimental Marine Biology and Ecology 381:S51–S64.

Taylor, D. M. 1967. Billion-dollar scallop find? Ocean Industry 2(12):20–24.

Vargo, G. A., C. A. Heil, K. A. Fanning, L. Kellie Dixon, M. B. Neely, K. Lester, D. Ault, S. Murasko, J. Havens, J. Walsh, and S. Bell. 2008. Nutrient availability in support of *Karenia brevis* blooms on the central west Florida shelf: what keeps *Karenia* blooming? Continental Shelf Research 28:73–98.

Wolff, M. 1988. Spawning and recruitment in the Peruvian scallop *Argopecten purpuratus*. Marine Ecology Progress Series 42:213–217.

Wolff, M., and J. Mendo. 2000. Management of the Peruvian bay scallop (*Argopecten purpuratus*) metapopulation with regard to environmental change. Aquatic Conservation: Marine and Freshwater Ecosystems 10:117–126.

Yang, H., and R. H Weisberg. 1999. Response of the west Florida shelf circulation to climatological wind stress forcing. Journal of Geophysical Research Oceans 104:5301–5320.

5

Genetic Identification of Chinook Salmon: Stock-Specific Distributions of Juveniles along the Washington and Oregon Coasts

David J. Teel*
National Oceanic and Atmospheric Administration, National Marine Fisheries Service, Northwest Fisheries Science Center, Conservation Biology Division, Post Office Box 130, Manchester, Washington 98353, USA

Brian J. Burke
National Oceanic and Atmospheric Administration, National Marine Fisheries Service, Northwest Fisheries Science Center, Fish Ecology Division, 2725 Montlake Boulevard East, Seattle, Washington 98112, USA

David R. Kuligowski
National Oceanic and Atmospheric Administration, National Marine Fisheries Service, Northwest Fisheries Science Center, Conservation Biology Division, Post Office Box 130, Manchester, Washington 98353, USA

Cheryl A. Morgan
Cooperative Institute for Marine Resources Studies, Hatfield Marine Science Center, Oregon State University, 2030 South Marine Science Drive, Newport, Oregon 97365, USA

Donald M. Van Doornik
National Oceanic and Atmospheric Administration, National Marine Fisheries Service, Northwest Fisheries Science Center, Conservation Biology Division, Post Office Box 130, Manchester, Washington 98353, USA

Abstract

We used microsatellite DNA data and genetic stock identification methods to delineate the temporal and spatial distributions of juvenile Chinook Salmon *Oncorhynchus tshawytscha* occupying coastal habitats extending from central Oregon to northern Washington. Juveniles were collected in trawl surveys conducted during spring, summer, and autumn over 15 years. Distributions (mean latitude and distance from shore) differed between yearling and subyearling life history types and between stocks; many of these differences were consistent across years. Yearlings were nearly all (98%) from Columbia River sources, and only 6% were naturally produced. In late May, yearlings from the lower Columbia and Willamette rivers were farther north than other yearlings, likely due to the early spring timing of their releases from hatcheries and subsequent out-migration from the Columbia River.

Subject editor: Kristina Miller, Fisheries and Oceans Canada, Pacific Biological Station, Nanaimo, British Columbia, Canada

*Corresponding author: david.teel@noaa.gov

However, yearling distributions in late June reflected known migration behaviors. Yearlings from interior Columbia and Snake River sources were farthest north by June, whereas yearlings from other stocks were more spread out in latitude. Subyearlings sampled in early summer were also largely from the Columbia River (98%), but greater percentages of subyearlings from coastal rivers were present during the fall (24%). In contrast to yearlings, natural production accounted for nearly one-third of subyearlings. Subyearlings of most stocks tended to remain relatively near their point of sea entry throughout the summer. Subyearlings from the Snake River fall-run stock and upper Columbia River summer–fall-run stock exhibited diverse distributions that included both southward and northward dispersal. Overall, distributions of Chinook Salmon stocks and life history types reflected differences in migration behavior but also reflected the influence of environmental factors and hatchery practices.

A long-held tenet of discerning fish biologists is that effective management and conservation of Pacific salmon *Oncorhynchus* spp. require knowledge of stock-specific migration and dispersal patterns (Rich 1939; Simon and Larkin 1972; Hartt 1980; Walters 1996). This requirement presents a challenge because salmon exhibit a great diversity of movement patterns, particularly during their ocean phase. Nonetheless, for many populations, the marine pathways of migrating adult fish are well known by fishery managers because adults are routinely sampled in fisheries conducted along much of the West Coast of North America (Cleaver 1969; Utter et al. 1987; Shaklee et al. 1999; Weitkamp and Neely 2002; Beacham et al. 2008; Weitkamp 2010). In many areas, the marine migrations of juvenile salmon have also been studied intensively as part of efforts to understand variability in salmon production and to determine why populations respond to environmental conditions in different ways. Most studies of juvenile Pacific salmon have focused on the distributions of fish during the first weeks and months after ocean entry because this period is characterized by highly variable growth and survival (Godfrey 1958; Hartt 1980; Mortensen et al. 2000; Miller et al. 2014). Knowledge of where fish are during this critical period is important since the physical and biological ocean conditions that are thought to determine year-class strength vary with latitude and also across the continental shelf (Pearcy 1992; Brodeur et al. 2000; Pearcy and McKinnell 2007).

Salmon tagging has long been the primary means for studying marine distributions during this early critical period. During initial research efforts, juvenile fish were captured and tagged at sea, and some were later recovered as adults when returning to spawn in their natal streams (Foerster 1955; Godfrey et al. 1975; Hartt and Dell 1986). Later, numerous ocean research surveys captured juveniles that had been coded-wire-tagged during hatchery rearing (Miller et al. 1983; Pearcy and Fisher 1988; Orsi and Jaenicke 1996; Morris et al. 2007; Trudel et al. 2009; Fisher et al. 2014). More recently, genetic stock identification methods have been applied to samples from many coastal areas to delineate the early marine distributions of both naturally and hatchery-produced juvenile salmon (Teel et al. 2003; Van Doornik et al. 2007; Tucker et al. 2009; Seeb et al. 2011; Beacham et al. 2012, 2014). The migrations of juvenile Chinook Salmon *O. tshawytscha* have been examined using genetic techniques, most extensively along the continental shelf off British Columbia and southeastern Alaska (Tucker et al. 2011, 2012) but also in Puget Sound, Washington (Rice et al. 2011), and off the southern Oregon and northern California coasts (Brodeur et al. 2004). Genetic data have also been used to study the distributions of juvenile Chinook Salmon in coastal areas along Oregon and Washington; however, those analyses were limited to a few selected genetic stocks (Teel 2004; Burke et al. 2013a).

Additional study of stock-specific distributions off the coasts of Oregon and Washington is warranted because those coastal areas are major rearing and migration habitats for juvenile Chinook Salmon (Miller et al. 1983; Pearcy and Fisher 1990; Peterson et al. 2010; Daly et al. 2012; Burke et al. 2013a). In an analysis of juvenile abundances along the coast from central California to the northern Gulf of Alaska, Fisher et al. (2007) found that the greatest catches of juvenile Chinook Salmon occurred from central Oregon to northern Washington. The region's Chinook Salmon productivity is centered in the Columbia River, which historically supported the largest runs in the world (Van Hyning 1973). Although widespread habitat losses, dam construction, and periods of sustained overfishing have greatly reduced the abundance of naturally produced Chinook Salmon in the Columbia River (Myers et al. 1998; Ford 2011), artificial production of this species is currently enormous, with more than 100 million juveniles released from the basin's hatcheries each year (Paquet et al. 2011).

In addition to their great abundance, Chinook Salmon in the region are characterized by high genetic, ecological, and life history diversity (Waples et al. 2001). Juvenile life histories are highly variable among and within populations, particularly with regard to the timing of downstream migration and the age at ocean entry. Two general ocean entry life history types of Chinook Salmon have been identified (Healey 1991): subyearlings and yearlings. Subyearlings enter the ocean during their first year (age 0) and have highly diverse downstream movement patterns (Rich 1920). In many areas, subyearlings move into estuaries and marine areas throughout the year, with temporally broad migratory peaks typically occurring in summer (Rich 1920; Reimers 1973; McCabe et al. 1986). Yearlings reside for a year in freshwater before moving seaward (at age 1) and typically migrate to the ocean as larger smolts during spring (Rich 1920; Weitkamp et al. 2012). Although the age of ocean entry is variable in most naturally spawning populations, yearling life histories are most common in populations with adults that return to freshwater in the spring (i.e., spring run; Waples et al. 2004). In contrast, subyearling types predominate in fall-run populations. Columbia River hatcheries focus on subyearling releases for fall-run populations and yearling culture strategies for spring runs. However, hatchery propagation of fall-run populations in the Snake River and summer-run populations in the upper Columbia River includes substantial production of both subyearlings and yearlings (Fisher et al. 2014).

Previous studies have shown that the marine migrations of the region's juvenile Chinook Salmon also differ between yearling and subyearling life history types (Miller et al. 1983; Fisher and Pearcy 1995; Trudel et al. 2009; Tucker et al. 2011). Yearlings disperse rapidly after entering the ocean. For example, yearlings from spring-run populations in the interior headwater tributaries of the Columbia and Snake rivers make very rapid northward migrations, and by late summer they are primarily found off the Alaska coast (Tucker et al. 2011; Fisher et al. 2014). However, the migrations of yearlings from spring-run populations in the lower Columbia and Willamette rivers and from summer-run sources in the upper Columbia River are much more variable; by the end of summer, yearlings from these stocks are widely distributed along the coast (Fisher et al. 2014). In contrast to the two yearling dispersal patterns, subyearlings from the Columbia River and adjacent coastal streams mostly remain south of Vancouver Island through autumn (Trudel et al. 2009; Tucker et al. 2011; Fisher et al. 2014). Although these three distinct migration types have been well documented, a genetic assessment of stock-specific juvenile distributions is lacking for marine areas immediately north and south of the Columbia River. This information gap is particularly notable for the region's subyearlings, which occupy coastal habitats in the area during their first summer at sea.

The objective of our study was to use genetic stock identification to delineate the temporal and spatial distributions of juvenile Chinook Salmon occupying coastal habitats from central Oregon to northern Washington. Juveniles were collected during 15 years of trawl surveys conducted in spring, summer, and autumn along a fixed grid that included both nearshore and offshore sampling stations. We report the first stock composition estimates for subyearling and yearling juveniles off Oregon and Washington, and we also present the proportions of naturally and hatchery-produced juveniles from specific stocks. Furthermore, we assessed whether the initial spatial distributions of yearlings (latitude and distance from shore) in spring and summer differed between stocks that are known to migrate rapidly northward and those that display more variable migrations. Samples collected in summer and autumn were used to examine the seasonal and spatial distributions of the region's more resident subyearlings and to search for stock-specific differences in habitat use.

METHODS

Ocean sampling and juvenile life history classification.—Juvenile Chinook Salmon were collected during annual juvenile salmon trawl surveys that were conducted along the Washington and Oregon coasts by National Ocean and Atmospheric Administration Fisheries (Figure 1; Brodeur et al. 2005; Peterson et al. 2010). Data used in the present study were from surveys conducted in late May of 2006–2012 and in late June and late September of 1998–2012. Trawls consisted of 30-min-long surface tows with a 264 Nordic rope trawl (see Krutzikowski and Emmett 2005 for a detailed description). Trawling occurred along seven to nine transects situated perpendicular to shore, ranging from northern Washington (48.2°N) to central Oregon (44.7°N; Figure 1). Sampling stations began as close to shore as possible (1–5 nautical miles offshore [1 nautical mile = 1.852 km]; ~30-m depth) and continued offshore (in about 5-nautical-mile increments) to just beyond the shelf.

Onboard the sampling vessel, juvenile Chinook Salmon were identified, measured (FL, mm), and frozen. In the laboratory, fish were remeasured and examined for markings indicative of hatchery origin (e.g., clipped adipose fin, coded wire tag [CWT], or other tag). Samples of fin tissue were taken for genetic analysis and were stored in 100% ethanol. For most trawls, tissues were collected from all fish captured. However, for several large catches (>100 fish), juveniles from all size-classes were subsampled to comprise the genetic sample. We used FL and month of ocean capture to classify juvenile life history type as either subyearling or yearling; size cutoffs were derived from known ages based on scale analysis and from tagged fish (Pearcy and Fisher 1990; Fisher and Pearcy 1995; Weitkamp et al. 2012). Fish larger than the yearling size category were assumed to be older (i.e., sea entry in earlier years) and were not included in analyses. Among fish that were captured in May, those ranging from 121 to 250 mm were classified as yearlings, and smaller individuals were considered subyearlings. Size cutoffs used to designate yearling fish were 141–280 mm for June captures and 251–400 mm for September captures. Subyearling and yearling CPUEs for each station were computed by dividing the number of captured subyearlings or yearlings by the number of kilometers that were towed to obtain the sample.

Genetic data collection.—Genomic DNA was isolated from fin tissue samples by using Wizard Genomic DNA Purification Kits (Promega Corp.) in accordance with the manufacturer's protocols. Polymerase chain reactions (PCRs) were then conducted to amplify 13 microsatellite DNA loci

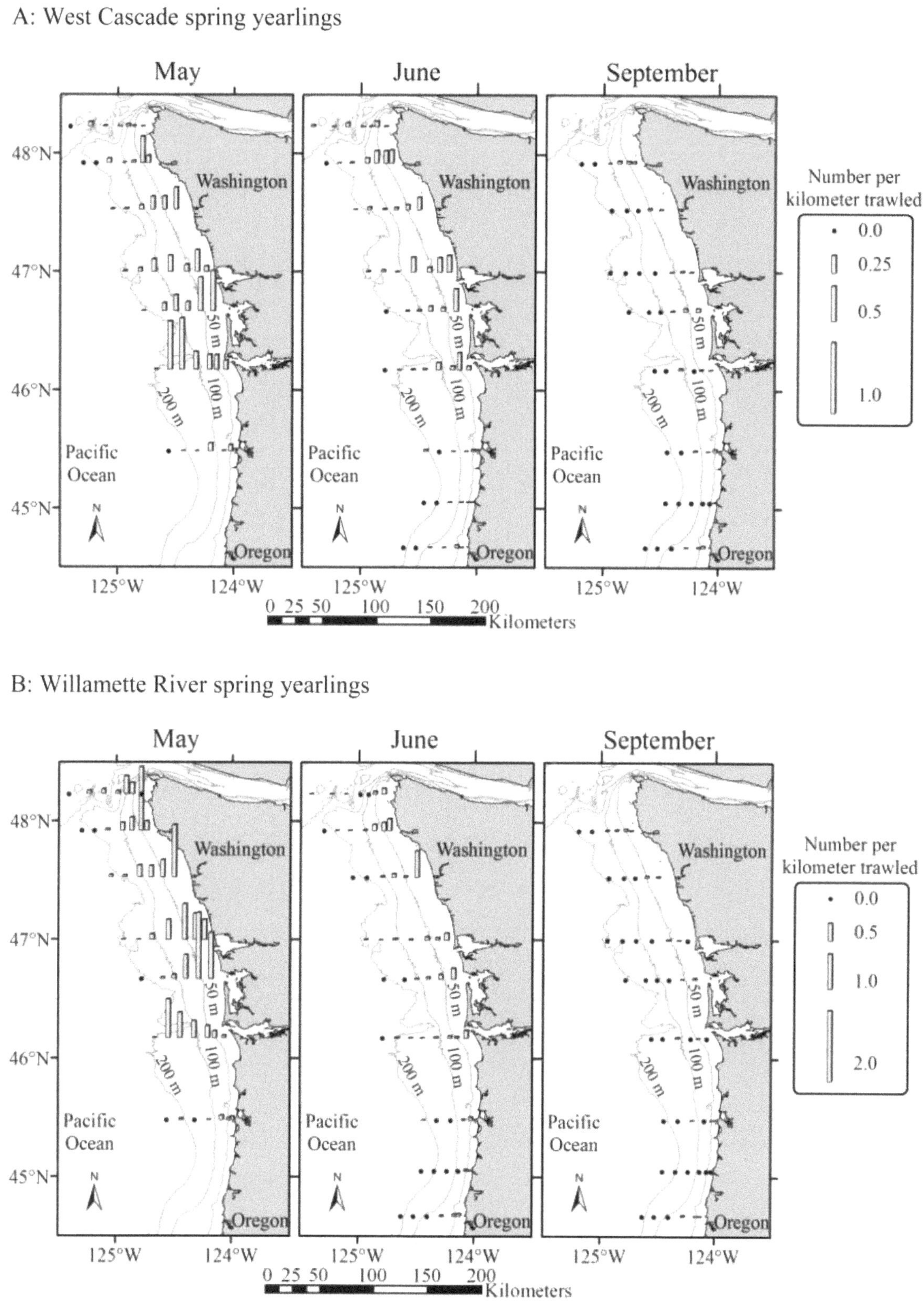

FIGURE 1. Maps showing the sampling transects and stock-specific CPUEs (fish/km trawled) for juvenile Chinook Salmon of each life history type sampled in May (2006–2012), June (1998–2012), and September (1998–2012): **(A)** West Cascade spring-run yearlings, **(B)** Willamette River spring-run yearlings, **(C)** mid-/upper Columbia River spring-run yearlings, **(D)** Snake River spring-run yearlings, **(E)** upper Columbia River summer–fall-run yearlings, **(F)** Snake River fall-run yearlings, **(G)** Washington coast subyearlings, **(H)** West Cascade fall-run subyearlings, **(I)** Spring Creek group fall-run subyearlings, **(J)** upper Columbia River summer–fall-run subyearlings, **(K)** Snake River fall-run subyearlings, and **(L)** Oregon coast subyearlings.

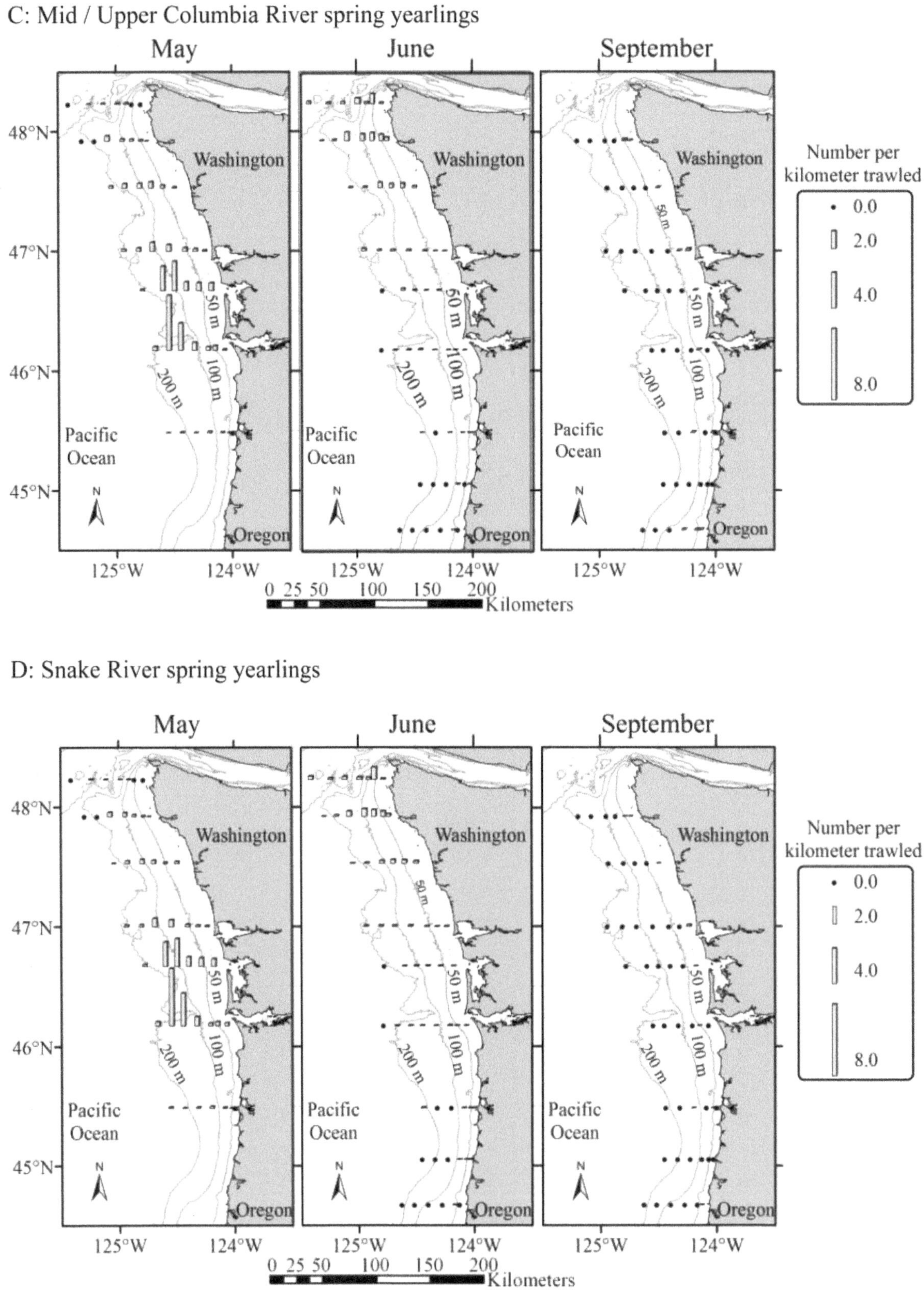

FIGURE 1. Continued.

E: Upper Columbia River summer/fall yearlings

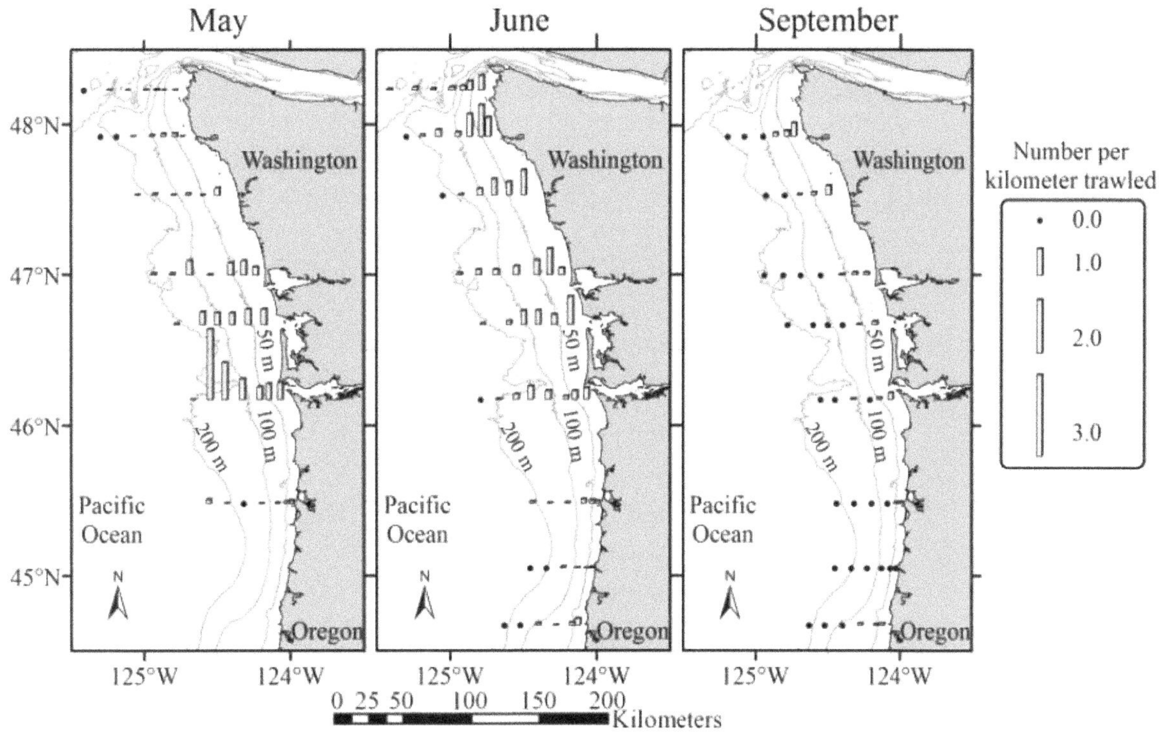

F: Snake River fall yearlings

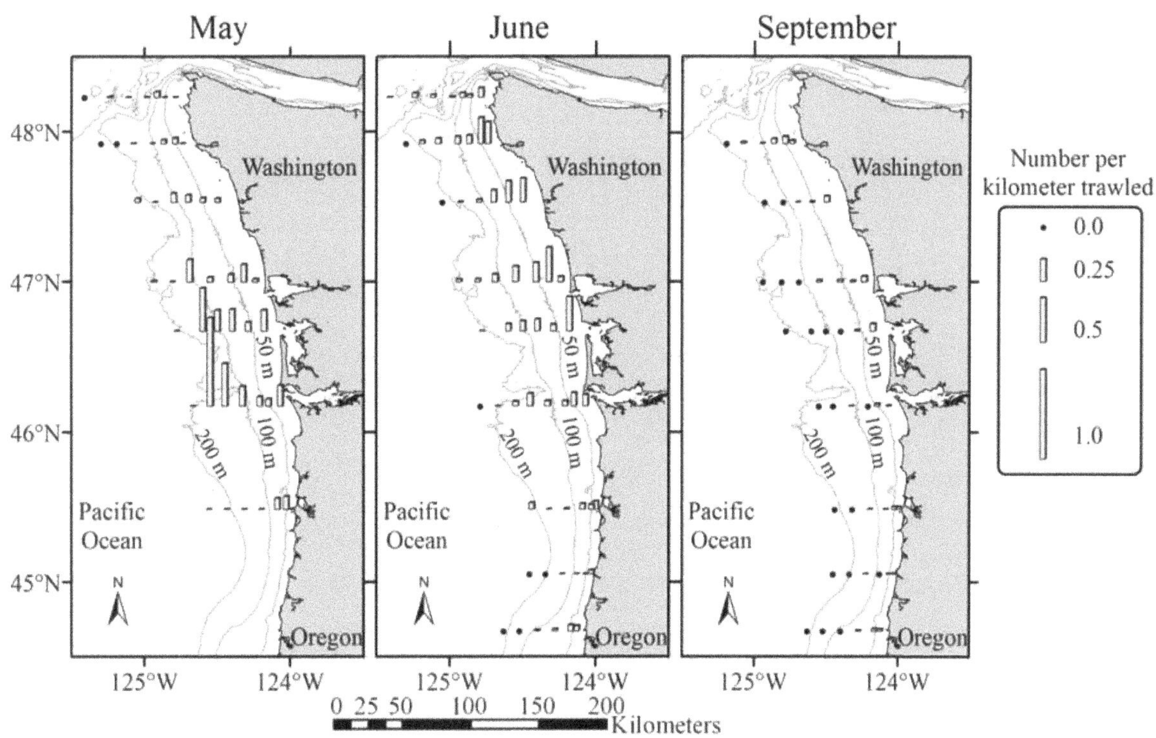

FIGURE 1. Continued.

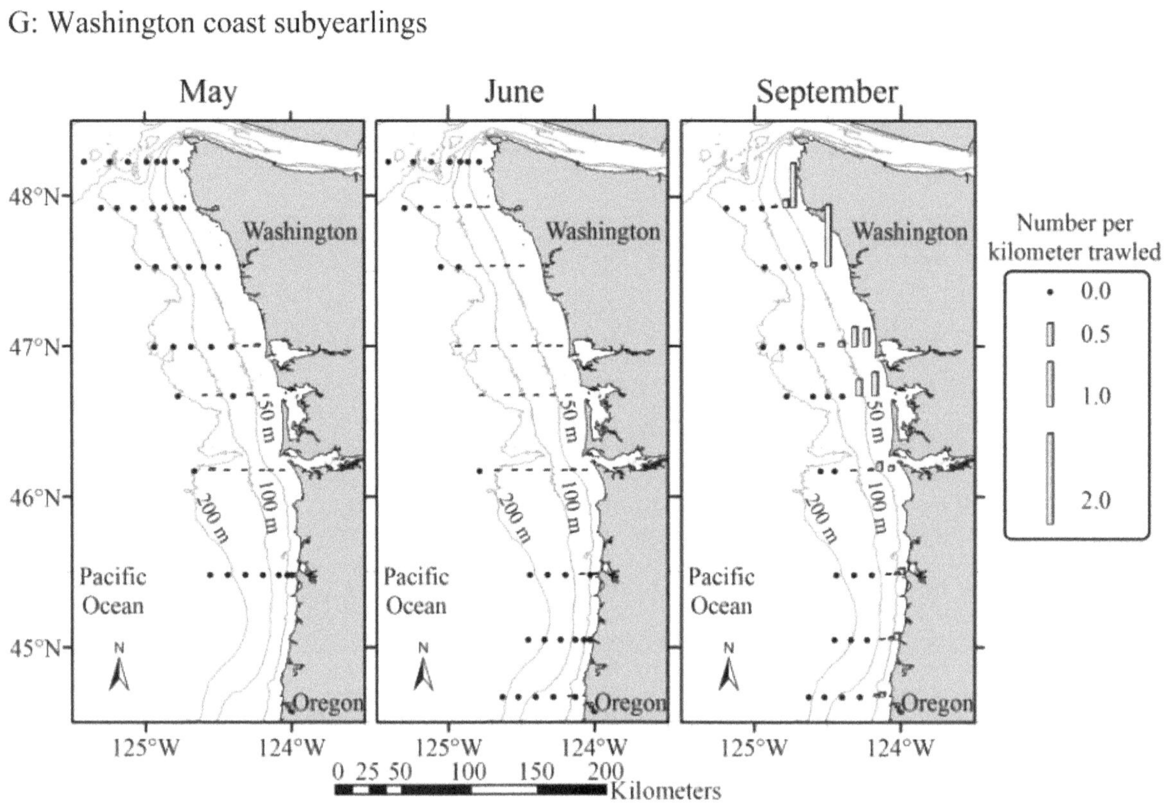

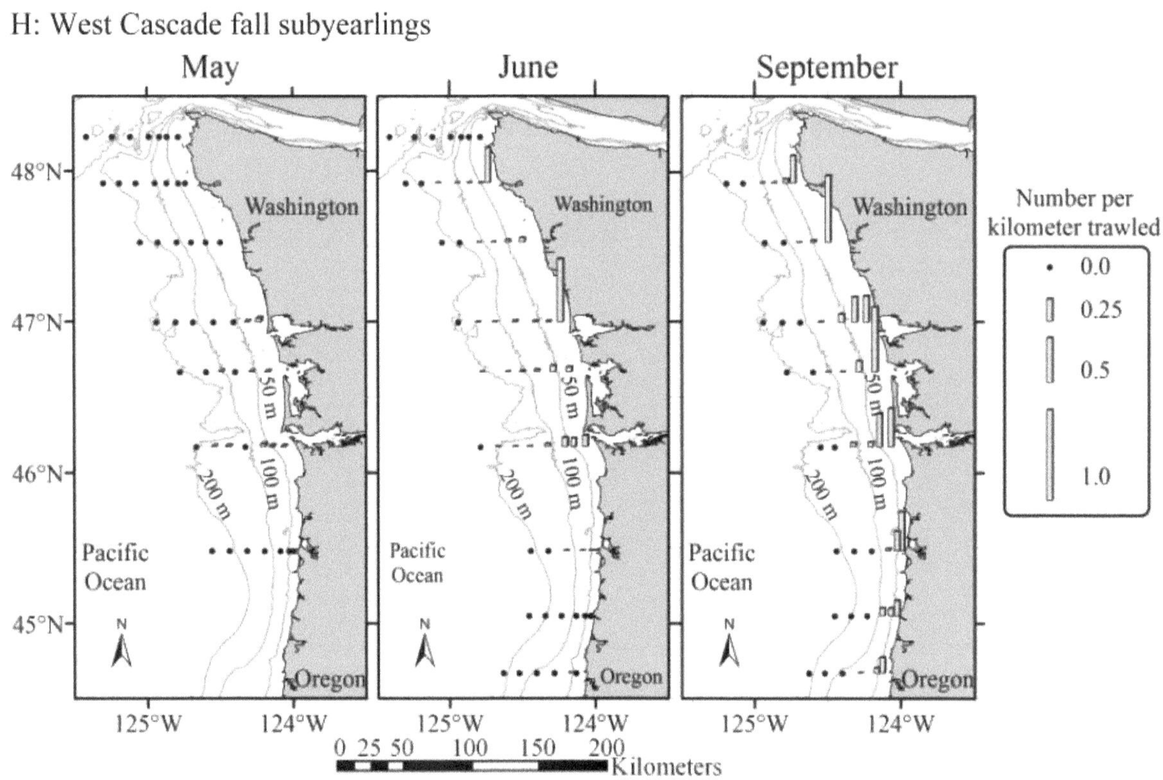

FIGURE 1. Continued.

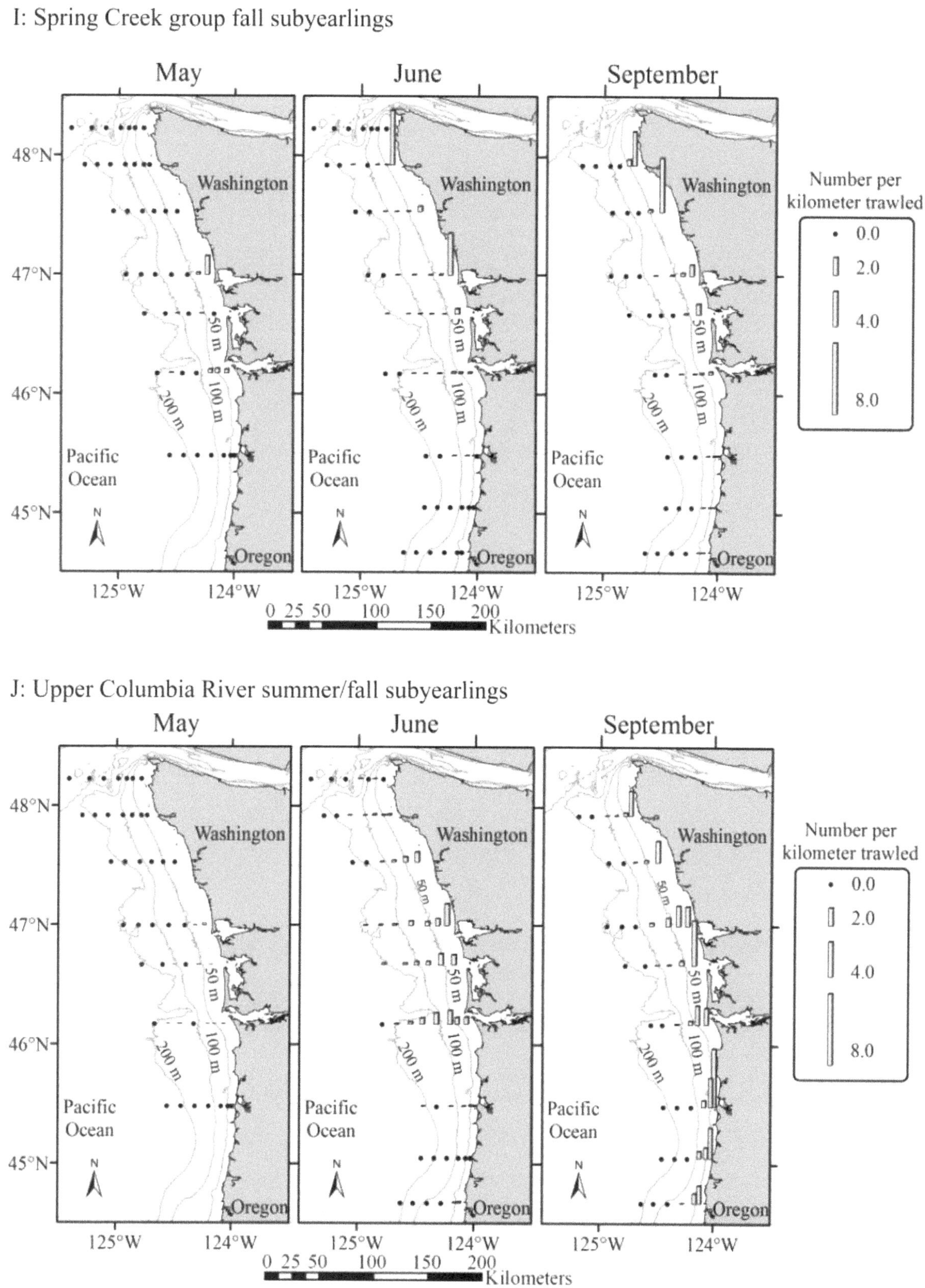

FIGURE 1. Continued.

K: Snake River fall subyearlings

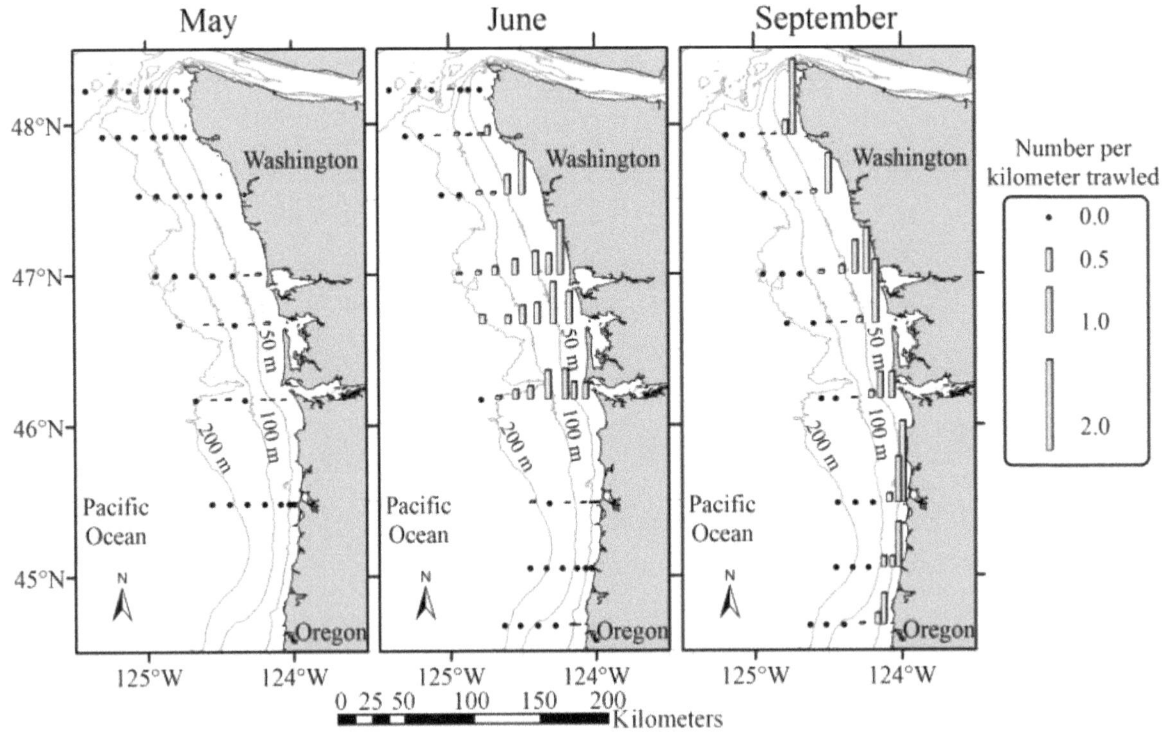

L: Oregon coast subyearlings

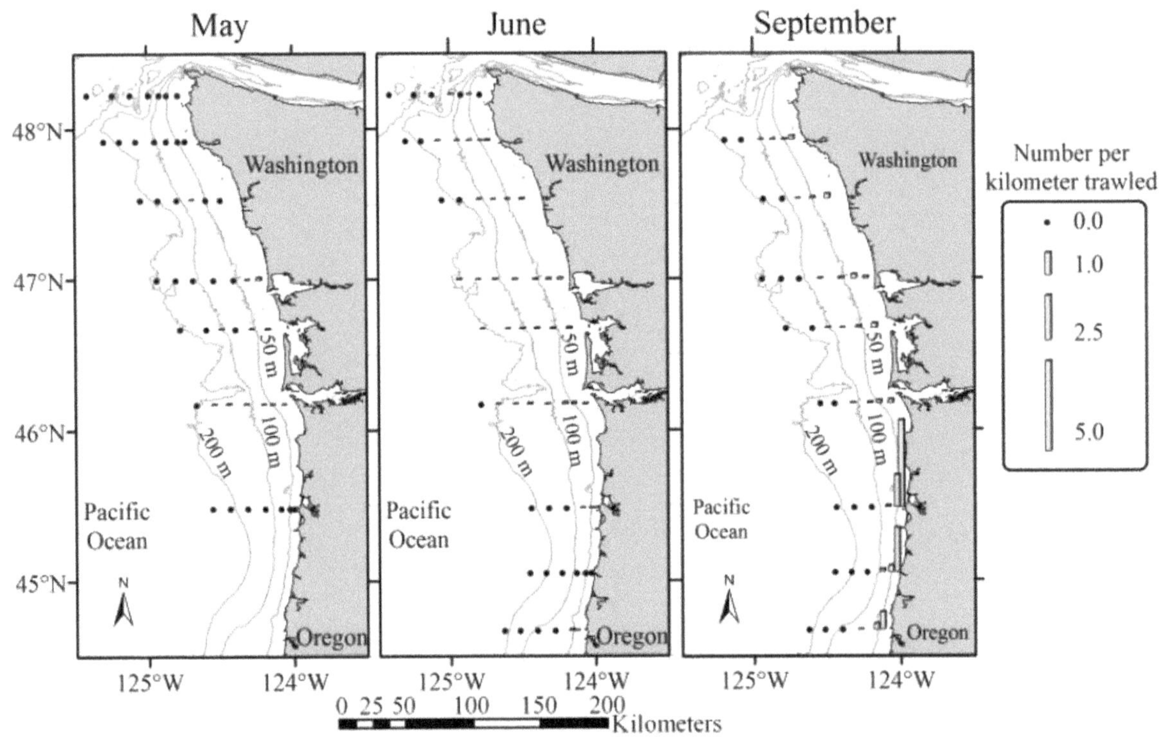

FIGURE 1. Continued.

that have been standardized for Chinook Salmon by several West Coast genetics laboratories (Seeb et al. 2007). The resulting PCR products were analyzed with Applied Biosystems, Inc. (ABI), 3100 and 3730 Genetic Analyzers, and genotypes were determined using ABI GeneScan and ABI Genotyper software.

Stock composition analysis.—Genetic stock identification analyses were conducted using the likelihood model of Rannala and Mountain (1997) as implemented by the genetic stock identification program ONCOR (Kalinowski et al. 2007). Two types of stock identification analysis were implemented to address different questions. First, ONCOR and a standard approach to genetic stock identification mixture modeling (Manel et al. 2005) were used to estimate the overall proportional stock compositions of samples grouped by month and by length-based juvenile life history type (i.e., subyearlings and yearlings). This approach uses a "baseline" of genotypic data from potential source populations (Milner et al. 1985) to estimate the most likely proportional composition of the set of mixture genotypes. The precision of stock composition results was estimated by bootstrapping baseline and mixture data 100 times (Kalinowski et al. 2007). Second, we used ONCOR and the genetic baseline to estimate the stock origins of individual fish. The individual fish stock assignments were used to estimate stock-specific CPUE per station for the analysis of spatial distributions and to estimate percentages of natural-origin fish (see below).

Genetic stock identification analyses were conducted by using a Chinook Salmon microsatellite DNA baseline that was compiled from the multi-agency databases described by Seeb et al. (2007), Moran et al. (2013), and Hess et al. (2014). Stock allocations were made to 141 individual Columbia River basin and coastal baseline populations and were summed to estimate the proportional contributions of 14 regional stocks (Table 1). The 14 stocks were generally congruent with the Chinook Salmon evolutionarily significant units (ESUs) identified under the U.S. Endangered Species Act (ESA; Myers et al. 1998; Table 1). However, due to a lack of appreciable genetic differentiation between several ESUs, estimates for juveniles from those areas were combined into larger reporting units. Results for the three Central Valley (California) ESUs were combined into a single reporting unit, as were results for the Mid-Columbia River ESU and Upper Columbia River ESU. In addition, preliminary results showed that the three southern Oregon and California coastal ESUs south of the study area were very minor contributors to our

TABLE 1. Chinook Salmon genetic baseline used for stock identification analysis in this study. Genetic stock or region, primary juvenile life history type, evolutionarily significant unit (ESU), number of baseline populations, and sample size are given (ESU status: 1 = endangered; 2 = threatened; 3 = species of concern; 4 = listing not warranted). Geographic areas for baseline populations in British Columbia (BC) are indicated. Genetic data are from Seeb et al. (2007), Moran et al. (2013), and Hess et al. (2014).

Genetic stock/region	Juvenile life history type	ESU (status)	Baseline populations	Sample size
Southern BC/Salish Sea	Subyearling[a]	South BC mainland, east Vancouver Island, Fraser River, and west Vancouver Island; and Puget Sound ESU (2)	60	8,169
Washington coast	Subyearling[a]	Washington Coast ESU (4)	7	767
West Cascade fall	Subyearling	Lower Columbia River ESU (2)	3	357
West Cascade spring	Yearling	Lower Columbia River ESU (2)	3	428
Spring Creek group fall	Subyearling	Lower Columbia River ESU (2)	4	384
Willamette River spring	Yearling	Upper Willamette River ESU (2)	5	559
Deschutes River fall	Subyearling	Deschutes River Summer–Fall ESU (4)	2	288
Upper Columbia River summer–fall	Subyearling[b]	Upper Columbia River Summer–Fall ESU (4)	4	706
Mid-/upper Columbia River spring	Yearling	Mid-Columbia River Spring ESU (4); and Upper Columbia River Spring ESU (1)	6	740
Snake River spring	Yearling	Snake River Spring–Summer ESU (2)	8	909
Snake River fall	Subyearling[b]	Snake River Fall ESU (2)	3	474
Oregon coast	Subyearling[a]	Oregon Coast ESU (4)	19	2,345
Southern Oregon/California coast	Subyearling[a]	Southern Oregon and Northern California Coastal ESU (4); Upper Klamath and Trinity Rivers ESU (3); and California Coast ESU (2)	8	1,119
Central Valley, California	Subyearling	Central Valley Fall ESU (3); Central Valley Spring ESU (2); and Sacramento River Winter ESU (1)	9	1,000
Total			141	18,245

[a]Spring-run populations in these regions primarily exhibit the yearling life history type.
[b]Some hatchery programs release yearlings from these stocks.

samples; we therefore combined estimates for those ESUs into a single regional stock. Similarly, results for the Puget Sound ESU were combined with results for the British Columbia sources north of the study area.

For the Lower Columbia River ESU, genetic estimates were made for three genetically distinct stocks (Myers et al. 2006; Seeb et al. 2007; Moran et al. 2013). Spring-run and fall-run populations originating in tributaries of the western Cascade Mountains were differentiated from each other and from a fall-run stock native to the Columbia River gorge (Table 1). Hereafter, we refer to the Columbia River gorge native fall-run stock as the "Spring Creek group" because the current distribution of the stock reflects the use (since 1901) of fish and eggs from Spring Creek National Fish Hatchery (NFH) to stock hatcheries and rivers throughout the lower Columbia River (Myers et al. 2006).

Some fish ($N = 319$) from our trawl surveys were initially classified as yearlings based on FL but were later reclassified as subyearlings. The reclassification of these fish (~3% of our total sample) was based on their individual genetic stock assignments (described below). The reclassified individuals were large fish that were assigned to the Central Valley stock and to the Spring Creek group fall-run stock in the lower Columbia River. Chinook Salmon from the Sacramento and San Joaquin rivers of California are thought to enter the ocean primarily as subyearlings, with out-migrations occurring throughout much of the year (Fisher 1994; Waples et al. 2004). Central Valley subyearlings that entered the ocean during winter or spring likely exceeded our size cutoffs after several months of marine growth. Spring Creek NFH releases large subyearling smolts in May of each year (Regional Mark Information System [RMIS]; www.rmpc.org), and these fish can exceed the subyearling size cutoffs when captured in ocean surveys. A few individuals that were reclassified (2 fish from the Central Valley; 48 fish from the Spring Creek group) were confirmed to be subyearlings based on hatchery release information from the CWTs.

Stock-specific catch per unit effort.—ONCOR was used to compute the relative probability of regional stock membership for each individual ocean-caught fish. In this procedure, an individual is "carved up" and allocated to each source in proportion to these relative probabilities (Manel et al. 2005). For each sampling station during a given survey, we summed the full individual assignment probability values for all subyearlings and for all yearlings to estimate the proportions of each stock in the catch. The resulting proportions were then applied to the total number of each life history type caught at the station to estimate stock-specific CPUEs.

Distribution maps.—Stock-specific and life-history-specific CPUEs for each station were averaged across years for each survey month to examine the average distributions. Only stations that were sampled during at least 25% of the years for each month were included (i.e., at least 2 of the 7 surveys conducted in May; at least 4 of the 15 surveys conducted in June; and at least 4 of the 15 surveys conducted in September). Charts were made using ArcMap 10.0 for each life history type and stock sampled in each month.

Spatial statistics.—Our sampling regime used fixed-location sampling sites that were partially designed to capture the full spatial distribution of particular stocks. Therefore, many trawls resulted in a CPUE of zero for select stocks (e.g., we sampled close to shore to catch certain stocks, knowing that other stocks might not be that close). Therefore, calculation of a simple mean for latitude or distance from shore would have resulted in biased estimates of location. Instead, we calculated the mean location (latitude or distance from shore) by weighting each sample by the CPUE for a given stock caught in that trawl. By design, trawls with zero catch of a given stock had no influence on the mean location for that stock. For this analysis of stock-specific spatial distributions by month and life history type, data were combined across all survey years.

For many of the Chinook Salmon stocks, the spatial distribution of juveniles was nonnormal, making traditional statistical tests inappropriate. We therefore chose to use permutation-based methods for all statistical tests. We first calculated the weighted mean latitude and weighted mean distance from shore for each stock individually (with subyearlings and yearlings considered separately) as described above. For each pair of stocks to be compared, we then calculated the difference in weighted mean latitude and distance from shore. To determine whether these differences were greater than one would expect if the two stocks' distributions had been drawn from the same underlying spatial distribution (i.e., our null hypothesis), we compared them to the differences generated from 5,000 permutations of the data. For each permutation, we randomly reassigned the two CPUE values to the two stocks (one value for each stock in the comparison) and recalculated the difference in weighted mean latitude and distance from shore between the two stocks. All P-values reported here represent the proportion of permutations for which the difference in weighted means was more extreme than the observed difference. Because we made multiple pairwise comparisons, significance was determined using Bonferroni-corrected probabilities. All stocks with sample sizes of 75 or more fish for a given month and life history type were included in the statistical tests.

For each stock and life history type included in statistical tests, we also characterized the interannual variability in spatial distributions. To estimate interannual variability in latitude or distance from shore during a given month, we calculated the weighted mean value for each year separately; interannual variability was then represented as the SD of the annual weighted means.

Estimates of natural-origin fish and coded-wire-tagged fish.—We used ONCOR to assign each individual fish to a most likely stock of origin (i.e., the stock with the highest relative probability). The numbers of marked fish (those with adipose fin clips or CWTs) and unmarked fish in the resulting set of fish for each stock were used to estimate the proportions of natural-origin fish. Not all hatchery fish are marked, and marking rates vary among hatcheries. Marking rates for juvenile Chinook Salmon released from hatcheries were compiled from the RMIS database (Table 2). To calculate the potential number of naturally produced fish in the catches, we (1) estimated the total number of hatchery fish (calculated as [number of marked fish assigned to a stock]/[marking rate for the stock]) and then (2) subtracted the estimated number of hatchery fish from the total number of fish in the sample. Estimation of the numbers of naturally produced fish was restricted to the most recent study years (2006–2012), when marking rates for several stocks were relatively high (>90%). To improve the overall accuracy of the estimates of natural-origin fish, estimates were not made for sample sizes smaller than 50 fish.

To assess the accuracy of our genetic estimates, we identified a subset of the juvenile Chinook Salmon that had received CWTs. Release data for each CWT numeric code, including species, run, fish age, date, hatchery, stock, and location, were obtained from the RMIS database. Based on this information, we classified each juvenile as originating from 1 of the 14 regional genetic stocks in our analysis. The CWT classification of each fish was then compared with the stock of origin (i.e., the stock with the highest relative probability) that was estimated using ONCOR and the fish's genotypic data. We also used the entire subset of fish with CWTs to create a mixed-stock sample of "known" origin to evaluate the accuracy of stock composition estimates produced by ONCOR.

RESULTS

Seasonal Abundance Trends in Catches

The CPUE varied among years, among seasons, and between life history types (Figure 2). For all years that included May sampling, the yearling CPUE was greatest in May, with a mean of 2.4 fish/km trawled. In all years, yearling CPUE in June was intermediate (mean = 0.9 fish/km), and September CPUE of yearlings was lowest (0.2 fish/km). The

TABLE 2. Average annual number of Chinook Salmon released from hatcheries in 2006–2012, months of releases, and percentages of hatchery fish that were marked at release (from the Regional Mark Information System; www.rmpc.org); numbers of unmarked and marked juveniles assigned to regional genetic stocks from sampling along the Washington and Oregon coasts (2006–2012); and estimated number of natural-origin juveniles, number of hatchery-origin juveniles, and percentage of natural-origin juveniles in samples based on marking rates of released fish.

	Hatchery releases			Juveniles in ocean catch				
Genetic stock/region	Average release (millions)	Peak month(s)	Percent marked	Unmarked	Hatchery marked	Estimated natural origin	Estimated hatchery origin	Percentage of catch estimated to be of natural origin
				Yearlings				
West Cascade spring	3.1	Feb–Mar	89.9	11	173	0[a]	184	0.0
Willamette River spring	5.5	Feb–Mar	98.5	48	405	42	411	9.3
Upper Columbia River summer–fall	2.8	Apr	92.9	13	570	0[a]	583	0.0
Mid-/upper Columbia River spring	8.3	Mar–Apr	97.5	95	706	77	724	9.6
Snake River spring	11.3	Mar–Apr	99.6	71	689	68	692	8.9
Snake River fall	0.9	Apr	97.9	8	205	4	209	1.9
Total yearlings	31.9			246	2,748	191	2,803	6.4
				Subyearlings				
Washington coast	10.7	May–Jun	89.1	69	62	61	70	46.6
West Cascade fall	16.1	Jun–Jul	90.1	56	45	51	50	50.5
Willamette River spring	1.8	Oct–Nov	95.6	53	6	53	6	89.8
Spring Creek group fall	24.2	Apr–May	89.1	58	476	0[a]	534	0.0
Deschutes River fall	<0.1	May	100.0	60	26	60	26	69.8
Upper Columbia River summer–fall	21.5	Jun–Jul	60.5	883	429	603	709	46.0
Snake River fall	4.1	May–Jun	65.2	315	268	172	411	29.5
Oregon coast	4.0	Jun–Jul	55.5	148	79	85	142	37.4
Total subyearlings	82.5			1,642	1,391	1,085	1,948	35.8

[a]The percentage of marked ocean-caught fish equaled or exceeded the hatchery marking rate; therefore, the estimated number of natural-origin juveniles was zero.

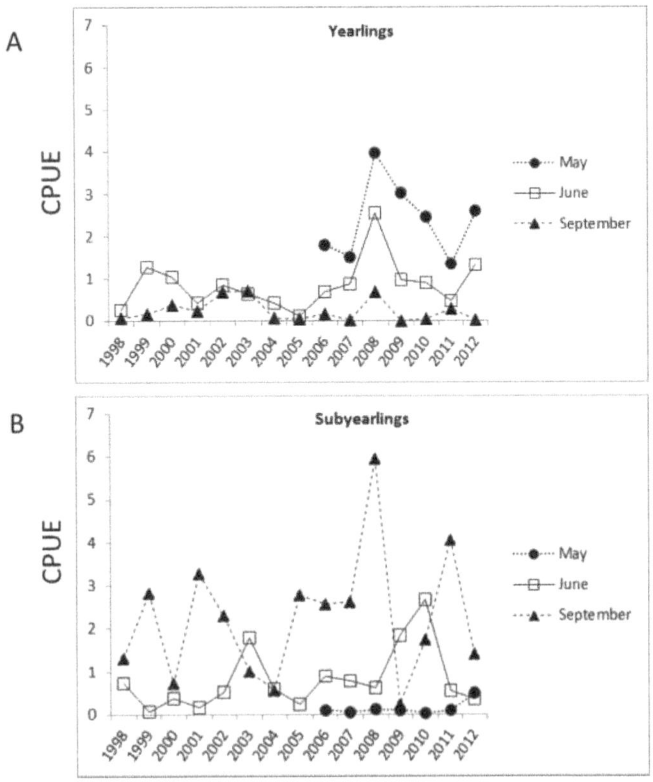

FIGURE 2. Chinook Salmon CPUE (fish/km trawled) for **(A)** yearlings and **(B)** subyearlings sampled along the Washington and Oregon coasts in May (2006–2012), June (1998–2012), and September (1998–2012).

CPUE pattern for subyearlings was opposite the yearling CPUE pattern in most years, with the lowest value observed in May (0.1 fish/km), an intermediate value in June (0.8 fish/km), and the highest value in September (2.2 fish/km). However, in 4 of the 15 sampling years, subyearling CPUE was greater in June than in September.

Comparison of Genetic Estimates and Stock Origins based on Coded Wire Tags

Genetic and CWT data were obtained for 2,223 sampled fish (Table 3). Overall, genetic assignments and CWT origins agreed for 82% of the individuals. Genetic assignment accuracy for the 14 genetic stocks ranged from 65% for the Snake River spring-run stock to 100% for the southern British Columbia/Salish Sea stock and the West Cascade (lower Columbia River) fall-run stock. Approximately 80% of the genetic allocations that did not agree with the CWT data were between two pairs of stocks from the interior Columbia River basin: (1) the upper Columbia River summer–fall run and Snake River fall run (54% of the misallocations); and (2) the mid-/upper Columbia River spring run and Snake River spring run (25% of the misallocations). The differences between genetic stock composition estimates and the composition based on CWT data averaged 0.3%. The greatest differences were observed for the Deschutes River fall run (0.8%) and the upper Columbia River summer–fall run (0.7%).

Stock Composition of Juveniles

Genetic data from a total of 9,703 sampled fish were used to estimate proportional stock composition of juvenile Chinook Salmon captured during coastal trawl surveys (Table 4; Figure 3). Overall, approximately 98% of the yearlings sampled in May were estimated to originate from Columbia River basin sources. The May yearlings were 82% spring-run fish from the basin, primarily from the mid-/upper Columbia River (28%), Snake River (27%), and Willamette River (19%) stocks. May yearlings included smaller percentages of upper Columbia River summer–fall-run (10%), West Cascade spring-run (8%), and Snake River fall-run (5%) fish. The contribution of all other stocks combined was approximately 3%.

Yearlings sampled in June were also predominately from the Columbia River basin (98%), but only 54% were from the basin's spring-run stocks. June percentages for the mid-/upper Columbia River spring run (21%), Snake River spring run (20%), Willamette River spring run (7%), and West Cascade spring run (7%) were lower than the corresponding May percentages. The greatest contributor to June yearlings was the upper Columbia River summer–fall stock (31%). The Snake River fall-run stock comprised 10% of the June sample of yearlings.

In addition to being the largest contributor to June yearling samples, the upper Columbia River summer–fall-run stock comprised the greatest percentage of June subyearlings (33%). Other stocks represented at high percentages in the subyearling samples included the Snake River fall run (30%) and Spring Creek group fall run (18%). Overall, greater than 98% of the subyearlings sampled in June were from Columbia River basin stocks.

Approximately 77% of September subyearlings were from Columbia River basin sources, with the largest percentages contributed by the upper Columbia River summer–fall-run (42%), Spring Creek group fall-run (15%), and Snake River fall-run (11%) stocks. Subyearlings from coastal Oregon (12%) and Washington (5%) were also present, as were smaller percentages of fish originating from systems south and north of the study area (1–3%).

Relatively few (n = 115) subyearlings were captured during 7 years of May sampling, and the majority (57%) of those fish belonged to the Spring Creek group fall stock. May subyearlings also included Columbia River basin spring-run fish (~40%). Yearlings were rare in September surveys; 161 samples were analyzed from 15 survey years. The greatest percentage contributions to September yearlings were from the upper Columbia River summer–fall run (44%), Snake River fall run (15%), West Cascade spring run (12%), and West Cascade fall run (9%).

TABLE 3. Number of juvenile Chinook Salmon with coded wire tags (CWTs) indicating origin, compared with genetic stock assignments to 14 stocks in samples collected along the Washington and Oregon coasts. Actual and genetic estimates of percentage stock composition for the 2,223 fish with CWTs are also shown (stocks: BCS = southern British Columbia/Salish Sea; WA = Washington coast; WCF = West Cascade fall run; WCS = West Cascade spring run; WS = Willamette River spring run; SCG = Spring Creek group fall run; DES = Deschutes River fall run; USF = upper Columbia River summer–fall run; MUS = mid-/upper Columbia River spring run; SS = Snake River spring run; SF = Snake River fall run; OR = Oregon coast; SOCA = southern Oregon/California coast; CV = Central Valley, California). Values in bold italics represent the numbers of individuals for which CWT origin and genetic stock assignment were in agreement.

Genetic stock	CWT origin													
assignment	BCS	WA	WCF	WCS	WS	SCG	DES	USF	MUS	SS	SF	OR	SOCA	CV
BCS	***3***	1						4	2		1		1	
WA		***47***						2	1			3		
WCF			***2***	2	1	1		1		1	1	2		
WCS				***57***	2				5			1		
WS				2	***80***			1		4				
SCG				1		***43***					3			
DES							***0***	5			6			
USF					3	1		***788***			104	4	2	1
MUS				4					***246***	59				
SS									42	***121***				
SF						2		112			***379***			
OR		6		1				1				***49***		
SOCA								1				2	***7***	
CV														***2***
Total	3	54	2	67	86	47	0	915	296	185	494	61	10	3
Actual %	0.1	2.4	0.1	3.0	3.9	2.1	0.0	41.2	13.3	8.3	22.2	2.7	0.4	0.1
Genetic %	0.1	2.3	0.5	3.4	3.6	2.0	0.8	40.5	13.7	7.5	22.0	2.6	0.5	0.1

Hatchery- and Natural-Origin Fish

Hatchery fish were predominant among juveniles sampled off the Washington and Oregon coasts from 2006 through 2012 (Table 2). Approximately 92% of yearlings and 46% of subyearlings were marked hatchery fish. For all stocks included in this analysis, hatchery releases of yearlings were marked at high rates (90–100%). The marked percentage of West Cascade spring-run yearlings in our ocean catches (94%) was higher than the hatchery release marking rate for the stock (90%). Similarly, 98% of the ocean-caught upper Columbia River summer–fall-run yearlings were marked, thus exceeding the stock's hatchery marking rate (93%). For these two stocks, the estimated percentage of hatchery fish was over 100%; we therefore concluded that the entire catch of these stocks was likely of hatchery origin. When catches of unmarked yearlings from other stocks were adjusted for hatchery releases of unmarked fish, the total estimated percentage of natural-origin yearlings in the study area was 6%. The highest percentages of natural-origin yearlings were contributed by the Willamette River, mid-/upper Columbia River, and Snake River spring-run stocks (9–10%).

Hatchery releases of subyearlings were marked at lower and more variable rates (56–100%) during the 7 years, making it more difficult to accurately assess the percentage of natural-origin subyearlings in our catches. After adjustment for hatchery marking rates, approximately 36% of the subyearlings sampled in our surveys were estimated to be of natural origin. Among the major contributors to subyearling catches, the percentages of natural-origin fish were 46% for subyearlings from the upper Columbia River summer–fall stock and 30% for subyearlings from the Snake River fall stock. However, for the Spring Creek group fall-run stock, the percentage of captured subyearlings that had marks was the same as the hatchery marking rate (89%); we therefore considered these fish to be entirely of hatchery origin.

Seasonal and Spatial Distributions of Stocks

Yearlings.—In May, yearlings from all of the stocks analyzed were primarily distributed off the mouth of the Columbia River and northward along the Washington coast (Table 5; Figure 1). Spring-run fish from the Willamette River stock were farther north than yearlings from interior basin stocks during May ($P \leq 0.01$; see Supplementary Table S.1 available online). The mean latitude of yearlings increased from May to June for all stocks, and the northward shift was most evident for spring-run yearlings from the mid-/upper Columbia River and Snake River ($P < 0.001$; Figure 4). By June, the interior

TABLE 4. Estimated percentage composition of 14 regional genetic stocks observed in samples of juvenile Chinook Salmon collected along the Washington and Oregon coasts (n = number of sampled fish). May samples were collected in 2006–2012; June and September samples were collected in 1998–2012. The range in parentheses below each estimate is the 95% confidence interval derived from 100 bootstrap resamplings of baseline and mixed-stock genotypes. Estimated composition values greater than 10% are shown in bold italics.

	May		June		September	
Stock	Yearlings (n = 1,896)	Subyearlings (n = 115)	Yearlings (n = 1,814)	Subyearlings (n = 1,752)	Yearlings (n = 161)	Subyearlings (n = 3,965)
Southern British Columbia/Salish Sea	0.7(0.5–1.8)	0.4(0.0–5.5)	0.3(0.3–1.2)	0.5(0.3–1.3)	2.5(0.6–7.2)	1.1(0.9–1.9)
Washington coast	0.6(0.2–0.9)	0.0(0.0–1.7)	0.1(0.0–0.4)	0.2(0.1–0.4)	1.2(0.0–3.5)	5.3(4.2–5.8)
West Cascade fall	1.0(0.6–1.9)	0.0(0.0–12.4)	1.7(1.1–2.5)	1.5(1.7–4.0)	9.3(2.5–14.3)	5.1(4.5–6.9)
West Cascade spring	7.6(6.3–9.5)	2.9(0.0–9.4)	6.5(5.4–8.2)	0.8(0.5–1.8)	***11.8***(6.8–19.5)	0.4(0.2–1.1)
Willamette River spring	***19.2***(16.4–20.4)	0.9(0.0–2.6)	6.8(5.4–7.6)	4.8(3.7–5.6)	7.3(1.9–11.6)	0.7(0.4–0.9)
Spring Creek group fall	0.0(0.0–0.0)	***57.4***(42.2–60.4)	0.0(0.0–0.1)	***17.8***(14.4–18.4)	0.0(0.0–2.0)	***15.0***(12.5–15.2)
Deschutes River fall	0.6(0.2–1.4)	0.0(0.0–0.0)	0.5(0.2–1.5)	5.0(3.0–6.7)	2.0(0.0–6.8)	2.8(2.1–4.1)
Upper Columbia River summer–fall	9.5(8.1–11.0)	0.9(0.0–5.1)	***31.0***(27.7–33.3)	***32.8***(29.1–36.6)	***43.8***(34.1–52.3)	***41.8***(36.4–42.8)
Mid-/upper Columbia River spring	***28.3***(23.8–29.9)	***13.5***(5.3–21.0)	***20.8***(17.1–21.7)	2.3(1.3–2.9)	1.4(0.0–3.5)	0.0(0.0–0.1)
Snake River spring	***27.3***(25.4–31.7)	***22.3***(11.0–27.8)	***20.2***(17.0–22.9)	3.0(2.1–3.9)	1.1(0.0–3.7)	0.1(0.0–0.1)
Snake River fall	4.8(3.3–5.7)	1.8(0.0–4.5)	***10.2***(8.6–12.7)	***30.4***(25.9–34.4)	***14.8***(8.7–22.4)	***10.6***(9.5–15.1)
Oregon coast	0.5(0.2–1.4)	0.0(0.0–1.3)	1.9(1.2–2.6)	0.6(0.1–1.5)	5.1(1.0–8.8)	***11.9***(10.8–13.6)
Southern Oregon/ California coast	0.0(0.0–0.2)	0.0(0.0–0.0)	0.0(0.0–0.4)	0.2(0.0–0.6)	0.0(0.0–2.2)	2.2(1.7–2.7)
Central Valley	0.0(0.0–0.1)	0.0(0.0–0.9)	0.0(0.0–0.1)	0.0(0.0–0.5)	0.0(0.0–3.1)	3.1(2.6–3.8)

spring-run stocks were farther north than other yearlings ($P < 0.001$; Table S.2). Upper Columbia River summer–fall-run and Snake River fall-run yearlings also shifted to the north from May to June ($P < 0.001$). For all yearlings, the spread in latitude was greater during June than during May, and this increase in variability was most evident for the upper Columbia River summer–fall stock and the Snake River fall stock ($P < 0.001$).

Yearlings were also closer to shore in June than in May (Table 5; Figure 1). The shift toward shore was most apparent for the interior basin stocks that were located farthest offshore in May ($P < 0.03$). Willamette River and West Cascade spring-run yearlings were also captured closer to shore, but the May-to-June comparisons were not significant for these stocks ($P = 0.29$ and 0.39, respectively).

By June, Willamette River spring-run yearlings were closer to shore than yearlings of the other stocks ($P < 0.01$), and the mid-/upper Columbia River spring-run stock was the farthest offshore, although the distance from shore was not significantly different from that of the Snake River spring-run yearlings ($P = 0.44$). Yearlings of the West Cascade spring run, upper Columbia River summer–fall run, and Snake River fall run were intermediate in location between the more-offshore interior basin spring-run stocks and the nearer-shore Willamette River spring-run stock during both May and June.

Subyearlings.—In June, subyearlings from the upper Columbia River summer–fall-run and Snake River fall-run stocks were primarily dispersed along the Washington coast (Table 5; Figure 1). By September, subyearlings from the two stocks were distributed farther south, and the fish were distributed broadly across the entire latitudinal range of the study area. The change in mean latitude was not significant for the upper Columbia River summer–fall ($P = 0.08$) or Snake River fall ($P = 0.18$) subyearlings. In contrast, June and September distributions were relatively similar for the Spring Creek group fall-run subyearlings; these subyearlings were farther north than upper Columbia River summer–fall-run and Snake River fall-run stocks during June ($P < 0.01$; Table S.3). Subyearlings of the Spring Creek group fall stock also had the most northerly distribution of any stock in September when coastal stocks contributed to our catches ($P \leq 0.01$; Table S.4). Subyearlings from the Oregon coast stock were primarily distributed off the Oregon coast, whereas Washington coast subyearlings were situated farther north ($P < 0.001$) and largely caught off the coast of Washington.

During the summer, the distributions of subyearlings shifted strongly toward shore, and the variability in distance from shore also decreased. The shift inshore from June to September was evident for subyearlings of the upper Columbia River summer–fall-run stock and the Snake River fall-run

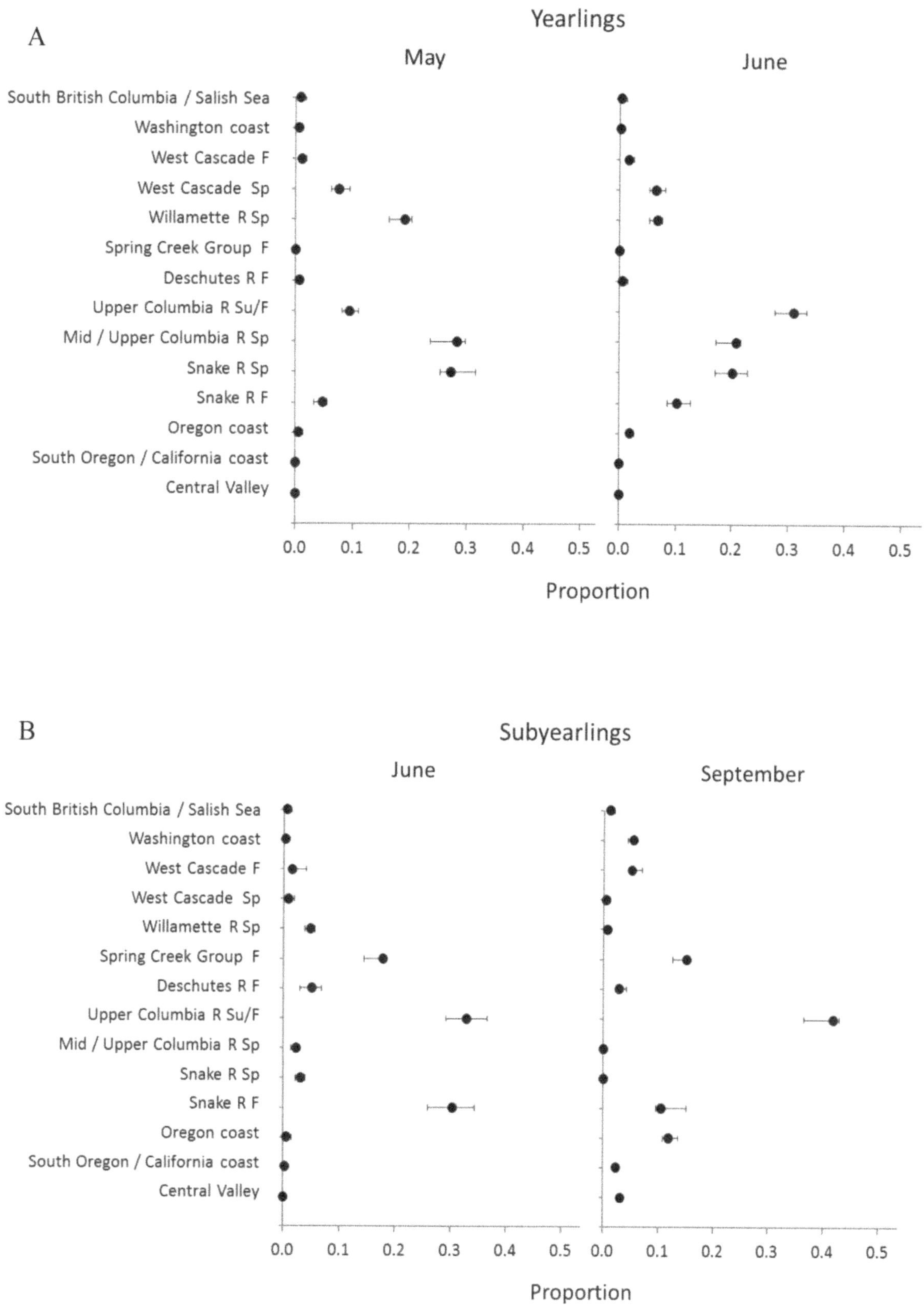

FIGURE 3. Estimated proportional stock composition (14 regional stocks) of **(A)** yearling and **(B)** subyearling Chinook Salmon sampled along the Washington and Oregon coasts in May (2006–2012), June (1998–2012), and September (1998–2012). Solid circles are the mean estimated proportions; horizontal lines are 95% confidence intervals derived from 100 bootstrap resamplings of baseline and mixed-stock genotypes (R = River, Sp = spring run, Su = summer run, F = fall run; May yearlings: n = 1,896; June yearlings: n = 1,814; June subyearlings: n = 1,752; September subyearlings: n = 3,965).

TABLE 5. Weighted mean latitude (°N) and distance from shore (nautical miles [nm]; 1 nautical mile = 1.852 km) for juvenile Chinook Salmon in samples collected along the Washington and Oregon coasts, presented for each life history type, stock, and sampling month. The variability metric is the weighted SD (spread) among annual means. For yearlings, bold italic values for a stock in June indicate a significant difference from the comparable May value. Similarly, for subyearlings, bold italic values for a stock in September indicate a significant difference from the comparable June value.

	Latitude (°N)		Distance from shore (nm)	
Stock	Mean	Spread	Mean	Spread
May yearlings				
West Cascade spring	46.72	0.40	13.25	53.58
Willamette River spring	47.03	0.45	10.74	38.10
Mid-/upper Columbia River spring	46.61	0.28	18.85	40.02
Snake River spring	46.55	0.24	18.92	39.11
Upper Columbia River summer–fall	46.49	0.21	15.62	62.07
Snake River fall	46.53	0.27	16.45	59.26
June yearlings				
West Cascade spring	47.02	0.54	8.72	27.13
Willamette River spring	47.12	0.73	6.78	15.50
Mid-/upper Columbia River spring	***47.68***	0.28	***13.45***	46.91
Snake River spring	***47.68***	0.37	***12.52***	42.88
Upper Columbia River summer–fall	***47.06***	***0.66***	***10.24***	39.82
Snake River fall	***46.92***	***0.71***	***9.78***	39.17
June subyearlings				
Willamette River spring	47.07	0.39	8.86	26.20
Spring Creek group fall	47.44	0.33	4.42	3.24
Upper Columbia River summer–fall	46.65	0.30	10.04	35.76
Snake River fall	46.76	0.27	10.92	42.32
September subyearlings				
Washington coast	47.10	0.71	5.40	5.07
West Cascade fall	46.44	0.84	4.75	6.02
Spring Creek group fall	47.53	0.26	5.14	1.94
Deschutes River fall	46.02	0.87	3.53	6.71
Upper Columbia River summer–fall	46.11	***0.94***	***4.04***	***7.39***
Snake River fall	46.31	***1.14***	***4.03***	***6.13***
Oregon coast	45.50	0.41	2.21	3.57
Southern Oregon/California coast	45.80	0.66	4.29	11.82
Central Valley	45.75	0.93	5.78	11.85

stock ($P < 0.001$). Overall, September subyearlings were primarily caught at the most inshore trawled station on each transect, and mean distances from shore ranged from 5.8 nautical miles for the Central Valley stock to 2.2 nautical miles for the Oregon coast stock.

Interannual variation.—Juvenile Chinook Salmon distributions varied among years, and several patterns were consistent among stocks (Tables 6, 7). Overall, the lowest levels of interannual variability in latitude were observed for yearlings sampled in May (range of SD among years = 0.19–0.27). Variability in latitude for yearlings was greater in June (SD = 0.37–0.77). For yearlings of spring-run stocks, the variability in distance from shore also increased from May (SD = 2.3–3.7) to June (SD = 3.1–4.1). However, interannual variability in distance from shore for summer–fall-run and fall-run yearlings was lower in June (SD = 2.4 and 2.8, respectively) than in May (SD = 4.8 and 3.8).

Among subyearlings, interannual variability in latitude increased from June to September for the Spring Creek group fall-run (SD = 0.59–0.76), upper Columbia River summer–fall-run (SD = 0.34–0.44), and Snake River fall-run (SD = 0.30–0.44) stocks (Table 7). In contrast, for those same three stocks, interannual variability in distance from shore decreased from June (SD = 2.4–6.3) to September (SD = 0.9–1.1).

Many of the patterns in mean latitude and distance from shore described above for yearlings and subyearlings were consistent among years (Figures 5, 6). For example, Willamette River spring-run yearlings were farther north and closer to shore in May than other stocks for all 6 years in which that

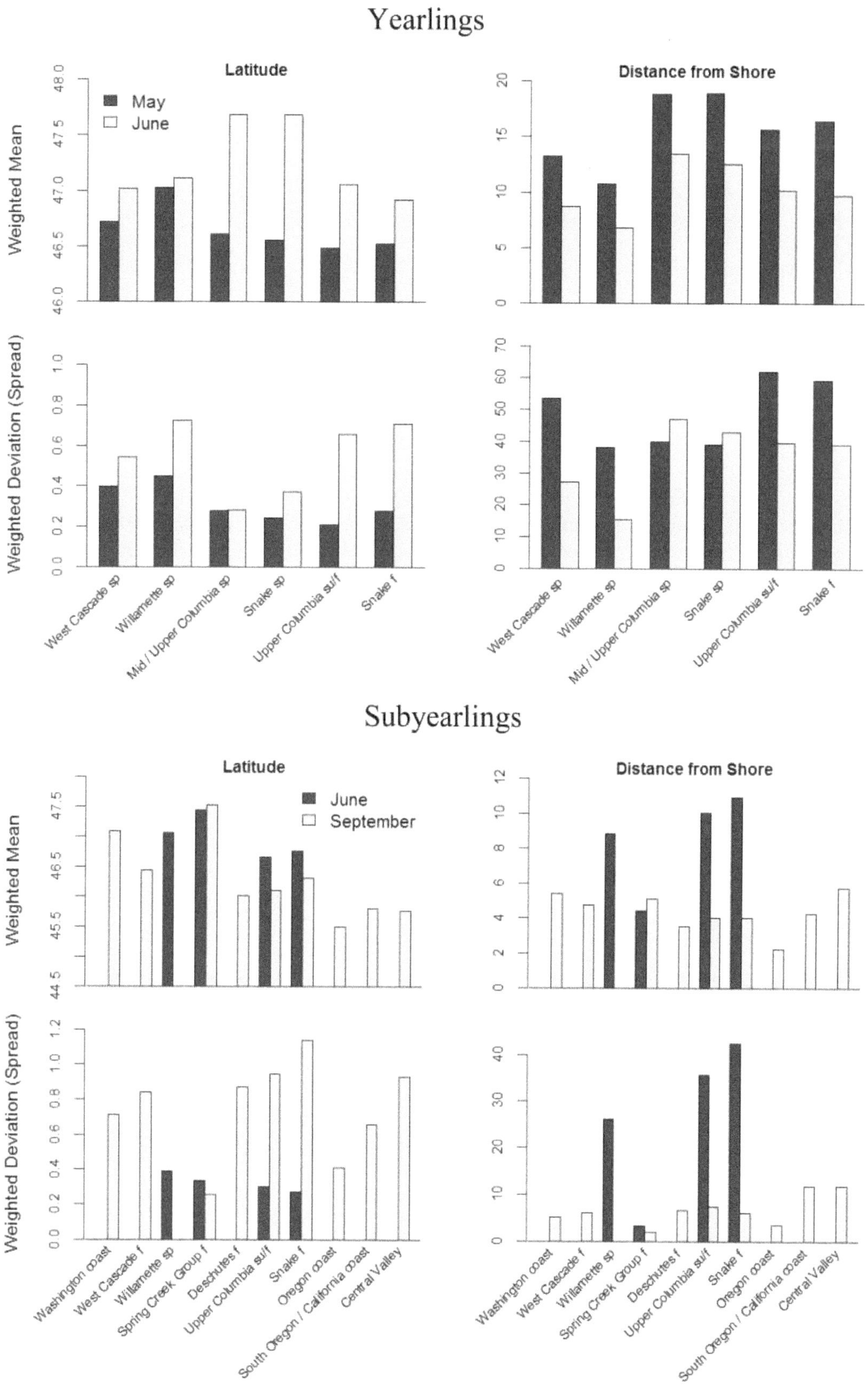

FIGURE 4. Weighted mean latitude (°N) and distance from shore (nautical miles [nm]; 1 nautical mile = 1.852 km) for yearling and subyearling Chinook Salmon of each stock sampled along the Washington and Oregon coasts in May, June, and September (sp = spring run; su = summer run; f = fall run). Variability is expressed as the weighted deviation (spread).

TABLE 6. Interannual variability in latitude (°N) and distance from shore (nautical miles [nm]; 1 nautical mile = 1.852 km) for yearling Chinook Salmon by sampling month and stock. Values were calculated by first estimating the weighted mean latitude or distance from shore for each year separately and then determining the SD of the annual values.

	May		June	
Stock	Latitude (°N)	Distance from shore (nm)	Latitude (°N)	Distance from shore (nm)
West Cascade spring	0.21	2.3	0.37	3.1
Willamette River spring	0.25	2.7	0.77	3.2
Mid-/upper Columbia River spring	0.27	3.5	0.45	3.7
Snake River spring	0.27	3.7	0.45	4.1
Upper Columbia River summer–fall	0.21	4.8	0.41	2.4
Snake River fall	0.19	3.8	0.47	2.8

stock was captured. During June, the mid-/upper Columbia River and Snake River spring-run yearlings were farther from shore than other stocks in 10 of 11 annual comparisons and were farther north than other stocks in all 11 years. However, although the mean latitude of spring-run yearlings increased from May to June in all 18 annual comparisons, this shift northward was observed in only 6 of 10 comparisons for yearlings of the upper Columbia River summer–fall and Snake River fall stocks.

Several distribution patterns were also consistent among years for subyearlings (Figure 6). The Spring Creek group fall-run stock was located farther north than the upper Columbia River summer–fall-run and Snake River fall-run stocks during both June and September in 14 of 15 annual comparisons. Upper Columbia River summer–fall-run and Snake River fall-run subyearlings shifted closer to shore in all 19 annual comparisons.

Yearlings and subyearlings from the same stock—Mean latitude and distance from shore were compared between yearlings and subyearlings from the same stock (Table 5; Figure 4). We were able to make this comparison for three of the stocks sampled in June. For the upper Columbia River summer–fall stock, yearlings were farther north than subyearlings ($P < 0.001$). Yearlings of the Snake River fall-run stock were also farther north than subyearlings of that stock, but the comparison was not significant ($P = 0.20$). Mean latitude was not significantly different between yearlings and subyearlings of the Willamette River spring-run stock ($P = 0.99$). For all three stocks, yearlings had a greater latitudinal spread than subyearlings (upper Columbia River summer–fall: $P < 0.001$; Snake River fall: $P < 0.001$; Willamette River spring: $P = 0.04$). In addition, Willamette River subyearlings were further offshore ($P = 0.01$) and more spread out in distance from shore ($P = 0.01$) than were the stock's yearlings.

Marked and unmarked juveniles from the same stock.—In September, marked Snake River fall subyearlings had a more northerly distribution than the stock's unmarked subyearlings

TABLE 7. Interannual variability in latitude (°N) and distance from shore (nautical miles [nm]; 1 nautical mile = 1.852 km) for subyearling Chinook Salmon by sampling month and stock. Values were calculated by first estimating the weighted mean latitude or distance from shore for each year separately and then determining the SD of the annual values. Stocks or months with no value had sample sizes less than 75 fish.

	June		September	
Stock	Latitude (°N)	Distance from shore (nm)	Latitude (°N)	Distance from shore (nm)
Washington coast			0.57	1.6
West Cascade fall			0.56	1.2
Willamette River spring	0.48	2.9		
Spring Creek group fall	0.59	6.3	0.76	0.9
Deschutes River fall			0.45	1.5
Upper Columbia River summer–fall	0.34	2.4	0.44	1.1
Snake River fall	0.30	2.9	0.44	1.0
Oregon coast			0.29	1.1
Southern Oregon/California coast			0.58	1.8
Central Valley			0.85	3.0

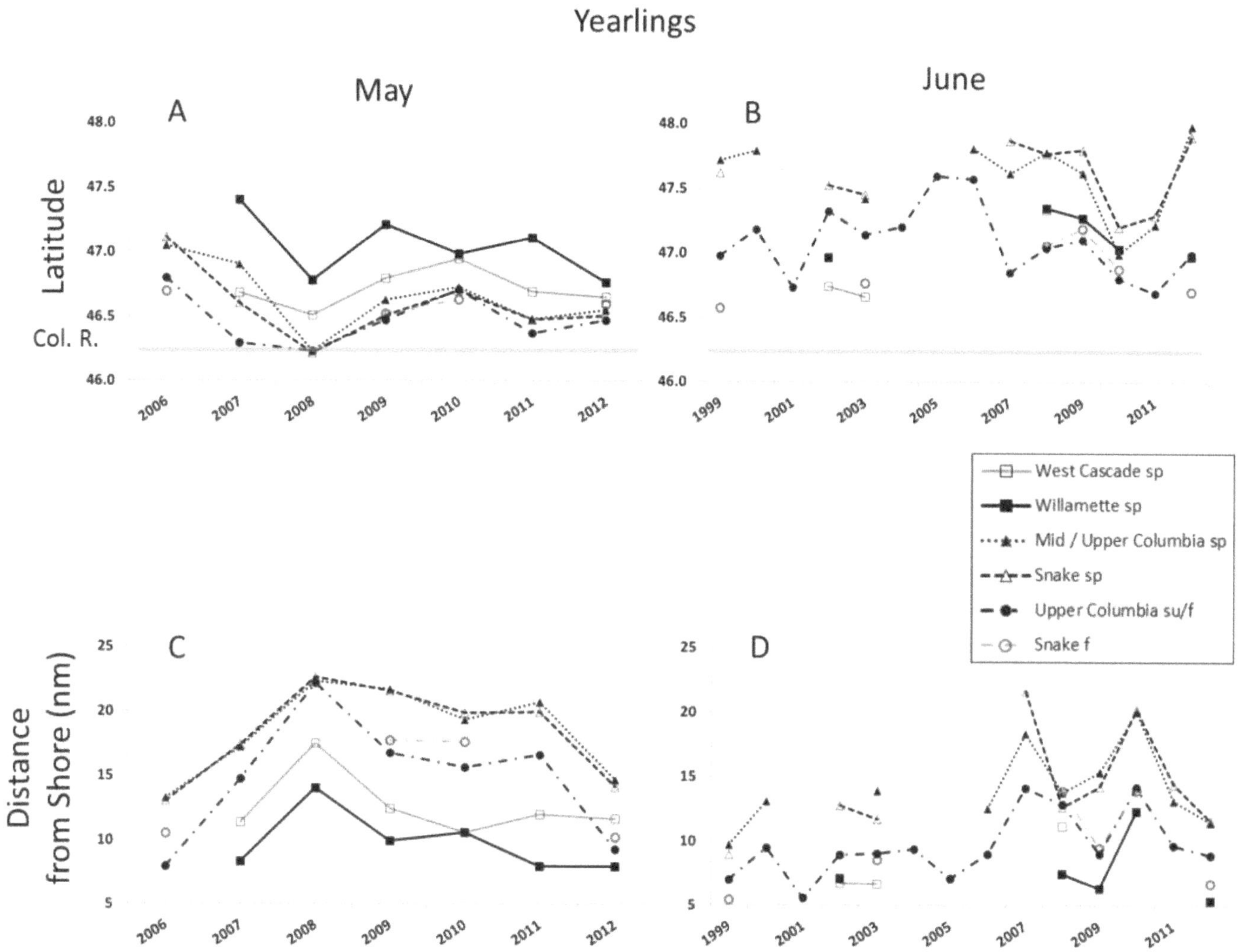

FIGURE 5. Annual weighted mean **(A)** latitude (°N) in May, **(B)** latitude in June, **(C)** distance from shore (nautical miles [nm]; 1 nautical mile = 1.852 km) in May, and **(D)** distance from shore in June for yearling Chinook Salmon representing various stocks sampled along the Washington and Oregon coasts (sp = spring run; su = summer run; f = fall run). Sampling months with fewer than 10 fish for a given stock were excluded from this analysis; none of the stocks sampled in June 1998 met the minimum sample size criterion. The gray horizontal line indicates the latitude of the Columbia River mouth.

($P = 0.03$). Latitude in June, distance from shore in June, and distance from shore in September did not significantly differ between the two groups. Similar comparisons for marked and unmarked upper Columbia River summer–fall subyearlings indicated no significant differences.

DISCUSSION

Chinook Salmon Stock Composition

Genetic estimates based on 15 years of sampling in coastal habitats off Washington and Oregon revealed that fish from Columbia River sources comprise approximately 98% of yearlings present during late spring and early summer. This finding is consistent with previous studies showing that most tagged yearlings within our sampling area are from hatcheries located in the Columbia River basin (Miller et al. 1983; Fisher and Pearcy 1995; Trudel et al. 2009; Fisher et al. 2014). In addition, we estimated that only 6% of yearlings in these marine habitats are naturally produced. In the seven most recent years of sampling, our catches included fewer than 200 naturally produced yearlings. Thus, our findings reflect the massive scale of contemporary yearling Chinook Salmon production by Columbia River hatcheries (Table 2) but also indicate the diminished abundance of natural populations in the Columbia and Snake rivers (Myers et al. 1998; Ford 2011). Moreover, taken together with the rarity of naturally produced Coho Salmon *O. kisutch* in the coastal ocean (Teel et al. 2003) and the patchy marine spatial distribution of Chinook Salmon and Coho Salmon yearlings (Peterson et al. 2010), our results highlight the difficulty of sampling ESA-protected Columbia River salmon populations in coastal habitats—yet those

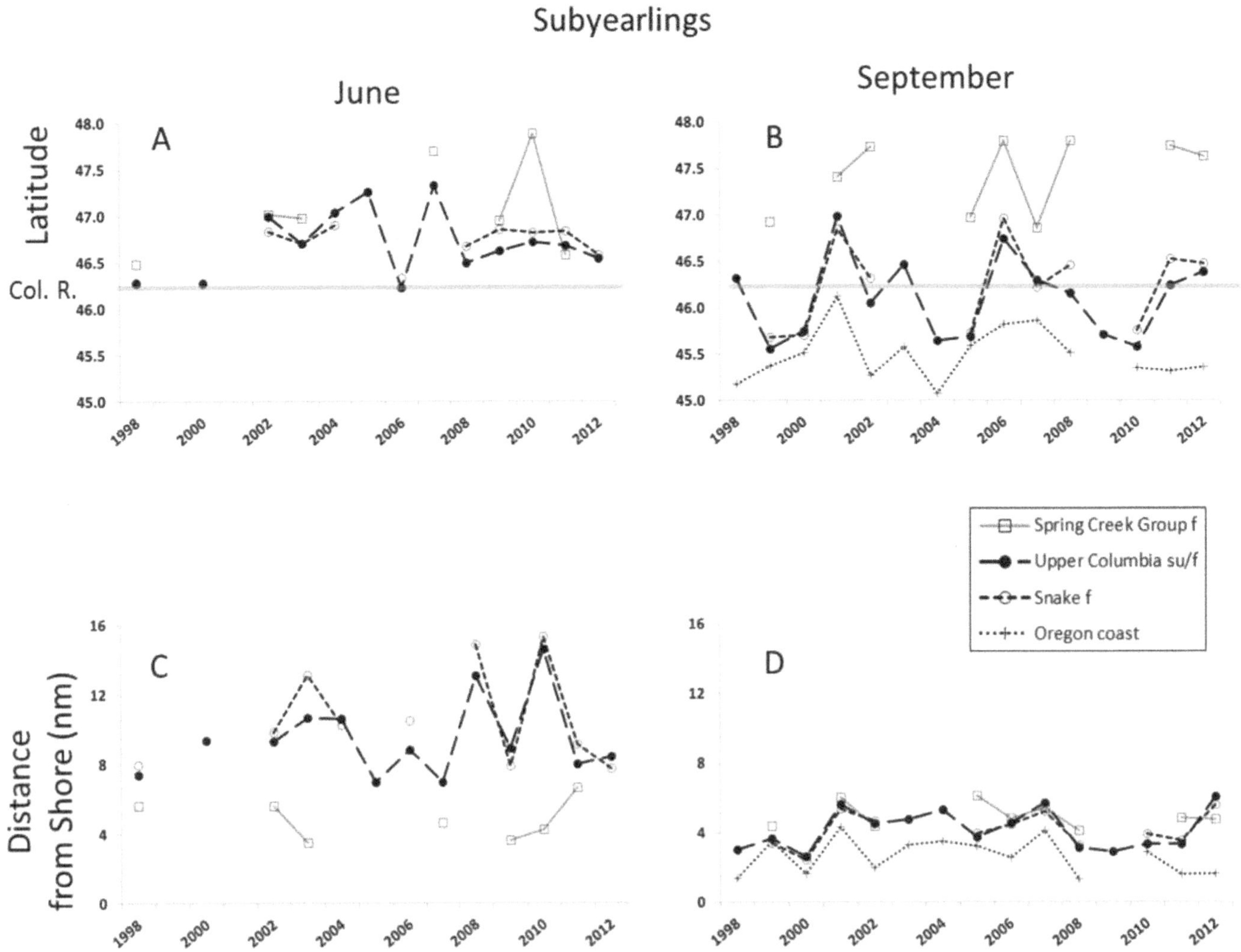

FIGURE 6. Annual weighted mean **(A)** latitude in June (°N), **(B)** latitude in September, **(C)** distance from shore (nautical miles [nm]; 1 nautical mile = 1.852 km) in June, and **(D)** distance from shore in September for subyearling Chinook Salmon of the four stocks that were most abundant in samples collected along the Washington and Oregon coasts (su = summer run; f = fall run). Sampling months with fewer than 10 fish for a given stock were excluded from this analysis; none of the stocks sampled in June 1999 or June 2001 met the minimum sample size criterion. The gray horizontal line indicates the latitude of the Columbia River mouth.

populations are frequently of greatest concern in marine growth and survival studies conducted in the region (Jacobson et al. 2008; Burla et al. 2010; Daly et al. 2012; Miller et al. 2014).

In contrast to the yearling composition, Chinook Salmon subyearlings sampled off the Oregon and Washington coasts comprised greater proportions of fish from coastal rivers and also included more fish that were naturally produced. In early summer, subyearlings were nearly all from the Columbia River, with substantial proportions from the upper Columbia River summer–fall-run (33%) and Snake River fall-run (30%) stocks. However, in autumn, when subyearlings were most abundant, nearly 25% of the subyearlings were from coastal sources and nearly half of those fish originated from northern and central Oregon coastal rivers. Small proportions of subyearlings from populations in coastal Washington, southern Oregon, and California also occupied the nearshore habitats. Unlike the sparseness of naturally produced yearlings off Oregon and Washington, we estimated that natural production accounted for more than one-third of the subyearlings. By far, the largest contribution of natural subyearlings was from the upper Columbia River summer–fall stock; this is likely due to the robust natural spawning of fall-run fish in the Hanford Reach, located upstream of the Snake River confluence (Dauble and Watson 1997).

It is worthwhile to note that although yearling life histories are predominant in spring-run populations of Chinook Salmon, substantial proportions of the fish that we classified as subyearlings were allocated to spring-run stocks. This result is not surprising with regard to lower Columbia River spring-run stocks (i.e., Willamette River and West Cascade), for which subyearling downstream movements and subsequent estuarine rearing are well documented (Craig and Townsend 1946; Hymer et al. 1992; Roegner et al. 2012; Teel et al. 2014). Less expected were our estimates of subyearling life histories for the Snake River and mid-/upper Columbia River spring runs, since their freshwater-returning adults almost all enter the ocean as yearlings (Waples et al. 2004; Hess et al. 2014). Among the interior basin spring-run fish that were smaller than our size-based thresholds for yearlings in May (>120 mm) and June (>140 mm), nearly all were without hatchery marks (data not shown), suggesting that they were naturally produced. Further study should include age determinations (e.g., analysis of scales or otoliths) to verify our life history classification of small, naturally produced fish from these stocks. Although we have verified our cut-offs for hatchery fish, additional data may show that our size-at-capture delineations are not appropriate for use with yearlings reared in low-growth environments of the interior Columbia and Snake rivers.

The lack of West Cascade fall-run subyearlings in our catches is intriguing and may be explained by an insufficiency of inshore sampling. Despite constituting nearly 25% of the subyearling hatchery production in the Columbia River basin, releases of the West Cascade fall stock contributed only 3% to our catches of hatchery subyearlings. Moreover, during summer, naturally produced fry and fingerlings of the West Cascade fall-run stock are predominant in shallow rearing areas throughout the lower Columbia River estuary (Roegner et al. 2012; Teel et al. 2014), yet the stock comprised only 5% of our ocean-caught subyearlings in September. Previous studies off the coasts of Oregon and Washington have documented the presence of small Chinook Salmon subyearlings (<130 mm) inshore of 4 km (Miller et al. 1983; Fisher and Pearcy 1995), including in shallow, sandy beach surf-zone habitats (Marin Jarrin et al. 2009). Stock-specific analyses were not conducted in those earlier studies. Nonetheless, we conjecture that small, naturally produced subyearlings of the West Cascade fall run are more abundant inshore of our sampling grid and likely north of the Columbia River, where our stations began 3–6 nautical miles offshore.

Seasonal Density Patterns

Although the densities of Chinook Salmon yearlings along the Washington and Oregon coasts fluctuated annually, strong seasonal shifts were apparent each year (Figure 2). The peak densities that we observed at the end of May occurred relatively soon after the spring emigration of yearlings from the Columbia River and adjacent coastal rivers (Rich 1920; Healey 1991; Weitkamp et al. 2012). At the end of June, the decreased yearling densities in the area likely reflected both mortality, which remains largely unquantified (Emmett and Krutzikowski 2008; Burke et al. 2013b), and the well-known migration of yearlings into more northerly regions (Orsi et al. 2000; Trudel et al. 2009; Tucker et al. 2011; Fisher et al. 2014). Although migration patterns and rates differed among stocks (Figure 1; Fisher et al. 2014), by autumn the density of yearlings in the region was low (Figure 2; Peterson et al. 2010). However, it is important to note that although many yearlings reside in coastal areas off Washington and Oregon rather briefly, it is a critical period of feeding and growth and is essential for determining the year-class strength of Columbia River spring-run Chinook Salmon (Tomaro et al. 2012; Burke et al. 2013b; Miller et al. 2014).

Subyearlings were prevalent in coastal areas during early summer and—in contrast to yearlings—they were also abundant off the Oregon and Washington coasts during autumn. In most years, subyearling densities increased during the summer, in part because the timing of ocean entry for subyearlings is very protracted and out-migration from the region's rivers is substantial throughout the summer (Rich 1920; Reimers 1973; Howell et al. 1985; McCabe et al. 1986; Roegner et al. 2012). Unlike yearlings, relatively few of the region's subyearlings migrate north of Washington during their first ocean summer (Tucker et al. 2011; Fisher et al. 2014). Interestingly, in 4 of the 15 years of sampling, subyearling densities were lower in September than in June, suggesting fluctuations in either recruitment or survival. However, a recent analysis of upper Columbia River summer–fall Chinook Salmon indicated that neither June subyearling density nor September subyearling density is a strong predictor of subsequent adult abundance (Miller et al. 2013). Nonetheless, Miller et al. (2013) found that survival is positively related to river plume size and also to the coastal ocean conditions experienced by subyearlings off Oregon and Washington during the summer.

Yearling Distributions

In late spring, differences in the initial distributions of yearlings from different stocks did not conform to the generalized migration patterns that have been previously documented for Columbia River stocks (Trudel et al. 2009; Fisher et al. 2014). We found that Willamette River and West Cascade spring-run juveniles, which comprised more than 25% of May yearlings, were farther north than yearlings from the interior basin spring-run stocks, which exhibit more rapid northward migrations. Clearly, distributions and stock abundances in the spring are influenced by (1) the timing of large hatchery releases of juveniles and (2) the timing of ocean entry. Large releases of Willamette River and West Cascade spring-run juveniles begin in February, well before the peak releases of yearlings from other parts of the basin (Table 2). The lower Columbia River yearlings enter the ocean primarily in April, earlier than yearlings from the interior Columbia River (Weitkamp et al., in press). Lower-river yearlings are also larger than interior basin yearlings after ocean entry (Burke et al. 2013a); as a result, they may be capable of faster swim speeds, which could also contribute to the more northward location of Willamette

River and West Cascade spring-run yearlings during the initial marine period.

Physical and environmental factors affect the marine migrations of Columbia River yearlings (Peterson et al. 2010; Burke et al. 2013a; McMichael et al. 2013) and likely influenced the distributions observed in our study. Our large catches along the Columbia River transect in May were mostly at stations farther offshore than our catches along the Washington coast. This pattern was very apparent for yearlings from the mid-/upper Columbia River and Snake River spring-run stocks, which were more offshore than other yearlings. By June, yearlings of these stocks as well as other stocks were distributed closer to shore. One explanation for these patterns is that currents associated with the Columbia River affect the early spatial distribution of yearlings by sweeping them offshore, which is followed by "corrective" movements toward shore as the fish swim northward. A change in movement direction associated with ocean currents was also recently documented by McMichael et al. (2013), who used acoustic telemetry to track yearling Chinook Salmon in marine habitats near the Columbia River mouth. In that study, yearlings initially migrated south in the river plume before reversing direction and moving northward. Behavioral compensation for ocean currents and other external factors is also supported by multiple recent modeling studies of salmon movements in the marine environment (Burke et al. 2014; Putman et al. 2014), which demonstrated that salmon alter their swim speed and direction to account for ocean currents. Moreover, the analysis by Burke et al. (2013a) showed that during their early ocean migration, Chinook Salmon yearlings from the Columbia River display stock-specific behavioral responses to local environmental factors (chlorophyll-*a* concentration and temperature) in addition to broad-scale geospatial cues (latitude and distance from shore).

By June, yearling distributions along the coast of Washington clearly reflected behavioral differences that characterized subsequent stock-specific migrations along the coastal shelf. Consistent with the CWT analysis of Fisher et al. (2014), we found that yearlings from the mid-/upper Columbia River and Snake River spring-run stocks were farthest north and also had the smallest latitudinal spread. After leaving our study area, yearlings from these stocks continue to migrate rapidly northward; they reach Alaska in the summer and move off the continental shelf by autumn (Orsi et al. 2000; Trudel et al. 2009; Tucker et al. 2011; Fisher et al. 2014). Our June data revealed a much different distributional pattern for lower Columbia River spring-run, upper Columbia River summer–fall-run, and Snake River fall-run yearlings. These stocks were significantly more spread out along the coast than the interior basin spring-run yearlings, and they included fish that migrated southward to habitats off the coast of Oregon. Moreover, although yearling densities off Oregon and Washington were low in autumn, the predominant sources of yearlings that remained in the region during autumn were the interior basin summer–fall-run and lower-river spring-run stocks. These findings are consistent with those of Fisher et al. (2014), who reported that coded-wire-tagged hatchery yearlings from these stocks dispersed in coastal areas from Oregon to Alaska.

Subyearling Distributions

Although subyearlings were concentrated near shore along the Oregon and Washington coasts at the end of summer, stock-specific distributions differed in latitude and inshore distance. For some stocks, these spatial differences were consistent with previous observations that Chinook Salmon subyearlings tend to remain relatively close to the point of sea entry throughout their first ocean summer (Tucker et al. 2011). For example, the movements of subyearlings from coastal rivers adjacent to the Columbia River were very limited: at the end of September, subyearlings from Washington coastal rivers were found along the Washington coast, and those from Oregon coastal rivers were mostly distributed along the Oregon shoreline. In addition, juveniles from Oregon coastal rivers remained much closer to shore than subyearlings from other stocks, including those from southern Oregon and California, which were also distributed along the Oregon coast in September.

However, other latitudinal differences among stocks are due to differing movement patterns, even between fall-run stocks with the same ocean entry location. The diverse marine migrations that have evolved in upper Columbia River summer–fall-run and Snake River fall-run Chinook Salmon include large proportions of subyearlings that migrate southward along the Oregon coast during their first summer at sea. By autumn, subyearlings from the two interior basin stocks were broadly distributed in nearshore habitats from central Oregon to northern Washington. Subyearlings of these stocks were the only groups in our analysis that had a mean shift to the south during the summer, exhibiting a distribution pattern very different from those of some other stocks. For example, among the subyearlings we studied, those of the Spring Creek group had the most northern mean latitude, with relatively little change throughout the summer. By September, Spring Creek group subyearlings had the smallest spread both in latitude and in distance from shore.

The distinctive marine distribution is one of several distinguishing life history characteristics of interior Columbia River fall-run Chinook Salmon (Howell et al. 1985; Myers et al. 1998). Unlike the Spring Creek group fall-run subyearlings, whose freshwater out-migration is concentrated in spring, fish from the upper river occupy estuarine habitats throughout the summer and have very protracted movements to the ocean (Roegner et al. 2012; Teel et al. 2014). However, the difference in ocean entry timing alone does not appear to explain the more southerly marine distribution of the upper-river fish during autumn. Claiborne et al. (2014) used otolith chemistry and structure to study a subset of the upper Columbia River summer–fall subyearlings from our study; those authors found that the latitudinal distribution in late September was not related to how long the fish had occupied marine habitats (i.e., even some of the early migrants were south of the Columbia River in September). Interestingly, the more southern early marine distribution of upper-river subyearlings does not provide insight into their subsequent coastal distributions as adults. For example, substantial proportions of upper Columbia River fall-run fish are intercepted in fisheries conducted off Alaska and northern British Columbia, whereas Spring Creek group fall-run fish are rarely caught north of Vancouver Island in southern British Columbia (Norris et al. 2000; Weitkamp 2010).

Accurate assessments of stock-specific marine distributions rely on accurate assignments of ocean-caught juveniles to their source stocks. A recent study concluded that reasonably accurate estimates of Chinook Salmon stock identities can be obtained with the microsatellite DNA loci and baseline database used in our study (Hess et al. 2014). Our allocations of fish with CWTs produced results that were consistent with the findings of Hess et al. (2014), who used a leave-one-out method to evaluate assignment accuracies. The accuracy of our assignments to the Columbia River (99%) was also similar to the 97% accuracy obtained by Tucker et al. (2011), who analyzed known-origin Chinook Salmon juveniles by using a different set of microsatellites and a different baseline of populations. However, within the Columbia River, assignment accuracy rates are diminished by a legacy of extensive hatchery stock transfers between regions (Myers et al. 1998, 2006; Hess et al. 2014). In our study, misallocations within interior basin genetic lineages likely inhibited our ability to detect potential differences in distribution (1) between Snake River spring-run yearlings and mid-/upper Columbia River spring-run yearlings and also (2) between Snake River fall-run and upper Columbia River summer–fall-run juveniles. This lack of power may have contributed to the near absence of significant differences in our comparisons of these two pairs of stocks. We anticipate that greater resolution for these stocks will be obtained through the identification of high-resolution single-nucleotide polymorphisms (Larson et al. 2014) and through recent efforts to implement parentage-based tagging at Columbia River hatcheries (Steele et al. 2013).

Conclusions

The juvenile Chinook Salmon distribution patterns we identified, many of which are consistent across years, illustrate inherent differences in migration behavior between life history types (age at sea entry) and among stocks. Our findings are in agreement with those of Fisher et al. (2014), who recently documented the differing dispersal of yearling Chinook Salmon from interior spring-run ESUs relative to yearlings from other Columbia River ESUs. Importantly, our study revealed that early marine distributions of subyearlings also differed among stocks, particularly between interior basin and lower-river stocks.

Distributions also appear to be influenced by extrinsic factors, such as currents or plume conditions, and by hatchery practices. For example, differences in the timing of hatchery releases may determine the sequence with which specific hatchery stocks occupy plume and coastal habitats, possibly influencing the match–mismatch between salmon and their ephemeral prey base (Scheuerell et al. 2009; Miller et al. 2014). Moreover, age at release is also under hatchery control, and we showed that yearlings and subyearlings from the same stock can have differing distributions. Of the stocks included in our analysis, the upper Columbia River summer–fall, Snake River fall, and Willamette River spring stocks have hatchery programs with large releases of both subyearlings and yearlings. The marine distributions of these stocks are therefore most affected by contemporary age-at-release management strategies.

It is hoped that the early marine distributions we have documented will be viewed in the context of ongoing efforts for Chinook Salmon conservation in the Pacific Northwest. Overall, the distribution patterns were largely consistent with the region's Chinook Salmon ESUs, which were delineated by using a synthesis of genetic, ecological, and life history data (Myers et al. 1998; Ford 2011). Moreover, age at ocean entry and ocean distribution patterns are among the variable traits that are considered important for the recovery of endangered and threatened salmon ESUs (Waples et al. 2001, 2004; Good et al. 2007). Such phenotypic and spatial diversity allows species and ESUs to access an array of marine habitats and therefore respond to both short-term and long-term environmental change. The framework developed under the ESA to evaluate salmonid population viability therefore includes trait diversity as a key parameter (McElhany et al. 2000). We recommend broadening those evaluations to incorporate metrics that describe among- and within-ESU variability in early marine distributions.

ACKNOWLEDGMENTS

This paper is dedicated to the memory of Robert L. Emmett (1955–2015), the chief scientist on many of our study's sampling surveys. We greatly benefited from his expertise, enthusiasm, and friendship. We greatly appreciate the many people who contributed to this project over the years, including the captains and crews who operated vessels and the many scientists who collected and processed samples. We especially thank Paul Bentley, Brian Beckman, Ric Brodeur, Cindy Bucher, Ed Casillas, Joe Fisher, Kurt Fresh, Susan Hinton, Kym Jacobson, Jessica Miller, Bill Peterson, Larissa Rohrbach, Tom Wainwright, Laurie Weitkamp, and Jen Zamon. Funding for this study was provided by the Bonneville Power Administration and the National Oceanic and Atmospheric Administration Fisheries. Reference to trade names does not imply endorsement by the National Marine Fisheries Service.

REFERENCES

Beacham, T. D., R. J. Beamish, J. R. Candy, C. Wallace, S. Tucker, J. H. Moss, and M. Trudel. 2014. Stock-specific migration pathways of juvenile Sockeye Salmon in British Columbia waters and in the Gulf of Alaska. Transactions of the American Fisheries Society 143: 1386–1403.

Beacham, T. D., J. R. Candy, C. Wallace, M. Wetklo, L. Deng, and C. MacConnachie. 2012. Microsatellite mixed-stock identification of Coho Salmon (*Oncorhynchus kisutch*) in British Columbia. Marine and Coastal Fisheries: Dynamics, Management, and Ecosystem Science [online serial] 4:85–100.

Beacham, T. D., B. Spilsted, K. D. Le, and M. Wetklo. 2008. Population structure and stock identification of Chum Salmon (*Oncorhynchus keta*) from British Columbia determined with microsatellite DNA variation. Canadian Journal of Zoology 86:1002–1014.

Brodeur, R. D., G. W. Boehlert, E. Casillas, M. B. Eldridge, J. H. Helle, W. T. Peterson, W. R. Heard, S. T. Lindley, and M. H. Schiewe. 2000. A coordinated research plan for estuarine and ocean research on Pacific salmon. Fisheries 25(6):7–16.

Brodeur, R. D., J. P. Fisher, R. L. Emmett, C. A. Morgan, and E. Casillas. 2005. Species composition and community structure of pelagic nekton off Oregon and Washington under variable oceanographic conditions. Marine Ecology Progress Series 298:41–57.

Brodeur, R. D., J. P. Fisher, D. J. Teel, R. L. Emmett, E. Casillas, and T. W. Miller. 2004. Juvenile salmonid distribution, growth, condition, origin, environmental and species associations in the northern California Current. U.S. National Marine Fisheries Service Fishery Bulletin 102:25–46.

Burke, B. J., J. J. Anderson, and A. M. Baptista. 2014. Evidence for multiple navigational sensory capabilities of Chinook Salmon. Aquatic Biology 20:77–90.

Burke, B. J., M. Liermann, D. J. Teel, and J. J. Anderson. 2013a. Environmental and geospatial factors drive juvenile Chinook Salmon distribution during early ocean migration. Canadian Journal of Fisheries and Aquatic Sciences 70:1167–1177.

Burke, B. J., W. T. Peterson, B. R. Beckman, C. Morgan, and E. A. Daly. 2013b. Multivariate models of adult Pacific Salmon returns. PLoS (Public Library of Science) ONE [online serial] 8(1).

Burla, M., A. M. Baptista, E. Casillas, J. G. Williams, and D. M. Marsh. 2010. The influence of the Columbia River plume on the survival of steelhead (*Oncorhynchus mykiss*) and Chinook Salmon (*Oncorhynchus tshawytscha*): a numerical exploration. Canadian Journal of Fisheries and Aquatic Sciences 67:1671–1684.

Claiborne, A. M., J. A. Miller, L. A. Weitkamp, D. J. Teel, and R. L. Emmett. 2014. Evidence for selective mortality in marine environments: the role of fish migration size, timing, and production type. Marine Ecology Progress Series 515:187–202.

Cleaver, F. C. 1969. Effects of ocean fishing on the 1961-brood fall Chinook Salmon from Columbia River hatcheries. Research Reports of the Fish Commission of Oregon, Portland.

Craig, J. A., and L. D. Townsend. 1946. An investigation of fish-maintenance problems in relation to the Willamette Valley project. U.S. Fish and Wildlife Service Special Scientific Report 33.

Daly, E. A., R. D. Brodeur, J. P. Fisher, L. A. Weitkamp, D. J. Teel, and B. R. Beckman. 2012. Spatial and trophic overlap of marked and unmarked Columbia River basin spring Chinook Salmon during early marine residence with implications for competition between hatchery and naturally produced fish. Environmental Biology of Fishes 94:117–134.

Dauble, D. D., and D. G. Watson. 1997. Status of fall Chinook Salmon populations in the mid-Columbia River, 1948–1992. North American Journal of Fisheries Management 17:283–300.

Emmett, R. L., and G. K. Krutzikowski. 2008. Nocturnal feeding of Pacific and Jack mackerel off the mouth of the Columbia River, 1998–2004: implications for juvenile salmon predation. Transactions of the American Fisheries Society 137:657–676.

Fisher, F. W. 1994. Past and present status of Central Valley Chinook Salmon. Conservation Biology 8:870–873.

Fisher, J., M. Trudel, A. Ammann, J. A. Orsi, J. Piccolo, C. Bucher, E. Casillas, J. A. Harding, B. MacFarlane, R. D. Brodeur, J. F. T. Morris, and D. W. Welch. 2007. Comparisons of the coastal distribution and abundance of juvenile salmon from central California to the northern Gulf of Alaska. Pages 31–80 *in* C. B. Grimes, R. D. Brodeur, L. J. Haldorson, and S. M. McKinnell, editors. The ecology of juvenile salmon in the northeast Pacific Ocean: regional comparisons. American Fisheries Society, Symposium 57, Bethesda, Maryland.

Fisher, J. P., and W. G. Pearcy. 1995. Distribution, migration, and growth of juvenile Chinook Salmon, *Oncorhynchus tshawytscha*, off Oregon and Washington. U.S. National Marine Fisheries Service Fishery Bulletin 93:274–289.

Fisher, J. P., L. A. Weitkamp, D. J. Teel, S. A. Hinton, M. Trudel, J. F. T. Morris, M. E. Thiess, R. M. Sweeting, J. A. Orsi, and E. V. Farley Jr. 2014. Early ocean dispersal patterns of Columbia River Chinook and Coho salmon. Transactions of the American Fisheries Society 143:252–272.

Foerster, R. E. 1955. The Pacific salmon (genus *Oncorhynchus*) of the Canadian Pacific coast, with particular reference to their occurrence in or near freshwater. International North Pacific Fisheries Commission Bulletin 1.

Ford, M. J., editor. 2011. Status review update for Pacific salmon and steelhead listed under the Endangered Species Act: Pacific Northwest. NOAA Technical Memorandum NMFS-NWFSC-113.

Godfrey, H. 1958. Comparisons of the index of return for several stocks of British Columbia salmon to study variations in survival. Journal of the Fisheries Research Board of Canada 15:891–900.

Godfrey, H., K. A. Henry, and S. Machidori. 1975. Distribution and abundance of Coho Salmon in offshore waters of the North Pacific Ocean. International North Pacific Fisheries Commission Bulletin 31.

Good, T. J., T. J. Beechie, P. McElhany, M. M. McClure, and M. H. Ruckelshaus. 2007. Recovery planning for Endangered Species Act-listed Pacific salmon: using science to inform goals and strategies. Fisheries 32:426–440.

Hartt, A. C. 1980. Juvenile salmonids in the oceanic ecosystem; the critical first summer. Pages 25–57 *in* W. J. McNeil and D. C. Himsworth, editors. Salmonid ecosystems of the North Pacific. Oregon State University Press, Corvallis.

Hartt, A. C., and M. B. Dell. 1986. Early oceanic migrations and growth of juvenile Pacific salmon and steelhead trout. International North Pacific Fisheries Commission Bulletin 46.

Healey, M. C. 1991. Life history of Chinook Salmon (*Oncorhynchus tshawytscha*). Pages 311–393 *in* C. Groot and L. Margolis, editors. Pacific salmon life histories. University of British Columbia Press, Vancouver.

Hess, J. E., J. M. Whiteaker, J. K. Fryer, and S. R. Narum. 2014. Monitoring stock specific abundance, run-timing, and straying of Chinook Salmon in the Columbia River using genetic stock identification (GSI). North American Journal of Fisheries Management 34:184–201.

Howell, P., K. Jones, D. Scarnecchia, L. LaVoy, W. Kendra, and D. Orrmann. 1985. Stock assessment of Columbia River anadromous salmonids, volume I. Bonneville Power Administration, Project 83-335, Portland, Oregon.

Hymer, J., R. Pettit, M. Wastel, P. Hahn, and K. Hatch. 1992. Stock summary reports for Columbia River anadromous salmonids. Bonneville Power Administration, Project 88-108, Portland, Oregon.

Jacobson, K. C., D. Teel, D. M. Van Doornik, and E. Casillas. 2008. Parasite associated mortality of juvenile Pacific salmon caused by the trematode *Nanophyetus salmincola* during early marine residence. Marine Ecology Progress Series 354:235–244.

Kalinowski, S. T., K. R. Manlove, and M. L. Taper. 2007. ONCOR: a computer program for genetic stock identification. Montana State University, Bozeman.

Krutzikowski, G. K., and R. L. Emmett. 2005. Diel differences in surface trawl fish catches off Oregon and Washington. Fisheries Research 71:365–371.

Larson, W. A., J. E. Seeb, C. E. Pascal, W. D. Templin, and L. W. Seeb. 2014. Single-nucleotide polymorphisms (SNPs) identified through genotyping-by-sequencing improve genetic stock identification of Chinook Salmon (*Oncorhynchus tshawytscha*) from western Alaska. Canadian Journal of Fisheries and Aquatic Sciences 71:698–708.

Manel, S., O. E. Gaggiotti, and R. S. Waples. 2005. Assignment methods: matching biological questions with appropriate techniques. Trends in Ecology and Evolution 20:136–142.

Marin Jarrin, J. R., A. L. Shanks, and M. A. Banks. 2009. Confirmation of the presence and use of sandy beach surf-zones by juvenile Chinook Salmon. Environmental Biology of Fishes 85:119–125.

McCabe, G. T. Jr., R. L. Emmett, W. D. Muir, and T. H. Blahm. 1986. Utilization of the Columbia River estuary by subyearling Chinook Salmon. Northwest Science 60:113–124.

McElhany, P., M. H. Ruckelshaus, M. J. Ford, T. Wainwright, and E. Bjorkstedt. 2000. Viable salmonid populations and the recovery of evolutionarily significant units. NOAA Technical Memorandum NMFS-NWFSC-42.

McMichael, G. A., A. C. Hanson, R. A. Harnish, and D. M. Trott. 2013. Juvenile salmonid migratory behavior at the mouth of the Columbia River and within the plume. Amimal Biotelemetry [online serieal] 1:14.

Miller, D. R., J. G. Williams, and C. W. Sims. 1983. Distribution, abundance and growth of juvenile salmonids off the coast of Oregon and Washington, summer 1980. Fisheries Research 2:1–17.

Miller, J. A., D. J. Teel, A. M. Baptista, and C. A. Morgan. 2013. Disentangling bottom-up and top-down effects on survival during early ocean residence in a population of Chinook Salmon (*Oncorhynchus tshawytscha*). Canadian Journal of Fisheries and Aquatic Sciences 70:617–629.

Miller, J. A., D. J. Teel, W. T. Peterson, and A. Baptista. 2014. Assessing the relative importance of local and regional processes on the survival of a threatened salmon population. PLoS (Public Library of Science) ONE [online serial] 9(6): e99814.

Milner, G. B., D. J. Teel, F. M. Utter, and G. A. Winans. 1985. A genetic method of stock identification in mixed populations of Pacific salmon. Marine Fisheries Review 47:1–8.

Moran, P., D. J. Teel, M. A. Banks, T. D. Beacham, M. R. Bellinger, S. M. Blankenship, J. R. Candy, J. C. Garza, J. E. Hess, S. R. Narum, L. W. Seeb, W. D. Templin, C. G. Wallace, and C. T. Smith. 2013. Divergent life-history races do not represent Chinook Salmon coast-wide: the importance of scale in quaternary biogeography. Canadian Journal of Fisheries and Aquatic Sciences 70:415–435.

Morris, J. F. T., M. Trudel, M. E. Thiess, R. M. Sweeting, J. Fisher, S. A. Hinton, J. A. Orsi, E. V. Farley Jr., and D. W. Welch. 2007. Stock-specific migrations of juvenile Coho Salmon derived from coded-wire tag recoveries on the continental shelf of western North America. Pages 81–104 *in* C. B. Grimes, R. D. Brodeur, L. J. Haldorson, and S. M. McKinnell, editors. The ecology of juvenile salmon in the northeast Pacific Ocean: regional comparisons. American Fisheries Society, Symposium 57, Bethesda, Maryland.

Mortensen, D. G., A. C. Wertheimer, S. Taylor, and J. Landingham. 2000. The relation between early marine growth of Pink Salmon, *Oncorhynchus gorbuscha*, and marine water temperature, secondary production, and survival to adulthood. U.S. National Marine Fisheries Service Fishery Bulletin 86:310–335.

Myers, J., C. Busack, D. Rawding, A. Marshall, D. Teel, D. M. Van Doornik, and M. T. Maher. 2006. Historical population structure of Pacific salmonids in the Willamette River and lower Columbia River basins. NOAA Technical Memorandum NMFS-NWFSC-73.

Myers, J. M., R. G. Kope, G. J. Bryant, D. Teel, L. J. Lierheimer, T. C. Wainwright, W. S. Grant, F. W. Waknitz, K. Neely, S. T. Lindley, and R. S. Waples. 1998. Status review of Chinook Salmon from Washington, Idaho, Oregon, and California. NOAA Technical Memorandum NMFS-NWFSC-35.

Norris, J. G., S. Hyun, and J. J. Anderson. 2000. Ocean distribution of Columbia River upriver bright fall Chinook Salmon stocks. North Pacific Anadromous Fish Commission Bulletin 2:221–232.

Orsi, J. A., and H. W. Jaenicke. 1996. Marine distribution and origin of pre-recruit Chinook Salmon, *Oncorhynchus tshawytscha*, in southeastern Alaska. U.S. National Marine Fisheries Service Fishery Bulletin 94: 482–497.

Orsi, J. A., M. V. Sturdevant, J. M. Murphy, D. G. Mortensen, and B. L. Wing. 2000. Seasonal habitat use and early marine ecology of juvenile Pacific salmon in southeastern Alaska. North Pacific Anadromous Fish Commission Bulletin 2:111–122.

Paquet, P., T. Flagg, A. Appleby, J. Barr, L. Blankenship, D. Campton, M. Delarm, T. Evelyn, D. Fast, J. Gislason, P. Kline, D. Maynard, L. Mobrand, G. Nandor, P. Seidel, and S. Smith. 2011. Hatcheries, conservation, and sustainable fisheries—achieving multiple goals: results of the Hatchery Scientific Review Group's Columbia River basin review. Fisheries 36:542–561.

Pearcy, W. G. 1992. Ocean ecology of North Pacific salmonids. University of Washington Press, Seattle.

Pearcy, W. G., and J. P. Fisher. 1988. Migrations of Coho Salmon, *Oncorhynchus kisutch*, during their first summer in the ocean. U.S. National Marine Fisheries Service Fishery Bulletin 86:173–195.

Pearcy, W. G., and J. P. Fisher. 1990. Distribution and abundance of juvenile salmonids off Oregon and Washington, 1981–1984. NOAA Technical Report NMFS 93.

Pearcy, W. G., and S. M. McKinnell. 2007. The ocean ecology of salmon in the northeast Pacific Ocean: an abridged history. Pages 7–30 *in* C. B. Grimes, R. D. Brodeur, L. J. Haldorson, and S. M. McKinnell, editors. The ecology of juvenile salmon in the northeast Pacific Ocean: regional comparisons. American Fisheries Society, Symposium 57, Bethesda, Maryland.

Peterson, W. T., C. A. Morgan, J. P. Fisher, and E. Casillas. 2010. Ocean distribution and habitat associations of yearling Coho (*Oncorhynchus kisutch*) and Chinook (*O. tshawytscha*) salmon in the northern California Current. Fisheries Oceanography 19:508–525.

Putman, N. F., M. M. Scanlan, E. J. Billman, J. P. O'Neil, R. B. Couture, T. P. Quinn, K. J. Lohmann, and D. L. G. Noakes. 2014. An inherited magnetic map guides ocean navigation in juvenile Pacific salmon. Current Biology 24:446–450.

Rannala, B., and J. L. Mountain. 1997. Detecting immigration by using multilocus genotypes. Proceedings of the National Academy of Sciences of the USA 94:9197–9201.

Reimers, P. E. 1973. The length of residence of juvenile fall Chinook Salmon in Sixes River, Oregon. Research Reports of the Fish Commission of Oregon 4(2).

Rice, C. A., C. M. Greene, P. Moran, D. J. Teel, D. R. Kuligowski, R. R. Reisenbichler, E. M. Beamer, J. R. Karr, and K. L. Fresh. 2011. Abundance, stock origin, and length of marked and unmarked juvenile Chinook Salmon in the surface waters of greater Puget Sound. Transactions of the American Fisheries Society 140:170–189.

Rich, W. H. 1920. Early history and seaward migration of Chinook Salmon in the Columbia and Sacramento rivers. U.S. Bureau of Fisheries Bulletin 37.

Rich, W. H. 1939. Local populations and migration in relation to the conservation of Pacific salmon in the western states and Alaska. Pages 45–50 *in* F. R. Moulton, editor. The migration and conservation of salmon. American Association for the Advancement of Science, Lancaster, Pennsylvania.

Roegner, G. C., R. McNatt, D. J. Teel, and D. L Bottom. 2012. Distribution, size, and origin of juvenile Chinook Salmon in shallow water habitats of the lower Columbia River and estuary, 2002–2007. Marine and Coastal Fisheries: Dynamics, Management, and Ecosystem Science [online serial] 4:450–472.

Scheuerell, M. D., R. W. Zabel, and B. P. Sandford. 2009. Relating juvenile migration timing and survival to adulthood in two species of threatened Pacific salmon (*Oncorhynchus* spp.). Journal of Applied Ecology 46:983–990.

Seeb, L. W., A. Antonovich, M. A. Banks, T. D. Beacham, M. R. Bellinger, S. M. Blankenship, M. R. Campbell, N. A. Decovich, J. C. Garza, C. M. Guthrie III, T. A. Lundrigan, P. Moran, S. R. Narum, J. J. Stephenson, K. T. Supernault, D. J. Teel, W. D. Templin, J. K. Wenburg, S. F. Young, and C. T. Smith. 2007. Development of a standardized DNA database for Chinook Salmon. Fisheries 32:540–552.

Seeb, L. W., J. E. Seeb, C. Habicht, E. V. Farley, and F. M. Utter. 2011. Single-nucleotide polymorphic genotypes reveal patterns of early juvenile migration of Sockeye Salmon in the eastern Bering Sea. Transactions of the American Fisheries Society 140:734–748.

Shaklee, J. B., T. D. Beacham, L. Seeb, and B. A. White. 1999. Managing fisheries using genetic data: case studies from four species of Pacific salmon. Fisheries Research 43:1–34.

Simon, R. C., and P. A. Larkin, editors. 1972. The stock concept in Pacific salmon. University of British Columbia, H. R. MacMillan Lectures in Fisheries, Vancouver.

Steele, C. A., E. C. Anderson, M. W. Ackerman, M. A. Hess, N. R. Campbell, S. R. Narum, and M. R. Campbell. 2013. A validation of parentage-based tagging using hatchery steelhead in the Snake River basin. Canadian Journal of Fisheries and Aquatic Sciences 70:1046–1054.

Teel, D. J. 2004. Genetic mixed-stock analysis of juvenile Chinook Salmon in coastal areas of western North America. Pages 47–48 *in* J. Irvine, L. Seeb, S. Urawa, N. Varnavskaya, and R. Wilmot, editors. Proceedings of an international workshop on application of stock identification in defining marine distribution and migration of salmon. North Pacific Anadromous Fish Commission, Technical Report 5, Vancouver.

Teel, D. J., D. L. Bottom, S. A. Hinton, D. R. Kuligowski, G. T. McCabe, R. McNatt, G. C. Roegner, L. A. Stamatiou, and C. A. Simenstad. 2014. Genetic identification of Chinook Salmon in the Columbia River estuary: stock-specific distributions of juveniles in shallow tidal freshwater habitats. North American Journal of Fisheries Management 34:621–641.

Teel, D. J., D. M. Van Doornik, D. R. Kuligowski, and W. S. Grant. 2003. Genetic analysis of juvenile Coho Salmon (*Oncorhynchus kisutch*) off Oregon and Washington reveals few Columbia River wild fish. U.S. National Marine Fisheries Service Fishery Bulletin 101:640–652.

Tomaro, L. M., D. J. Teel, W. T. Peterson, and J. A. Miller. 2012. When is bigger better? Early marine residence of middle and upper Columbia River spring Chinook Salmon. Marine Ecology Progress Series 452:237–252.

Trudel, M., J. P. Fisher, J. A. Orsi, J. F. T. Morris, M. E. Thiess, R. M. Sweeting, A. Fergusson, and D. W. Welch. 2009. Distribution and migration of juvenile Chinook Salmon derived from coded wire tag recoveries along the continental shelf of western North America. Transactions of the American Fisheries Society 138:1369–1391.

Tucker, S., M. Trudel, D. W. Welch, J. R. Candy, J. F. T. Morris, M. E. Thiess, C. Wallace, and T. D. Beacham. 2011. Life history and seasonal stock-specific ocean migration of juvenile Chinook Salmon. Transactions of the American Fisheries Society 140:1101–1119.

Tucker, S., M. Trudel, D. W. Welch, J. R. Candy, J. F. T. Morris, M. E. Thiess, C. Wallace, and T. D. Beacham. 2012. Annual coastal migration of juvenile Chinook Salmon: static stock-specific patterns in a highly dynamic ocean. Marine Ecology Progress Series 449:245–262.

Tucker, S., M. Trudel, D. W. Welch, J. R. Candy, J. F. T. Morris, M. E. Thiess, C. Wallace, D. J. Teel, W. Crawford, E. V. Farley Jr., and T. D. Beacham. 2009. Seasonal stock-specific migrations of juvenile Sockeye Salmon along the West Coast of North America: implications for growth. Transactions of the American Fisheries Society 138:1458–1480.

Utter, F., D. Teel, G. Milner, and D. McIsaac. 1987. Genetic estimates of stock compositions of 1983 Chinook Salmon harvests off the Washington coast and the Columbia River. U.S. National Marine Fisheries Service Fishery Bulletin 85:13–23.

Van Doornik, D. M., D. J. Teel, D. R. Kuligowski, C. A. Morgan, and E. Casillas. 2007. Genetic analyses provide insight into the early ocean stock distribution and survival of juvenile Coho Salmon (*Oncorhynchus kisutch*) off the coasts of Washington and Oregon. North American Journal of Fisheries Management 27:220–237.

Van Hyning, J. M. 1973. Factors affecting the abundance of fall Chinook Salmon in the Columbia River. Research Reports of the Fish Commission of Oregon 4(1).

Walters, C. 1996. Information requirements for salmon management. Pages 61–68 *in* D. J. Stouder, P. A. Bisson, and R. J. Naimon, editors. Pacific salmon and their ecosystems: status and future options. Chapman and Hall, New York.

Waples, R. S., R. G. Gustafson, L. A. Weitkamp, J. M. Myers, O. W. Johnson, P. J. Busby, J. J. Hard, G. J. Bryant, F. W. Waknitz, K. Neely, D. Teel, W. S. Grant, G. A. Winans, S. Phelps, A. Marshall, and B. Baker. 2001. Characterizing diversity in salmon from the Pacific Northwest. Journal of Fish Biology 59(Supplement A):1–41.

Waples, R. S., D. J. Teel, J. M. Myers, and A. R. Marshall. 2004. Life-history divergence in Chinook Salmon: historic contingency and parallel evolution. Evolution 58:386–403.

Weitkamp, L. A. 2010. Marine distributions of Chinook Salmon from the West Coast of North America determined by coded wire tag recoveries. Transactions of the American Fisheries Society 139:147–170.

Weitkamp, L. A., P. B. Bentley, and M. N. C. Litz. 2012. Seasonal and interannual variation in juvenile salmonids and associated fish assemblage in open waters of the lower Columbia River estuary, USA. U.S. National Marine Fisheries Service Fishery Bulletin 110:426–450.

Weitkamp, L. A., and K. Neely. 2002. Coho Salmon (*Oncorhynchus kisutch*) ocean migration patterns: insight from marine coded-wire tag recoveries. Canadian Journal of Fisheries and Aquatic Sciences 59:1100–1115.

Weitkamp, L. A., D. J. Teel, M. Liermann, S. A. Hinton, D. M. Van Doornik, and P. J. Bentley. In press. Stock-specific size and timing at ocean entry of Columbia River juvenile Chinook Salmon and steelhead: implications for early ocean growth. Marine and Coastal Fisheries: Dynamics, Management, and Ecosystem Science [online serial]. DOI: 10.1080/19425120.2015.1047476.

6

Factors Influencing Daily Growth in Young-of-the-Year Winter Flounder along an Urban Gradient Revealed Using Hierarchical Linear Models

Brian K. Gallagher,[1]* Lyndie A. Hice,[2] Anne E. McElroy, Robert M. Cerrato, and Michael G. Frisk

School of Marine and Atmospheric Sciences, Stony Brook University, Stony Brook, New York 11794-5000, USA

Abstract

Growth during early life history plays a key role in the recruitment dynamics of marine fishes; however, the effects of environmental stressors on growth are often difficult to quantify. In this study, increment widths from sagittal otoliths were used as a proxy for daily growth in 102 young-of-the-year Winter Flounder *Pseudopleuronectes americanus* collected over a 2-year period from three sites in Long Island, New York. We hypothesized that we would observe different growth patterns among bays due to an environmental gradient driven primarily by contaminant loadings and environmental stressors in our study sites. Hierarchical linear models were utilized to associate individual attributes (ontogeny, condition, and gene expression) to daily growth patterns during each year. As expected, daily growth generally displayed a negative relationship with age and daily average temperature, although the effect of temperature was much more variable. Out of 14 individual attributes, the settlement date, the age at capture, the condition indices Fulton's *K* and hepatosomatic index, and the expression of genes associated with immune response (pleurocidin), contaminant exposure (cytochrome P5401A), and glucose and glycogen metabolism (glycerol-3-phosphate dehydrogenase) were observed to significantly and consistently affect growth. The results provide evidence of differential growth based on the date of settlement and condition, and the molecular indicators of stress suggest that growth is also influenced by habitat quality. There were significantly different relationships between individual attributes and growth among bays, but these did not always reflect the proposed environmental gradient. Together, the results suggest that anthropogenic stressors likely play a role in growth and recruitment processes in Long Island bays and indicate that growth is both spatially and temporally dynamic at multiple scales. Furthermore, this study highlights the utility of hierarchical linear models in analyzing complex daily growth data in juvenile fish, which may be applicable to other species.

Growth during early life history is recognized as an important determinant of recruitment success in fish populations (Houde 2008). Moreover, growth rates become increasingly important if selective processes reduce mortality in larger or faster-growing individuals relative to their smaller, slower-growing counterparts (Sogard 1997). This differential mortality may be due to decreased predation risk, increased resilience to starvation and environmental change, or some combination of

Subject editor: Richard Brill, Virginia Institute of Marine Science, Gloucester Point

*Corresponding author: bgallagh@umces.edu

[1]Present address: Chesapeake Biological Laboratory, University of Maryland Center for Environmental Sciences, Post Office Box 38, Solomons, Maryland 20688, USA.

[2]Present address: Delaware National Estuarine Research Reserve, Dover, Delaware 19901, USA.

these factors as individuals become larger (van der Veer et al. 2000). By these mechanisms, even minute changes in growth rates during early life history can substantially impact survivability and recruitment strength (Houde 1987).

Winter Flounder *Pseudopleuronectes americanus* is a historically important species that once supported a robust commercial and recreational fishery in the inshore bays of Long Island, New York, until a precipitous decline of local populations occurred in the 1990s. Winter Flounder in this region are presently considered to be at record low abundances (Nuttall et al. 2011; Sagarese and Frisk 2011). In spite of a commercial moratorium in federal waters and recreational catch limits in New York, Winter Flounder populations have shown little to no signs of recovery in Long Island waters. In the most recent stock assessment, spawning stock biomass estimates in the southern New England and Middle Atlantic Bight, which includes Long Island populations, ranged from 16% to 24% of target levels, and recent recruitment has been consistently lower than the long-term average (NEFSC 2011). Furthermore, microsatellite analyses of Winter Flounder from Long Island bays have revealed an alarmingly high incidence of inbreeding, which is likely the result of a very low effective population size in these bays (O'Leary et al. 2013). Previous studies have proposed that inadequate recruitment of young-of-the-year (age-0) fish is limiting recovery in this region (Socrates and Colvin 2006; Yencho 2009), although the specific mechanisms underlying recruitment limitation in Long Island bays have not been extensively studied. Moreover, daily mortality in Long Island bays has been observed to be higher than in Connecticut and Rhode Island estuaries (Yencho 2009; L.A.H., unpublished data), and identifying the factors that influence growth may provide insight into the potential causes of these high rates of mortality.

Throughout their range, adult Winter Flounder enter estuaries from the continental shelf during the winter and early spring to spawn (Klein-MacPhee 2002). Juveniles are typically retained within the estuaries where they were spawned and spend the first 2 years of life in these inshore habitats. Thus, the environmental conditions of these inshore habitats are considered to be important determinants of growth and survival in juvenile Winter Flounder (Goldberg et al. 2002). For example, previous studies in other estuarine systems have demonstrated that habitat suitability for age-0 Winter Flounder is driven primarily by dynamic and sometimes simultaneous changes in temperature, dissolved oxygen, salinity, and food availability (Meng et al. 2001; Manderson et al. 2002). The timing of settlement, wherein pelagic larvae metamorphose into demersal, laterally compressed juveniles, also influences the environmental conditions that juveniles are exposed to and thus may affect long-term growth and other vital rates (Sogard et al. 2001; Manderson et al. 2003).

Growth in Winter Flounder would be expected to decline with increasing age, and studies in both field (Sogard and Able 1992; Phelan et al. 2000) and laboratory (Keller and Klein-MacPhee 2000) settings have indicated that high temperatures (typically > 21°C) negatively influence growth in Winter Flounder. However, the effect of temperature may diminish later in the growing season as fish grow larger (Sogard et al. 2001). Long Island bays are near the southern extremity of the range of Winter Flounder, which may suggest that fish in this region are more likely to experience long periods of unfavorably high temperatures (Manderson 2008; Able et al. 2014). Many watersheds in Long Island are also highly developed and have been historically impacted by nutrient and contaminant loading (Bopp et al. 1993). Exposure to contaminants, such as metals, hydrocarbons, and synthetic organic compounds, can lead to alterations in metabolism, hormone production, condition, and liver function in fish (Polgar et al. 1985; McArdle et al. 2004). The excessive loading of nutrients from sewage or septic wastewater, particularly nitrogen, can lead to eutrophication and subsequent hypoxic conditions that degrade fish habitat, especially when aided by thermal stratification (Nixon 1995). Thus, age-0 Winter Flounder in Long Island are likely subjected to numerous local stressors and may be especially vulnerable when these stressors combine with unfavorable temperatures and dissolved oxygen levels during summer months.

The objectives of this study were threefold: (1) to identify which individual attributes significantly and consistently affected patterns in otolith increment width (hereafter referred to as daily growth) in separate models for 2010 and 2011, (2) to construct and compare multivariable models to select the best model for each year, and (3) within the selected models, to examine and compare relationships between age, temperature, and individual attributes and daily growth in each study site. Central to this framework was the testing of 14 variables related to the ontogeny (age at capture, settlement date), condition (Fulton's *K*, hepatosomatic index [HSI], RNA/DNA ratio), and expression of genes associated with contaminant exposure (cytochrome P4501A [CYP1A], vitellogenin [VTG]), immune response (pleurocidin, hepcidin II, complement C3, phospholipase A2), and metabolic stress (glutamate decarboxylase, glucokinase, glycerol 3-phosphate dehydrogenase [GPDH]) within individual fish. We conducted an exploratory analysis, utilizing hierarchical linear models to investigate which of these factors influenced otolith-derived estimates of daily growth while controlling for age, daily average temperature, and location. Ultimately, we sought to provide insight into how individual variation in these ontogenetic, physiological, and molecular traits control patterns of growth, and possibly recruitment processes, in these vulnerable populations of juvenile fish.

METHODS

Study sites.—The sites evaluated in this study were Jamaica, Moriches, and Shinnecock bays (Figure 1), each of which are shallow inshore bays along the south shore of Long Island, New York. Although separated by less than 200 km,

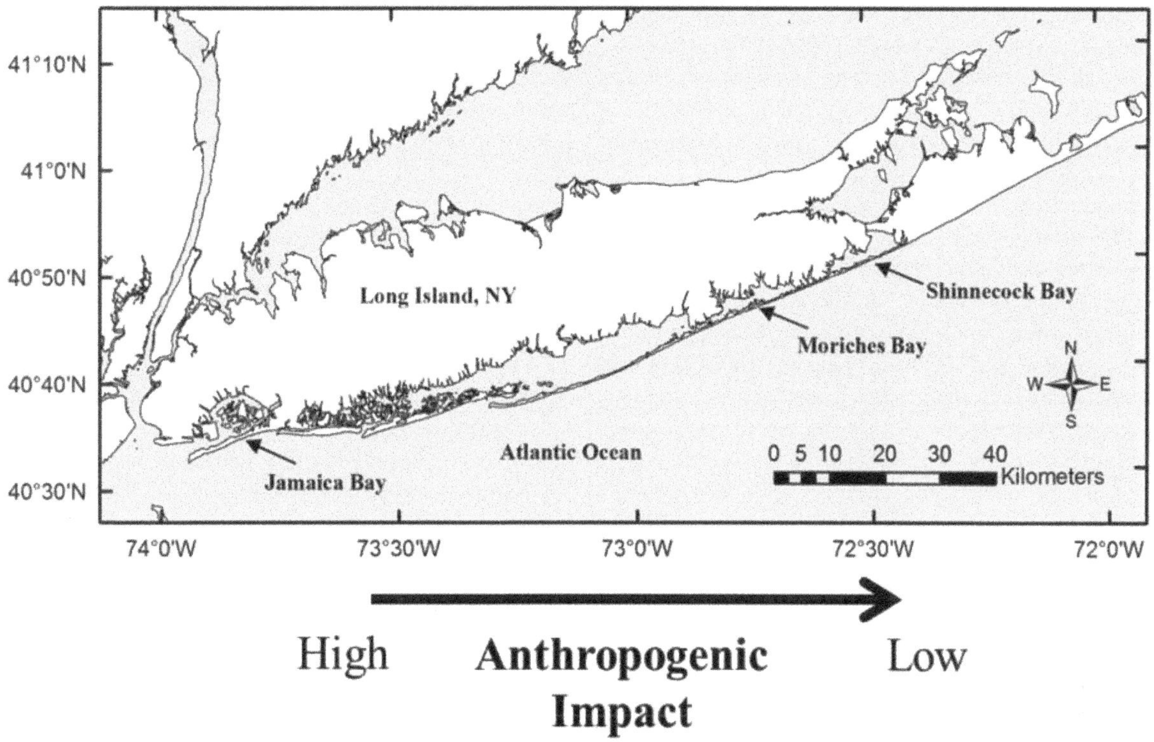

FIGURE 1. Map showing the locations of the three sites sampled in Long Island, New York. An arrow denoting a gradient of anthropogenic impact reflects that Jamaica and Moriches bays are more impacted sites, while Shinnecock Bay is comparatively less impacted, with no direct sewage effluent and a lower human population density in its watershed. Sample locations consisted of areas within the bays known to support juvenile Winter Flounder and were randomly sampled (Frisk et al. 2013).

these sites lay along a decreasing west-to-east gradient of human population density (LIRPC 2010) and environmental conditions, primarily associated with differences in nutrient input and loadings of organic and inorganic pollutants (Bopp et al. 1993; Hartig et al. 2002). Analysis of priority pollutants in fine-grain sediments sampled from 2000 to 2006 as part of the U.S. Environmental Protection Agency's National Coastal Assessment generally supports a west–east gradient, with average concentrations of organic (PAHs and PCBs) and inorganic (Pb, Cu, Cd, and Hg) pollutants decreasing by at least a factor of five from west to east (USEPA 2006; McElroy et al. 2015). A previous study of Atlantic Silverside *Menidia menidia* in Long Island detected a significant longitudinal trend in sex ratios, such that female-biased sex ratios were more prevalent in more urbanized western embayments, which reflected a west–east gradient in municipal wastewater discharge (Duffy et al. 2009). In addition, environmental data collected from our three study sites in the summer of 2011 suggest that there is a gradient in bottom-water hypoxia, with the percentage of dissolved oxygen measurements below 2.3 mg/L being 23.0, 6.0, and 0.6% at Jamaica, Moriches, and Shinnecock bays, respectively (McElroy et al. 2015). Consequently, we anticipated that growth patterns in age-0 Winter Flounder would vary in accordance with this environmental gradient, with greater reductions in growth due to stress expected in more impacted western sites.

Fish collections and basic metrics.—Winter Flounder were randomly sampled as a part of a larger study during a biweekly beam trawl survey from May through October in 2010 and 2011 (Frisk et al. 2013). Fish were flash frozen on dry ice after capture and brought back to the laboratory and stored in freezers at −80°C until they were processed. The total length (to nearest millimeter) and weight (whole body weight; to nearest 0.01 g) were recorded for each individual, and these measures were subsequently used to calculate condition indices. Fish from Shinnecock, Moriches, and Jamaica bays were included in this analysis if they were caught during time periods that allowed for 20 or more days of increment width data and corresponding age and temperature data to be collected (see below).

Otolith processing and analysis.—The left and right sagittal otoliths of each individual were removed and affixed to clean, labeled slides using heat-activated Crystalbond 509 adhesive. The left otolith was then polished whole, in the sagittal plane, using 30-μm and 3-μm lapping paper. The left otolith was selected for age and growth estimation because it has been observed to have a stronger relationship to somatic growth than the right otolith (Sogard 1991). After polishing, each otolith was immersed in Cargille Type A immersion oil and four coincident photographs were taken at 100× magnification using a Nikon DX1200C digital camera connected to a Nikon Eclipse 80i compound microscope. These images were merged to improve the depth of field and then analyzed using Image-Pro 6.0 software (Media Cybernetics 2006). An axis was placed along the postrostrum after the method of Yencho (2009). The axis started approximately 10 increments away from the formation of the secondary growth primordium because metamorphosis is known to be rapid and typically occurs in conjunction with the formation of secondary growth centers in this species (Sogard 1991). The axis was extended to the otolith edge so that increments could be counted back to the settlement period (Figure A.1 in the appendix). Age was estimated in each individual by counting the number of daily growth rings present, and the width between each increment was measured. Daily increment width is a valid proxy for daily growth in age-0 Winter Flounder in Long Island because otolith size (total width of the postrostral axis) and fish size have been observed to be linearly and strongly correlated ($n = 126$, $r^2 = 0.77$, $P < 0.001$ in 2010; $n = 119$, $r^2 = 0.81$, $P < 0.001$ in 2011; data not shown). Settlement date was back-calculated by subtracting the number of daily rings from the date of capture in ordinal days (day number).

Environmental data.—The modeling procedure required water temperature data in order to relate growth and temperature histories; however, environmental data were only continuously collected from the end of June through August of 2011 with sondes (Hydrolab MS5; Hach Hydromet, Loveland, Colorado) in each bay. To complete the summer temperature histories for each site, the sonde data from 2011 was combined with water temperature data obtained from a variety of sources and converted to daily averages for 2010 and 2011. In Jamaica Bay, daily bottom temperatures for all of May–August were obtained by combining conductivity, temperature, and depth sensor data collected by the New York City Department of Environmental Protection (NYCDEP 2010, 2011) and YSI Model 85 probe (Yellow Springs, Ohio) data obtained during the beam trawl survey used to collect fish. Temperature in Moriches Bay were recorded every 6 min by a tide and temperature gauge (Sea Bird Electronics SBE-26) at Smith Point, and data for all of May–August during both years was extracted from LIShore (http://www.lishore.org; Thomas Wilson, Stony Brook University, personal communication). Continuous data from Shinnecock Bay were available from the beginning of June to the end of August in both 2010 (Christopher Gobler, Stony Brook University, unpublished data) and 2011 (Bradley Peterson, Stony Book University, unpublished data). There were gaps in the Jamaica Bay data set varying from 5 to 8 d in length, with one 14-d gap in 2011, and temperature during these time periods was inferred using linear interpolation. The resulting temperature data sets for both 2010 and 2011 spanned from June to August in Shinnecock Bay and from May to August in Jamaica and Moriches bays. For each fish, daily growth measurements were subsequently aligned with the corresponding age and daily average temperature during these time periods.

Condition indices and gene expression data.—Data detailing individual condition, gene expression, and microsatellite profiles at the time of collection were developed through other components of a larger study of Long Island Winter Flounder (Frisk et al. 2013). Fulton's K was calculated by dividing the body weight by the cube of total length for each individual and multiplying by 100, while HSI was calculated as the ratio of liver weight (to nearest 0.001 g) to body weight. Differential mRNA expression was examined in liver tissue from individuals to identify cellular pathways associated with the expression of two genes related to contaminant exposure (CYP1A and VTG), four associated with immune response (pleurocidin, complement C3, hepcidin II, and phospholipase A2), and three associated with glucose and glycogen metabolism (glucokinase, glutamate decarboxylase, GPDH). The raw data from the gene expression analysis spanned several orders of magnitude, so values were $\log_e$ transformed for all subsequent analyses. Full details of RNA isolation, cDNA synthesis, and quantitative PCR are provided in McElroy et al. (2015). Descriptions of all individual-level variables can be found in Table 1.

Hierarchical linear models.—Otolith increment widths, while useful in reconstructing growth patterns (Brothers et al. 1976; Campana and Neilson 1982), typically result in unbalanced, autocorrelated repeated-measures data (Campana and Jones 1992), which are often difficult to analyze from a statistical standpoint. In juvenile fish, growth variation can be attributed to factors operating over large scales, such as climatic and hydrographic regimes, as well as responses to local environmental conditions (Houde 1989; Sogard et al. 2001). Thus, factors affecting daily growth may be considered hierarchically structured (Figure 2). The three-level hierarchical linear models (HLMs) used in this study employed iterative maximum likelihood methods to estimate parameters, as well as variance and covariance components, in nested linear regression equations (Raudenbush and Bryk 2002). The hierarchical framework allowed data to be structured across three levels: level-1 factors that affected daily growth, level-2 factors that affected individual fish, and level-3 factors that affected all fish within each bay.

In HLMs, the levels are nested (i.e., growth increments within fish within bays) so that variation at one level is taken into account when calculating the variance components at any other level. In addition, level-2 and level-3 coefficients are estimated to explain variability in group means rather than individual level-1 responses, which addresses the correlation of error within groups when working with hierarchically structured data (McMahon and Diez 2007). One previous study utilized a hierarchical nonlinear modeling framework to investigate trends in annual size-at-age data in adult fish (Schaalje et al. 2002); however, no research, to our knowledge, has applied HLMs to describe daily growth patterns in juvenile fish.

TABLE 1. List of the 14 candidate level-2 variables, with their category and a brief description. The last two columns denote how each variable performed during the first stage of the exploratory analysis each year. The abbreviation NS means the variable was not significant in at least four of the six models tested, and PI (shown in bold italics) means that the variable was considered potentially important because it had a *P*-value <0.01 in at least four of the six models tested; the parentheses denote the number of times each variable was significant for that year.

Variable	Type	Definition and interpretation	2010	2011
RNA:DNA	Condition (short term)	The RNA:DNA ratio is used as an index of the amount of protein synthesized per cell as a proxy of recent tissue growth (Buckley et al. 1999).	NS (2/6)	NS (3/6)
Hepatosomatic index (HSI)	Condition (long term)	Relationship between liver weight and body weight. Increased liver weight may be an indicator of increased energy reserves in fish in good condition due to increased RNA, protein, or lipid content. Alternatively, liver weight has been shown to increase when exposed to contaminants, possibly due to increased protein synthesis or lipid storage (Khan 2010).	***PI (6/6)***	NS (2/6)
Fulton's *K*	Condition (long term)	An index of condition that assumes that heavier fish at a given length have a more robust body form and thus are in better condition (Amara et al. 2007; Khan 2010).	***PI (5/6)***	NS (2/6)
Age at capture	Ontogeny	Individual age when captured, estimated by otolith increment counts.	***PI (4/6)***	***PI (4/6)***
Settlement date	Ontogeny	Estimated date when settlement occurred as determined by subtracting age from date of capture (reported in ordinal days).	***PI (4/6)***	***PI (4/6)***
Cytochrome P4501A (CYP1A)	Contaminant exposure	A biomarker commonly used to evaluate exposure to planar aromatic hydrocarbons, such as chlorinated hydrocarbons like PCBs, dioxins, and furans, as well as other compounds associated with urban runoff (Stegeman and Hahn 1994).	NS (2/6)	***PI (4/6)***
Vitellogenin (VTG)	Contaminant exposure	Elevated expression of VTG in male or juvenile fish is a commonly used biomarker of exposure to estrogen or chemicals that mimic estrogen in wild-caught fish (Heppell et al. 1995).	***PI (4/6)***	NS (1/6)
Pleurocidin	Immune response	Gene codes for the antimicrobial peptide pleurocidin, an innate molecule effective at killing gram-positive and -negative bacteria (Cole et al. 1997).	***PI (4/6)***	NS (3/6)

(*Continued on next page*)

TABLE 1. Continued.

Variable	Type	Definition and interpretation	2010	2011
Complement C3	Immune response	Elevated C3 expression could indicate a generally higher level of pathogen exposure and inflammation in fish from more urban areas.	NS (1/6)	NS (2/6)
Hepcidin II	Immune response	A peptide hormone produced by the liver, known to have antimicrobial activity.	NS (2/6)	NS (3/6)
Phospholipase A2-like	Immune response	An important catalyst within the arachidonic acid pathway, leading to the formation of inflammatory and thrombogenic compounds (Pruzanski and Vadas 1991).	NS (2/6)	NS (2/6)
Glucokinase-like	Glucose and glycogen metabolism	Glucokinase is upregulated in response to hypoxia to increase anaerobic glucose metabolism.	NS (2/6)	NS (1/6)
Glutamate decarboxylase-like	Glucose and glycogen metabolism	Rate-limiting enzyme in gamma-aminobutyric acid synthesis (a neurotransmitter with a primary signaling role in the pancreas).	NS (2/6)	NS (2/6)
Glycerol 3-phosphase dehydrogenase-like (GPDH)	Glucose and glycogen metabolism	Glycerol 3-phosphate dehydrogenase catalyzes a key step in the breakdown of glucose for energy production. Upregulation could indicate elevated energy demand (Baumgarner et al. 2012)	NS (1/6)	***PI (4/6)***

The general form of a three-level HLM is provided in this section to fully conceptualize the hierarchical structure of this analysis. The notation is modified from Raudenbush et al. (2011), and regression coefficients are indicated by α, β, and γ, independent variables by x, y, and z, and random error terms by u, v, and w. At level 1, daily incremental growth G_{ijk} for increment i in fish j collected in bay k is represented as follows:

$$G_{ijk} = \alpha_{0jk} + \sum_{p=1}^{P} \alpha_{pjk} x_{pijk} + u_{ijk}, \tag{1}$$

where α_{pjk} are level-1 regression coefficients, x_{pjk} are level-1 independent variables, and the random error term u_{ijk} is assumed to be $N(0, \sigma^2)$. In the current study, age and temperature associated with each otolith increment were taken to be the level-1 independent variables.

At level 2, individual attributes, such as measurements of ontogeny, condition, and gene expression, were allowed to influence the level-1 coefficients:

$$\alpha_{pjk} = \beta_{0pk} + \sum_{q=1}^{Q} \beta_{qpk} y_{qjk} + v_{pjk}, \tag{2}$$

where β_{qpk} are level-2 regression coefficients, y_{qjk} are level-2 independent variables, and the random error terms for individual j in bay k are v_{pjk}.

The final level, level 3, was used to represent bay-level characteristics:

$$\beta_{qpk} = \gamma_{0pk} + \sum_{r=1}^{R} \gamma_{rqp} z_{rk} + w_{qpk}, \tag{3}$$

where γ_{rqp} are level-3 regression coefficients, z_{rk} are level-3 independent variables, and the random error terms for bay k are w_{qpk}. In the present study, level-3 variables were categorical variables denoting different bays.

If equation (3) is substituted into equation (2), and then the terms from the resulting equation are substituted into equation

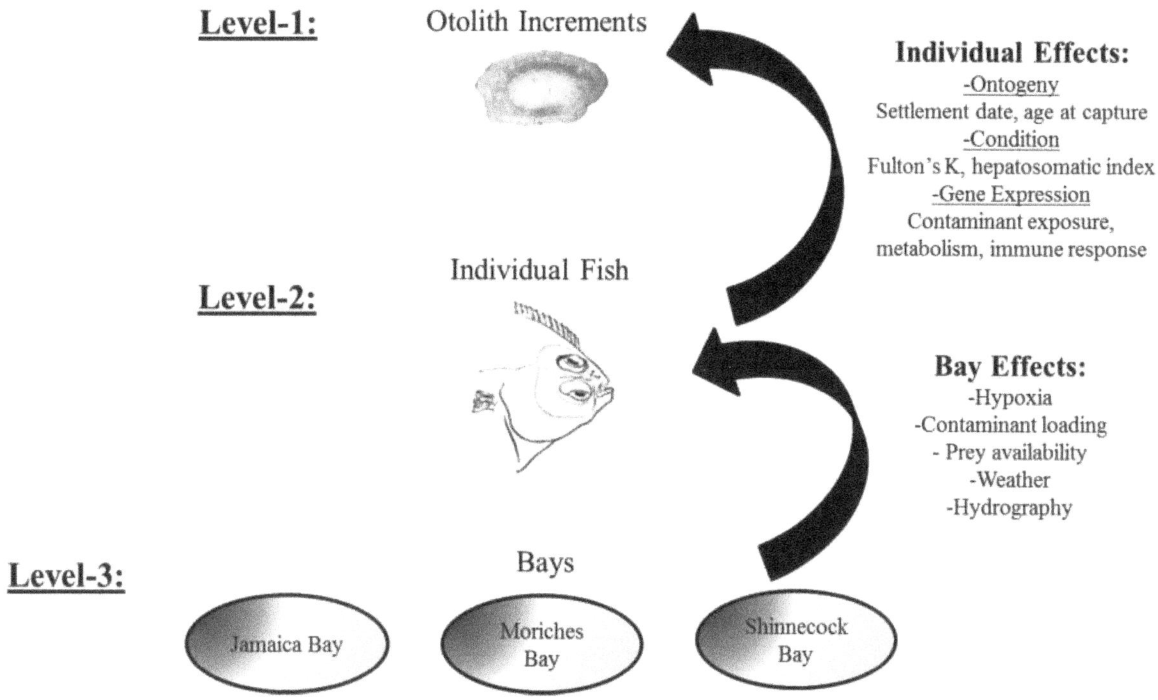

FIGURE 2. Conceptual diagram of the hierarchical structure of daily growth data in age-0 Winter Flounder. The data are broken up into three levels, which are nested within one another. The effects of these variables on growth move up the hierarchy (arrows) such that the characteristics of each bay influence the individual attributes of fish, which in turn affect the daily growth patterns reflected in otolith increments.

(1), the result is a linear mixed-effects model in terms of γ_{rqp}. This final mixed model incorporates fixed and random factors on all three levels, in which parameters are estimated simultaneously. In this study, full maximum likelihood was used for estimation so that the deviance of models with different fixed factors could be directly compared (Kreft and de Leeuw 1998).

Model selection.—An exploratory analysis was conducted using HLM7 (Raudenbush et al. 2011) in order to determine which independent variables and model structures best predicted daily growth in separate data sets for 2010 and 2011. To do so, the analysis was broken up into two steps. The level-1 independent variables, age and temperature, were retained throughout the selection process. This resulted in an intercept term, an age slope, and a temperature slope affecting daily growth (as described in the previous section). An interaction term between age and temperature did not substantially improve fit during either year and was thus not used in subsequent models. In addition, all level-2 variables were scaled and centered to their grand mean throughout the selection process.

Stage 1: Initially, there were 14 candidate explanatory variables at level 2 related to ontogeny, condition, contaminant exposure, immune response, and metabolism (Table 1) that were each tested independently in order to trim the number of explanatory variables. Each model incorporated a single level-2 variable affecting one of the three level-1 coefficients and level-3 site descriptors influencing the level-2 variable in one of two possible ways (i.e., through the intercept or slope). Random error terms were assigned to the intercept terms on level 2 (β_{0pk}) and level 3 (γ_{0pk}) in every model (i.e., random intercept model). Dummy variables denoting Shinnecock and Jamaica bays were used as level-3 variables because including all three sites resulted in collinearity. Thus, six models (i.e., three level-1 coefficients × two level-2 coefficients) were tested for each level-2 variable for each year, yielding 196 models overall. Variables progressed to the next stage of the analysis if they consistently had a significant influence on growth, such that the effect on level-1 coefficients had a P-value < 0.01 in four or more out of the six model runs conducted for each variable for each year. Six variables satisfied this criterion in 2010 while four variables did so in 2011.

Stage 2: In the final stage, various model configurations were compared to select the best models for each year; these models included different combinations of the selected level-2 variables affecting the level-1 coefficients (e.g., the intercept, age slope, and temperature slope). In addition, each level-2 effect was allowed to be influenced by site on level 3. The Akaike information criterion (AIC) was used to compare the relative fit of all model structures tested until the model with the lowest AIC was determined, selecting a single model for each year without creating over-parameterized models.

Model performance.—The performance of level-2 variables throughout stage 2 was reported as the proportion of total model runs that had a P-value < 0.05. Akaike weights (AIC_w) were used to evaluate the performance of models of varying complexity (i.e., number of parameters estimated), given the data (Wagenmakers and Farrell 2004). The AIC_w is estimated by first determining the difference between the AIC of the best model (minAIC) and that of all the candidate models (i):

$$\Delta_i(\mathrm{AIC}) = \mathrm{AIC}_i - \mathrm{minAIC}.$$

The AIC_w is computed for all models with the following equation:

$$\mathrm{AIC}w_i = \frac{\exp\{-\frac{1}{2}\Delta_i(\mathrm{AIC})\}}{\sum_{k=1}^{n}\exp\{-\frac{1}{2}\Delta_k(\mathrm{AIC})\}},$$

where n is the total number of models being compared. The AIC_w can be interpreted as a conditional probability and provides an estimate of how likely a model is relative to all candidate models (Wagenmakers and Farrell 2004).

Model performance was also investigated by constructing level-1 residual box plots for the model with the smallest AIC for each year (hereafter referred to as the selected model). These residuals represented the deviations, at level 1, of daily growth from the corresponding predicted values for each individual fish in the 2010 and 2011 models. Box plots for individuals were displayed in order of decreasing median increment width, with outliers and extreme values included. Intraclass correlations, which represent the proportion of total variance that was attributed to each level, were also calculated for the selected model for each year. Finally, model performance was evaluated by comparing observed growth-at-age data to predicted growth-at-age trajectories in each bay during each year. Ideally, the predicted trajectories would provide a realistic representation of the data and would fall within the range of observed growth values at a given age.

The selected models for 2010 and 2011 were analyzed graphically to illustrate the effects of each level-1 and level-2 variable on daily growth and to evaluate how these effects varied between the three sample sites. For each variable, we plotted the intercept, slope, and range of values at each site while holding all other variables constant at their mean. Contrasts were performed to (1) examine whether each slope was significantly different from 0 and (2) test whether the slopes of each relationship were significantly different among the three study sites.

RESULTS

Daily Growth

In total, otoliths from 57 fish were analyzed in 2010, consisting of 14 fish from Jamaica Bay, 21 from Moriches Bay, and 22 from Shinnecock Bay. In 2011, otoliths from 45 individuals were examined, including 17 individuals from Jamaica Bay, 16 from Moriches Bay, and 12 from Shinnecock Bay.

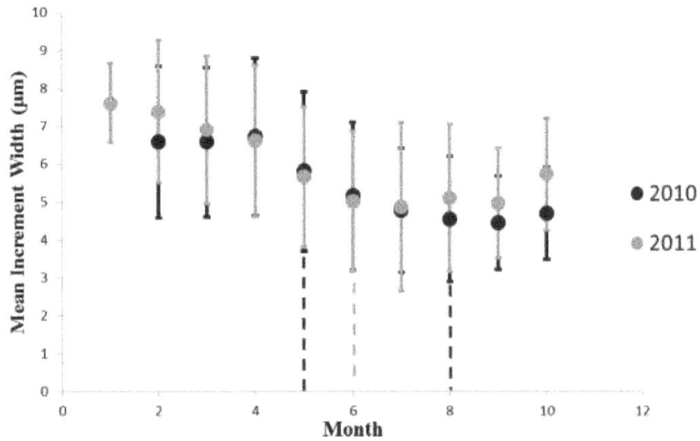

FIGURE 3. Plot of average otolith increment widths for all three sites in 2010 (black) and 2011 (gray), with error bars representing one standard deviation. The black dashed lines denote the time period analyzed in Moriches and Jamaica bays, while the gray dashed line denotes the start of the time period analyzed in Shinnecock Bay (which also ended in August), when daily growth was declining on average.

However, HLMs treat the structure of the variance in a manner that allows otolith increment widths to be considered independent when estimating degrees of freedom at level 1, such that 3,851 data points were used in the 2010 model and 2,158 data points were used in 2011. Otolith increment widths during the study period ranged from 1.96 to 15.01 µm in 2010 and from 2.12 to 12.53 µm in 2011. Average monthly otolith increment widths among all sites (Figure 3) throughout the whole growing season were variable and displayed similar trends during both years. Growth was highest early in the season, from February to April, and was followed by decreases in growth during the late spring through the summer months (i.e., those evaluated in the HLM analysis), before increasing again in the fall. Fish collected in our three study sites over the entire sampling period averaged 65.4 mm in length (range: 28–135 mm) and 116 d in age (55–238 d) in 2010, while averaging 59.9 mm in length (24–99 mm) and 101 d in age (38–185 d) in 2011.

Hierarchical Linear Models

Model selection.—Stage 1: Model selection at this stage resulted in a reduction of the potential level-2 explanatory variables. In 2010, six explanatory variables satisfied the criterion of having four or more effects on daily growth with a P-value < 0.01 among the six runs for each variable: settlement date, age at capture, Fulton's K, HSI, and expression of pleurocidin and VTG. In 2011, four variables (settlement date, age at capture, and expression of CYP1A and GPDH) met the criterion.

Stage 2: The proportion of significant effects ($P < 0.05$) of level-2 variables in all potential 2010 models ranged from 0.98 for pleurocidin to 0.29 for Fulton's K in 2010. In 2011, the proportion of significant level-2 variables ranged from 0.80 for GPDH to 0.31 for CYP1A (Table 2). A total of 73

TABLE 2. Table summarizing the performance of each level-2 variable tested in stage 2 of model selection. The Models row shows the number of models tested, the Significant row the number of model runs that were significant at the $P < 0.05$ level, and the Proportion row the proportion of significant model runs. Variables listed with an asterisk did not pass the stage-1 criterion and were not analyzed in stage 2.

Measure	Year	Age at capture	Settlement date	Fulton's K	HSI	Pleurocidin	VTG	CYP1A	GPDH
Models	2010	57	57	48	48	48	47	*	*
Significant	*	49	52	14	20	47	31	*	*
Proportion	*	0.86	0.91	0.29	0.42	0.98	0.66	*	*
Models	2011	38	38	*	*	*	*	35	35
Significant	*	29	27	*	*	*	*	11	28
Proportion	*	0.76	0.71	*	*	*	*	0.31	0.80

models in 2010 and 43 in 2011 were compared using AIC to determine the selected model for each year. The selected model in 2010 did not include VTG expression as a level-2 variable and resulted in five variables at level 2. Level 3 (site) influenced terms on level 2 in seven places, yielding a 25-parameter model. In 2011, the best model retained all four level-2 variables that passed through stage 1 of model selection, and level-3 variables influenced level-2 terms in six places, resulting in a 22-parameter model. Akaike weights based on the number of parameters estimated in all model configurations in both years (Table 3) generally showed that the selected models in both years were much more probable than the other model structures examined.

2010 Selected-model contrasts.—A summary of all terms in the selected model for 2010 can be found in Table A.1 in the appendix. The intercept term was highly significant and positive in the selected model for 2010 ($\gamma_{000} = 7.703$; $P = 0.003$), and this was the only term in the model which was not influenced by site. The age slope term was significantly negative in Shinnecock ($\gamma_{101} = -0.019$; $P < 0.001$), Moriches ($\gamma_{100} = -0.023$; $P < 0.001$), and Jamaica ($\gamma_{102} = -0.039$; $P < 0.001$) bays, and the relationship in Jamaica Bay was significantly different from the other bays ($P < 0.020$). The temperature slope term was negative in Shinnecock ($\gamma_{201} = -0.101$; $P < 0.001$) and Moriches ($\gamma_{200} = -0.041$; $P < 0.001$) bays but was positive in Jamaica Bay ($\gamma_{202} = 0.040$; $P = 0.128$), and all three sites were significantly different from one another ($P < 0.010$). Settlement date was negatively related to daily growth in Shinnecock ($\gamma_{011} = -0.005$; $P = 0.003$), Moriches ($\gamma_{010} = -0.062$; $P < 0.001$), and Jamaica ($\gamma_{012} = -0.002$; $P = 0.459$) bays, and Moriches Bay was significantly different than the other sites ($P < 0.001$). Age at capture significantly affected daily growth in Moriches Bay ($\gamma_{110} = -1.148$; $P = 0.001$) but not in Shinnecock ($\gamma_{111} = 0.254$; $P = 0.595$) or Jamaica ($\gamma_{112} = 0.160$; $P = 0.842$) bays in 2010, and Moriches Bay had a significantly different relationship than Shinnecock Bay ($P = 0.020$). The expression of pleurocidin had a varying relationship with daily growth in Shinnecock ($\gamma_{121} = 0.006$; $P = 0.098$), Moriches ($\gamma_{120} = -0.003$; $P < 0.001$), and Jamaica ($\gamma_{122} = -0.018$; $P < 0.001$) bays, and each site was significantly different from one another ($P < 0.020$). The condition indicator HSI was negatively related to daily growth in Shinnecock ($\gamma_{211} = -0.010$; $P = 0.431$), Moriches ($\gamma_{210} = -0.019$; $P = 0.002$), and Jamaica ($\gamma_{212} = -0.087$; $P < 0.001$) bays, and this effect was significantly different in Jamaica Bay relative to the other sites ($P < 0.001$). The relationships between daily growth and Fulton's K were not significant at any site ($P > 0.080$), and no site was significantly different from any other ($P > 0.100$).

TABLE 3. Akaike weights (AIC_w) for all models from stage 2 of the exploratory analysis, organized by the number of parameters estimated (i.e., model complexity).

Year	Parameters	Models	Sum AIC_w	Mean AIC_w
2010	28	5	0.119	0.0238
	25	6	0.865	0.1441
	24	8	0.004	0.0005
	21	9	0.012	0.0014
	18	17	<0.001	<0.0001
	15	28	<0.001	<0.0001
2011	22	8	0.983	0.1228
	18	10	0.003	0.0003
	16	1	<0.001	0.0010
	15	20	0.015	0.0007
	12	6	<0.001	0.0002

2011 Selected-model contrasts.—A summary of all terms in the selected model for the 2011 data can be found in Table A.1 in the appendix. In 2011, the intercept term was again highly significant and positive ($\gamma_{000} = 8.875$; $P < 0.001$), and this was once again the only term in the model not influenced by site. The age slope was also significantly negative in Shinnecock ($\gamma_{101} = -0.041$; $P < 0.001$), Moriches ($\gamma_{100} = -0.042$; $P < 0.001$), and Jamaica ($\gamma_{102} = -0.065$; $P < 0.001$) bays, and Jamaica Bay once again had a different slope than the other sites ($P < 0.001$). Similar to 2010, the temperature slope term was negative in Shinnecock ($\gamma_{201} = -0.108$; $P < 0.001$) and Moriches ($\gamma_{200} = -0.009$; $P = 0.589$) bays but positive in Jamaica Bay ($\gamma_{202} = 0.019$; $P = 0.552$), but

Shinnecock Bay was significantly different than the other two sites ($P < 0.010$). Settlement date positively affected daily growth in Shinnecock ($\gamma_{011} = 0.697$; $P = 0.037$) and Moriches ($\gamma_{010} = 0.694$; $P = 0.018$) bays but was negative in Jamaica Bay ($\gamma_{012} = 0.388$; $P = 0.487$), and none of these relationships were significantly different from one another ($P > 0.080$). Age at capture was positively related to daily growth in Shinnecock ($\gamma_{111} = 0.013$; $P < 0.001$), Moriches ($\gamma_{110} = 0.017$; $P < 0.001$), and Jamaica ($\gamma_{112} = 0.026$; $P < 0.001$) bays, but no sites differed from one another ($P > 0.050$). The effect of CYP1A expression on daily growth varied in its direction in Shinnecock ($\gamma_{211} = -0.012$; $P = 0.495$), Moriches ($\gamma_{210} = 0.044$; $P < 0.001$), and Jamaica ($\gamma_{212} = -0.025$; $P = 0.038$) bays, and Moriches Bay was significantly different than the other two sites ($P < 0.010$). In 2011, GPDH differentially affected daily growth in Shinnecock ($\gamma_{121} = 0.001$; $P = 0.514$), Moriches ($\gamma_{120} = -0.008$; $P = 0.025$), and Jamaica ($\gamma_{122} = -0.007$; $P = 0.072$) bays, and Shinnecock Bay had a significantly different relationship than Moriches Bay ($P < 0.010$).

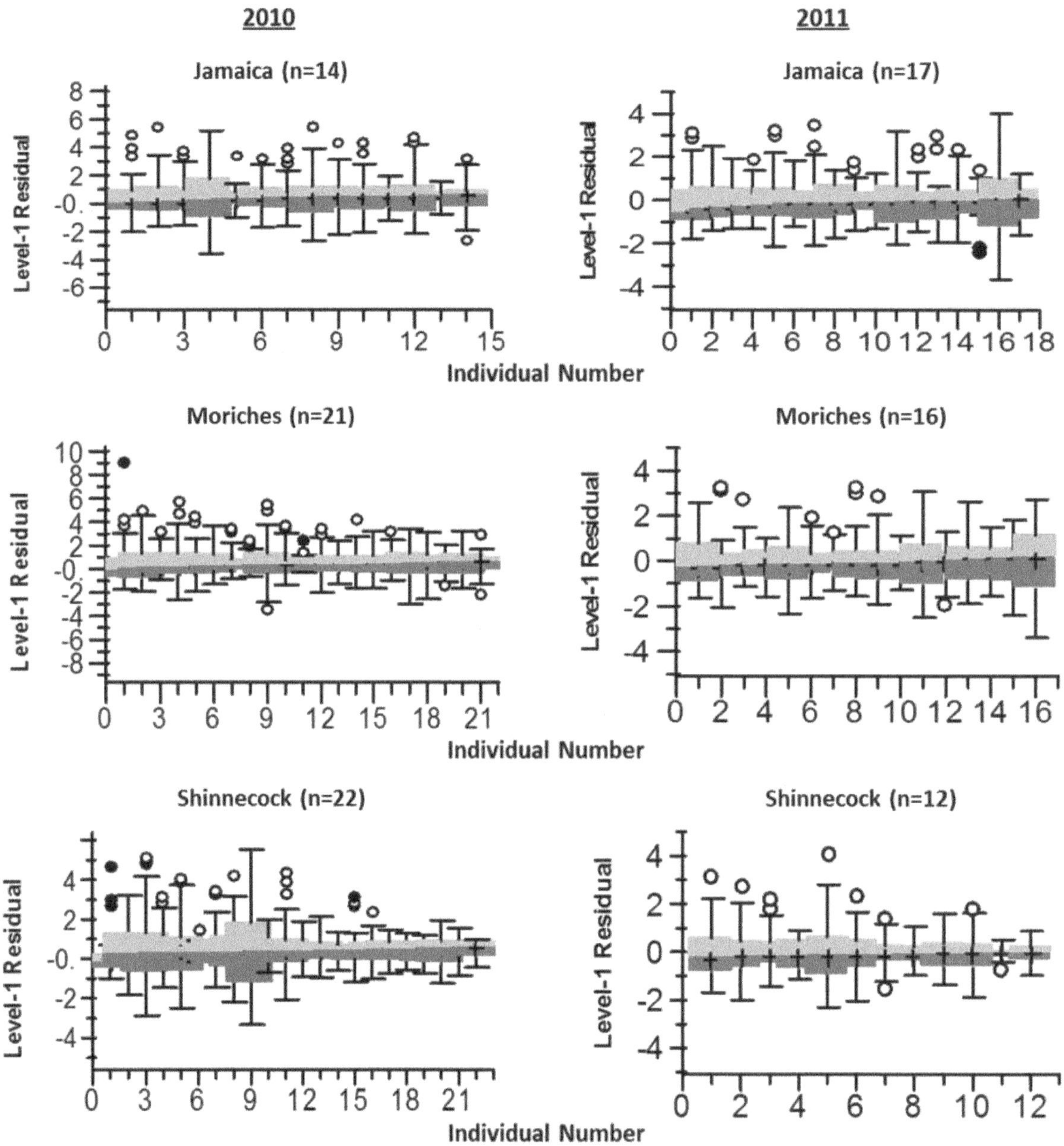

FIGURE 4. Level-1 residual daily growth box plots for individual fish (sample sizes for each site in parentheses) in the selected models in 2010 (left column) and 2011 (right column). The gray-shaded boxes indicate the 25th, 50th, and 75th percentiles for individual fish, the error bars represent the 10th and 90th percentiles, the white dots denote outliers, and the black dots denote extreme values. Note the different scales of the *y*-axes (i.e., magnitude of variation) for each bay.

TABLE 4. Intraclass correlation (ICC) estimates denoting the proportion of total variance attributed to level 1, level 2, and level 3 in the selected models for 2010 and 2011.

Level	2010	2011
Level-1 ICC	0.3782	0.4422
Level-2 ICC	0.5281	0.5576
Level-3 ICC	0.0937	0.0001

Model performance.—In general, box plots of level-1 residuals did not appear to display a strong trend in relation to the median observed increment width (Figure 4), and residuals were well centered around 0. However, there were differences in the range and the number of outliers and extreme values in each model for each year. In the selected models for both years, the intraclass correlation estimates indicated that over 90% of the total variance was attributed to level 1 and level 2 (Table 4), which suggests that the majority of variability in growth occurs at the daily growth (temperature and age) and individual fish (condition, gene expression, etc.) levels. Within each model, settlement date only had a significant effect in Shinnecock and Moriches bays, while gene expression and (or) condition metrics effects were significant in Moriches and Jamaica bays, the two more impacted sites in this study.

The majority of growth-at-age trajectories performed reasonably well during both years, with predicted values generally falling within the range of observed values (Figure A.2 in the appendix). There was a tendency, however, for the model to underestimate growth at early ages (<50 d), and this was particularly evident in Shinnecock Bay in 2010. In addition, there was one individual in Jamaica Bay (ID number 63) that had a consistently low predicted growth trajectory, eventually falling to 0 at an age of ~110 d. This individual fell well below the range of observed values, but this may be attributed to the fact that all values of its level-2 variables were below average (and were thus negative) and also because of the comparatively late age and high temperature at which it entered the model.

The graphical analyses of the selected models in 2010 and 2011 reflected the results of the contrasts, wherein relationships between increment width and variables of interest were displayed as being significantly different from 0 or not. The effect of age was significantly negative during both years, while temperature did not have the strong negative effect on daily growth that was expected (Figures 5, 6). In addition, all level-2 variables, with the exception of Fulton's K in 2010, had significant effects on increment width in at least one of the study sites.

DISCUSSION

Daily growth in age-0 Winter Flounder was analyzed using HLMs and showed significant relationships with ontogenetic factors, environmental conditions, and indices of health and condition. The selected models for 2010 and 2011 each contained a different set of several level-2 variables, thereby supporting previous studies that have demonstrated that relationships between growth and environmental conditions in this species are driven by multiple factors (Manderson et al. 2002). Interestingly, temperature did not have a strong direct effect on growth over the range of observed values (13–27°C; Figure A.3 in the appendix). Age at capture was negatively related to growth for Moriches Bay in 2010 and was positively related in all three locations in 2011. Settlement date was negatively related to daily growth in Shinnecock Bay and Moriches Bay but was positively related to daily growth in 2011 in both bays. Indicators of fish health (HSI in 2010) and immunological and metabolic stress (pleurocidin in 2010, CYP1A and GPDH in 2011) only had significant effects on daily growth in more impacted sites (Jamaica and Moriches bays). Our findings suggest that the optimum conditions for growth are spatially and temporally dynamic and that daily growth cannot be reduced to a single factor; instead, growth patterns are a complex response to a highly variable environment.

The switch from a negative to a positive relationship between settlement date and growth suggests that annual variation in environmental conditions influences the success of recruits settled during different periods of a given year. In general, early settlement corresponds with cooler temperatures, late settlement would favor higher temperatures, and both would have an impact on the bioenergetics and growth of juveniles (Bullock 1955). However, the association of daily growth with temperature was generally weak, suggesting other factors likely have an important influence on the role settlement date plays in determining growth. Previous research has explained lower growth rates in juvenile Winter Flounder with higher temperature (Meng et al. 2000) and a greater range of temperature (Meise et al. 2003); however, the former also found a positive relationship with temperature, suggesting additional factors are likely involved. Other factors that could have additive effects on growth include higher temperatures resulting in higher feeding rates (Fonds et al. 1992) or increased prey abundance (Franz and Tanacredi 1992). Alternatively, older and larger individuals that settled early in the season may have advantages in terms of resiliency to stressful environmental conditions and predation (Johnson and Hixon 2010) or in terms of outcompeting smaller conspecifics (Shepherd and Cushing 1980). These observations highlight the trade-offs that age-0 Winter Flounder encounter, and the optimal settlement date likely changes annually depending on the temperature, food resources, size reached by a given date, and ability to survive unfavorable conditions, such as hypoxia. These results also suggest that adult spawning behaviors that result in multiple cohorts of young of year in the same year-class (Yencho 2009) may have historically enabled Long Island Winter Flounder to produce viable year-classes in the face of uncertain environmental conditions.

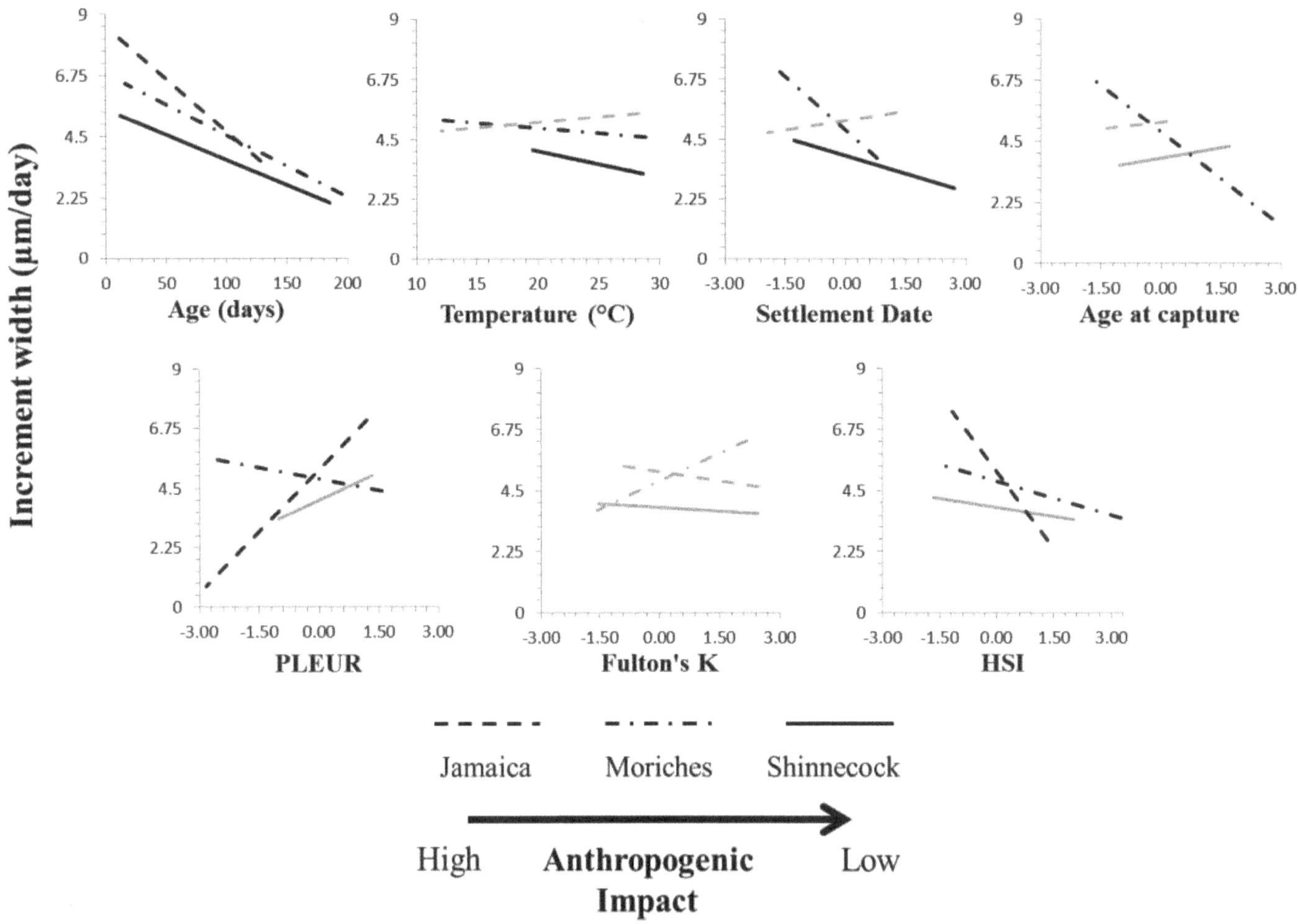

FIGURE 5. Plots of site-specific effects of level-1 and level-2 variables on the predicted otolith growth (y-axis) relationship in the selected 2010 model. Solid lines correspond to fish from Shinnecock Bay, hatched lines correspond to fish from Moriches Bay, and dashed lines correspond to fish from Jamaica Bay. Black lines denote slopes that are significantly different than 0 ($P < 0.05$), while lines in gray are not significant; PLEUR = pleurocidin.

The selected models during both years also contained five other variables: Fulton's *K*, HSI, GPDH, pleurocidin, and CYP1A. We expected that the condition indices (HSI, Fulton's *K*) would display a positive relationship with daily growth (Bolger and Connolly 1989), while the gene expression metrics (pleurocidin, CYP1A, and GPDH) would be more complex, with the potential for a mixed response depending on exposure to contaminants and potential compensatory responses, such as heightened immune response or mobilization of energy reserves. Fulton's *K* was included in the final 2010 model selected but did not have significant slopes for any site in 2010. The other condition index examined, HSI, had a significant negative relationship with daily growth for the more impacted Jamaica and Moriches bays in 2010. The HSI may be reflective of the energetic status and ration of individual fish, and it is known to be negatively affected by exposure to metals in age-0 Winter Flounder (Pereira et al. 1993).

Cytochrome P4501A mRNA expression, enzymatic activity, or protein levels are a common biomarker of exposure to organic pollutants, responding to many contaminants found in wastewater and urban runoff, such as many polycyclic aromatic hydrocarbons and a variety of planar chlorinated hydrocarbons, including some PCBs, dioxins, and furans (Stegeman and Hahn 1994; Schlenk et al. 2008). However, as reactive metabolites produced through CYP1A-mediated metabolism can have toxic consequences, and as induction has been linked mechanistically to a variety of developmental deficits, induction of CYP1A is sometimes considered a biomarker of toxic effect. The data presented here is among the first to link a physiological meaningful response (growth) to CYP1A induction. The fact that the relationship between CYP1A and growth is negative in Jamaica Bay Winer Flounder and positive in those from Moriches Bay, however, suggests that other factors may also be influencing CYP1A expression and its relationship to growth.

Glycerol 3-phosphate dehydrogenase is an enzyme involved in glycolysis whose expression has been associated with the mobilization of glycogen reserves in European Flounder *Platichthys flesus* exposed to hypoxia (Jørgensen and Mustafa 1980). The negative relationship between daily growth and GPDH observed in Moriches Bay in 2011 may therefore

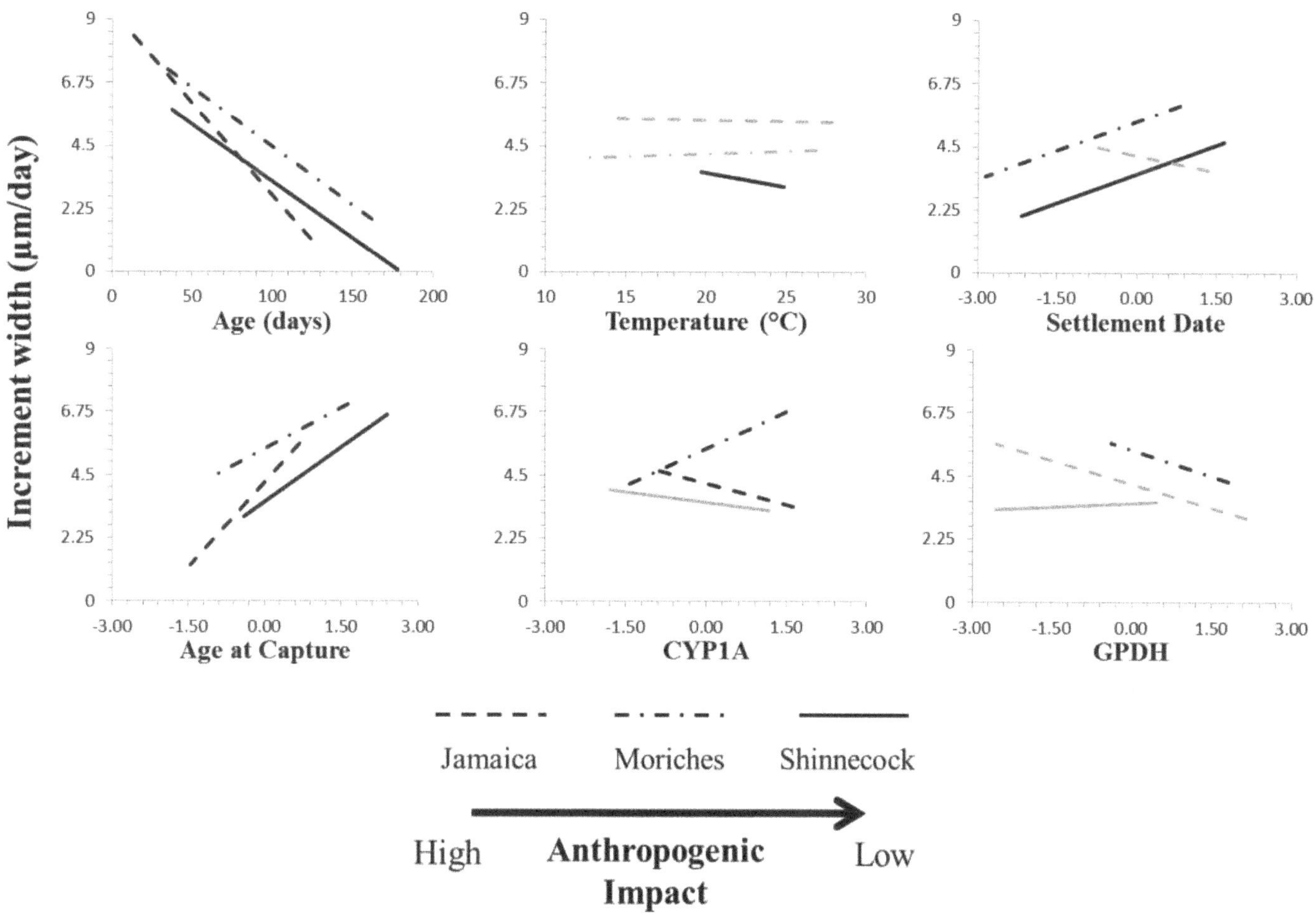

FIGURE 6. Plots of site-specific effects of level-1 and level-2 variables on the predicted otolith growth (*y*-axis) relationship in the selected 2011 model. Solid lines correspond to fish from Shinnecock Bay, hatched lines correspond to fish from Moriches Bay, and dashed lines correspond to fish from Jamaica Bay. Black lines denote slopes that are significantly different than 0 ($P < 0.05$), while lines in gray are not significant.

be the result of exposure to low dissolved oxygen in these fish, although this relationship was not significant in Jamaica Bay, another site known to experience intermittent hypoxic and hyperoxic conditions (Frisk et al. 2013). Pleurocidin is a peptide isolated from the mucous secretions of Winter Flounder that possesses antimicrobial properties (Cole et al. 1997). In 2010, pleurocidin mRNA expression was significant in the model, but the relationship was positive for Jamaica Bay and negative in Moriches Bay. The difference in the direction of this effect is consistent with adaptive (immune stimulation) and maladaptive (immune suppression) responses occurring in Jamaica and Moriches bays, respectively (Romany et al. 2015). The significant relationships between growth and pleurocidin expression in Moriches and Jamaica bays in 2010 may reflect the potential importance of immune status in controlling growth patterns in age-0 Winter Flounder, especially in more impacted sites.

Compensatory growth may explain the increased daily growth rates that were observed in fish with a reduced condition (HSI in 2010) and an increased gene expression associated with environmental stress (pleurocidin in 2010, CYP1A and GPDH in 2011). Although compensatory growth is likely an important driver of growth in wild fish, there is a paucity of field-based research on the subject (Ali et al. 2003). Laboratory experiments have confirmed that compensatory growth in age-0 Winter Flounder is possible. Bejda et al. (1992) exposed three treatments of fish to low, high, or fluctuating dissolved oxygen levels for 10–11 d, and growth rates were significantly lower in conditions of low or fluctuating oxygen. Following initial treatments, all individuals were placed in higher oxygen conditions for 5 weeks and individuals from the hypoxic treatment grew significantly faster than those from the other groups. This suggests that while prolonged exposure to hypoxic conditions can suppress growth rates, these same fish may be able to alter their consumption rates or reallocate energy reserves to increase short-term growth and compensate for adverse environmental conditions. These experiments may provide an explanation for some of the unexpected growth patterns observed in Jamaica Bay and Moriches Bay, the two most impacted sites. However, compensatory growth is not without trade-offs and it has been associated with decreased survival (Munch and Conover 2003) and various short- and long-term costs (Mangel and Stamps 2001; Mangel and Munch 2005). The populations of Winter Flounder in this study are in a depleted state, and previous research suggests they are suffering from a population bottleneck occurring during the first year of life (Yencho 2009; Frisk et al. 2013); the potential for compensatory growth may be another indicator of a highly stressed population.

Distinct responses in growth to attributes related to health (HSI) and stress (gene expression metrics) in Jamaica and Moriches bays may be due in part to population-specific responses. O'Leary et al. (2013) detected weak, but significant, fine-scale genetic differences in age-0 Winter Flounder in Long Island bays. Similarly, recent and historical studies indicate that Winter Flounder display diversity in migratory and spawning behaviors and the existence of resident individuals (Poole 1966; Sagarese and Frisk 2011; Frisk et al. 2014). Black et al. (1988) documented that PCB contamination can be passed down from mother to offspring in Winter Flounder from New England, and it may be possible for this process to elicit phenotypic and genetic changes over many generations. Over time, this may result in fish from more anthropogenically impacted sites becoming either adapted or acclimated to withstand higher contaminant loadings (Wirgin and Waldman 2004). Understanding how different local populations of Winter Flounder respond and adapt to such long-term stressors is also important for rebuilding the species in Long Island.

Hierarchical linear models provided a useful framework to evaluate complex relationships between important traits, such as growth, and a myriad of potential explanatory variables. The model structure and statistical technique provided a means to analyze data measured across various scales, including longitudinal daily growth data (Chambers and Miller 1995; McMahon and Diez 2007). This type of data structure is likely to arise in many situations when investigators are interested in evaluating growth in juvenile fish, especially if repeated measurements are taken from fish collected from multiple locations. The HLMs performed reasonably well, capturing growth in age-0 Winter Flounder. In some individuals, very high growth early in the season was not always captured by the models, especially in Shinnecock Bay in 2010. Future research exploring alternative model structures and additional explanatory variables may help better capture individuals with high growth early in the season. Our study would have benefited from more consistent and longer-term measurements of temperature, dissolved oxygen, and salinity, as these and other potentially important level-1 and level-2 factors have been observed to influence growth in age-0 Winter Flounder. A combination of controlled laboratory experiments and additional field-based research is needed to explore the temporal variation in the influence of potential drivers of growth, the timing of settlement patterns, and the relationship between somatic growth rate and otolith increment width. In addition, a better understanding of compensatory growth and its triggers is important for understanding observed growth trajectories in impacted sites.

In conclusion, daily growth patterns in age-0 Winter Flounder are complex, being related to settlement patterns, condition, and gene expression associated with contaminant exposure, immune response, and energy metabolism. The dynamic response of growth with settlement date and other level-1 and level-2 factors suggests that variability in the timing of spawning within bays may also play an important role in controlling these processes. A full life cycle research approach is needed to link spawning periodicity, larval stage duration, timing of settlement, and recruitment success in order to elucidate drivers of Winter Flounder population dynamics and recovery in Long Island bays (Frisk et al. 2014).

ACKNOWLEDGMENTS

We would like to thank Hannes Baumann for his assistance in developing a protocol for processing otoliths and using ImagePro software. We would also like to thank Christopher Gobler, Bradley Peterson, and Tom Wilson from the School of Marine and Atmospheric Sciences at Stony Brook University for sharing environmental data collected in Shinnecock and Moriches bays during the study period, which was critical for the success of this analysis. In addition, we are grateful for the comments by two anonymous reviewers, which improved the quality of this paper.

We thank the Saltonstall–Kennedy Grant Program through the National Marine Fisheries Service (award number NA10NMF4270202) for funding, and work during the summer of 2012 was supported in part by the Undergraduate Research and Creative Activities Program at Stony Brook University.

REFERENCES

Able, K., T. Grothues, J. Morson, and K. Coleman. 2014. Temporal variation in Winter Flounder recruitment at the southern margin of their range: is the decline due to increasing temperatures? ICES Journal of Marine Science 71:2186–2197.

Ali, M., A. Nicieza, and R. Wootton. 2003. Compensatory growth in fishes: a response to growth depression. Fish and Fisheries 4:147–190.

Amara, R., T. Meziane, C. Gilliers, G. Hermel, and P. Laffargue. 2007. Growth and condition indices in juvenile Sole *Solea solea* measured to assess the quality of essential fish habitat. Marine Ecology Progress Series 351:201.

Baumgarner, B. L., C. P. Riley, M. S. Sepulveda, P. B. Brown, J. L. Meyer, and J. Adamec. 2012. Increased expression of GAPDH protein is not indicative of nitrosative stress or apoptosis in liver of starved Rainbow Trout (*Oncorhynchus mykiss*). Fish Physiology and Biochemistry 38:319–327.

Bejda, A. J., B. A. Phelan, and A. L. Studholme. 1992. The effect of dissolved oxygen on the growth of young-of-the-year Winter Flounder, *Pseudopleuronectes americanus*. Environmental Biology of Fishes 34:321–321.

Black, D. E., D. K. Phelps, and R. L. Lapan. 1988. The effect of inherited contamination on egg and larval Winter Flounder, *Pseudopleuronectes americanus*. Marine Environmental Research 25:45–62.

Bolger, T., and P. Connolly. 1989. The selection of suitable indices for the measurement and analysis of fish condition. Journal of Fish Biology 34:171–182.

Bopp, R. F., H. J. Simpson, S. N. Chillrud, and D. W. Robinson. 1993. Sediment-derived chronologies of persistent contaminants in Jamaica Bay, New York. Estuaries 16:608–616.

Brothers, E. B., C. P. Mathews, and R. Lasker. 1976. Daily growth increments in otoliths from larval and adult fishes. U.S. National Marine Fisheries Service Fishery Bulletin 74:1–8.

Buckley, L., E. Caldarone, and T. L. Ong. 1999. RNA–DNA ratio and other nucleic acid-based indicators for growth and condition of marine fishes. Pages 265–277 *in* J. P. Zehr and M. Voytek, editors. Molecular ecology of aquatic communities. Kluwer Academic Publishers, Dordrecht, The Netherlands.

Bullock, T. H. 1955. Compensation for temperature in the metabolism and activity of poikilotherms. Biological Reviews 30:311–342.

Campana, S. E., and C. M. Jones. 1992. Analysis of otolith microstructure data. Canadian Special Publication of Fisheries and Aquatic Sciences 117:73–100.

Campana, S. E., and J. D. Neilson. 1982. Daily growth increments in otoliths of Starry Flounder (*Platichthys stellatus*) and the influence of some environmental variables in their production. Canadian Journal of Fisheries and Aquatic Sciences 39:937–942.

Chambers, R. C., and T. J. Miller. 1995. Evaluating fish growth by means of otolith increment analysis: special properties of individual-level longitudinal data. Pages 155–175 *in* D. H. Secor, J. M. Dean, and S. E. Campana, editors. Recent developments in fish otolith research. University of South Carolina Press, Columbia.

Cole, A. M., P. Weis, and G. Diamond. 1997. Isolation and characterization of pleurocidin, an antimicrobial peptide in the skin secretions of Winter Flounder. Journal of Biological Chemistry 272:12008–12013.

Duffy, T. A., A. E. McElroy, and D. O. Conover. 2009. Variable susceptibility and response to estrogenic chemicals in *Menidia menidia*. Marine Ecology Progress Series 380:245–254.

Fonds, M., R. Cronie, A. Vethaak, and P. Van der Puyl. 1992. Metabolism, food consumption and growth of Plaice (*Pleuronectes platessa*) and Flounder (*Platichthys flesus*) in relation to fish size and temperature. Netherlands Journal of Sea Research 29:127–143.

Franz, D. R., and J. T. Tanacredi. 1992. Secondary production of the amphipod *Ampelisca abdita Mills* and its importance in the diet of juvenile Winter Flounder (*Pleuronectes americanus*) in Jamaica Bay, New York. Estuaries 15:193–203.

Frisk, M. G., A. Jordaan, and T. J. Miller. 2014. Moving beyond the current paradigm in marine population connectivity: are adults the missing link? Fish and Fisheries 15:242–254.

Frisk, M. G., A. E. McElroy, M. D. Fast, and L. A. Hice. 2013. Restoring Long Island's Winter Flounder fishery: influence of natural and anthropogenic factors on health, fitness and recruitment success. Final Report submitted to National Marine Fisheries Service Saltonstall-Kennedy Grant Program Office, Washington, D.C.

Goldberg, R., B. Phelan, J. Pereira, S. Hagan, P. Clark, A. Bejda, A. Calabrese, A. Studholme, and K. W. Able. 2002. Variability in habitat use by young-of-the-year Winter Flounder, *Pseudopleuronectes americanus*, in three northeastern US estuaries. Estuaries 25:215–226.

Hartig, E. K., V. Gornitz, A. Kolker, F. Mushacke, and D. Fallon. 2002. Anthropogenic and climate-change impacts on salt marshes of Jamaica Bay, New York City. Wetlands 22:71–89.

Heppell, S. A., N. D. Denslow, L. C. Folmar, and C. V. Sullivan. 1995. Universal assay of vitellogenin as a biomarker for environmental estrogens. Environmental Health Perspectives 103:9.

Houde, E. D. 1987. Fish early life dynamics and recruitment variability. Pages 17–29 *in* R. D. Hoyt, editor. 10th Annual larval fish conference. American Fisheries Society, Symposium 2, Bethesda, Maryland.

Houde, E. D. 1989. Subtleties and episodes in the early life of fishes. Journal of Fish Biology 35:29–38.

Houde, E. D. 2008. Emerging from Hjort's shadow. Journal of Northwest Atlantic Fisheries Science 41:53–70.

Johnson, D., and M. Hixon. 2010. Ontogenetic and spatial variation in size-selective mortality of a marine fish. Journal of Evolutionary Biology 23:724–737.

Jørgensen, J. B., and T. Mustafa. 1980. The effect of hypoxia on carbohydrate metabolism in Flounder (*Platichthys flesus L.*)—I. Utilization of glycogen and accumulation of glycolytic end products in various tissues. Comparative Biochemistry and Physiology Part B 67:243–248.

Keller, A. A., and G. Klein-MacPhee. 2000. Impact of elevated temperature on the growth, survival, and trophic dynamics of Winter Flounder larvae: a mesocosm study. Canadian Journal of Fisheries and Aquatic Sciences 57:2382–2392.

Khan, R. 2010. Influence of sediment contaminated with untreated pulp and paper mill effluent on Winter Flounder, *Pleuronectes americanus*. Archives of Environmental Contamination and Toxicology 58:158–164.

Klein-MacPhee, G. 2002. Righteye flounders: family *Pleuronectidae*. Pages 560–587 *in* B. B. Collette and G. Klein-MacPhee, editors. Bigelow and Schroeder's fishes of the Gulf of Maine, 3rd edition. Smithsonian Institution Press, Washington, D.C.

Kreft, I., and J. De Leeuw. 1998. Introducing multilevel modeling. Sage Publications, Thousand Oaks, California.

LIRPC (Long Island Regional Planning Council). 2010. Summary of census data for the Long Island region. LIRPC, Syosset, New York.

Manderson, J., J. Pessutti, C. Meise, D. Johnson, and P. Shaheen. 2003. Winter Flounder settlement dynamics and the modification of settlement patterns by post-settlement processes in a NW Atlantic estuary. Marine Ecology Progress Series 253:253–267.

Manderson, J., B. Phelan, C. Meise, L. Stehlik, A. Bejda, J. Pessutti, L. Arlen, A. Draxler, and A. Stoner. 2002. Spatial dynamics of habitat suitability for the growth of newly settled Winter Flounder *Pseudopleuronectes americanus* in an estuarine nursery. Marine Ecology Progress Series 228:227–239.

Manderson, J. P. 2008. The spatial scale of phase synchrony in Winter Flounder (*Pseudopleuronectes americanus*) production increased among southern New England nurseries in the 1990s. Canadian Journal of Fisheries and Aquatic Sciences 65:340–351.

Mangel, M., and S. B. Munch. 2005. A life-history perspective on short-and long-term consequences of compensatory growth. American Naturalist [online serial] 166:E155–E176.

Mangel, M., and J. Stamps. 2001. Trade-offs between growth and mortality and the maintenance of individual variation in growth. Evolutionary Ecology Research 3:583–593.

McArdle, M. E., A. E. Mcelroy, and A. A. Elskus. 2004. Enzymatic and estrogenic responses in fish exposed to organic pollutants in the New York–New Jersey (USA) Harbor Complex. Environmental Toxicology and Chemistry 23:953–959.

McElroy, A. E., L. A. Hice, M. G. Frisk, S. L. Purcell, N. C. Phillips, and M. D. Fast. 2015. Spatial patterns in markers of contaminant exposure, glucose and glycogen metabolism, and immunological response in juvenile Winter Flounder (*Pseudopleuronectes americanus*). Comparative Biochemistry and Physiology Part D: Genomics and Proteomics 14:53–65.

McMahon, S. M., and J. M. Diez. 2007. Scales of association: hierarchical linear models and the measurement of ecological systems. Ecology Letters 10:437–452.

Media Cybernetics. 2006. Image-Pro Plus version 6.0 for windows. Media Cybernetics, Silver Spring, Maryland.

Meise, C., D. Johnson, L. Stehlik, J. Manderson, and P. Shaheen. 2003. Growth rates of juvenile Winter Flounder under varying environmental conditions. Transactions of the American Fisheries Society 132:335–345.

Meng, L., C. Gray, B. Taplin, and E. Kupcha. 2000. Using Winter Flounder growth rates to assess habitat quality in Rhode Island's coastal lagoons. Marine Ecology Progress Series 201:287–299.

Meng, L., J. C. Powell, and B. Taplin. 2001. Using Winter Flounder growth rates to assess habitat quality across an anthropogenic gradient in Narragansett Bay, Rhode Island. Estuaries 24:576–584.

Munch, S. B., and D. O. Conover. 2003. Rapid growth results in increased susceptibility to predation in *Menidia menidia*. Evolution 57:2119–2127.

Nixon, S. W. 1995. Coastal marine eutrophication: a definition, social causes, and future concerns. Ophelia 41:199–219.

NEFSC (Northeast Fisheries Science Center). 2011. Southern New England/Mid-Atlantic Winter Flounder stock assessment. NEFSC, Woods Hole, Massachusetts.

Nuttall, M., A. Jordaan, R. Cerrato, and M. Frisk. 2011. Identifying 120 years of decline in ecosystem structure and maturity of Great South Bay, New York using the Ecopath modelling approach. Ecological Modelling 222:3335–3345.

NYCDEP (New York City Department of Environmental Protection). 2010. New York Harbor water quality report. NYCDEP, New York.

NYCDEP (New York City Department of Environmental Protection). 2011. New York Harbor water quality report. NYCDEP, New York.

O'Leary, S. J., L. A. Hice, K. A. Feldheim, M. G. Frisk, A. E. McElroy, M. D. Fast, and D. D. Chapman. 2013. Severe inbreeding and small effective number of breeders in a formerly abundant marine fish. PLoS (Public Library of Science) ONE [online serial] 8:e66126.

Pereira, J., R. Mercaldo-Allen, C. Kuropat, D. Luedke, and G. Sennefelder. 1993. Effect of cadmium accumulation on serum vitellogenin levels and hepatosomatic and gonadosomatic indices of Winter Flounder (*Pleuronectes americanus*). Archives of Environmental Contamination and Toxicology 24:427–431.

Phelan, B., R. Goldberg, A. Bejda, J. Pereira, S. Hagan, P. Clark, A. Studholme, A. Calabrese, and K. Able. 2000. Estuarine and habitat-related differences in growth rates of young-of-the-year Winter Flounder (*Pseudopleuronectes americanus*) and Tautog (*Tautoga onitis*) in three northeastern US estuaries. Journal of Experimental Marine Biology and Ecology 247:1–28.

Polgar, T., J. Summers, T. Cummins, K. Rose, and D. Heimbuch. 1985. Investigation of relationships among pollutant loadings and fish stock levels in northeastern estuaries. Estuaries 8:125–135.

Poole, J. C. 1966. Growth and age of Winter Flounder in four bays of Long Island. New York Fish and Game Journal 13:206–220.

Pruzanski, W., and P. Vadas. 1991. Phospholipase A2—a mediator between proximal and distal effectors of inflammation. Immunology Today 12:143–146.

Raudenbush, S. W., A. S. Bryk, Y. F. Cheong, R. T. Congdon, and M. du Toit. 2011. HLM 7: hierarchical linear and nonlinear modeling. Scientific Software International, Lincolnwood, Illinois. Available: http://www.ssicentral.com/hlm/index.html. (April 2015).

Raudenbush, S. W., and A. S. Bryk. 2002. Hierarchical linear models: applications and data analysis methods. Sage Publications, Thousand Oaks, California.

Romany, J. S., A. E. McElroy, and M. D. Fast. 2015. Measures of immune system status in young-of-the-year Winter Flounder *Pseudopleuronectes americanus*. Journal of Fish Biology 86:148–161.

Sagarese, S. R., and M. G. Frisk. 2011. Movement patterns and residence of adult Winter Flounder within a Long Island estuary. Marine and Coastal Fisheries: Dynamics, Management, and Ecosystem Science [online serial] 3:295–306.

Schaalje, G. B., J. L. Shaw, and M. C. Belk. 2002. Using nonlinear hierarchical models for analyzing annulus-based size-at-age data. Canadian Journal of Fisheries and Aquatic Sciences 59:1524–1532.

Schlenk, D., M. Celander, E. P. Gallagher, S. George, M. James, S. W. Kullman, P. van den Hurk, and K. Willett. 2008. Biotransformation in fishes. Toxicology of Fishes:153–234.

Shepherd, J., and D. Cushing. 1980. A mechanism for density-dependent survival of larval fish as the basis of a stock-recruitment relationship. ICES Journal of Marine Science 39:160–167.

Socrates, J., and G. Colvin. 2006. A study of the Striped Bass in the marine district of New York State. New York State Department of Environmental Conservation, East Setauket.

Sogard, S. M. 1991. Interpretation of otolith microstructure in juvenile Winter Flounder (*Pseudopleuronectes americanus*): ontogenetic development, daily increment validation, and somatic growth relationships. Canadian Journal of Fisheries and Aquatic Sciences 48:1862–1871.

Sogard, S. M. 1997. Size-selective mortality in the juvenile stage of teleost fishes: a review. Bulletin of Marine Science 60:1129–1157.

Sogard, S. M., and K. W. Able. 1992. Growth variation of newly settled Winter Flounder (*Pseudopleuronectes americanus*) in New Jersey estuaries as determined by otolith microstructure. Netherlands Journal of Sea Research 29:163–172.

Sogard, S. M., K. W. Able, and S. M. Hagan. 2001. Long-term assessment of settlement and growth of juvenile Winter Flounder (*Pseudopleuronectes americanus*) in New Jersey estuaries. Journal of Sea Research 45:189–204.

Stegeman, J. J., and M. E. Hahn. 1994. Biochemistry and molecular biology of monooxygenases: current perspectives on forms, functions, and regulation of cytochrome P450 in aquatic species. Aquatic Toxicology: Molecular, Biochemical, and Cellular Perspectives 87:206.

USEPA (U.S. Environmental Protection Agency). 2006. National Estuary Program coastal condition report. EPA, EPA842/13-06/001, Washington, D.C.

van der Veer, H. W., R. Berghahn, J. M. Miller, and A. D. Rijnsdorp. 2000. Recruitment in flatfish, with special emphasis on North Atlantic species: progress made by the flatfish symposia. ICES Journal of Marine Science 57:202–215.

Wagenmakers, E. J., and S. Farrell. 2004. AIC model selection using Akaike weights. Psychonomic Bulletin and Review 11:192–196.

Wirgin, I., and J. R. Waldman. 2004. Resistance to contaminants in North American fish populations. Mutation Research/Fundamental and Molecular Mechanisms of Mutagenesis 552:73–100.

Yencho, M. 2009. Abundance, Mortality, Age and Growth of Young of the Year Winter Flounder (*Pseudopleuronectes americanus*) in Two Locations on Long Island. Master's thesis. Stony Brook University, Stony Brook, New York.

Appendix: Parameter Estimates, Model Performance, and Supplementary Information

TABLE A.1. Summary of the effect of level-1 and level-2 variables in the selected models for 2010 and 2011. For both years, the gene expression metrics (pleurocidin, CYP1A, and GPDH) were $\log_e$ transformed and all level-2 variables were standardized to their grand means. If the variable was included in the selected model for a particular year, the parameter estimates and standard errors are displayed. The *P*-value is based on the ratio of the parameter estimate to its standard error (*t*-ratio) to test whether it was significantly different than 0. The *P*-values for contrasts by site are also included, which demonstrate whether the effect of the variable was distinct in fish from one site relative to those from another. For the site contrasts, the abbreviations are as follows: S = Shinnecock Bay, M = Moriches Bay, and J = Jamaica Bay.

	2010 selected model							2011 selected model						
	Level 1		Level 2					Level 1		Level 2				
Coefficient	Age	Temperature	Age at capture	Settlement date	Fulton's *K*	HSI	Pleuro-cidin	Age	Temperature	Age at capture	Settlement date	CYP1A	GPDH
Type	Fixed	Fixed	Random	Fixed	Random	Fixed	Fixed	Fixed	Fixed	Fixed	Random	Fixed	Fixed
Approximate df	3,776	3,776	48	3,776	48	3,776	3,776	2,094	2,094	2,094	39	2,094	2,094
Shinnecock slope	−0.019	−0.101	0.254	−0.005	−0.091	0.006	−0.010	−0.041	−0.108	0.017	0.697	−0.012	0.001
Shinnecock SE	0.002	0.015	0.474	0.002	0.388	0.004	0.013	0.005	0.019	0.002	0.322	0.017	0.002
P-value	<0.001	<0.001	0.595	0.003	0.815	0.098	0.431	<0.001	<0.001	<0.001	0.037	0.495	0.514
Moriches slope	−0.023	−0.040	−1.148	−0.015	0.684	−0.003	−0.019	−0.042	−0.009	0.013	0.694	0.044	−0.008
Moriches SE	0.002	0.010	0.335	0.001	0.382	0.001	0.006	0.005	0.017	0.003	0.282	0.010	0.004
P-value	<0.001	<0.001	0.001	<0.001	0.080	<0.001	0.002	<0.001	0.589	<0.001	0.018	<0.001	0.025
Jamaica slope	−0.039	0.040	0.160	0.003	−0.225	0.018	−0.087	−0.065	0.019	0.026	−0.388	−0.025	−0.007
Jamaica SE	0.006	0.026	0.794	0.004	0.443	0.003	0.011	0.008	0.032	0.006	0.553	0.012	0.004
P-value	<0.001	0.128	0.842	0.459	0.613	<0.001	<0.001	<0.001	0.552	<0.001	0.487	0.038	0.072
S–M contrast	0.004	−0.061	1.402	0.010	−0.775	0.009	0.009	0.001	−0.099	0.004	0.003	−0.055	0.009
P-value	0.098	0.001	0.020	<0.001	0.161	0.013	0.527	0.848	<0.001	0.324	0.994	0.006	0.021
J–M contrast	−0.016	0.080	1.308	0.018	−0.909	0.021	−0.068	−0.023	0.028	0.014	−1.082	−0.069	0.001
P-value	0.015	0.004	0.136	<0.001	0.127	<0.001	<0.001	0.015	0.436	0.059	0.089	<0.001	0.901
J–S contrast	−0.020	0.141	−0.094	0.008	−0.134	0.012	−0.077	−0.025	0.127	0.009	−1.085	−0.013	−0.008
P-value	0.002	<0.001	0.919	0.052	0.821	0.019	<0.001	0.010	0.001	0.165	0.098	0.532	0.056

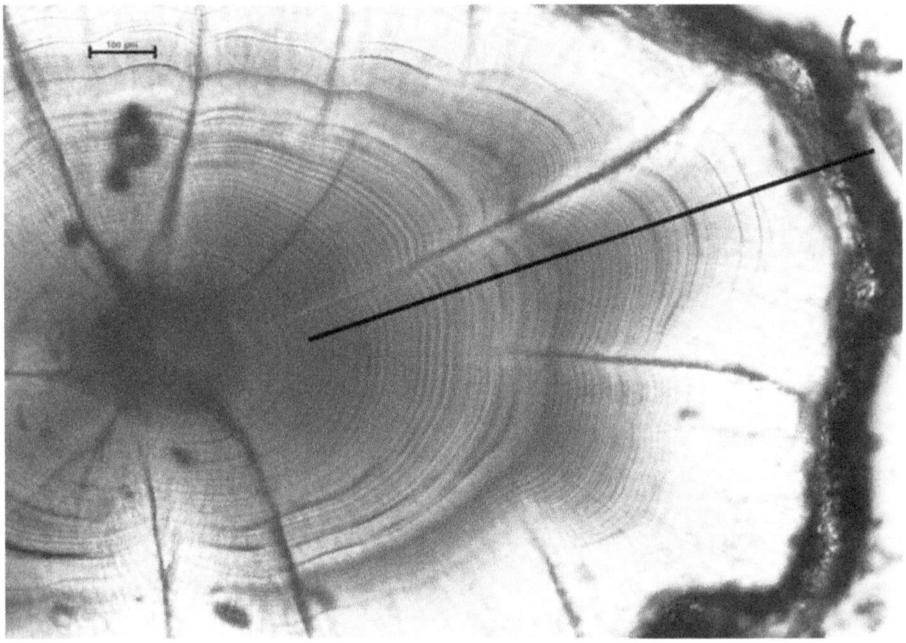

FIGURE A.1. Sample image of the left sagittal otolith of an age-0 Winter Flounder, with an axis (black line) placed along the postrostrum. Increments were counted and measured along this axis for each individual, starting roughly 10 increments away from the secondary growth center. The scale bar in the top left corner is for 100 μm. Note that the increments toward the edge of the otolith are less visible, but these were identified by zooming in and manipulating the brightness and contrast of the image.

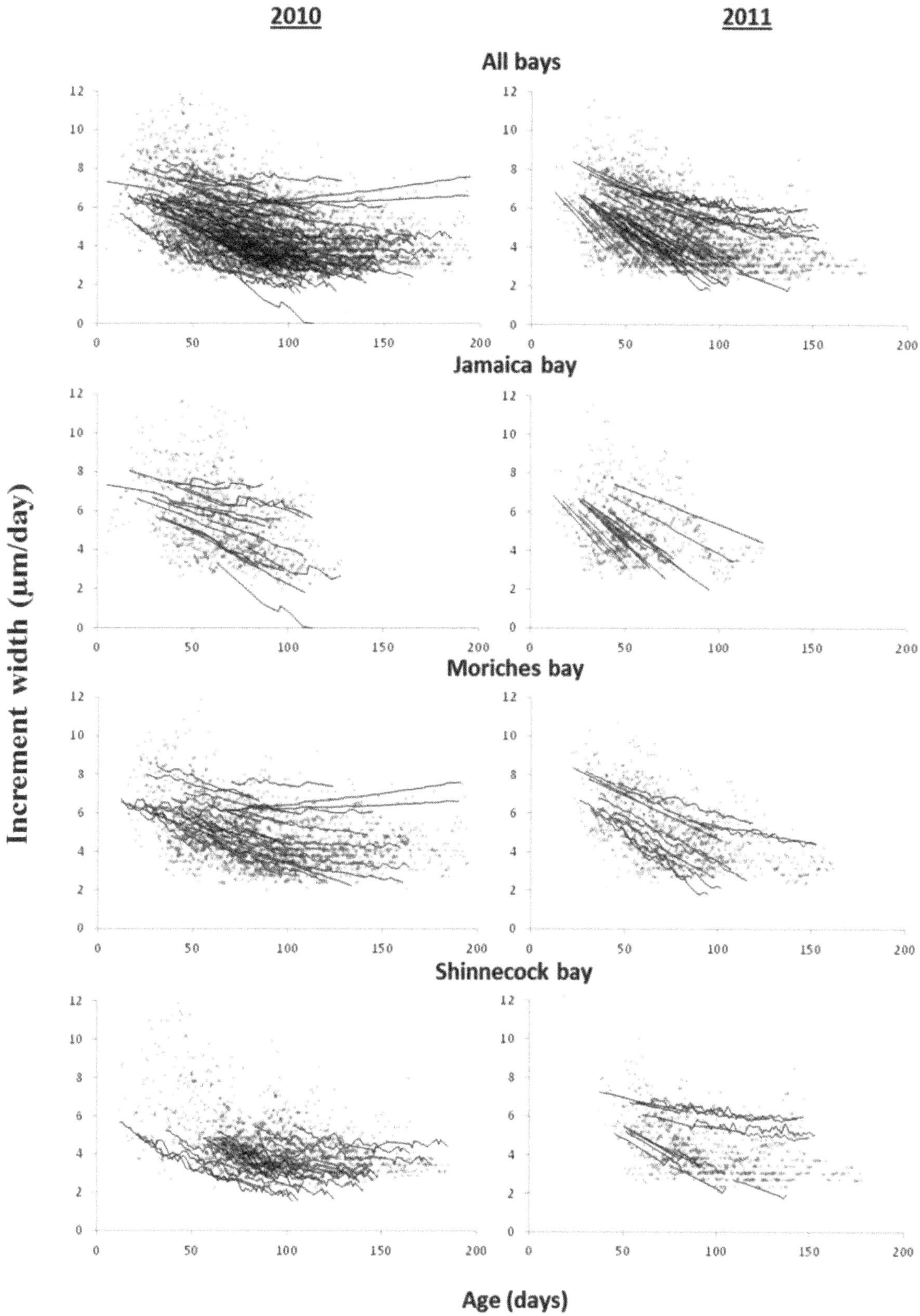

FIGURE A.2. Plots comparing the observed daily growth-at-age data (open circles) and predicted growth-at-age trajectories (black lines) in each bay during 2010 and 2011.

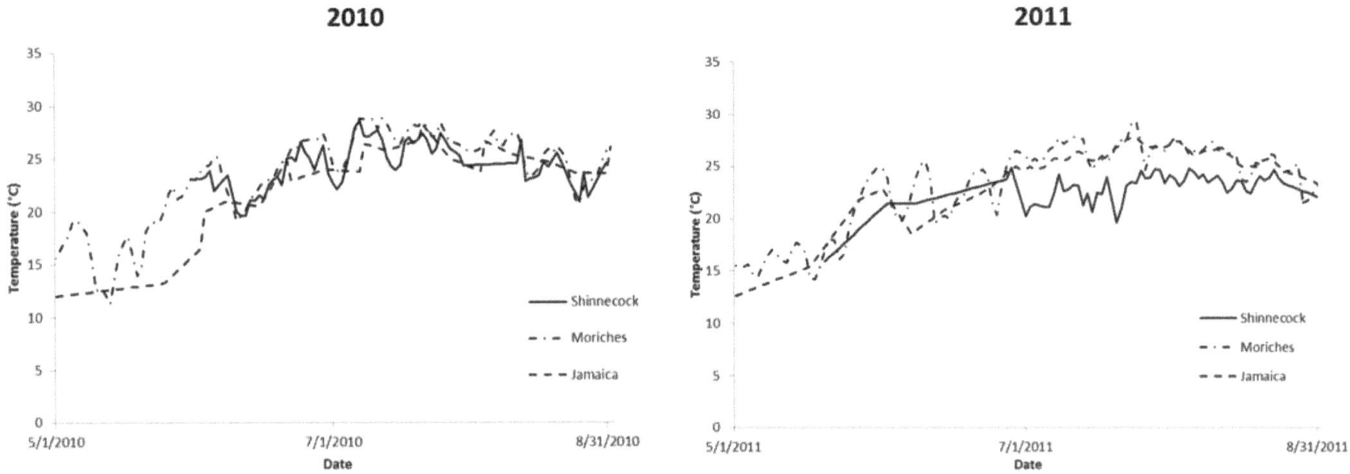

FIGURE A.3. Average daily temperature data for Shinnecock Bay (solid lines), Moriches Bay (hatched lines), and Jamaica Bay (dashed lines) for both 2010 and 2011. The temperature data in Jamaica Bay had gaps that were filled in using linear interpolation, and the Shinnecock Bay data only spans from June to August during both years.

7

Estuarine Residency and Migration of Southern Flounder Inferred from Conventional Tag Returns at Multiple Spatial Scales

J. Kevin Craig*

National Marine Fisheries Service, Southeast Fisheries Science Center, Beaufort Laboratory, Beaufort, North Carolina 28516, USA

William E. Smith[1]

North Carolina Division of Marine Fisheries, 3441 Arendell Street, Morehead City, North Carolina 28557, USA

Frederick S. Scharf

Department of Biology and Marine Biology, University of North Carolina Wilmington, 601 South College Road, Wilmington, North Carolina 28403, USA

James P. Monaghan[2]

North Carolina Division of Marine Fisheries, 3441 Arendell Street, Morehead City, North Carolina 28557, USA

Abstract

An improved understanding of the spatial structure and movements of harvested populations can promote more efficient management of marine resources. Conventional tagging is a valuable approach to study the movements of marine fishes due to its relatively low expense and the typically broad spatial extent over which movements can be characterized. We present the findings of multiple tag return studies initiated in the estuaries of North Carolina during the past two decades to better understand habitat residency and migration patterns of Southern Flounder *Paralichthys lethostigma*, an economically important marine flatfish in the southeastern USA. Tag return data indicated large-scale (>50 km) movements of relatively large fish in the fall, which were presumably associated with offshore winter spawning migrations. Nearly all Southern Flounder that demonstrated large-scale movement were recovered to the south of the system in which they were tagged, suggesting that the spawning activity of fish using North Carolina estuaries may be concentrated mostly off the southeastern U.S. continental shelf. Tag returns from within multiple estuarine systems during the spring and summer were in close proximity to release sites (typically < 1 km), suggesting limited movement during estuarine residency. Recaptures in the spring of fish tagged the previous summer or fall were also in close proximity to release sites, in some cases within the same estuarine creek, indicating limited movement of fish overwintering in the estuary as well. Our findings reveal saltatory movement dynamics of Southern Flounder characterized by limited movement during estuarine residency and large-scale movements in the fall associated with spawning migrations. Our synthesis of several tag return studies across multiple spatial scales should contribute to a better alignment of Southern Flounder management with their spatial dynamics.

Subject editor: Patrick Sullivan, Cornell University, Ithaca, New York

*Corresponding author: kevin.craig@noaa.gov

[1]Present address: U.S. Fish and Wildlife Service, Bay-Delta Fish and Wildlife Office, 650 Capitol Mall, Sacramento, California 95814, USA.

[2]Deceased.

There is increasing recognition that the stock definition of harvested marine resources is more complex than the traditional view of geographically distinct units with homogenous vital rates that are reproductively isolated from adjacent populations (Stephenson 1999; Quinn 2003; Cadrin and Secor 2009; Hamilton et al. 2011). Many fish populations exhibit complex spatial patterns and movements that generate spatial heterogeneity in demography, population biomass, and realized rates of exploitation (Goethel et al. 2011; Hamilton et al. 2011; Berger et al. 2012). Despite growing recognition that the movements of fish populations can have important implications for understanding harvested ecosystems, spatial dynamics have been largely ignored in the assessment and management of marine resources (Secor 2005).

In many cases, the stock boundaries of harvested populations are based on jurisdictional or management convenience rather than on the life history and ecology of the species under consideration (Cope and Punt 2011; Berger et al. 2012). For example, mismatches between the spatial structure of marine populations and the stock definitions used for management were identified for about one third of harvested stocks in the northeastern Atlantic Ocean (Stephenson 2002). As a result, many stock assessment models violate the basic assumption that emigration and immigration rates are negligible, potentially confounding these demographic rates with other model parameters. Failure to match the spatial scale of fisheries stock assessments to ecologically relevant processes that influence population dynamics can hinder the achievement of fishery harvest and conservation objectives (Fay et al. 2011; Hamilton et al. 2011; Taylor et al. 2011; Ying et al. 2011).

Adequately addressing the spatial aspects of harvested populations is limited more by the lack of reliable field observations on movement dynamics than by the availability of appropriate modeling tools. Conventional tagging studies are a useful approach for understanding movement dynamics because they are relatively inexpensive, can cover a large spatial extent, and generate direct information on individual fish movements (as opposed to indirect information from various molecular and chemical markers). Such studies are particularly useful for defining the appropriate geographic scale at which to monitor, assess, and manage marine resources (e.g., Bacheler et al. 2009; Cadrin and Secor 2009). In addition, recent advances in stock assessment modeling allow tagging data to be incorporated directly into population dynamic models to simultaneously estimate movement, exploitation rate, and the spatial distribution of population biomass (McGarvey et al. 2010; Goethel et al. 2011; Taylor et al. 2011; Hulson et al. 2013).

Southern Flounder *Paralichthys lethostigma* is a harvested marine flatfish that is distributed along the U.S. southern Atlantic and Gulf of Mexico coasts from southern Virginia to the Yucatan peninsula in northern Mexico. Adult Southern Flounder spawn in offshore waters from November to March (Stokes 1977; Miller et al. 1984; Burke et al. 1991), with pelagic larvae recruiting to estuarine habitats at 1–2 months of age. After settlement, Southern Flounder remain in estuaries for the first 2–3 years of life before moving into ocean waters (Powell and Schwartz 1977; Stokes 1977; Wenner et al. 1990). Maturation is typically attained by age 2 (Midway and Scharf 2012), with immature individuals thought to overwinter in the estuary while mature individuals emigrate from estuaries to offshore spawning grounds in the fall. After spawning, mature fish are believed to return to nearshore or estuarine waters in late spring. Southern Flounder support valuable commercial and recreational fisheries, with landings concentrated primarily in estuarine waters. In the southeastern USA, the majority of commercial landings are harvested in North Carolina (>45.4 million kg landed since 1972, ~1.13 million kg annually; http://portal.ncdenr.org/web/mf/statistics/comstat). The fisheries are managed independently by individual states, and North Carolina is presently the only state in the region with a comprehensive management plan for Southern Flounder. A 2009 assessment of the North Carolina Southern Flounder fishery concluded that the stock was overfished (low spawning biomass) and that overfishing (high fishing mortality) was occurring. An important assumption of this stock assessment is that there is no net migration of Southern Flounder outside of North Carolina state waters.

Studies of the movements and habitat use of Southern Flounder have focused mostly on recently settled and young-of-the-year (age-0) fish in estuarine habitats (Powell and Schwartz 1977; Burke et al. 1991; Burke 1995; Glass et al. 2008; Nañez-James et al. 2009; Taylor et al. 2010; Froeschke et al. 2013a, 2013b; Furey and Rooker 2013). Collectively, these studies suggest some level of habitat specificity, with settlement patterns potentially related to the physical transport processes affecting larvae (e.g., Taylor et al. 2010), and habitat selection of age-0 fish driven by several abiotic and biotic factors (e.g., Burke et al. 1991; Walsh et al. 1999; Froeschke et al. 2013a; Furey and Rooker 2013). Estuarine movements of larger age-0 and subadult fish appear to be fairly localized, with fish moving less than 10 km over periods of at least 1 month (Furey et al. 2013). Recent studies using otolith microchemistry suggest more extensive use of low-salinity habitats than previously thought (Lowe et al. 2011; Farmer et al. 2013; Nims and Walther 2014). Despite indications of limited movement, there is little evidence for discrete populations within the Gulf of Mexico or Atlantic basins, though molecular and otolith morphometric studies indicate high levels of genetic divergence between southern Atlantic Ocean and Gulf of Mexico populations (Blandon et al. 2001; Anderson and Karel 2012; Anderson et al. 2012; Midway et al. 2014; Wang et al. 2015). The movements of larger, older (>age-1) Southern Flounder, as well as the migratory patterns associated with maturation, are poorly understood.

There are few published studies of Southern Flounder movement and no quantitative analysis of large-scale movements associated with spawning. Our primary objective was to quantify the spatial scale of Southern Flounder movements in order to better understand the stock structure of this species and its relevance to the management unit assumed for this stock. Our secondary objective was to better understand the localized movements of Southern Flounder during the period of estuarine residency. We investigated the movements of Southern Flounder using three conventional tagging data sets collected from multiple estuarine ecosystems in North Carolina. First, we quantified the directionality of Southern Flounder movements, the total distance moved, and the occurrence of large-scale movements (>50 km) in relation to fish size, season, year, and estuary based on tag return data for fish tagged in the Neuse and New River estuaries during 2005 and 2006. We then integrated these results with similar analyses of prior tagging studies conducted in multiple North Carolina estuarine systems during the 1980s and 1990s and also with data from a small-scale study conducted in a single marsh creek from 2003 to 2005. We interpret our results in terms of their relevance to the spatial dynamics of Southern Flounder and other similar flatfish species, as well as to recent management decisions based on the most current assessment of the stock supporting the fishery in North Carolina waters.

METHODS

Study Systems

Paired tagging studies using identical methodologies were conducted in the Neuse and New River estuaries in 2005 and 2006 (Figure 1). The Neuse River estuary is a large (~75 km in length) riverine estuary located in the central portion of the North Carolina coast that drains into the southwestern corner of Pamlico Sound. The New River estuary is located in a separate drainage basin along the southern portion of the coast and

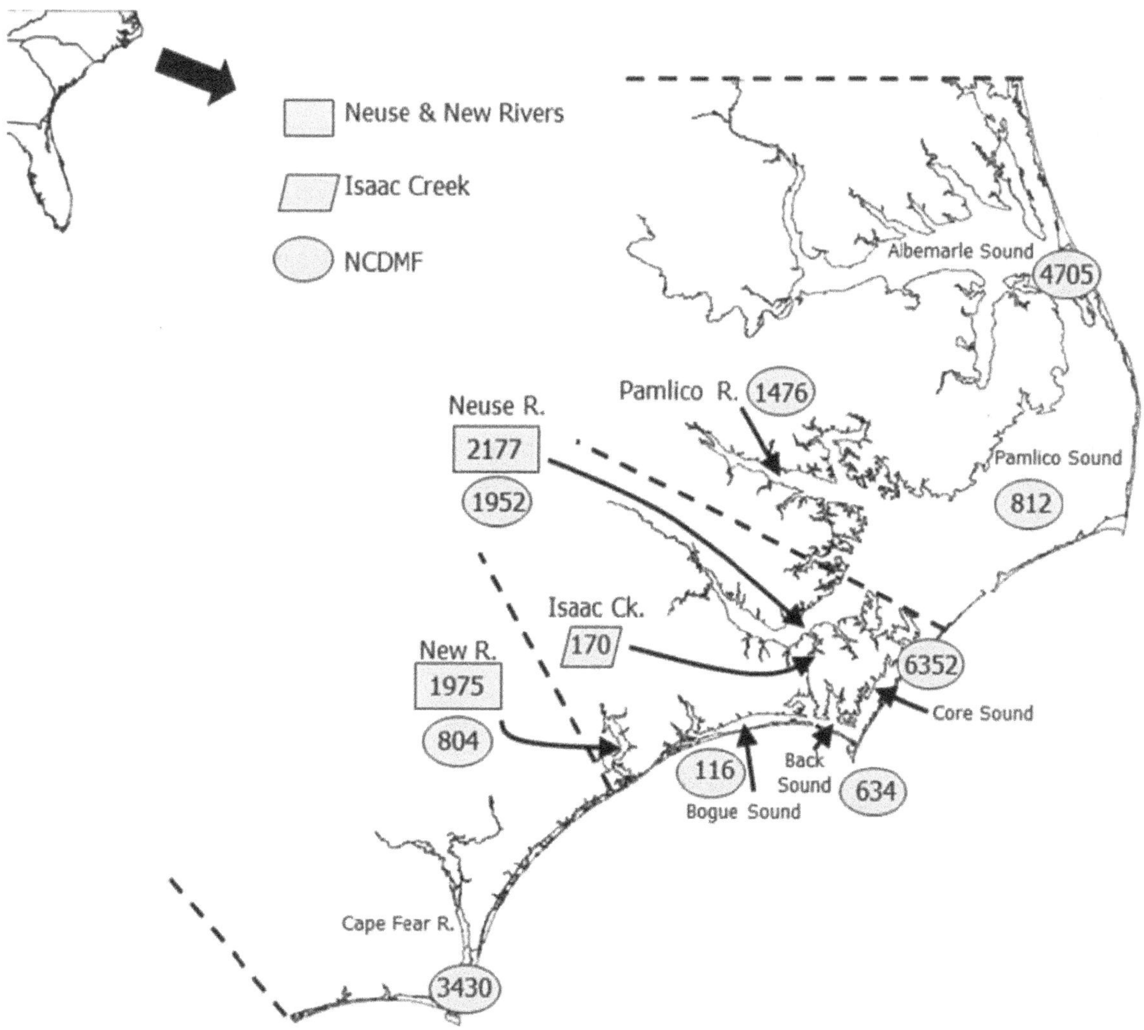

FIGURE 1. Release areas for Southern Flounder tagged in North Carolina waters from three tagging data sets. The numbers in the shaded shapes reflect the total number of fish tagged and released in the Neuse and New rivers (2005–2006), by the North Carolina Division of Marine Fisheries (NCDMF; 1980–1982, 1988–1995), and in Isaac Creek (2003–2005). The dashed lines delineate northern, central, and southern North Carolina waters.

extends over 25 km before emptying into the coastal ocean. Both estuaries range from oligohaline to polyhaline along their length, have slow flushing and high nutrient retention rates, and are considered moderately to severely eutrophic (Mallin et al. 2000). Earlier tagging of Southern Flounder in the 1980s and 1990s was conducted in inshore estuarine waters from Albemarle Sound to the Cape Fear River estuary as part of a long-term tagging program by the North Carolina Division of Marine Fisheries (NCDMF) (Figure 1). These data included riverine estuaries as well as fish tagged in the larger bays and sounds throughout the state. A small-scale tagging study was conducted in Isaac Creek, a small (2.5 km in length) marsh creek adjacent to the Neuse River estuary, from 2003 to 2005.

Tagging Methodology

Neuse and New River study.—Tagging was conducted during the 2005 and 2006 fishing seasons (March to November) in the Neuse River estuary and the New River estuary in collaboration with commercial gill-net fishermen (Figure 1; Table 1). Over the 2-year period, approximately 2,000 Southern Flounder were tagged in each system. Fish were retrieved from overnight sets of 14-cm stretched-mesh gill nets, held in insulated coolers for less than 15 min, tagged with an external orange cinch-up tag (Floy Tag, Seattle), measured for total length (TL), and released within 200 m of the capture location. Tags were inserted into the dorsal region of the caudal peduncle with a stainless steel canula and cinched to allow a remaining space of about 10–15 mm for growth. Each tag was labeled with a cash reward (US$5 or $50), a unique identification number, and contact information. Some undersized fish (≤356 mm) were tagged in the Neuse River, while all tagged fish in the New River were of harvestable size. Over the 2-year period, only 20 Southern Flounder (<1%) were recaptured more than once (all in the Neuse River). These fish were measured and re-released. Only the first recapture was retained in the statistical analysis.

Tagging study by NCDMF.—Southern Flounder were tagged as part of a long-term tagging program for paralichthid flounders conducted by the NCDMF from 1980 to 1982 and from 1988 to 1995 (Figure 1; Table 1). Fish were captured primarily in commercial pound nets and trawls from January to December, with 87% of fish captured between April and November. Each fish was held in a holding tank onboard the vessel, tagged with a laminated internal anchor tag inserted into the abdominal cavity, and then returned to the water after a 15–20-min recovery period. Tags were labeled with a $2 reward prior to June 1990 and a $5 reward or cap thereafter. From 1980 to 1982, Southern Flounder greater than 200 mm were tagged, while from 1988 to 1995 all fish greater than 160 mm were tagged. As a result, many more small fish were tagged in the NCDMF study than in the Neuse and New River study (85% ≥ 356 mm; Table 1). Over the 11-year period, 20,281 fish were tagged, with most of the tagging conducted in southern Albemarle Sound, the Pamlico River estuary, Core Sound, the Neuse River estuary, and the Cape Fear River estuary (Figure 1).

Isaac Creek study.—From March 2003 to May 2005, 170 Southern Flounder were tagged in Isaac Creek (Figure 1; Table 1). These fish were also captured from overnight sets of gill nets as part of a larger study of the piscivore community in shallow estuarine creeks (Kirby-Smith et al. 2003; J. K. Craig, unpublished data). Gill nets were set weekly to biweekly from March 2003 to April 2004 and monthly thereafter. Captured fish were tagged between the dorsal pterygiophores with 63.5-mm T-bar anchor tags (Floy Tag). Tags were printed with a unique identifying number and contact information. Initially, tags were not labeled with a reward, but a $5 reward was added to the tags midway through the study.

Data Analysis

We conducted separate but parallel analyses of the three data sets (Neuse and New rivers, NCDMF, Isaac Creek) rather than a single integrated analysis due to inherent differences in the data and the types of analyses that could be conducted. In some cases, response variables were defined differently across data sets based on the spatial scale and the resolution of the

TABLE 1. Southern Flounder tagged in North Carolina waters from three tag–recapture studies (Neuse and New rivers, 2005–2006; NCDMF, 1980–1982 and 1988–1995; and Isaac Creek, 2003–2005). Mean total length (TL) is given in millimeters, with the standard deviation given in parentheses. Recap is the percentage of tagged fish that were recaptured, with the number of recaptures given in parentheses. Days at large is given as the median, with the range given in parentheses. Recap location is the percentage of recaptured Southern Flounder that were recaptured within the same system where they were tagged.

Study	Year	Number tagged	Tagging period	Mean TL	Recap	Days at large	Recap location (%)
Neuse River	2005	1,018	Apr–Aug	367 (30.3)	25.0 (255)	7 (1–463)	93.7
	2006	1,159	Apr–Oct	372 (41.6)	24.4 (283)	21 (1–444)	88.0
New River	2005	1,011	Jun–Nov	387 (34.3)	24.3 (246)	19 (1–958)	89.8
	2006	964	May–Oct	390 (29.0)	52.4 (505)	26 (1–432)	96.1
NCDMF	1980–1982, 1988–1995	20,281	Jan–Dec	291 (66.6)	5.7 (1,156)	43 (1–2,429)	78.9
Isaac Creek	2003–2005	170	Jan–Dec	345 (69.8)	8.8 (15)	98.5 (12–206)	80.0

tagging information, and not all predictor variables were available for all data sets. The NCDMF data were unbalanced with respect to season, system, and year and required some degree of pooling across one or more predictors, while the Neuse and New River data were relatively balanced with respect to these factors. Sample sizes from the Isaac Creek study were small (Table 1) and not sufficient to statistically detect the effects of multiple predictor variables; therefore, only a descriptive analysis of these data are presented. Our analytical approach focused on the Neuse and New River study and then used the two other data sets to corroborate our main results and provide additional insights beyond those available in the Neuse and New River data.

Response variables.—Tag returns from the Neuse and New rivers, the NCDMF tagging, and Isaac Creek were divided into two categories and analyzed separately for the three data sets. The first category included fish that were recovered prior to December 1 in the same calendar year they were tagged. These fish represented 86% of all tag returns and are referred to as the "prewinter analysis." The second category included fish that were recovered after having the opportunity to join the winter offshore spawning migration (after December 1 of the calendar year of tagging). Emigration of mature adults from estuaries to the coastal ocean is thought to be completed by December 1 in most years (L. Hollensead and F. S. Scharf, University of North Carolina Wilmington, unpublished data). These fish represented 14% of all tag returns and are referred to as the "overwinter analysis."

Three response variables were developed that reflected different aspects of the movement dynamics of Southern Flounder: (1) the total distance moved, (2) the direction of movement, and (3) a binomial variable indicating whether a fish moved more than 50 km (overwinter analysis only). The total distance moved (km) was calculated in ArcGIS (ESRI 2011) as the shortest distance between tagging and recovery locations that did not cross land. The direction of movement was defined differently depending on the spatial scale and resolution of particular data sets. For the prewinter analysis of Neuse and New River data, the direction of movement was categorized as upstream, downstream, or stationary within the study system (i.e., Neuse or New rivers) because movement out of the river of tagging was generally not observed prior to winter. Stationary was defined as a tag recovery location that was < 1 km from the tagging location. Movement upstream or downstream could not be determined for most fish in the prewinter analysis of the NCDMF data because tagging and recovery locations were defined only broadly and tagging was conducted mainly in the larger bays and sounds rather than in particular river systems. Based on initial tag returns, Monaghan (1992) hypothesized that Southern Flounder begin to move south within the Albemarle–Pamlico estuarine system prior to moving through southern inlets to the coastal ocean. Therefore, the direction of movement was categorized as north, south, or stationary for the prewinter analysis of the NCDMF data. Stationary was defined as a tag recovery occurring within the system of tagging (Figure 1). For the overwinter analysis of both data sets, the direction of movement was categorized as north or south of the study system based on whether the recapture location was north or south of a latitudinal line bisecting the tagging location. Fish that moved >50 km were determined from the total distance moved. All overwinter recoveries of fish that had moved >50 km occurred outside of the estuarine system where they were tagged, and no fish that had left the estuarine system of tagging moved < 50 km. Most of these fish were recovered in other estuarine systems or coastal waters downstream of the tagging location, while a few fish were recovered near inlets in the larger Pamlico Sound and were presumed to be migrating offshore to spawn.

Statistical analysis.—Multiple regression and multinomial logistic regression were used to test the effects of predictor variables on the distance moved, direction of movement, and probability of moving >50 km. For the prewinter analysis of Neuse and New River data, TL at tagging, season of tagging (summer and fall), system (Neuse River or New River), and year (2005 or 2006) were tested for their effects on the total distance moved and the direction of movement. Summer was defined as April–July and fall was defined as August–November. Unbalanced sample sizes with respect to system and season across the 11 years of tagging by the NCDMF precluded the analysis of year and system effects. Therefore, data were pooled across years and systems, and the effects of TL at tagging and season of tagging were tested for their effects on the total distance moved and the direction of movement. Both analyses were conditioned on the days at large of individual fish because we assumed greater movement was possible with increasing days at large. In addition, models of the Neuse and New River data were conditioned on monthly fishing mortality because nearly all returns were from commercial fishermen, and we assumed that the level of fishing mortality would influence the probability of tag return. The monthly fishing mortality rate was estimated separately for the Neuse and New rivers during the 2005 and 2006 fishing seasons using a tag–recapture model applied to these same tag return data (Craig and Rice 2008; Smith et al. 2009). Monthly fishing mortality rates were not available for the NCDMF tagging data set.

For the overwinter analysis of Neuse and New River data, the effects of predicted TL at the beginning of the spawning season (November 1), system, and year were tested for their effects on the probability of moving > 50 km using binomial logistic regression. Due to small sample sizes, only the effect of length on overwinter movement was tested for the NCDMF data set. For both data sets, the length at the beginning of the spawning season was predicted using the length at tagging and the von Bertalanffy growth model parameters estimated in the 2009 North Carolina stock assessment (Takade-Heumacher and Batsavage 2009) and a growth model based on mark–recapture data (Fabens 1965). Southern Flounder growth is sexually dimorphic, with females (maximum observed size = 835 mm TL) reaching a much larger asymptotic length than males (maximum observed size = 495 mm TL) (Takade-Heumacher and Batsavage 2009). Because males and females have different growth schedules and possibly migratory patterns, fish predicted to be less than 356 mm TL on November 1 were removed from the overwinter analysis of both data sets. These smaller fish could have been mature males that might have joined the offshore spawning migration at relatively small sizes or immature females that overwintered in the estuaries. At lengths > 356 mm TL at least 85% of Southern Flounder are estimated to be female (see Figure 4 in Takade-Heumacher and Batsavage 2009); therefore, the overwinter results apply primarily to female fish.

All possible models, including intercept-only models, were compared using the Bayesian information criterion (BIC), and the model with the lowest BIC was selected as the best model. We chose BIC (rather than Akaike information criterion) because it penalizes model complexity more severely and, hence, is more conservative in terms of the predictor variables retained in the best model (Burnham and Anderson 2002). The weight of evidence in support of alternative models was compared using BIC weights. Normal, Poisson, and negative binomial error distributions were tested for models of the distance moved; the negative binomial provided the best fit and was used for all models. Each of the best models' regression coefficients were analyzed for significance using *t*-tests ($\alpha = 0.05$). All statistical analyses were conducted in R 3.1.0 (R Core Development Team 2012).

RESULTS

Recaptures of Southern Flounder ranged from 5.7% to 52.4% depending on the system, year, and data set (Table 1). Most recaptured Southern Flounder were recovered within the same system where they were tagged (79–96%). The mean size of tagged Southern Flounder was similar across systems and years, except for fish tagged by the NCDMF, which were about 20% smaller on average than those tagged in the Neuse and New rivers. Based on the examination of otolith annuli, most recovered Southern Flounder were age 1 and nearly all were less than age 2. The median days at large was similar for fish tagged in the Neuse and New rivers (19–26 d), with the exception of the Neuse River in 2005 (7 d). For fish tagged statewide by the NCDMF, the median days at large was 43 d, with fish recaptured within the year of tagging at large for a median of 20 d, while fish recaptured the following year were at large for a median of 280 d. The median days at large for recaptures of fish tagged in Isaac Creek was 99 d, with those recaptured within the creek at large for up to 206 d.

Prewinter Analysis

Southern Flounder were resident within local riverine and estuarine systems in North Carolina for large portions of the year. Across the 2 years and the two river systems, 69% of Southern Flounder tagged and recovered within the Neuse River or New River did not move considerable distances (i.e., fish were recaptured < 1.0 km from the tagging location). Fish that did move > 1.0 km were recovered at relatively moderate distances from the tagging location, with median distances moved of 3.6 km (range = 1.0–19.5 km) in the Neuse River and 4.2 km (range = 1.0–24.6 km) in the New River. The best model to explain the distance moved prior to winter included only the season of tagging as an explanatory variable (BIC weight = 0.26; Table 2). The null model and models with only a year or system effect were closely ranked (BIC weights = 0.24, 0.18, and 0.15, respectively), however, suggesting the predictor variables were not highly informative about factors affecting the distances Southern Flounder moved within estuaries. Fish moved 0.16 ± 0.01 km/d (mean ± standard error) during the summer season (April to July) and 0.24 ± 0.05 km/d during the fall season (August to November) (Figure 2A).

Similar to the Neuse and New rivers, fish tagged by the NCDMF also showed little movement during the prewinter period. Nearly 80% of the fish were recovered in the same area as the tagging location, and of those that had clearly moved from the tagging location, the median distance moved was only 14 km. Similar to the Neuse and New River study, the best model to predict the distance moved also included only season as an explanatory variable (BIC weight = 0.66; Table 2). On average fish moved 0.33 ± 0.00 km/d during the summer and 0.99 ± 0.11 km/d during the fall (Figure 2B).

The best model of the direction of Southern Flounder movements in the Neuse and New rivers during the prewinter period included the effects of TL, season, and system of tagging (BIC weight = 0.71; Table 3); year was not included. These effects were primarily driven by the downstream movements of relatively large Southern Flounder in the fall (Figure 3). Relative to remaining stationary, Southern Flounder were 2–3 times more likely to move downstream in the fall than in the summer, and relatively large fish (≥450 mm TL) were more than twice as likely to move downstream than were relatively small fish (≤375 mm). Also, downstream movements were slightly more likely to be observed in the

TABLE 2. Results from the best multiple linear regression models of the total distance moved from the prewinter analysis of Southern Flounder, fit separately for the Neuse and New River data set and the NCDMF statewide data set. For both season and TL (mm), model inputs refer to the time of tagging. The *P*-values are defined as follows: three asterisks indicate < 0.001 and two asterisks indicate < 0.01. The "X" indicates terms that were removed during the BIC model selection process.

Data set	*N*	Intercept	TL	Season	System	Year
Neuse and New rivers	789	6.3***	X	1.1**	X	X
NCDMF	728	8.5***	X	0.9***	X	X

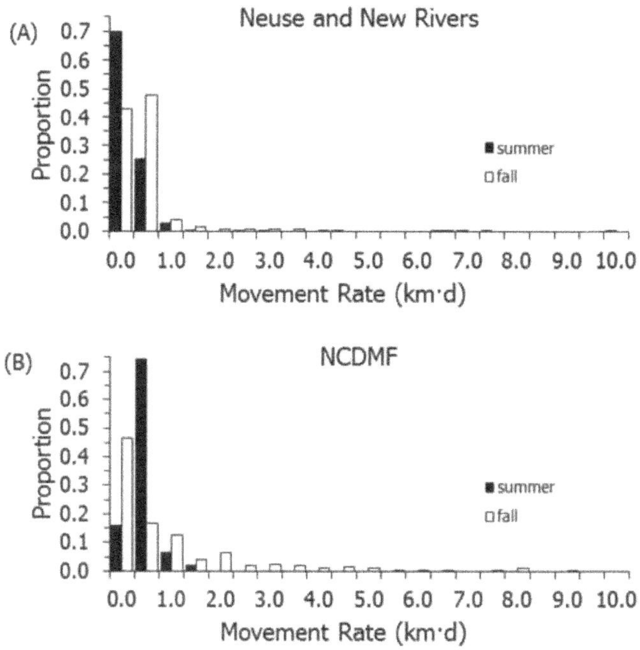

FIGURE 2. Movement rate of prewinter Southern Flounder tagged and recaptured in (**A**) the Neuse and New rivers (2005–2006) and (**B**) by the NCDMF (1980–1982, 1988–1995). Data were pooled across systems and years for each data set.

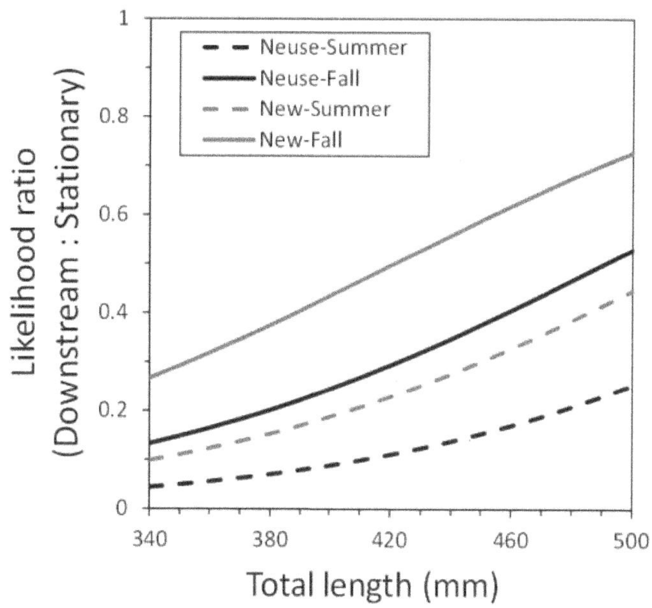

FIGURE 3. Multinomial logistic regression model predictions of the probability ratio of recovering tagged Southern Flounder after downstream movement within the Neuse and New River system (2005–2006). Probability ratios compare the probability of moving downstream to the probability of remaining stationary. Model predictions of upstream movement are not shown because upstream movement parameters were not significant (see Table 3).

New River (66% of moves were downstream) than in the Neuse River (50% of moves were downstream).

In the NCDMF data set, the best model explaining movement direction north or south of the tagging location also included a seasonal effect, but no other effects were significant (BIC weight = 0.69; Table 3). Similar to the Neuse and New River data set, most directional movement was to the south. In contrast to the Neuse and New River data set, however, slightly more southerly movement occurred in the summer (64% of recaptured fish) than in the fall (57% of recaptured fish). The more southerly movement of NCDMF-tagged fish in the summer was probably due to the much greater time at large for NCDMF-tagged fish in the summer (118 ± 94 d [mean ± standard deviation]) than in the fall (22 ± 20 d). As a result, recaptured fish tagged in the fall were less likely to show strong directionality in movement because they had only been at large for a limited time.

In Isaac Creek, 12 of the 170 Southern Flounder tagged were recaptured in the creek prior to winter after 12–154 d at large (Table 1; Figure 4). The average distance moved by fish recaptured within the creek was 0.22 km (range = 0.02–0.40 km). There was no evidence of directional movement at the scale of the creek, with four fish recaptured toward the head of the creek, three recaptured toward the mouth, and the remainder showing no directional movement relative to the tagging location.

Overwinter Analysis

The probability that Southern Flounder tagged in the Neuse and New rivers were recaptured > 50 km from the tagging

TABLE 3. Results from the best multinomial logistic regression models of movement direction during the prewinter period. Movement direction was categorized as upstream, downstream, or stationary (tag recovery < 1 km from tagging location) for the Neuse and New River data set and north, south, or stationary (tag recovery within system of tagging) for the NCDMF data set. For both season and TL, model inputs refer to the time of tagging. Probabilities of upstream or downstream movement are referenced to the probability of remaining stationary as a probability ratio. The *P*-values are defined as follows: three asterisks indicate < 0.001 and ns = not significant. The "X" indicates terms that were removed during the BIC model selection process.

Data set	*N*	Response	Intercept	TL (mm)	Season	System	Year
Neuse and New rivers	789	Upstream	−4.5 (ns)	0.003 (ns)	1.3 (ns)	0.89 (ns)	X
		Downstream	−7.7 (ns)	0.012***	1.2***	0.87***	X
NCDMF	663	North	−1.3***	X	0.3***	X	X
		South	−2.6***	X	1.5***	X	X

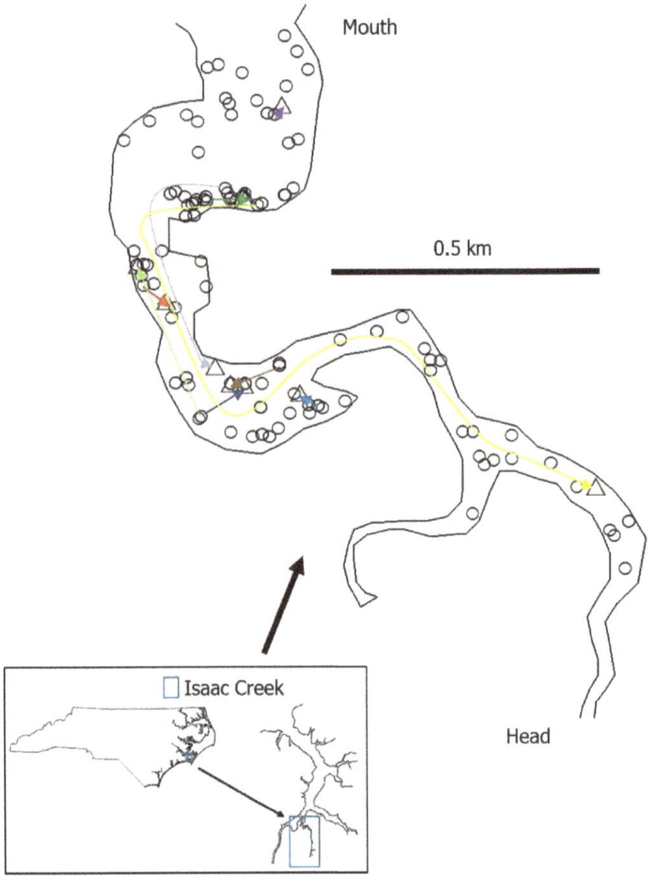

FIGURE 4. Distribution of Southern Flounder tagged and recaptured in Isaac Creek, North Carolina (2003–2005). Each line connects the tagging location (circles) to the recovery location (triangles) for fish recaptured within the creek. Colors indicate individual recaptures that could be uniquely identified.

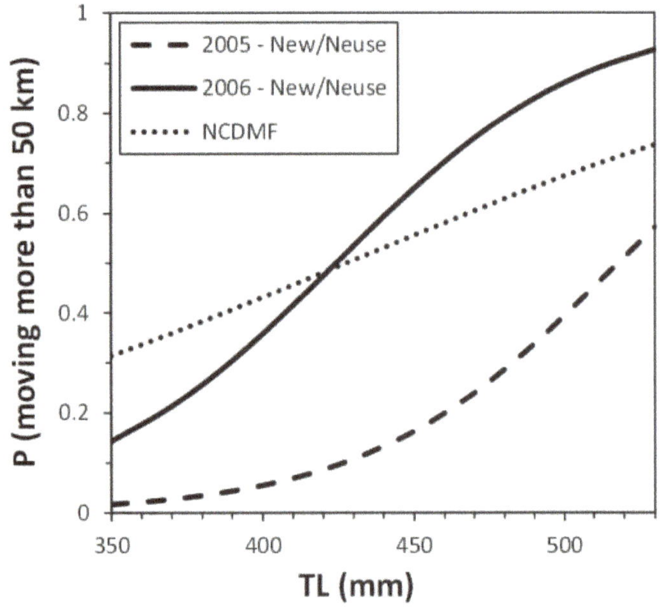

FIGURE 5. Regression model predictions of the probability that Southern Flounder moved more than 50 km from the location of release as a function of total length (TL; mm) for fish tagged in the Neuse and New rivers (2005–2006) and in coastal waters by the NCDMF (1980–1982, 1988–1995).

location was related to TL and year (BIC weight of best model = 0.54; Table 4; Figure 5); system was excluded from the best model. Southern Flounder ≥ 450 mm TL were more than three times as likely to emigrate from the system where they were tagged than were fish ≤ 375 mm TL. Also, fish tagged in 2006 were more likely to be recovered outside of the Neuse and New rivers (34.6%) than were those tagged in 2005

TABLE 4. Results from the best logistic regression models of Southern Flounder emigration from the estuarine system of tagging (tag recovery >50 km from the tagging location) based on the overwinter analysis of fish tagged in the Neuse and New rivers and statewide by the NCDMF. The *P*-values are defined as follows: one asterisk indicates < 0.05 and ns = not significant. The "X" indicates terms that were removed during the BIC model selection process.

Data set	*N*	Intercept	TL (mm)	System	Year
Neuse and New rivers	75	−12.5*	0.024*	X	2.3*
NCDMF	58	−4.4 (ns)	0.010 (ns)	X	X

(20.4%). Due to disparate returns among systems and years, only TL could be tested for its effect on the probability of emigrating for overwintering fish tagged by the NCDMF. The probability of emigrating increased with increasing TL, as for the Neuse and New River data set, but the effect was marginally insignificant ($P = 0.10$; Table 4; Figure 5) and the null model was a better model (BIC weight = 0.65) than the model including TL (BIC weight = 0.35).

All Southern Flounder that left inshore estuarine waters and were recaptured > 50 km from the tagging location were recovered to the south of where they were tagged (Figure 6). Southern Flounder tagged in the New River were recaptured from coastal waters south of the New River to northern Florida, while those tagged in the Neuse River were recaptured in coastal North Carolina waters to the south and in South Carolina. Pooled over the three data sets, 44% of overwintering fish that had moved > 50 km were recaptured across North Carolina state lines in waters to the south. Most of these fish (73.9%) were recaptured in South Carolina, while the remainder were recaptured in Georgia and north Florida.

There was no apparent relationship between tagging location and recapture location. For example, tag returns from state waters to the south (South Carolina, Georgia, and Florida) in the NCDMF data were from fish tagged as far south as the Cape Fear River and as far north as the Pamlico River and the Pamlico Sound in North Carolina. The most rapid movement speeds of emigrating Southern Flounder were > 1 km/d, with the fastest fish moving 5–8 km/d. However, the distance moved was not strongly correlated with the days at

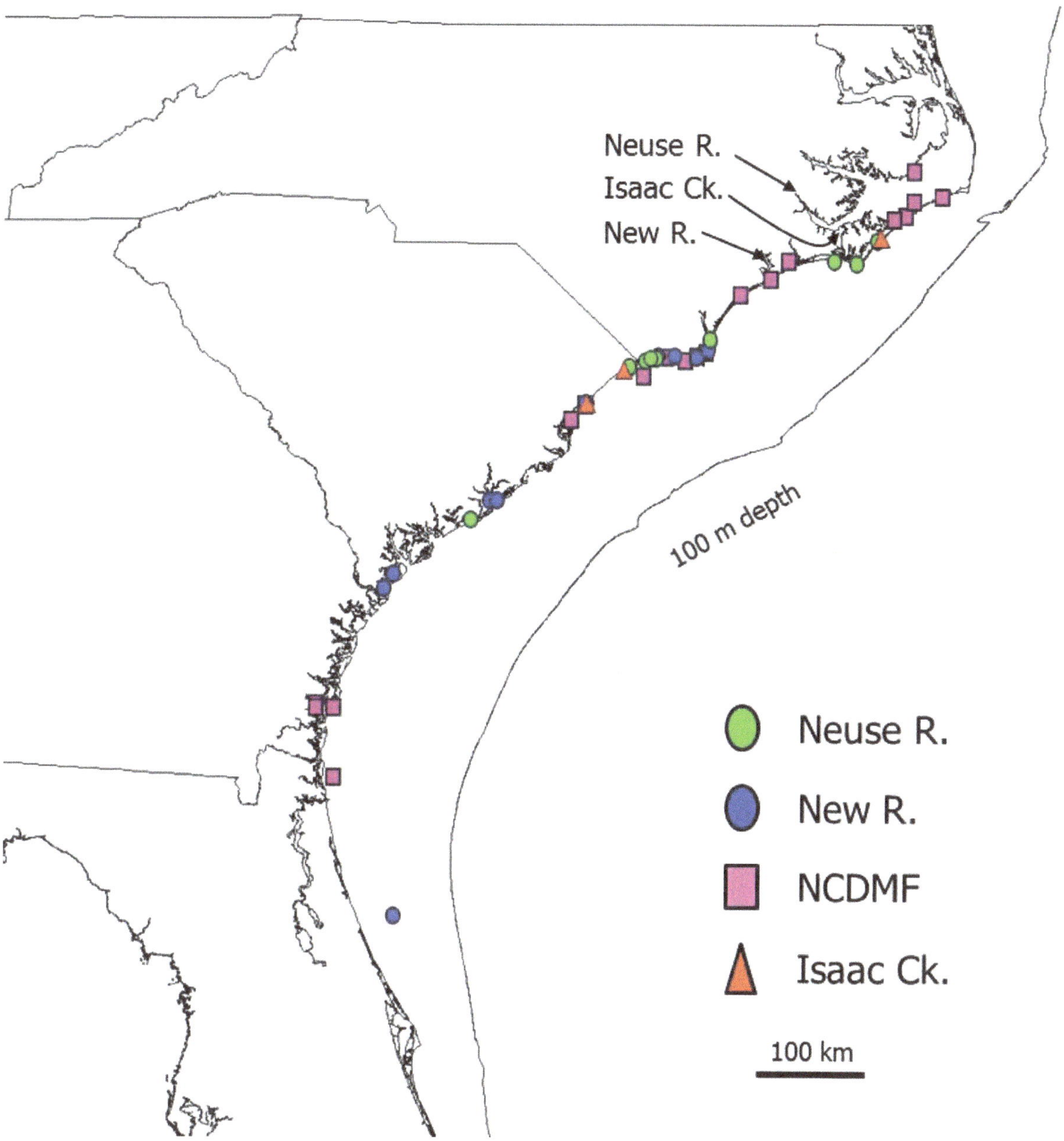

FIGURE 6. Recapture locations for Southern Flounder from three tagging studies: (1) Neuse and New rivers (2005–2006), (2) NCDMF (1980–1982, 1988–1995), and (3) Isaac Creek (2003–2005). Each symbol represents the recapture location of individual fish that had moved greater than 50 km from the tagging location and were recaptured after the overwinter period. The 100-m depth contour approximates the continental shelf–slope break.

large ($r^2 = 0.002$, $P = 0.55$). In fact, the days at large for fish that moved the greatest distances (603–676 km) ranged from 77 to 958 d, and those fish that moved the shortest distances (50–100 km) were at large for 81–647 d. While the data were too limited to evaluate temporal trends, tag returns from the NCDMF data (tagging from 1980 to 1982 and from 1986 to 1995) occurred from 1982 to 1994, with about an equal number of returns in the 1980s (11 recaptures) and the 1990s (15 recaptures).

Recaptures of overwintering fish tagged in Isaac Creek were consistent with the results from the Neuse and New River and NCDMF data sets. Two fish tagged in Isaac Creek in late summer (August) and fall (October) were recaptured in the creek during winter (December; 116 d at large) and during the following summer (June; 206 d at large), suggesting either high site fidelity or overwintering at the scale of the creek. Of the three fish recaptured outside of the creek after being at large for 81–150 d, one was recaptured in Core Sound between two estuarine inlets and two were recaptured in South Carolina waters (Figure 6), consistent with the southerly movement of fish tagged in the Neuse and New rivers and by the NCDMF.

DISCUSSION

Migration Patterns of Southern Flounder

Within the U.S. southern Atlantic Ocean, Southern Flounder appear to make consistent large-scale movements to the south that are most likely associated with migration from inshore estuarine nursery habitats to offshore winter spawning locations. Although findings from conventional tag returns cannot provide definitive insights on individual fish behaviors or migration routes, the collective results from multiple tag return studies indicate that large-scale movement in a southerly direction during fall is a general feature of the life history of Southern Flounder along the U.S. southern Atlantic coast. Many other flatfish species undergo extensive seasonal migrations associated with spawning and feeding, which is common among marine fishes (Secor 2015). Indeed, several aspects of the large-scale migration dynamics of flatfishes, including temporal synchronization, consistent migratory pathways, site fidelity, and complex larval transport processes between offshore spawning areas and inshore nurseries, have been described for selected species (Harding et al. 1978; Bailey and Picquelle 2002; Hunter et al. 2003; Loher and Seitz 2006).

Within North Carolina, the evidence for large-scale southern migration during fall that was observed during the more recent work in the New and Neuse River estuaries aligns well with past observations from the 1980s and 1990s during NCDMF tagging efforts. Furthermore, within the Atlantic range of Southern Flounder, tagging studies conducted in Georgia (Music and Pafford 1984) and South Carolina (Wenner et al. 1990) also demonstrated patterns of large-scale movement to the south, suggesting that offshore spawning activities could be concentrated close to the southern end of the species' Atlantic distribution. The fact that we observed an increased likelihood of downstream movement within the estuary during fall by larger individuals, coupled with an increased probability of large-scale (> 50 km) movement for larger individuals, also suggests that these movements were linked to maturation and spawning behavior. The strong size dependence of Southern Flounder female maturity was recently confirmed (Midway and Scharf 2012), and smaller, immature fish are suspected to primarily overwinter in estuarine systems rather than participate in offshore migrations.

The locations of offshore spawning sites for Southern Flounder are currently unknown, but spawning is suspected to be centered on the outer continental shelf along the southeastern U.S. Atlantic coast. Although sparse, both historic and recent ichthyoplankton data collected off the North Carolina coast support a hypothesized spawning location on the outer continental shelf (Smith et al. 1975; Walsh 2007). In addition, the absence of adult Southern Flounder aggregations in shallow neritic (depth < 40 m) habitats during recent winter scuba surveys also suggests deeper spawning habitats (Watterson and Alexander 2004; Tucker 2011). The combined weight of evidence, thus, points to the outer continental shelf as the most likely spawning site for Southern Flounder. The distribution of conventional tag returns suggests that spawning may be concentrated in more southern outer-shelf habitats, with northward advection of some individuals during the transport of egg and larval stages across the shelf. Depth-specific, cross-shelf circulation patterns may combine with Gulf Stream transport to determine larval pathways for several winter-spawning fishes, including flatfishes, along the southeastern U.S. Atlantic coast (Miller et al. 1984; Kraus and Musick 2001; Hare and Govoni 2005; Walsh 2007).

Estuarine Movements of Southern Flounder

Southern Flounder resided in North Carolina estuaries for extended periods between late spring and early fall, with tagged fish remaining in the New or Neuse rivers for up to 166 d. During this residency period, the vast majority of fish were recaptured less than 10 km from their release site, with many recoveries occurring within 1 km of the release locations. The NCDMF tag returns revealed similar patterns of limited estuarine movement until fall, with nearly 80% of all tag returns occurring within the system of tagging. During spring and summer, most fish were recaptured within the same system in which they were released, with most fish recaptured less than 20 km from their release site. Southern Flounder appear to demonstrate similar localized movement patterns during summer and early fall as was observed for the closely related Summer Flounder *Paralichthys dentatus* occupying U.S. mid-Atlantic estuaries. Both Sackett et al. (2008) and Capossela et al. (2013) detected some Summer Flounder individuals residing in estuarine systems for > 200 d (mean residence times: 86 d in New Jersey, 130 d in Virginia) using acoustic telemetry. In both of these studies, Summer Flounder resided mainly in higher salinity habitats, with Sackett et al. (2008) noting a habitat shift toward the ocean inlet in late summer prior to offshore emigration. Fish demonstrated highly localized movement within the estuary, with large-scale (hundreds of meters) movements uncommon until fish were beginning to emigrate. To our knowledge, no telemetry studies of Southern Flounder habitat use have been conducted at the whole-estuary scale; however, Furey et al. (2013) used a fine-scale positioning system to quantify Southern Flounder habitat use within a confined area of a Texas estuary. They found that individuals also tended to remain within relatively small areas for extended periods but that some individuals did show considerable movement (e.g., > 8 km over ~8 d) within their study area. The strong seasonal and size effects on within-estuary movements that we observed in the present study were most likely associated with the transition to offshore habitats. However, we also noted that Southern Flounder tended to show greater within-estuary movements during fall, including upstream movements not associated with emigration. We suspect that these movements may have been in response to episodic changes in temperature and barometric pressure that occur more frequently during fall or in response to shifts in prey fish distributions.

At a finer spatial scale, the tag return data from Isaac Creek provided additional evidence of limited movement by Southern Flounder when occupying estuarine habitats. Nearly all recaptures occurred in the creek, in some cases after over 100 d at large. For these fish, the average distance between release and relocation sites was 0.22 km, with most recaptures occurring < 0.1 km from the release site. Even after presumably overwintering in the creek, one fish was located only 1.15 km from its release position, providing evidence for habitat fidelity during the overwinter period at the scale of a small marsh creek. Indeed, at the whole-estuary scale, fish suspected of overwintering in the estuary were relocated in the spring in close proximity to their fall release sites (all postwinter relocations in the New and Neuse rivers were < 20 km from the release site). These included mainly smaller individuals that were putatively immature and thus had likely foregone offshore emigration. Our findings support the possibility that individual Southern Flounder often remain confined in relatively small areas throughout estuarine residency but may demonstrate high activity levels within those areas (e.g., Sackett et al. 2008; Furey et al. 2013).

Potential Limitations of Conventional Tagging Results

The use of conventional tag return data means that our conclusions about Southern Flounder movement are inferred only from release and recapture locations and time at large. The data provide no direct evidence that fish participated in spawning or that spawning occurred over the outer continental shelf, nor do they provide direct evidence that fish recaptured close to their estuarine release locations demonstrated limited movement during their time at large. We observed that all fish recovered > 50 km from their tagging location after the overwinter period were recaptured in areas south of the tagging area and inferred southerly movement most likely associated with offshore winter spawning. Alternatively, the potential exists for higher recapture or reporting probabilities to the south of the tagging location. The much higher landings of Southern Flounder in North Carolina waters relative to other states, particularly in the northern area (Figure 7), suggests this possibility is unlikely, and we know of no evidence for regional differences in reporting probabilities. Furthermore, several previous tag return studies for this species obtained similar results pertaining to large-scale movements (Music and Pafford 1984; Wenner et al. 1990). In addition, because most fish that had moved large distances to the south were at large for longer periods, any increase in the probability of tag loss with time at large would create a negative bias for detecting southerly migration. Similarly, our conclusions related to limited within-estuary movement are supported by recent telemetry studies for both Southern Flounder (Furey et al.

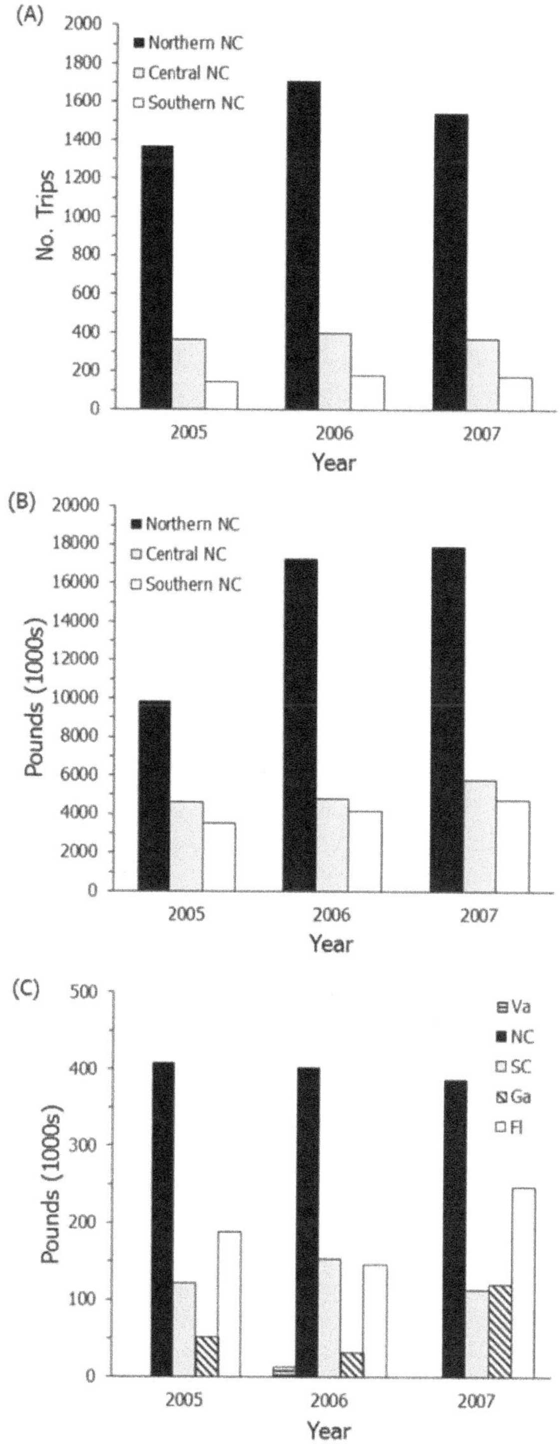

FIGURE 7. Fishing effort and landings for Southern Flounder in North Carolina and U.S. southern Atlantic Ocean waters (2005–2007), showing **(A)** the number of commercial trips and **(B)** commercial landings reported by the NCDMF in different regions of the state and **(C)** the recreational landings by state along the U.S. southern Atlantic coast obtained from the U.S. Marine Recreational Information Program (www.st.nmfs.noaa.gov/recreational-fisheries). State abbreviations are as follows: Va = Virginia, NC = North Carolina, SC = South Carolina, Ga = Georgia, and FL = Florida. Delineations for northern, central, and southern regions of North Carolina are shown in Figure 1.

2013) and the closely related Summer Flounder (Sackett et al. 2008; Capossela et al. 2013). While it is possible that Southern Flounder demonstrate more extensive movement during estuarine residency, which if coupled with strong site fidelity could lead to high rates of recapture near the original release location, a more parsimonious conclusion for our observations is that movement is limited.

Management Considerations

The Southern Flounder stock is currently categorized as "depleted" in North Carolina waters, a status that is believed to be related to a long period of elevated commercial harvest rates beginning in the early 1990s. Outside of North Carolina, only Florida produces any considerable commercial landings (~200,000–300,000 lb annually; Florida Fish and Wildlife Conservation Commission 2015) of Southern Flounder along the southeastern U.S. Atlantic coast. However, recreational landings have increased steadily during the past three decades in all states. Total annual recreational catch of Southern Flounder for southern Atlantic states has averaged more than 0.45 million kg since 2010, more than doubling since the 1980s (National Oceanic and Atmospheric Administration 2015). While South Carolina, Georgia, and Florida each enforce recreational bag and size limits for Southern Flounder in state waters, the regulations are liberal relative to those in North Carolina. Additionally, given that the fishery resource has had only moderate historical importance in those states, none has developed a comprehensive fishery management plan for Southern Flounder. In contrast, North Carolina has recently imposed even stricter management measures in both commercial and recreational fishery sectors intended to lower Southern Flounder harvest rates and reduce interactions with protected species, most notably sea turtles (Cheloniidae; NCDMF 2013). Continued growth of recreational fisheries and any increased participation in estuarine and nearshore commercial fisheries in other states may necessitate the development of more comprehensive management plans in the near future. In that event, there may be an increased need for interstate cooperation through a federal fishery management plan under the auspices of the Atlantic States Marine Fisheries Commission. The cooperative management of the closely related Summer Flounder, a species that demonstrates several life history features similar to Southern Flounder and is harvested by multiple states in the U.S. mid-Atlantic region, could serve as a useful model.

The management of Southern Flounder in North Carolina is based on the assumption that the North Carolina fishery represents a unit stock (Takade-Heumacher and Batsavage 2009; NCDMF 2013). This is primarily due to a lack of available information about migration patterns and stock structure within the U.S. southern Atlantic region. Throughout the range of Southern Flounder, their life history includes offshore spawning during winter, larval transport to estuarine nurseries, and rapid growth during estuarine residency, followed by maturity and spawning-related migration to ocean habitats (Stokes 1977; Wenner et al. 1990; Burke et al. 1991). This basic life history model has also included a return to estuarine habitats by adult Southern Flounder in the spring following spawning, and the unit stock assumption would necessitate that adults emigrating from North Carolina in the fall would return to North Carolina waters in the spring. However, findings from this study, as well as observations from previous conventional tag return studies for Southern Flounder conducted in Georgia (Music and Pafford 1984) and South Carolina (Wenner et al. 1990), demonstrate that most fish that return to inshore waters may do so within the jurisdiction of a neighboring state to the south. Although the available data do not yet allow any definitive conclusions about Southern Flounder migration pathways and behavior, the consistent and considerable movement to waters located south of the tagging location implies that there is high potential for extensive mixing within the southeastern U.S. Atlantic region.

Recently, genetic analysis of Southern Flounder stocks demonstrated clear separation between Gulf of Mexico and southeastern U.S. Atlantic populations (Anderson et al. 2012; Wang et al. 2015). However, stock structure within the southeastern U.S. Atlantic basin remains poorly understood. Analysis of otolith morphometrics was able to detect the Gulf of Mexico–Atlantic separation but only low levels of shape disparities among regions within the Atlantic basin, suggestive of a high level of mixing (Midway et al. 2014). Similarly, using fingerprints from both mitochondrial DNA and amplified fragment length polymorphisms, Wang et al. (2015) detected the presence of only weak genetic structure among possible subpopulations of Southern Flounder along the southeastern U.S. Atlantic coast. If Southern Flounder from North Carolina mix extensively with individuals from other areas within the southeastern U.S. Atlantic region, this has several important implications for effective management of the species. For instance, if fish that emigrate from North Carolina estuaries reenter estuarine waters in another part of the range, they are effectively removed from the harvestable stock in North Carolina. Alternatively, immigrants from other areas may replenish local stocks in North Carolina that are heavily exploited. Importantly, any attempts to estimate spawning stock biomass, a key biological reference point used in management, would be biased if the movements of fish outside of North Carolina state waters were ignored. While our findings suggest that a broad exchange of individuals within the U.S. southern Atlantic region is likely, a more refined understanding of Southern Flounder migration dynamics will be necessary to promote effective fishery management of this important resource species.

ACKNOWLEDGMENTS

We thank the commercial fishermen who participated in this study: H. Bogey, P. Darna, S. Notargiacomo, B. Padgett,

J. Padgett, and B. Sanderford. We thank Z. Tait, J. Williams, and J. Leonard for help with tagging and data processing. Funding for this project was provided by grants to J. A. Rice and J. K. Craig (FEG-05-15), and to F. S. Scharf (FEG-05-16), from the North Carolina Sea Grant Fishery Resource Grant Program. We thank B. Hooper, D. Skinner, C. Batsavage, and K. West of the North Carolina Division of Marine Fisheries and the North Carolina State University Center for Marine Science and Technology for logistical support. We thank N. Bacheler, C. Taylor, and two anonymous reviewers for comments on the manuscript. This work is dedicated to James P. Monaghan (deceased) for his influence on JKC and his contributions to the understanding and management of flatfish populations in North Carolina. The findings and conclusions in this paper are those of the author(s) and do not necessarily represent the views of the National Oceanic and Atmospheric Administration nor any of its subagencies or of the North Carolina Division of Marine Fisheries.

REFERENCES

Anderson, J. D., and W. J. Karel. 2012. Population genetics of Southern Flounder with implications for management. North American Journal of Fisheries Management 32:656–662.

Anderson, J. D., W. J. Karel, and A. C. S. Mione. 2012. Population structure and evolutionary history of Southern Flounder in the Gulf of Mexico and western Atlantic Ocean. Transactions of the American Fisheries Society 141:46–55.

Bacheler, N. M., L. M. Paramore, J. A. Buckel, and J. E. Hightower. 2009. Abiotic and biotic factors influence the habitat use of an estuarine fish. Marine Ecology Progress Series 377:263–277.

Bailey, K. M., and S. J. Picquelle. 2002. Larval distribution of offshore spawning flatfish in the Gulf of Alaska: potential transport pathways and enhanced onshore transport during ENSO events. Marine Ecology Progress Series 236:205–217.

Berger, A. M., M. L. Jones, Y. Zhao, and J. R. Bence. 2012. Accounting for spatial population structure at scales relevant to life history improves stock assessment: the case for Lake Erie Walleye *Sander vitreus*. Fisheries Research 115–116:44–59.

Blandon, I. R., R. Ward, T. L. King, W. J. Karel, and J. P. Monaghan. 2001. Preliminary genetic population structure of Southern Flounder, *Paralichthys lethostigma*, along the Atlantic coast and Gulf of Mexico. U.S. National Marine Fisheries Service Fishery Bulletin 99:671–678.

Burke, J. S. 1995. Role of feeding and prey distribution of Summer and Southern flounder in selection of estuarine nursery habitats. Journal of Fish Biology 47:355–366.

Burke, J. S., J. M. Miller, and D. E. Hoss. 1991. Immigration and settlement pattern of *Paralichthys dentatus* and *P. lethostigma* in an estuarine nursery ground, North Carolina, USA. Netherlands Journal of Sea Research 27:393–405.

Burnham, K. P., and D. R. Anderson. 2002. Model selection and multimodel inference: a practical information-theoretic approach. Springer-Verlag, New York.

Cadrin, S. X., and D. H. Secor. 2009. Accounting for spatial population structure in stock assessment: past, present, and future. Pages 405–426 *in* R. J. Beamish and B. J. Rothschild, editors. The future of fisheries science in North America. Springer Science, New York.

Capossela, K. M., M. C. Fabrizio, and R. W. Brill. 2013. Migratory and within-estuary behaviors of adult Summer Flounder (*Paralichthys dentatus*) in a lagoon system of the southern mid-Atlantic Bight. U.S. National Marine Fisheries Service Fishery Bulletin 111:189–201.

Cope, J. M., and A. E. Punt. 2011. Reconciling stock assessment and management scales under conditions of spatially varying catch histories. Fisheries Research 107:22–38.

Craig, J. K., and J. A. Rice. 2008. Estuarine residency, movements, and exploitation of Southern Flounder (*Paralichthys lethostigma*) in North Carolina. North Carolina Sea Grant, Fishery Resource Grant 05-FEG-15, Final Report, Raleigh.

ESRI (Environmental Systems Research Institute). 2011. ArcGIS desktop: release 10. ESRI, Redlands, California.

Fabens, A. J. 1965. Properties and fitting of the von Bertalanffy growth curve. Growth 29:265–289.

Farmer, T. M., D. R. DeVries, R. A. Wright, and J. E. Gagnon. 2013. Using seasonal variation in otolith microchemical composition to indicate Largemouth Bass and Southern Flounder residency patterns across an estuarine salinity gradient. Transactions of the American Fisheries Society 142:1415–1429.

Fay, G., A. E. Punt, and A. D. M. Smith. 2011. Impacts of spatial uncertainty on performance of age structure-based harvest strategies for Blue Eye Trevalla (*Hyperoglyphe antarctica*). Fisheries Research 110:391–407.

Florida Fish and Wildlife Conservation Commission. 2015. Commercial fisheries landings in Florida. Available: http://myfwc.com/research/saltwater/fishstats/commercial-fisheries/landings-in-florida. (March 2015).

Froeschke, B. F., G. W. Stunz, M. M. R. Robillard, J. Williams, and J. T. Froeschke. 2013a. A modeling and field approach to identify essential fish habitat for juvenile Bay Whiff (*Citharichthys spilopterus*) and Southern Flounder (*Paralichthys lethostigma*) within the Aransas Bay complex, TX. Estuaries and Coasts 36:881–892.

Froeschke, B. F., P. Tissot, G. W. Stunz, and J. T. Froeschke. 2013b. Spatiotemporal predictive models for juvenile Southern Flounder in Texas estuaries. North American Journal of Fisheries Management 33:817–828.

Furey, N. B., M. A. Dance, and J. R. Rooker. 2013. Fine-scale movements and habitat use of juvenile Southern Flounder *Paralichthys lethostigma* in an estuarine seascape. Journal of Fish Biology 82:1469–1483.

Furey, N. B., and J. R. Rooker. 2013. Spatial and temporal shifts in suitable habitat of juvenile Southern Flounder (*Paralichthys lethostigma*). Journal of Sea Research 76:161–169.

Glass, L. A., J. R. Rooker, R. T. Kraus, and G. J. Holt. 2008. Distribution, condition, and growth of newly settled Southern Flounder (*Paralichthys lethostigma*) in the Galveston Bay Estuary, TX. Journal of Sea Research 59:259–268.

Goethel, D. R., T. J. Quinn, and S. X. Cadrin. 2011. Incorporating spatial structure in stock assessment: movement modeling in marine fish population dynamics. Reviews in Fisheries Science 19:119–136.

Hamilton, S. L., J. R. Wilson, T. Ben-Horin, and J. E. Caselle. 2011. Utilizing spatial demographic and life history variation to optimize sustainable yield of a temperate sex-changing fish. PLoS (Public Library of Science) ONE [online serial] 6(9):e24580.

Harding, D., J. H. Nichols, and D. S. Tungate. 1978. The spawning of plaice (*Pleuronectes platessa* L.) in the Southern Bight. Rapports et Procès-Verbaux des Réunions / Conseil Permanent International pour l'Exploration de la Mer 172:102–113.

Hare, J. A., and J. J. Govoni. 2005. Comparison of average larval fish vertical distributions among species exhibiting different transport pathways on the southeast United States continental shelf. U.S. National Marine Fisheries Service Fishery Bulletin 103:728–736.

Hulson, P. F., T. J. Quinn, D. H. Hanselman, and J. N. Ianelli. 2013. Spatial modeling of Bering Sea Walleye Pollock with integrated age-structured assessment models in a changing environment. Canadian Journal of Fisheries and Aquatic Sciences 70:1402–1416.

Hunter, E., J. D. Metcalfe, and J. D. Reynolds. 2003. Migration route and spawning area fidelity by North Sea plaice. Proceedings of the Royal Society of London B 270:2097–2103.

Kirby-Smith, W. W., M. E. Lebo, and R. B. Herrmann. 2003. Importance of water quality to nekton habitat use in a North Carolina branch estuary. Estuaries and Coasts 26:1480–1485.

Kraus, R. T., and J. A. Musick. 2001. A brief interpretation of Summer Flounder, *Paralichthys dentatus*, movements and stock structure with new tagging data on juveniles. Marine Fisheries Review 63:1–6.

Loher, T., and A. Seitz. 2006. Seasonal migration and environmental conditions of Pacific Halibut *Hippoglossus stenolepis*, elucidated from pop-up archival transmitting (PAT) tags. Marine Ecology Progress Series 317:259–271.

Lowe, M. R., D. R. DeVries, R. A. Wright, S. A. Ludsin, and B. J. Fryer. 2011. Otolith microchemistry reveals substantial use of freshwater by Southern Flounder in the northern Gulf of Mexico. Estuaries and Coasts 34:630–639.

Mallin, M. A., J. M. Burkholder, L. B. Cahoon, and M. H. Posey. 2000. The North and South Carolina coasts. Marine Pollution Bulletin 41:56–75.

McGarvey, R., A. J. Linnane, J. E. Feenstra, A. E. Punt, and J. M. Matthews. 2010. Integrating recapture-conditioned movement estimation into spatial stock assessment: a South Australian lobster fishery application. Fisheries Research 105:80–90.

Midway, S. R., S. X. Cadrin, and F. S. Scharf. 2014. Southern Flounder (*Paralichthys lethostigma*) stock structure inferred from otolith shape analysis. U.S. National Marine Fisheries Service Fishery Bulletin 112:326–338.

Midway, S. R., and F. S. Scharf. 2012. Histological analysis reveals larger size at maturity for Southern Flounder with implications for biological reference points. Marine and Coastal Fisheries: Dynamics, Management, and Ecosystem Science [online serial] 4:628–638.

Miller, J. M., J. P. Reed, and L. J. Pietrafesa. 1984. Patterns, mechanisms and approaches to the study of migrations of estuarine-dependent fish larvae and juveniles. Pages 209–225 *in* J. D. McCleave, G. P. Arnold, J. J. Dodson, and W. H. Neill, editors. Mechanisms of migration in fishes. Plenum, New York.

Monaghan, J. P. 1992. Tagging studies of Southern Flounder (*Paralichthys lethostigma*) and Gulf Flounder (*Paralichthys albigutta*) in North Carolina. North Carolina Department of Environment and Natural Resources, Division of Marine Fisheries, Project F-29, Study 3B, Completion Report, Morehead City.

Music, J. L., and J. L. Pafford. 1984. Population dynamics and life history aspects of major marine sportfishes in Georgia's coastal waters. Georgia Department of Natural Resources, Coastal Resources Division, Contribution Series 38, Brunswick.

Nañez-James, S. E., G. W. Stunz, and S. A. Holt. 2009. Habitat use patterns of newly settled Southern Flounder, *Paralichthys lethostigma*, in Aransas–Copano Bay, Texas. Estuaries and Coasts 32:350–359.

National Oceanic and Atmospheric Administration. 2015. Recreational fisheries statistics. Available: http://www.st.nmfs.noaa.gov/recreational-fisheries/index. (March 2015).

NCDMF (North Carolina Division of Marine Fisheries). 2013. North Carolina fishery management plan for Southern Flounder, Amendment 1. North Carolina Department of Environment and Natural Resources, Morehead City.

Nims, M. K., and B. D. Walther. 2014. Contingents of Southern Flounder from subtropical estuaries revealed by otolith chemistry. Transactions of the American Fisheries Society 143:721–731.

Powell, A. B., and F. J. Schwartz. 1977. Distribution of paralichthid flounders (Bothidae: *Paralichthys*) in North Carolina estuaries. Chesapeake Science 18:334–339.

Quinn, T. J. 2003. Ruminations on the development and future of population dynamics models in fisheries. Natural Resource Modeling 16:341–392.

R Core Development Team. 2012. R: a language and environment for statistical computing. R Foundation for Statistical Computing, Vienna, Austria. Available: http://www.R-project.org/. (March 2015).

Sackett, D. K., K. W. Able, and T. M. Grothues. 2008. Habitat dynamics of Summer Flounder *Paralichthys dentatus* within a shallow USA estuary, based on multiple approaches using acoustic telemetry. Marine Ecology Progress Series 364:199–212.

Secor, D. H. 2015. Migration ecology of marine fishes. Johns Hopkins University Press, Baltimore, Maryland.

Secor, D. H. 2005. Fish migration and the unit stock: three formative debates. Pages 17–44 *in* S. X. Cadrin, K. D. Friedland, and J. R. Waldman, editors. Stock identification methods: applications in fishery science. Elsevier Academic Press, San Diego, California.

Smith, W. E., F. S. Scharf, and J. E. Hightower. 2009. Fishing mortality in North Carolina's Southern Flounder fishery: direct estimates of F from a tag-return experiment. Marine and Coastal Fisheries: Dynamics, Management, and Ecosystem Science [online serial] 1:283–299.

Smith, W. G., J. D. Sibunka, and A. Wells. 1975. Seasonal distributions of larval flatfishes (Pleuronectiformes) on the continental shelf between Cape Cod, Massachusetts, and Cape Lookout, North Carolina, 1965–1966. NOAA Technical Report NMFS SSRF-691.

Stephenson, R. L. 1999. Stock complexity in fisheries management: a perspective of emerging issues related to population sub-units. Fisheries Research 43:247–249.

Stephenson, R. L. 2002. Stock structure and management structure: an ongoing challenge for ICES. ICES Marine Science Symposia 215:305–314.

Stokes, G. M. 1977. Life history studies of Southern Flounder (*Paralichthys lethostigma*) and Gulf Flounder (*Paralichthys albigutta*) in the Aransas Bay area of Texas. Texas Parks and Wildlife Department, Technical Series 25, Austin.

Takade-Heumacher, H., and C. Batsavage. 2009. Stock status of North Carolina Southern Flounder (*Paralichthys lethostigma*). North Carolina Department of Environment and Natural Resources, Division of Marine Fisheries, Morehead City.

Taylor, J. C., J. M. Miller, L. J. Pietrafesa, D. A. Dickey, and S. W. Ross. 2010. Winter winds and river discharge determine juvenile Southern Flounder (*Paralichthys lethostigma*) recruitment and distribution in North Carolina estuaries. Journal of Sea Research 64:15–25.

Taylor, N. G., M. K. McAllister, G. L. Lawson, T. Carruthers, and B. A. Block. 2011. Atlantic Bluefin Tuna: a novel multistock spatial model for assessing population biomass. PLoS (Public Library of Science) ONE [online serial] 6(12):e27693.

Tucker, C. R. 2011. Reproductive analysis of Southern and Gulf flounders (*Paralichthys lethostigma* and *P. albigutta*) in South Carolina based on scuba surveys. Master's thesis. College of Charleston, Charleston, South Carolina.

Walsh, H. J. 2007. Distribution of fall/winter-spawned larval fish in relation to hydrographic fronts on the North Carolina shelf: implications for larval transport mechanisms. Master's thesis. North Carolina State University, Raleigh.

Walsh, H. J., D. S. Peters, and D. P. Cyrus. 1999. Habitat utilization by small flatfishes in a North Carolina estuary. Estuaries 22:803–813.

Wang, V. H., M. A. McCartney, and F. S. Scharf. 2015. Population genetic structure of Southern Flounder inferred from multilocus DNA profiles. Marine and Coastal Fisheries: Dynamics, Management, and Ecosystem Science [online serial] 7:220–232.

Watterson, J. C., and J. L. Alexander. 2004. Southern Flounder escapement in North Carolina. North Carolina Department of Environment and Natural Resources, Grant F-73, Segments 1–3, Final Performance Report, Morehead City.

Wenner, C. A., W. A. Roumillat, J. E. Moran, M. B. Maddox, L. B. Daniel, and J. W. Smith. 1990. Investigations on the life history and population dynamics of marine recreational fishes in South Carolina: part 1. South Carolina Marine Resources Research Institute, Charleston.

Ying, Y., Y. Chen, L. Lin, and T. Gao. 2011. Risks of ignoring fish population spatial structure in fisheries management. Canadian Journal of Fisheries and Aquatic Sciences 68:2101–2120.

8

Estimating the Dependence of Spawning Frequency on Size and Age in Gulf of Mexico Red Snapper

C. E. Porch*

National Marine Fisheries Service, Southeast Fisheries Science Center, Sustainable Fisheries Division, 75 Virginia Beach Drive, Miami, Florida 33149-1099, USA

G. R. Fitzhugh

National Marine Fisheries Service, Southeast Fisheries Science Center, Panama City Laboratory, 3500 Delwood Beach Road, Panama City, Florida 32408, USA

E. T. Lang[1] and H. M. Lyon

Riverside Technology Contracting for National Marine Fisheries Service, Panama City Laboratory, 3500 Delwood Beach Road, Panama City, Florida 32408, USA

B. C. Linton[2]

National Marine Fisheries Service, Southeast Fisheries Science Center, Sustainable Fisheries Division, 75 Virginia Beach Drive, Miami, Florida 33149-1099, USA

Abstract

In 2011, a large multivessel survey was conducted to provide nearly synoptic sampling of Red Snapper *Lutjanus campechanus* throughout their reproductive season in the U.S. Gulf of Mexico. A total of 2,487 Red Snapper were caught with a female : male ratio that was approximately 1:1. The ovaries of 1,002 females were histologically examined. Females (n = 391) were found with spawning markers (postovulatory follicles and hydrated oocytes) throughout the study area, but primarily in outer shelf waters. Statistical models were developed to quantify and test the dependence of the proportion of females bearing spawning markers (spawning fraction) on female length and age, time of year, depth, gear type (vertical line or longline), or region (east or west of the Mississippi River). Most of the variance in spawning fraction was explained by the time of year; spawning fractions were generally low in spring, peaked in midsummer, and declined by fall. There was also strong statistical evidence of a positive relationship between spawning fraction and either age or length. The effects of region and gear type were not significant once time of year and size or age were accounted for. These results demonstrate the need to account for differences in the time of year and age structure of the population when the productivity of populations of Red Snapper are compared. For example, productivity has been hypothesized to be greater in the western Gulf than in the eastern Gulf, as evidenced by regional patterns of egg and larval abundance. Our results suggest that this regional difference is not due to any intrinsic difference in the biology of the fish, but simply a consequence of there being more large, old Red Snapper in the western Gulf. Recent stock assessments have indicated that Red Snapper

Subject editor: Patrick Sullivan, Cornell University, Ithaca, New York

*Corresponding author: clay.porch@noaa.gov

[1]Present address: Louisiana Department of Wildlife and Fisheries, 2000 Quail Drive, Baton Rouge, Louisiana 70808, USA.

[2]Present address: National Marine Fisheries Service, Northeast Fisheries Science Center 166 Water Street, Woods Hole, Massachusetts 02543, USA.

are increasing in abundance and there is a need to continue monitoring to detect any possible compensation in reproduction.

Many tropical and subtropical fishes exhibit indeterminate fecundity and spawn multiple times over an extended period that may last much of the year. Several studies have demonstrated that the number of eggs produced during a spawning event (batch fecundity) increases with age or body length at a faster rate than does body weight (Cooper et al. 2013; Hixon et al. 2014; He et al. 2015). Moreover, evidence is emerging that spawning frequency also increases with age and body size (LaPlante and Schultz 2007; Mehault et al. 2010; Fitzhugh et al. 2011; Cooper et al. 2013; Klibansky and Scharf 2013). As a result, egg production in tropical and subtropical species is probably less correlated to mature (spawning) biomass than has been observed for species from higher latitudes (Fitzhugh et al. 2011; Lowerre-Barbieri et al. 2011; Klibansky and Scharf 2013).

Red Snapper *Lutjanus campechanus* are indeterminate spawners that inhabit the tropical and subtropical waters of the southeastern United States and Mexico. They are long lived (>50 years: Wilson and Nieland 2001) and considered to have a periodic life history (Winemiller and Rose 1992), yet maturity can occur at age 2 (SEDAR 2013). Full reproductive potential has been estimated to occur by about 12–15 years of age (Goodyear 1995; Woods 2003; Kulaw 2012). The Red Snapper population in the northern Gulf of Mexico has been harvested commercially since the mid 19th century and currently supports one of the most important recreational fisheries in the world. Perhaps not surprisingly, the assessment and management of this species has been highly controversial (Shipp and Bortone 2009). Even the U.S. Congress has weighed in on the matter, holding congressional hearings and sponsoring several bills specifically addressing Red Snapper (e.g., Senate Bill 157 and Congressional Bill H.R. 3099).

One area of active debate concerns the level of abundance to which the Red Snapper stock must rebuild, which depends partly on the perceived relationship between egg production and age or size, and which is tied directly into fisheries management by law through the Magnuson–Stevens Fishery Conservation and Management Act (MSFCMA). Recent stock assessments (Porch 2004, 2007) modeled Red Snapper egg production in the Gulf of Mexico as the product of batch fecundity and maturity at age. Porch et al. (2007) acknowledged the need to account for age or size dependence in the number of annual spawns. Until recently, however, the available data were insufficient to determine whether such a relationship existed, let alone to quantify it.

In 2011 a congressionally supported supplemental survey, in which Red Snapper was one of the target species, was conducted throughout the northern Gulf of Mexico (Campbell et al. 2012). Nearly 2,500 Red Snapper were sampled from multiple vessels operating throughout the Red Snapper spawning season (April through October) and across all known spawning areas, resulting in the most spatially and temporally extensive survey of Red Snapper spawning ever conducted. Nevertheless, the survey was not perfectly synoptic in that all areas and depths were not sampled on the same dates. Previous work has suggested that the spawning intensity of Red Snapper and other species varies considerably during the course of the season and in different locations (Woods 2003; Porch et al. 2007; Brown-Peterson et al. 2009; Lowerre-Barbieri et al. 2011). For this reason, the perceived relationship between spawning frequency and age or size is likely to vary depending on when and where reproductive samples are taken (Lowerre-Barbieri et al. 2011). For example, since older Red Snapper tend to be caught in deeper water, sampling deep and shallow waters at different times of the spawning season could alter the apparent relationship between spawning frequency and age.

The objective of this paper was to use the samples obtained from the supplemental survey to develop unbiased estimates of the relationship between spawning frequency and body length or age. The proposed modeling framework attempted to "standardize" these estimates by explicitly accounting for the effects of any factors that may have varied systematically during the sampling.

METHODS

Field survey and laboratory processing.—Red Snapper were caught by hook-and-line gear (vertical line and longline) deployed from Dry Tortugas, Florida, through Brownsville, Texas, and from the inner shelf (9 m) to a depth of about 180 m according to a stratified random design as described in Campbell et al. (2012). The time of day of all catches were recorded as the gear was retrieved. All of the Red Snapper caught were measured for TL and FL (mm), weighed (kg), and dissected to extract otoliths and gonads. Gonads were sexed and macroscopically staged while on the vessel, and ovaries were frozen for further processing in the laboratory.

Red Snapper were aged from sectioned otoliths as described in Allman and Fitzhugh (2007) and Allman et al. (2012). Frozen ovaries were weighed (nearest 0.1 g) and subsamples were taken from the right posterior lobe and placed into 10% neutral buffered formalin (NBF) for histological preparation (hematoxylin and eosin-y stain; Mass Histology Services, Worcester, Massachusetts). Histological examination determined the presence–absence of spawning markers in females (postovulatory follicles and/or hydrated oocytes: Hunter and Macewicz 1985; Murua et al. 2003). During selected cruises, a sample of 50 Red Snapper

TABLE 1. Number of Red Snapper females and fraction bearing histological spawning markers by age-class from the 2011 congressional supplemental survey.

Age-class	Number of females	Fraction with spawning markers
1	2	0.00
2	67	0.16
3	78	0.19
4	174	0.38
5	224	0.34
6	157	0.42
7	137	0.51
8	45	0.42
9	25	0.56
10+	83	0.57

ovaries were weighed and tissue subsamples fixed in fresh formalin at sea prior to freezing the remainder of the ovaries, for the purpose of quality assessment. A comparison test of histology sections from the paired freshly preserved and frozen–preserved ovarian tissue subsamples was conducted to determine whether spawning markers could be readily identified after freezing.

Models of spawning fraction.—We defined spawning fraction as the proportion of females bearing spawning markers. This definition is similar to usage elsewhere (Priede and Watson 1993; Murua et al. 2003; Lowerre-Barbieri et al. 2009; Kurita et al. 2010; Kurita 2012), but differs in that we refer to the proportion of all sampled females in the survey without further distinction of active or mature females. In later calculations (see section below), we address the conversion of spawning fraction to daily probabilities of spawning. This separation of steps recognizes that various factors may govern the duration of spawning markers, which subsequently can affect daily probabilities and the estimated total number of spawns (Priede and Watson 1993; Kurita 2012).

Tabulations of the 2011 survey results appeared to confirm previous findings that the proportion of females with spawning markers (spawning fraction) increases with age or size (Table 1) and varies with the time of year (Table 2). However, the size and age of the Red Snapper in the sample also tended to vary with the time of year, and somewhat older fish on average were sampled near the peak of the spawning season. Accordingly, the apparent relationship between spawning fraction and age (or length) may be confounded to some unknown degree by the relationship between spawning fraction and time of year, and it is therefore necessary to model the effects of age and time of year simultaneously in order to disentangle them. The situation is further complicated by observations of variations in spawning fraction by depth, gear type (vertical line or longline), and region (east or west of the Mississippi River).

Models are often used to help distinguish the response of a measure to an explanatory variable of interest from its response

TABLE 2. Summary of Red Snapper females from the 2011 congressional supplemental survey. Within each approximate 2-week period, the initial date of sampling is listed. The numbers of sampling dates and females and fraction of females bearing histological spawning markers are indicated, along with the mean and range of age. The survey covered a total period of 198 d.

Start date (month/day)	Number of dates	Number of females	Fraction with spawning markers	Mean age (years)	Age range (years)
4/9	9	58	0.02	6	2–14
4/19	9	52	0.00	8	2–31
5/3	8	70	0.21	7	3–14
5/17	9	87	0.28	5	2–14
5/31	12	90	0.63	6	2–29
6/15	6	39	0.62	9	2–34
6/29	1	2	1.00	21	8–34
7/13	8	58	0.79	8	3–28
7/26	10	123	0.67	6	2–22
8/9	10	58	0.62	7	3–21
8/23	7	58	0.66	5	1–15
9/10	7	34	0.29	6	2–24
9/20	9	173	0.23	5	2–25
10/4	12	79	0.10	6	2–22
10/19	4	11	0.09	7	4–9

to other variables of perhaps less interest. This is particularly true where the sampling is unbalanced in the sense that more measurements are made for some levels of the explanatory variables than for others. In the present case we are particularly interested in distinguishing real changes in spawning fraction with age or length from perceived changes that result by sampling different stages under different average conditions as illustrated in Table 3. If the model is approximately correct, then important biophysical relationships should be reflected by parameter estimates that are statistically different from the corresponding null model (where the explanatory variable is assumed to have no effect) and substantially increase the fraction of the variance explained by the model.

The standard approach for modeling dependent variables that can have only two possible outcomes (e.g., success or failure to detect spawning markers) is a binomial regression. The dependent variable is treated as the outcome of a Bernoulli trial such that the total number of successes from a series of like trials is approximately binomially distributed. The maximum likelihood estimate for the probability of observing a success (p) is therefore obtained by minimizing the negative log-likelihood expression:

$$-\ln L = -\sum_{i}^{N} o_i \ln(p) + (1 - o_i) \ln(1 - p), \qquad (1)$$

where i denotes one of N observations, o is an indicator that takes on a value of 1 for a "success" (spawning markers present) and a

TABLE 3. Plausible biophysical and sampling-related explanations for perceived trends in spawning frequencies of Red Snapper by stage, depth, region, gear, and season. If the model is approximately correct, biophysical causes for changes in spawning frequency should manifest as parameter estimates that are statistically different from the null model.

Independent variable	Biophysical causes	Sampling issues
Stage (age or length)	Body cavity size and/or energy reserves increase with stage.	Older, larger stages sampled under more favorable conditions (depth, region, gear, season).
Depth	Spawning conditions improve with depth.	Stages that spawn more frequently are sampled more often at depth.
Region	Spawning conditions differ between regions (east and west).	Stages that spawn more frequently are disproportionately sampled between regions.
Gear	Spawning behavior makes some fish more vulnerable to one gear than another.	Stages that spawn more frequently are more vulnerable to one gear than another.
Season	Spawning conditions vary consistently through the year.	Stages that spawn more frequently are disproportionately sampled through the year.

value of 0 for a "failure" (spawning markers absent), and p is the probability of observing a spawning marker.

One of the most common forms of binomial regression is the logistic regression, which assumes that the probability p of observing a success (presence of spawning markers) can be well approximated by a logistic function of a linear combination of explanatory variables x:

$$p(x) = 1/\left(1 + e^{\alpha + \beta \bullet x}\right). \tag{2}$$

Equation (2) can be linearized by the logit transformation, $\ln[p/(1-p)]$ and falls into the class of generalized linear models handled by many standard statistical packages. Probit regression is similar except it assumes $p(x)$ follows the cumulative standard normal distribution. The two approaches tend to give similar results, but logistic regressions may be somewhat more robust to outliers inasmuch as the logistic function has thicker tails than the probit.

The logistic model above implies that the probability of observing an event changes monotonically with the value of the explanatory variable x, increasing to a maximum of one if β is negative and decreasing to a minimum of zero otherwise. It seems reasonable to expect a monotonic relationship between the probability of spawning (spawning fraction) and length or age, but even the largest, oldest females may not be in spawning condition at all times. For this reason it is advantageous to incorporate a scale parameter that allows the maximum spawning fraction to be something less than one, in which case the appropriate model would take the form

$$p(x) = \pi/\left(1 + e^{\alpha + \beta \bullet x}\right). \tag{3}$$

We allow for the possibility that the relationship between spawning fraction and age or length may vary across regions and gear types by allowing π, α, and β to vary categorically with each of the four combinations of gear and region (effectively gear–region–stage interaction terms). The relationship between spawning fraction and depth, if present, is assumed to be monotonic. This would be consistent with observations that there is an ontogenetic shift in the distribution of Red Snapper, and older, large fish tend to move to deeper water (Patterson 2007; Mitchell et al. 2014). If the cause of this shift is related to spawning behavior, then one might expect spawning frequency to increase with depth.

Field observations suggest that Red Snapper spawning is seasonal, as has also been demonstrated for Pouting *Trisopterus luscus*, and may be common among other species (Alonso-Fernández and Saborido-Rey 2011; Lowerre-Barbieri et al. 2011). A seasonal pattern would imply that the relationship between spawning fraction and time of year is dome-shaped rather than monotonic. A dome-shaped relationship could be mimicked by the use of categorical variables that represent discrete intervals of time, say 1 month, in which case our model would take the form:

$$p(s, d, r, g, m) = \pi_{grm}/\left(1 + e^{\alpha_{grm} + \beta_{gr}s + \delta d}\right), \tag{4}$$

where s is a continuous variable representing the stage of life (here either age or length), d is a continuous variable representing depth, m is a categorical variable representing the month, g is a categorical variable representing gear type (longline or vertical line), r is a categorical variable representing region (east or west of the Mississippi River), π_{grm} is the scaling parameter, which may vary categorically by gear, region, and month (absent a gear or regional effect, $\pi_{grt} = \pi_t$), α_{grt} is the linear intercept parameter, which may vary categorically by gear, region and month (absent a gear or regional effect, $\alpha_{grt} = \alpha_t$), β_{gr} is the slope parameter for age or length, which may vary categorically by gear and region (absent a gear and regional effect, $\beta_{gr} = \beta$), and δ is the slope parameter for depth.

An important drawback of the model described by equation (4) is the large number of parameters associated with the time intervals (months), some of which will be poorly estimated if few samples are available for that month (as was the case for some months in our Red Snapper example). Alternatively, one can use a functional form to represent the dependence of spawning fraction on time of year. We chose to use the flexible gamma function:

$$p(t|s) = \left(\frac{t}{\mu}\right)^{\mu/(\kappa+\gamma s)} e^{(\mu-t)/(\kappa+\gamma s)}, \tag{5}$$

where t is the fraction of the year elapsed beginning on January 1, μ is the mode (time of peak spawning) and the expression $\kappa + \gamma s$ represents the dispersion coefficient (which is allowed to increase linearly with age or length to accommodate the possibility of age or size dependence in the duration of the spawning season). Note that this form of the gamma density has been divided by the value at the mode such that the maximum value is always 1.0.

The full model used in this paper is given by the product of equation (4) (without the categorical month effects) and equation (5); i.e.,

$$p(s,d,r,g,t) = \frac{e^{(\mu-t)/(\kappa+\gamma s)}}{1+e^{\alpha_{gr}+\beta_{gr}s+\delta d}}\left(\frac{t}{\mu}\right)^{\mu/(\kappa+\gamma s)}. \tag{6}$$

This model allows the duration of the spawning season to vary with age or length, but assumes that it is the same across regions, gear types, and depth. For completeness, we note that all of the models posed in equations (1)–(6) treat the explanatory variables s (length or age) and d (depth) as though they are measured without error. In the case of length and depth, the measurements were made by scientific observers and the error is negligible for all practical purposes. The age of the fish is subject to reader error, but Allman et al. (2005) suggested that this error is relatively small and unbiased for Red Snapper.

Statistical estimation and comparisons of alternative models.—Equation (6) does not fall into the class of general linear models owing to the incorporation of the scaling parameters π and the gamma function of t. Nevertheless, the parameters are easily estimated by numerical minimization of the likelihood expression given by equation (1), where p_i is understood to be the probability of observing a spawning marker predicted by equation (6) given the values of the explanatory variables s, t, d, g, and r associated with sample i. In this study the numerical minimization was accomplished using Excel Solver and AD Model Builder.

Statistical comparisons among alternative models were made using Akaike's information criterion (AIC; Akaike 1973):

$$\text{AIC} = -2\ln L + 2n,$$

where n is the total number of parameters estimated and L is the measure of goodness of fit (e.g., likelihood function) being maximized. The AIC attempts to identify the most parsimonious explanation of the data by balancing the relative improvement in model fit against the number of parameters required to achieve that fit. The "best" model is considered to be the one with the lowest AIC. A "rule of thumb" is that differences in AIC (ΔAIC) of less than 2 constitute weak evidence that one model is better than another, differences between 3 and 10 are regarded as moderate evidence, and differences greater than 10 are regarded as strong evidence (Burnham and Anderson 2002). A pseudo-R^2 statistic was also computed as the fraction of the variance explained by the regression:

$$R^2 = 1 - \frac{\sum_i^N (o_i - \hat{p}_i)^2}{\sum_i^N (o_i - \bar{o})^2},$$

where $\bar{o}$ is the mean of the observations and $\hat{p}$ is the maximum likelihood estimate of the probability that a fish will have spawning markers.

Conversion to total number of spawns.—The average spawning fraction during the course of a year for age or length s is obtained as

$$\bar{p}(s) = p(s)\int_{t=0}^{1} p(t|s)dt. \tag{7}$$

In cases where $p(t|s)$ is estimated to be essentially 0 at $t = 1$,

$$\int_{t=0}^{1} p(t|s)dt \cong \int_{t=0}^{\infty} p(t|s)dt = \frac{\Gamma(\mu/(\kappa+\gamma s+1)\kappa^{\mu/(\kappa+\gamma s)+1}}{(\mu/e)^{\mu/(\kappa+\gamma s)}}, \tag{8}$$

where the right hand side of equation (8) is simply the inverse of the gamma density evaluated at the mode. In cases where spawning occurs throughout the year so that $p(t|s) > 0$ at $t = 1$, the integral in equation (8) can be evaluated numerically.

The prevalence of hydrated oocytes and postovulatory follicles (histological spawning markers) can be detected over a period of time and may be influenced by temperature and the diel spawning pattern. Following Priede and Watson (1993), the average spawning fraction can be converted to a daily probability of spawning P:

$$\bar{P}(s) = \frac{24}{T_M}\bar{p}(s) \tag{9}$$

where T_M is the duration in hours that spawning markers can be detected. The expected number of spawns during the course of a year is therefore obtained from multiplying equation (9) by the number of days as follows:

$$N(s) = 365\frac{24}{T_M}\frac{\Gamma(\mu/(\kappa+\gamma s+1)\kappa^{\mu/(\kappa+\gamma s)+1}}{(\mu/e)^{\mu/(\kappa+\gamma s)}}p(s). \qquad (10)$$

Final oocyte maturation and hydration have been estimated to begin during morning hours (0830 hours) and remain evident throughout the day (to about 1800 hours, or 10 h duration: Jackson et al. 2006). Field observations and spontaneous spawning in tanks have indicated that ovulation occurs in mid to late afternoon, peaking at about 1600 hours with spawning commonly occurring before sunset around 2000 hours in summer months (Papanikos et al. 2003, 2008; Jackson et al. 2006). While postovulatory follicles have been noted to appear as early as 1100 hours, new postovulatory follicles are not thought to be common until after 1600 hours as ovulation peaks and postovulatory follicle duration is not thought to exceed about 24 h in the Gulf of Mexico during summer (Nieland et al. 2002; Jackson et al. 2006). Together, these findings indicate that T_M is about 34 h.

RESULTS

Survey Summary

The 2011 congressional supplemental survey was conducted over 846 sea-days (April to October) and included 1,171 longline (four vessels) and 1,939 vertical-line (two vessels) stations (Campbell et al. 2012). Histological analyses were conducted on 1,002 females, of which 992 were aged. Based on comparison of the 50 paired quality control histology sections (freshly fixed versus initially frozen) it was visually apparent that freezing reduced the ability to resolve oocyte details such as germinal vesicles, follicular membranes, lipid vacuoles, and yolk globules. However, spawning markers (hydrated oocytes, postovulatory follicles) could be distinguished in the frozen samples. Comparison of the leading oocyte stage (primary growth, cortical alveolar, vitellogenesis, and oocyte maturation including hydration) yielded 90% agreement and presence–absence of postovulatory follicles yielded 82% agreement.

Females with spawning markers ($n = 391$) were distributed throughout the U.S. Gulf survey area, predominantly along the outer shelf at depths between 15 and 158 m (Figure 1). The distribution of females with or without spawning markers overlapped spatially; females without markers were caught at an average depth of 60 m and females with markers were caught at an average depth of 62 m. These depths center on a discrete depth zone (50–100 m) that encompasses the high-relief paleoreefs found on the outer shelf of the southeastern United States (Koenig et al. 2000).

Modeling the Spawning Fraction

Stepwise model building exercises were developed beginning with the null model ($p = \pi$), which assumed all fish have

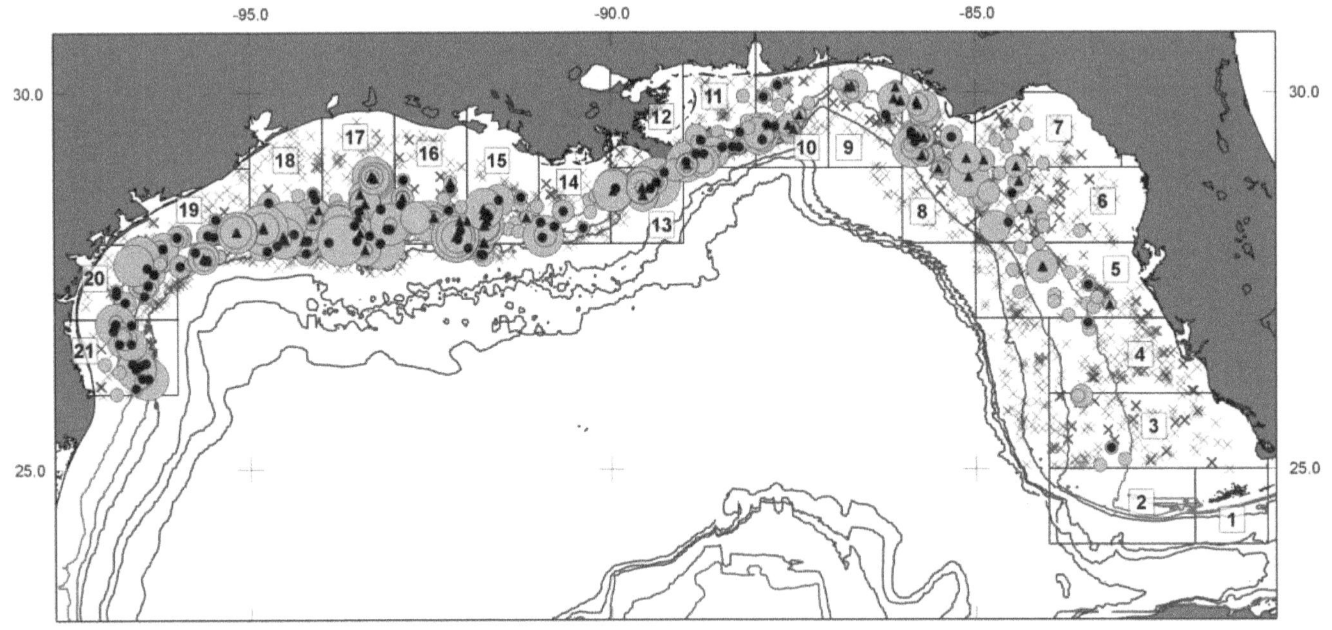

FIGURE 1. Distribution of sampling stations with zero (× symbol) and positive catches of Red Snapper during the 2011 supplemental survey (grey circle; progressively increasing size of circle indicates number landed per set: 1–4, 5–10, 10–13, 14–17, 18+). Solid dark symbols indicate capture locations of females with spawning markers (round symbols: bottom longline gear, triangles: bandit gear). The 50-, 100-, 200-, 1,000-, 1,300-, and 2,000-m isobaths are indicated. Statistical subareas are denoted from 1 to 21 within U.S. Gulf of Mexico waters. Numbers along horizontal and vertical borders of figure are longitude and latitude, respectively. Note that symbols are for graphical illustration and do not reflect scale of area fished per station.

the same probability of being found in spawning condition. In the first step, the logistic age or length effects were added to the null model. The corresponding AIC values were decreased by more than 40 and about 6% of the residual variance was explained (Table 4). Accordingly, the statistical evidence for both the age and length effects was strong and these terms were retained in all subsequent models.

In the second step, the spawning fraction was allowed to vary with time of year to accommodate the biological dynamics of the spawning season. Initially, the duration of the

TABLE 4. Length-based and age-based models for Red Snapper spawning fractions developed during the stepwise model building procedure (based on the binomial regression of equation 6). The shaded column highlights the final models. The use of t' indicates the model allows the seasonal effect to vary with age or length. Subscripts refer to gear or region (subscripts in parentheses identify the gear–region combination when gear and region effects are estimated simultaneously); l = length, a = age, t = time elapsed, d = depth, g = gear type, r = region (see text for explanation), NA = not applicable.

Statistic	P	$p(l)$	$p(l,t)$	$p(l,t')$	$p(l,t,d)$	$p(l,t,g)$	$p(l,t,r)$	$p(l,t',d,r,g)$
				Length-based models				
μ			0.536	0.535	0.537	0.538	0.542	0.541
κ			0.024	0.003	0.025	0.026	0.026	0.015
γ				0.000037				0.00002
δ					0.474			0.671
$\pi_{1\,(1,1)}$	0.390	0.547	0.866	0.858	0.886	0.775	0.986	0.987
$\alpha_{1\,(1,1)}$		3.295	9.084	12.762	6.311	9.803	63.044	43.042
$\beta_{1\,(1,1)}$		−0.008	−0.027	−0.042	−0.025	−0.030	−0.180	−0.130
$\pi_{2\,(2,1)}$						0.899	0.928	1.000
$\alpha_{2\,(2,1)}$						150.899	3.451	−6.253
$\beta_{2\,(2,1)}$						−0.436	−0.009	−0.040
$\pi_{1,2}$								0.822
$\alpha_{1,2}$								6.096
$\beta_{1,2}$								−0.027
$\pi_{2,2}$								0.884
$\alpha_{2,2}$								108.642
$\beta_{2,2}$								−0.334
AIC	1,339	1,283	1,000	992	986	999	989	981
r^2	NA	0.06	0.30	0.31	0.32	0.31	0.32	0.33
				Age-based models				
Statistic	p	$p(a)$	$p(a,t)$	$p(a,t')$	$p(a,t,d)$	$p(a,t,g)$	$p(a,t,r)$	$p(a,t',d,r,g)$
μ			0.538	0.535	0.537	0.539	0.542	0.540
κ			0.026	0.012	0.025	0.026	0.025	0.018
γ				0.0020				0.0010
δ					0.330			0.276
$\pi_{1\,(1,1)}$	0.388	0.563	0.940	0.878	0.920	0.896	1.000	1.000
$\alpha_{1\,(1,1)}$		1.889	1.696	1.547	1.170	1.653	6.808	6.658
$\beta_{1\,(1,1)}$		−0.513	−0.650	−0.861	−1.055	−0.636	−2.633	−3.189
$\pi_{2\,(2,1)}$						0.934	0.952	1.000
$\alpha_{2\,(2,1)}$						0.222	1.773	1,406.203
$\beta_{2\,(2,1)}$						−0.491	−0.573	−364.045
$\pi_{1,2}$								0.970
$\alpha_{1,2}$								0.620
$\beta_{1,2}$								−0.704
$\pi_{2,2}$								0.881
$\alpha_{2,2}$								753.651
$\beta_{2,2}$								−202.841
AIC	1,327	1,283	1,008	1,007	995	1,011	995	993
R^2	NA	0.05	0.29	0.29	0.30	0.29	0.31	0.32

spawning season was assumed to be independent of age or length (i.e., $\gamma = 0$). In that case the AIC values were further reduced by over 200 and about 30% of the residual variance was explained. Accordingly, the statistical evidence for a seasonal effect on the spawning fraction was strong and these terms were retained in all subsequent models.

In the third step, the potential variation in the duration of the spawning season with age or length was investigated by estimating the parameter γ. The resulting AIC values suggested little evidence for a variation in season duration with age (ΔAIC < 3) and only moderate evidence for a variation in season duration with length (ΔAIC < 10). Moreover, the percentage of variation explained by the model was negligibly improved.

Finally, the effects of incorporating the covariates depth, gear, and region were examined by adding the corresponding parameters to the seasonal model derived from step 2 (one covariate at a time). The model fits suggested little evidence for gear effects (ΔAIC < 3), but strong evidence for either a regional or depth effect (ΔAIC > 10). None of these models, however, contributed to explaining a substantial fraction of the variance. Even the full model incorporating all parameters did not substantially improve the explanatory power of the regression (Figure 2). For the length-based analysis, the r^2 for the full model was 33% compared with 30% for the length + season model. For the age-based analysis, the r^2 for the full model was 32% compared with 29% for the age + season model. Accordingly, the final model included only age or length and time of year.

The spawning season was estimated in the final model to occur primarily from early April through late October, with a peak in July (Figure 3). The spawning fraction was estimated to increase rapidly with length or age (Figure 4), starting at very low values for fish under 300 mm (age 1) and beginning to level off around 500 mm (age 6–7). There were no obvious trends in the residuals (difference between model fit and observed data) to suggest that the model was not fitting the oldest (or youngest) age-classes (Figure 2). The asymptotic estimates of the variance of the parameters and associated correlation matrix (obtained from the Hessian via the application in AD Model Builder) are shown in Table 5.

Conversion to Total Number of Spawns

The expected number of spawns per year was computed according to equation (10). The duration of the spawning season did not vary with age or size in the base models (the terms not being significantly different from zero); therefore, the integral of $p(t|s)$ was a constant equal to 0.295 for all ages and equal to 0.288 for all lengths. The value of $p(s)$ for the base model followed the logistic functions specified by the estimates for the parameters π, α, and β in Table 4. The value of T_M, the time during which the prevalence of hydrated oocytes and postovulatory follicles (histological spawning markers) in Red Snapper can be detected, was set to 34 h (see discussion below). The resulting ogive indicates that 2-year-old Red Snapper spawn an average of 29 times per year, while the largest, oldest Red Snapper spawn an average of about 71 times per year (Figure 5).

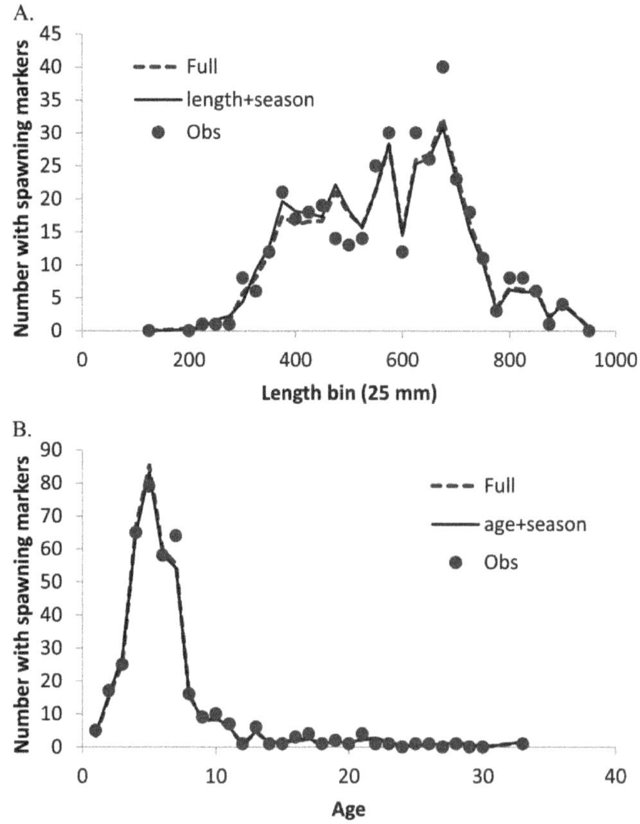

FIGURE 2. Observed number of Red Snapper with spawning markers compared with the corresponding predictions of the final model (length + season, or age + season) and full model (all covariates included). Panels show the results for (A) length and (B) age.

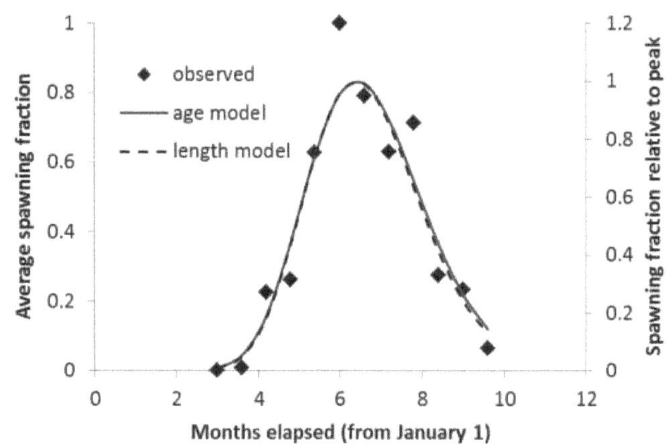

FIGURE 3. Average observed spawning fraction of Red Snapper (primary axis) compared with model predictions of relative proportion of females with spawning markers (secondary axis) by time of year for age-based and length-based models.

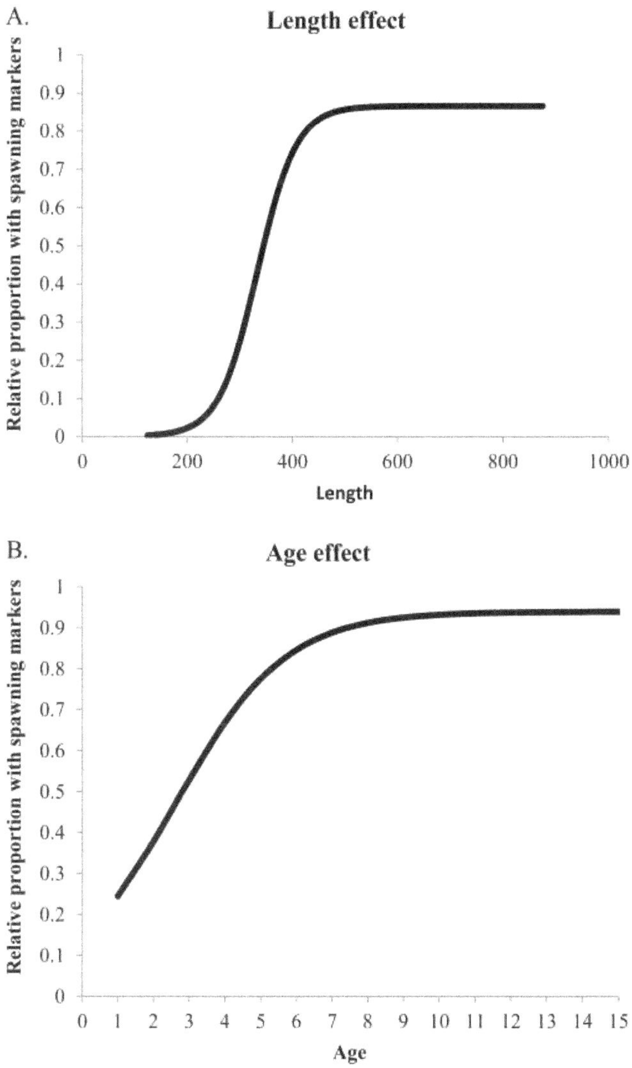

FIGURE 4. Predicted relationships between spawning fraction of Red Snapper (proportion of females with spawning markers) and (**A**) length (mm) or (**B**) age (years).

TABLE 5. Correlation matrix for final models to determine Red Snapper spawning fractions.

Statistic	Estimate	SE	Correlation coefficients π	α	β	μ	κ
			Length-based model				
π	0.8660	0.0315	1.0				
α	−0.2723	0.1086	0.2	1.0			
β	9.0840	3.5484	−0.2	−1.0	1.0		
μ	0.5363	0.0048	−0.3	0.2	−0.2	1.0	
κ	0.0244	0.0019	−0.5	0.2	−0.2	0.2	1.0
			Age-based model				
π	0.9399	0.0409	1.0				
α	−0.6504	0.1740	0.6	1.0			
β	1.6962	0.6135	−0.4	−0.9	1.0		
μ	0.5381	0.0050	0.0	0.1	0.0	1.0	
κ	0.0255	0.0019	−0.1	0.1	0.1	0.1	1.0

DISCUSSION

Our estimates of spawning fraction and related quantities such as spawning frequency are intended to reflect the population of sampled Red Snapper females regardless of whether they are active or not. Others have expressed these quantities in terms of active females only (Priede and Watson 1993; Stratoudakis et al. 2006; Kurita 2012). The two sets of metrics will lead to equivalent perceptions of the target population as long as the units are consistent. For example, the total number of spawning events in the population can be computed equivalently as either the product of spawning frequency per female and the number of females or the spawning frequency per active female and the number of active females. More importantly, neither set of metrics necessarily represented the average values observed in the population at large owing to the size selection of the gear and spatial variations in the availability of different size- and age-classes of the target species. It is possible in principle to adjust our estimates of spawning fraction to account for variations in selection and availability, but these are not as well understood for Gulf of Mexico Red Snapper as they are for a few other species (see Lowerre-Barbieri et al. 2009). Alternatively, population-level estimates of spawning frequency may be obtained by multiplying the estimated spawning frequency at size or age by independent estimates of the size or age structure of the female population (as might be obtained by a stock assessment). In this sense the procedure is similar to multiplying an age–length key by size composition data to determine the age structure of a population.

The estimates of spawning frequency (number of spawns per year) for Red Snapper from this study increase with age from 29 events per year for 2-year-old fish to about 71 per year for fish age 10 and older, corresponding to interspawning intervals ranging between 6.5 and 2.5 d. Not surprisingly, this range brackets previous estimates of interspawning intervals

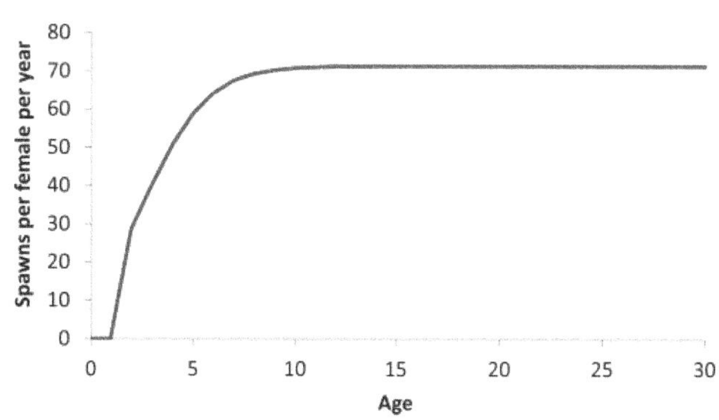

FIGURE 5. Estimated number of Red Snapper spawns at age (years) from the final model, $p(a,t')$, in Table 4.

that were averaged across multiple age-classes (3.97–5.95 d: Collins et al. 1996; 3.4–4.2 d: Woods 2003; 4.3 d: Brown-Peterson et al. 2009) and is consistent with the frequency of spontaneous spawning observed under experimental aquaculture conditions (Papanikos et al. 2008). There was no evidence that spawning frequency decreased for the oldest fish sampled, but even our oldest fish (34 years) was considerably younger than the maximum age of Red Snapper (over 50 years). While it is possible that senescence may occur at some point, it is unlikely to be an important factor to consider in stock assessments as such old fish are very rare in the population.

More importantly, the fact that younger mature Red Snapper do not spawn as often as older mature fish implies that they contribute less to the reproductive potential of the population than might be inferred from their maturity alone (Figure 6; Porch 2004). For this reason, the per-capita fecundity of each age- or size-class of Red Snapper should be modeled as the product of spawns per female per year and batch fecundity rather than as the product of maturity and batch fecundity. The effect of replacing maturity with spawns per female on the perception of the per-capita fecundity of young fish is somewhat mitigated by the rather low batch fecundity of younger Red Snapper (Figure 7a). This is true even when one accounts for the fact that young Red Snapper tend to be much more abundant than larger, older fish. If, for example, the per-capita fecundity at age is multiplied by the equilibrium age structure associated with the maximum sustainable yield (SEDAR 2013), one finds that fish between 2 and 4 years old account for only 3% of the total egg production based on spawning frequency compared with 5% based on maturity (Figure 7b).

The seasonality of spawning was in general agreement with previous studies. Spawning begins in late April or early May, reaches a peak in midsummer, and declines by fall. It is possible that some spawning may occur outside the time period sampled during this study (April–October) when conditions are favorable; however, this has not been detected during several prior studies spanning decades (Collins et al. 1996; Woods 2003). The data also hint that the spawning season of 2- and 3-year-old Red Snapper is shorter than that of older Red Snapper; no 2- and 3-year-old fish were found in spawning condition before late May and very few after August. The season durations predicted by the models where the parameter γ was estimated are consistent with this observation (Figure 8), but the statistical evidence was not strong and little additional variance was explained. It is possible, therefore, that this apparent trend is spurious. Nevertheless, the matter is of sufficient importance to merit further sampling. As Wright and Trippel (2009) pointed out, an increase in the duration of the spawning season with age has implications beyond merely increasing the total number of annual spawning opportunities. By expanding the duration of the spawning season, older fish may increase their chance of spawning during favorable

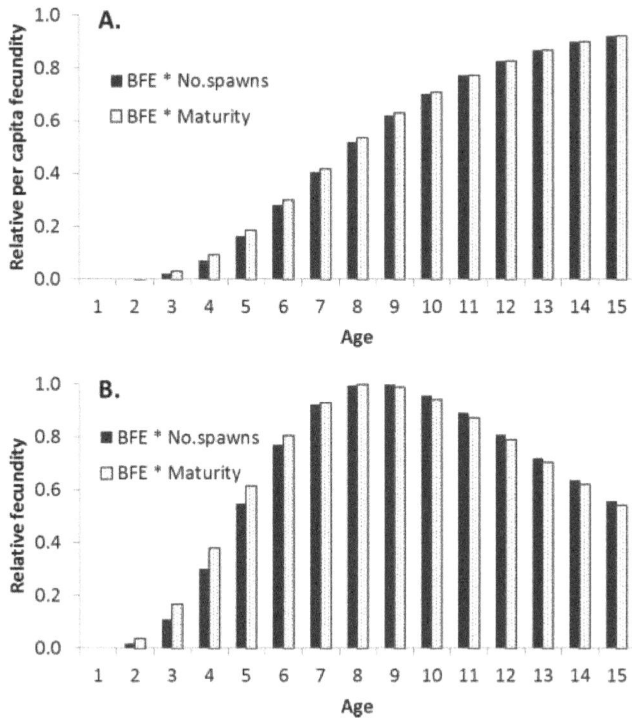

FIGURE 7. Trends in (**A**) relative per-capita fecundity at age (years) and (**B**) relative population fecundity at age (combined fecundity of all members in the age-class) in Red Snapper when the calculated number of spawns per female per year is used (solid bars) compared with the trends when spawning frequency is assumed to be proportional to maturity (open bars). The label "BFE * No. Spawns" refers to the calculation where age-specific estimates of batch fecundity are multiplied by age-specific estimates of the number of spawns per year. Similarly, the label "BFE * Maturity" refers to the calculation where age-specific estimates of batch fecundity are multiplied by age-specific estimates of maturity. The relative (population) fecundity is obtained by multiplying the per-capita fecundity at age by the relative abundance of each age-class at the equilibrium level associated with fishing at the maximum sustainable yield.

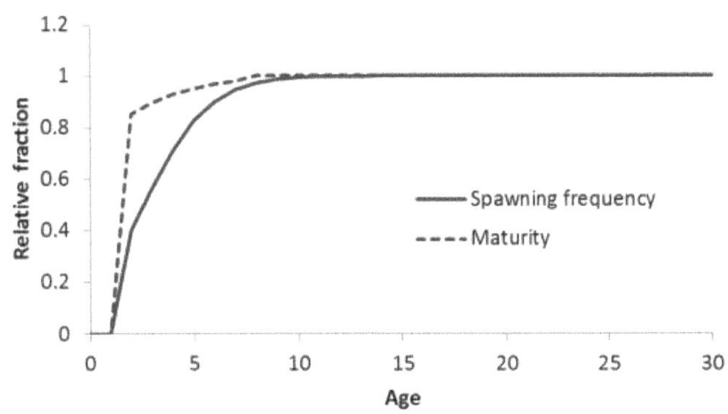

FIGURE 6. Comparison of maturity of Red Snapper and relative spawning frequency at age (years). Female maturity was based upon the presense of vitellogenic or maturing oocytes during the peak reproductive months of June, July, and August. The label on the vertical axis, relative fraction, refers to the value for maturity at age or spawning frequency at age divided by their respective maximum values (to put the two measures on the same scale).

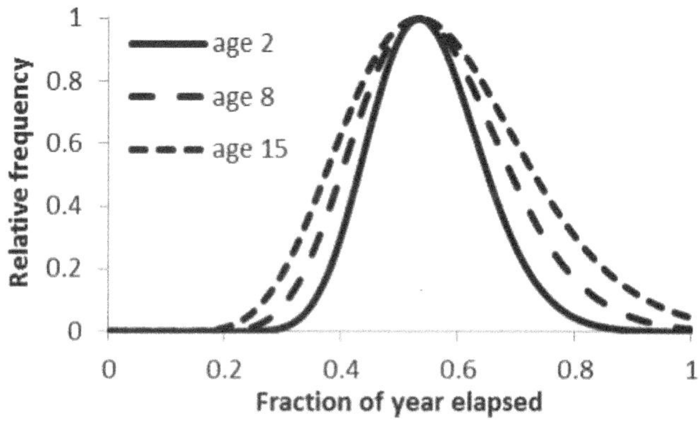

FIGURE 8. Estimated variation in the duration of the Red Snapper spawning season with age (i.e., ages 2, 8, and 15 years) corresponding to the final model, $p(a,t')$, in Table 4.

conditions, thereby reducing the variance and increasing the mean of progeny survival.

Spawning frequency was also estimated to increase with depth and to vary east and west of the Mississippi River. However, while the AIC indicated strong support for including either depth or region in the model, the support for including both simultaneously was weak. This implies that the two effects may be somehow correlated. The average depth sampled in the west (63 m) was slightly deeper than in the east (57 m), so it is possible that perceived regional differences may be partly attributable to differences in average depth. Nevertheless, it is important to point out that depth and region together contributed to explain less than 3% of the total variance and it remains possible that the estimated trends are spurious. This seems particularly likely in the case of the regional model where spawning frequency in the east was estimated to drop to essentially zero for fish below 400 mm largely as a consequence of having very few samples of fish below that size.

The existence of regional differences in the productivity of Red Snapper could have important implications for the management of Red Snapper. The highest spawning frequencies in our samples were exhibited by females caught in areas of the western Gulf associated with outer shelf banks and reef tracts (the western Louisiana shelf and central to the southern Texas shelf) which also happen to be the areas with the greatest concentration of Red Snapper eggs and larvae (Lyczkowski-Shultz and Hanisko 2007). The results of our study indicate that the regional effects were small, implying that the apparently higher spawning frequencies observed in these areas is mostly a reflection of the greater average size of fish and the time of year when the samples are taken, rather than any intrinsic differences in individual fish behavior. It is possible of course that spawning frequency does vary over different spatial and temporal scales owing to differences in ambient conditions. Nevertheless, it is clear that any comparisons must also account for possible differences in local population age structure and the time of year that the collections are made. A population that appears to enjoy a higher per-capita spawning frequency may simply have a higher proportion of older, larger fish. Similarly, the differences in the apparent spawning frequency of Red Snapper caught on longlines and vertical lines were attributable primarily to the differences in the size and age of fish caught by those gears.

The reproductive tissues used in this study were initially frozen, as opposed to the more common practice of preserving them first in 10% buffered formalin (but see Young et al. 2003). This occurred because the survey utilized commercial vessels working over extended days at sea with contracted samplers and, early in the study design, there were concerns about the routine use of formalin under such conditions. We tested the ability to discern spawning markers from frozen samples by comparing the results from blind reads of paired subsamples (one subsample initially frozen, one subsample initially fixed as fresh tissue) from 50 females. The results indicated 90% agreement for the leading oocyte stage and 82% agreement for presence–absence of postovulatory follicles, suggesting the ability to detect spawning markers may be slightly lower for frozen samples than for those fixed initially in formalin. Young et al. (2003) encountered similar sampling circumstances, but reported no difference between samples that were frozen and those that were fixed in formalin in terms of the classification of oocyte stage, presence or absence of postovulatory follicles, or degree of atresia. In any case, we have no reason to expect that the degree of underestimation we observed would vary in time and space. Therefore, while it is possible that spawning frequency could be somewhat higher than estimated here, the statistical inferences and estimated relative trends in spawning frequency would not be affected.

Previous studies have demonstrated a clear relationship between the batch fecundity of Red Snapper and size or age (Collins et al. 1996; Woods 2003; Porch et al. 2007; Kulaw 2012). The present study has further demonstrated that spawning frequency also increases with size or age and is consistent with the idea that it is important to conserve old ages in the population (Hixon et al. 2014). What remains is to determine how egg quality may vary with age and other factors. For example, egg quality and fertilization success in Red Snapper are known to be related to adult nutrition (Papanikos et al. 2003). Moreover, the abundance of Red Snapper has been increasing rapidly and is expected to continue increasing over the next few years. This raises the possibility that surplus energy for growth and reproduction may be reduced (Goodwin et al. 2006; McBride et al. 2015). Thus, there is a need to continue monitoring Red Snapper reproductive ecology while remaining mindful of important temporal and spatial scales.

ACKNOWLEDGMENTS

Thanks are due to Melissa Cook, the sampling team at the National Marine Fisheries Service (NMFS) Pascagoula Laboratory, and the charter vessel crews. Beverly Barnett and Hanna Lang organized and proofed the data from the biological collections. Robert Allman and the NMFS Panama City aging team provided the age data. Nikolai Klibansky, Matthew Campbell, and two anonymous reviewers gave many helpful comments that improved the manuscript.

REFERENCES

Akaike, M. 1973. Information theory and an extension of the maximum likelihood principle. Pages 267–281 *in* B. Petrov and F. Csaki, editors. Proceedings of the 2nd international symposium on information theory. Akademia Kiado, Budapest.

Allman, R., B. Barnett, H. Trowbridge, L. Goetz, and N. Evou. 2012. Red Snapper (*Lutjanus campechanus*) otolith ageing summary for collection years 2009–2011. Southeast Data Assessment and Review, SEDAR31-DW05, North Charleston, South Carolina.

Allman, R. J., and G. R. Fitzhugh. 2007. Temporal age progressions and relative year-class strength of Gulf of Mexico Red Snapper. Pages 311–328 *in* W. F. Patterson III, J. H. Cowan Jr., G. R. Fitzhugh, and D. L. Nieland, editors. Red Snapper ecology and fisheries in the U. S. Gulf of Mexico. American Fisheries Society, Symposium 60, Bethesda, Maryland.

Allman, R. J., G. R. Fitzhugh, K. J. Starzinger, and R. A. Farsky. 2005. Precision of age estimation in Red Snapper (*Lutjanus campechanus*). Fisheries Research 73:123–133.

Alonso-Fernández, A., and F. Saborido-Rey. 2011. Maternal influence on the variation of the reproductive cycle of *Trisopterus luscus* (Gadidae). Ciencias Marinas 37:619–632.

Brown-Peterson, N., K. M. Burns, and R. M. Overstreet. 2009. Regional differences in Florida Red Snapper reproduction. Proceedings of the Gulf and Caribbean Fisheries Institute 61:149–155.

Burnham, K. P., and D. R. Anderson 2002. Model selection and multimodal inference: a practical information-theoretic approach, 2nd edition. Springer-Verlag, New York.

Campbell, M. D., A. Pollack, T. Henwood, J. Provaznik, and M. Cook. 2012. Summary report of the Red Snapper (*Lutjanus campechanus*) catch during the 2011 congressional supplemental sampling program. Southeast Data Assessment and Review, SEDAR31-DW17, North Charleston, South Carolina.

Collins, L. A., A. G. Johnson, and C. P. Keim. 1996. Spawning and annual fecundity of the Red Snapper (*Lutjanus campechanus*) from the northeastern Gulf of Mexico. Pages 174–188 *in* F. Arreguin-Sanchez, J. L. Munro, M. C. Balgos, and D. Pauly, editors. Biology, fisheries and culture of tropical groupers and snappers. International Center of Living Aquatic Resources Management, Conference Proceedings 48, Manila.

Cooper, W. T., L. R. Barbieri, M. D. Murphy, and S. K. Lowerre-Barbieri. 2013. Assessing reproductive potential in species with indeterminate fecundity: effects of age truncation and size-dependent reproductive timing. Fisheries Research 138:31–41.

Fitzhugh, G. R., K. W. Shertzer, G. T. Kellison, and D. M. Wyanski. 2011. Review of size- and age-dependence in batch spawning: implications for stock assessment of fish species exhibiting indeterminate fecundity. U.S. National Marine Fisheries Service Fishery Bulletin 110:413–425.

Goodwin, N. B., A. Grant, A. L. Perry, N. K. Dulvey, and J. D. Reynolds. 2006. Life history correlates of density dependent recruitment in marine fishes. Canadian Journal of Fisheries and Aquatic Sciences 63:494–509.

Goodyear, C. P. 1995. Red Snapper in U.S. waters of the Gulf of Mexico. Southeast Fisheries Center, Contribution MIA-95/96-05, Miami.

He, X., J. C. Field, S. G. Beyer, and S. M. Sogard. 2015. Effects of size-dependent relative fecundity specifications in fishery stock assessments. Fisheries Research 165:54–62.

Hixon, M. A., D. W. Johnson, and S. M. Sogard. 2014. BOFFFFs: on the importance of conserving old-growth age structure in fishery populations. ICES Journal of Marine Science 71:2171–2185.

Hunter, J. R., and B. J. Macewicz. 1985. Measurement of spawning frequency in multiple spawning fishes. NOAA Technical Report NMFS 36:79–94.

Jackson, M. W., D. L. Nieland, and J. H. Cowan 2006. Diel spawning periodicity of Red Snapper *Lutjanus campechanus* in the northern Gulf of Mexico. Journal of Fish Biology 68:695–706.

Klibansky, N., and F. S. Scharf. 2013. Size-dependent and temporal variability in batch number and fecundity of Red Porgy, a protogynous, indeterminate spawner, in the U.S. South Atlantic. Marine and Coastal Fisheries: Dynamics, Management and Ecosystem Science [online serial] 5:39–52.

Koenig, C. C., F. C. Coleman, C. B. Grimes, G. R. Fitzhugh, K. M. Scanlon, C. T. Gledhill, and M. Grace. 2000. Protection of fish spawning habitat for the conservation of warm-temperate reef-fish fisheries of shelf-edge reefs of Florida. Bulletin of Marine Science 66:593–616.

Kulaw, D. 2012. Habitat- and region-specific reproductive biology of female Red Snapper (*Lutjanus campechanus*) in the Gulf of Mexico. Master's thesis. Louisiana State University, Baton Rouge.

Kurita, Y. 2012. Revised concepts for estimation of spawning fraction in multiple batch spawning fish considering temperature-dependent duration of spawning markers and spawning time frequency distribution. Fisheries Research 117–118:121–129.

Kurita, Y., Y. Fujinami, and M. Amano. 2010. The effect of temperature on the duration of spawning markers—migratory-nucleus and hydrated oocytes and postovulatory follicles—in the multiple-batch spawner Japanese flounder (*Paralichthys olivaceus*). U.S. National Marine Fisheries Service Fishery Bulletin 109:79–89.

LaPlante, L. H., and E. T. Schultz. 2007. Annual fecundity of tautog in Long Island Sound: size effects and long-term changes in a harvested population. Transactions of the American Fisheries Society 136:1520–1533.

Lowerre-Barbieri, S. K., K. Ganias, F. Saborido-Rey, H. Murua, and J. R. Hunter. 2011. Reproductive timing in marine fishes: variability, temporal scales, and methods. Marine and Coastal Fisheries: Dynamics, Management and Ecosystem Science [online serial] 3:71–91.

Lowerre-Barbieri, S. K., N. Henderson, J. Llopiz, S. Walters, J. Bickford, and R. Muller. 2009. Defining a spawning population (Spotted Seatrout, *Cynoscion nebulosus*) over temporal, spatial, and demographic scales. Marine Ecology Progress Series 394:231–245.

Lyczkowski-Shultz, J., and D. S. Hanisko. 2007. A time series of observations on Red Snapper larvae from SEAMAP surveys, 1982–2003: seasonal occurrence, distribution, abundance and size. Pages 3–23 *in* W. F. Patterson III, J. H. Cowan Jr., G. R. Fitzhugh, and D. L. Nieland, editors. Red Snapper ecology and fisheries in the U. S. Gulf of Mexico. American Fisheries Society, Symposium 60, Bethesda, Maryland.

McBride, R. S., S. Somarakis, G. R. Fitzhugh, A. Albert, N. A. Yaragina, M. J. Wuenschel, A. Alonso-Fernández, and G. Basilone. 2015. Energy acquisition and allocation to egg production in relation to fish reproductive strategies. Fish and Fisheries 16:23–57.

Mehault, S., R. Dominguez-Petit, S. Cervino, and F. Saborido-Rey. 2010. Variability in total egg production and implications for management of the southern stock of European Hake. Fisheries Research 104:111–122.

Mitchell, W. A., G. T. Kellison, N. M. Bacheler, J. C. Potts, C. M. Schobernd, and L. F. Hale. 2014. Depth-related distribution of post juvenile Red Snapper in southeastern U.S. Atlantic Ocean waters: ontogenetic patterns and implications for management. Marine and Coastal Fisheries: Dynamics, Management and Ecosystem Science [online serial] 6:143–155.

Murua, M., G. Kraus, F. Saborido-Rey, P. R. Witthames, A. Thorsen, and S. Junquera. 2003. Procedures to estimate fecundity of marine fish species in relation to their reproductive strategy. Journal of Northwest Atlantic Fisheries Science 33:33–34.

Nieland, D. L., G. Thomas, and C. A. Wilson. 2002. Age, growth, and reproduction of Spotted Seatrout in Barataria Bay, Louisiana. Transactions of the American Fisheries Society 131:245–259.

Papanikos, N., R. P. Phelps, D. A. Davis, A. Ferry, and D. Maus. 2008. Spontaneous spawning of captive Red Snapper, *Lutjanus campechanus*, and dietary lipid effect on reproductive performance. Journal of the World Aquaculture Society 39:324–338.

Papanikos, N., R. P. Phelps, K. Williams, A. Ferry, and D. Maus. 2003. Egg and larval quality of natural and induced spawns of Red Snapper, *Lutjanus campechanus*. Fish Physiology and Biochemistry 28:487–488.

Patterson, W. F. III. 2007. A review of movement in Gulf of Mexico Red Snapper: implications for population structure. Pages 245–261 *in* W. F.

Patterson III, J. H. Cowan Jr., G. R. Fitzhugh, and D. L. Nieland, editors. Red Snapper ecology and fisheries in the U.S. Gulf of Mexico. American Fisheries Society, Symposium 60, Bethesda, Maryland.

Porch, C. E. 2004. Batch-fecundity and maturity estimates for the 2004 assessment of Red Snapper in the Gulf of Mexico, revised. Southeast Data Assessment and Review, SEDAR7-AW-5, North Charleston, South Carolina.

Porch, C. E. 2007. An assessment of the Red Snapper fishery in the U.S. Gulf of Mexico using a spatially-explicit age-structured model. Pages 355–384 *in* W. F. Patterson III, J. H. Cowan Jr., G. R. Fitzhugh, and D. L. Nieland, editors. Red Snapper ecology and fisheries in the U.S. Gulf of Mexico. American Fisheries Society, Symposium 60, Bethesda, Maryland.

Porch, C. E., G. R. Fitzhugh, M. S. Duncan, L. A. Collins, and M. W. Jackson. 2007. Modeling the dependence of batch fecundity on size and age for use in stock assessments of Red Snapper in U.S. Gulf of Mexico waters. Pages 229–244 *in* W. F. Patterson III, J. H. Cowan Jr., G. R. Fitzhugh, and D. L. Nieland, editors. Red Snapper ecology and fisheries in the U.S. Gulf of Mexico. American Fisheries Society, Symposium 60, Bethesda, Maryland.

Priede, I. G., and J. J. Watson. 1993. An evaluation of the daily egg production method for estimating biomass of Atlantic Mackerel (*Scomber scombrus*). Bulletin of Marine Science 53:891–911.

SEDAR (Southeast Data Assessment and Review). 2013. Gulf of Mexico Red Snapper stock assessment report, SEDAR 31. SEDAR, North Charleston, South Carolina.

Shipp, R. L., and S. A. Bortone. 2009. A perspective of the importance of artificial habitat on the management of Red Snapper in the Gulf of Mexico. Reviews in Fisheries Science 17:41–47.

Stratoudakis, Y., M. Bernal, K. Ganias, and A. Uriarte. 2006. The daily egg production method: recent advances, current applications and future challenges. Fish and Fisheries 7:35–57.

Wilson, C. A., and D. L. Nieland. 2001. Age and growth of Red Snapper, *Lutjanus campechanus*, from the northern Gulf of Mexico off Louisiana. U.S. National Marine Fisheries Service Fishery Bulletin 99: 653–664.

Winemiller, K. O., and K. A. Rose. 1992. Patterns of life-history diversification in North American fishes: implications for population regulation. Canadian Journal of Fisheries and Aquatic Sciences 49:2196–2218.

Woods, M. K. 2003. Demographic differences in reproductive biology of female Red Snapper (*Lutjanus campechanus*) in the northern Gulf of Mexico. Master's thesis. University of South Alabama, Mobile.

Wright, P. J., and E. A. Trippel. 2009. Fishery-induced demographic changes in the timing of spawning: consequences for reproductive success. Fish and Fisheries 10:283–284.

Young, J., A. Drake, M. Brickhill, J. Farley, and T. Carter. 2003. Reproductive dynamics of Broadbill Swordfish, *Xiphias gladius*, in the domestic longline fishery off eastern Australia. Marine and Freshwater Research 54:315–332.

9

Genetic Stock Structure of *Terapon jarbua* in Taiwanese Waters

Shang-Yin Vanson Liu
Department of Oceanography and Department of Marine Biotechnology and Resources, National Sun Yat-Sen University, Kaohsiung 80424, Taiwan

I-Hsiang Huang
Department of Oceanography, National Sun Yat-Sen University, Kaohsiung 80424, Taiwan

Min-Yun Liu
Taiwan Ocean Research Institute, National Applied Research Laboratories, Kaohsiung 852, Taiwan

Hung-Du Lin
Affiliated School of National Tainan First Senior High School, Tainan 701, Taiwan

Feng-Yu Wang
Taiwan Ocean Research Institute, National Applied Research Laboratories, Kaohsiung 852, Taiwan

Te-Yu Liao*
Department of Oceanography, National Sun Yat-Sen University, Kaohsiung 80424, Taiwan

Abstract

Over the past century, stock identification, which is crucial for fisheries stock assessments, has relied heavily on intraspecies variations to differentiate management units. In recent years, however, DNA approaches have shed additional light on some aspects of the natural history and ecology of species and stocks. The Crescent Grunter *Terapon jarbua* is widely distributed across the Indo-Pacific region. It inhabits coastal waters with sandy substrates and tends to aggregate in estuaries. In the present study, we collected 96 individuals of *T. jarbua* from 10 locations around the shoreline of Taiwan as well as near Guangdong, China. A concatenated sequence (1,753 bp) of mtDNA (cytochrome *c* oxidase I and cytochrome *b*) was obtained from all individuals. We discovered two genetically distinct clades (lineages A and B) with different historical demographies occurring sympatrically except at the Yilan collection site, which was dominated by lineage A haplotypes. Connectivity within this region is high according to F_{ST} and AMOVA tests. The genetic variation between the two clades is far below the interspecies threshold for the genus *Terapon* (0.007 versus 0.3156). Therefore, we suggest that they be considered different genetic stocks from a fisheries management point of view and that future stock reassessment should be conducted based on the genetic information provided in this study. Further large-scale sampling is needed to understand the mechanisms that drive genetic partitioning on regional (Northwest Pacific) and global (Indo-Pacific) scales.

Subject editor: Kristina Miller, Pacific Biological Station, Nanaimo, British Columbia

*Corresponding author: swp0117@gmail.com

Many marine ecosystems are being degraded by ongoing catastrophic impacts caused by human activities such as overfishing and pollution. The former is currently considered the main factor accounting for the accelerated shrinkage of populations and extinction of species (Jackson et al. 2001; Worm et al. 2006). According to Worm et al. (2006), the fisheries of all currently harvested taxa will collapse by the year 2048 if current exploitation rates persist. Drastic declines in population size due to overfishing could result in a loss of genetic diversity, which may in turn impair species' ability to survive in changing environments. This could produce a vicious cycle leading to failure to recover, even smaller population sizes, and the eventual collapse of marine ecosystems (Worm et al. 2006; Cano et al. 2008).

Stock identification is an interdisciplinary science for studying fisheries stocks, through which appropriate scales of harvesting and monitoring are developed to maintain sustainable fisheries and protect endangered species (Begg et al. 1999). The definition of a fish stock has varied with the approach to identifying and classifying stocks and evolved with management needs (Begg and Waldman 1999). Early definitions relied on distinct internal dynamics and fishery characteristics (Russell 1931). Over the past 100 years, they have strongly depended on different methodological approaches designed to evaluate the homogeneity of populations, e.g., measurements of meristic and reproductive traits as well as geographic distributions. However, these methods can overlook stock structure when dealing with cryptic species/linages that cannot be distinguished by biological and morphological traits. Without proper methodologies for evaluating stock structure, multiple subpopulations of any given species have generally been managed as a single homogeneous unit. Additionally, due to the enormous size of the ocean and the lack of apparent physical barriers, local populations were considered open systems until a few decades ago (review in Swearer et al. 2002). However, recent evidence has shown that environmental and landscape features can greatly affect and shape population structure at relatively small scales in terms of the life history diversification and local adaptation of aquatic organisms (Storfer et al. 2007). Molecular genetics has made remarkable contributions to our understanding of populations, providing novel insight into previously inaccessible aspects of biological disciplines, e.g., new species descriptions (Allen and Erdmann 2012; Liu et al. 2013) as well as species and management unit identification (Liu et al. 2013; Lim et al., in press). This is especially true for specimens lacking morphological and biological differences among populations (Viñas et al. 2014) or having no morphological information, e.g., shark fins and fish filets (Cutarelli et al. 2014). Genetic markers have also been successfully applied to discriminating the stock structure of fishes (Wu et al. 2014; Singh et al., in press). There is an urgent need to conduct research of this type since species might be genetically heterogeneous over their ranges.

The Crescent Grunter *Terapon jarbua* is widely distributed across the Indo-Pacific region, from the Red Sea to Fiji and Japan (Vari 2001). It inhabits coastal waters with sandy substrates and is euryhaline, with juveniles commonly being found in brackish and fresh waters. It feeds mainly on small fish and invertebrates. Its planktonic larval duration is relatively short and stable, averaging 25.1 d (Lavergne 2012). The *Terapon jarbua* occurs along the coast of Taiwan, especially in estuaries, and is a popular game fish caught mainly for food. Larvae and juveniles enter estuaries in great abundance during May and November for feeding (Miu et al. 1990). Once fully grown, they return to deeper waters away from estuaries for spawning. Although the commercial fishery is minor in terms of total amount, *T. jarbua* is a high-priced species in Taiwan, with adults being caught by long lines and trawlers at depths shallower than 50 m. Its reproduction, body deformation caused by warm water discharged from nuclear power plants, and linear alkylbenzene sulfonate toxicity have been investigated for Taiwanese waters (Miu et al. 1990; Wang and Huang 1995; Chang et al. 2010). However, studies focusing on genetic stock structure have not yet been conducted.

In the present study, the cytochrome *b* (cyt *b*) and cytochrome *c* oxidase I (COI) regions of mtDNA were used as molecular markers to examine the level of genetic differentiation and the historical demography of *T. jarbua* in Taiwanese waters for future management and conservation purposes.

METHODS

Sample collection and DNA extraction.—A total of 96 *T. jarbua* were collected from 10 locations around the shoreline of Taiwan, including New Taipei City (TP), Yilan County (YL), Hua-lian City (HL), Hua-lian County (HFB), Tai-Tung County (TT), Kaohsiung County (KH), Chiayi County (CI), Taichung County (TC), Hsinchu County (HC), Penghu County (PH), and a site near Guangdong, China (RP) (Figure 1). Sample sizes from each locality ranged from 5 to 10 individuals. Genomic DNA was extracted from muscle tissue or fin clips using commercial DNA extraction kits (Genomics BioSci and Tech, Taiwan). DNA extracts were diluted in tris–EDTA buffer and stored at −20°C until amplification by polymerase chain reaction (PCR).

Genetic diversity and phylogenetic analyses.—Complete cyt *b* and partial COI genes were amplified with the following primers: COI: FishF1 = 5′-TCAACCAACCACAAAGACATTGGCAC-3′, FishR1 = 5′-ACTTCAGGGTGACCGAAGAATCAGAA-3′ (Ward et al. 2005); cyt *b*: Thr33 = 5′-TAACCTTCAGCCTCCTGCTTACA-3′, Glu31 = 5′-GTGACTTGAAAAACCACCGTT-3′ (Davis et al. 2012). Polymerase chain reactions were run in 30-μL amounts containing 10–40 ng template DNA, 3 μL 10× buffer, 0.2 mM dNTPs, 1.5 mM $MgCl_2$, 10 mM of each primer, and 0.2 units of *Taq* polymerase (MDbio, Taipei). The thermocycling profile consisted of initial denaturation at 94°C for 2 min followed by 35 cycles of denaturation at 94°C for 30 s, annealing at 48°C for

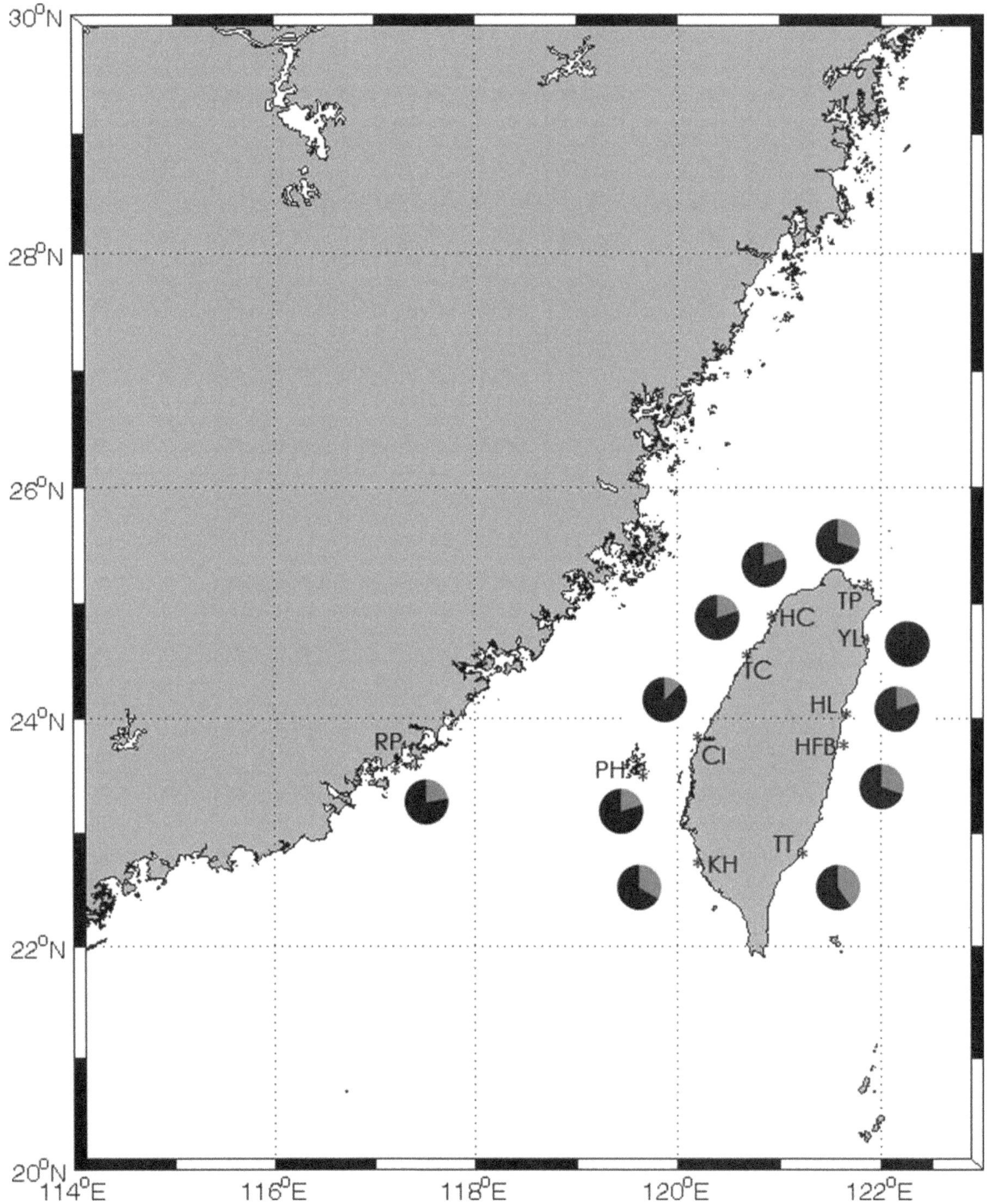

FIGURE 1. Locations sampled for *Terapon jarbua* in the coastal areas of Taiwan. Location codes are given in Table 1. Within the pie charts, black indicates the percentage of lineage A fish and gray the percentage of lineage B fish.

30 s, extension at 70°C for 40 s, and a final extension at 72°C for 2 min. The purification of PCR products and sequencing reactions was done by Genomics BioSci and Tech. Nucleotide sequences of forward and reverse strands were determined using an ABI 3730XL automated sequencer (Applied Biosystems, Carlsbad, California). Sequences were assembled and edited using Sequencher version 4.2 software (Gene Codes, Ann Arbor, Michigan). The sequences obtained in this study were deposited in GenBank under accession numbers KP204162–KP204259 and KP152133–KP152230 for COI and cyt *b*, respectively. Sequences were aligned using CLUSTAL W (Thompson et al. 1994), followed by manual editing with

Sequencher 4.2 (Gene Codes). We concatenated the sequences of two loci for all individuals, and the analyses performed in this study were based on this data set. Arlequin 3.5 (Excoffier and Lischer 2010) was used to analyze genetic diversity indexes and index of population differentiation (Φ_{st}). Unique haplotypes were quantified and the genetic diversity of populations calculated. Analysis of molecular variance (AMOVA; Excoffier et al. 1992) was performed under two hypothetical groupings and the proportions of variations among regions (Φ_{ct}), among populations within regions (Φ_{sc}), and within populations (Φ_{st}) were estimated. One of the groupings was based on the difference in coastal habitats between Taiwan's west coast, which is primarily characterized by muddy or sandy bottoms (RP, KH, PH, HC, TC, CI and TP), and its east coast, which is mainly rocky or reef habitat (TT, HFB, HL and YL). Another grouping included these two areas but also considered a possible genetic break between China and Taiwan, caused by either the currents of the Taiwan Strait or the emergence of the continental shelf under the Taiwan Strait during the Pleistocene Epoch (Wang 1999). Therefore, in this latter grouping RP was separated from the west coast group. Median-joining parsimony analysis was done with NETWORK (Bandelt et al. 1999) on the nucleotide sequence matrix of the 96 concatenated sequences using the program's default settings.

The general time reversible model plus the gamma rate heterogeneity model (GTR+G) was selected by MEGA 6 (Tamura et al. 2013) as the best-fitting substitution model based on the Bayesian information criterion. Maximum likelihood tree searching was conducted in RAxML-HPC BlackBox under CIPRES Science Portal version 3.1. Sequences of closely related species (*T. theraps* and *T. puta)* were downloaded from GenBank for estimating interspecies Kimura two-parameter (K2P) genetic distance with MEGA 6 (Tamura et al. 2013).

Additionally, Bayesian phylogenetic reconstructions were created by Mr. Bayes 3.12 (Ronquist and Huelsenbeck 2003) with the K2P model and default priors. Subsequently, two duplicate runs of three heated and one cold Markov chain–Monte Carlo (MCMC) chains were established, and each was initiated from a random tree and run for 1,000,000 generations. The convergence diagnostic was applied and the stop probability was set to 0.01. Trees were then sampled every 100 generations and a consensus tree built for all trees with the exclusion of the first 25% of sampled trees to allow for sufficient burn-in.

Historical demography and molecular dating.—The significance of the *F*-statistics for population comparisons was assessed using 1,000 permutations. In addition, a pairwise mismatch distribution comprised of the pairwise differences among all haplotypes was performed for the historical demographic test of samples of the two major clades based on the deep divergence found in the phylogenetic and minimum-spanning trees. The distribution is usually multimodal when data comply with a demographic equilibrium. In contrast, a unimodal distribution may indicate a recent demographic expansion (Slatkin and Hudson 1991).

The software BEAST version 1.7.5 (Drummond and Rambaut 2007) was used to estimate the time to the most recent common ancestor (TMRCA) using an MCMC Bayesian approach. Divergence rates of approximately 2.0% and 1.2% per million years for cyt *b* and COI, respectively, were used to estimate the absolute TMRCA values (Horne et al. 2012; Taillebois et al. 2013). All analyses were performed using the GTR+G model of nucleotide substitution. Two independent MCMC chains were run for 100 million generations, with sampling every 10,000 generations and 10% burn-in of the posterior samples. The effective sampling size parameter was found to exceed 200, implying that there was acceptable mixing and sufficient sampling. The analysis was run five times to test the stability and convergence of the MCMC chains in plots of posterior log likelihoods in Tracer version 1.6 (Rambaut et al. 2014).

RESULTS

A 1,753-bp concatenated sequence of mtDNA COI (612 bp) and cyt *b* (1,141 bp) was analyzed for 96 individuals obtained from 11 locations (Figure 1) in Taiwanese waters. The nucleotide composition was as follows: 24% adenine, 28.7% thymine, 31.6% cytosine, and 15.7% guanine. In total, 106 variable sites and 45 parsimoniously informative sites were found. Among the 96 individuals sequenced, 63 unique haplotypes were identified. Nucleotide diversity (π) ranged from 0.0013 ± 0.0009 (mean $\pm$ SD) to 0.0079 ± 0.0045, and haplotype diversity (h) ranged from 0.8571 ± 0.1371 to 1.0000 ± 0.0625 among locations (Table 1).

The phylogenetic analysis revealed two clades among the 96 individuals, herein named lineage A and lineage B, both having strongly supporting values on the nodes (99 maximum likelihood, 0.99 Bayesian support; Figure 2). Both lineages were found at all locations except YL, where we observed only the lineage A haplotype. The general topology of the median-joining tree corresponded with the maximum likelihood tree, and the two lineages were separated by nine mutation steps. Lineage A showed star-like branches (Figure 3). Pairwise θ estimates based on concatenated sequences ranged from −0.09184 to 0.17288, and only two comparisons (TC with HFB and TC with RP) showed marginally significant genetic structure (Table 2). The results of the AMOVA comparisons of genetic variation revealed that the variance was mainly from within the population level and that there was no significant genetic structure under these two groupings ($P > 0.05$; Table 3). Mismatch distribution tests on lineage A and lineage B failed to reject the hypothesis of the sudden-expansion model (Figure 3). However, the topologies of the distributions differ, with lineage A being unimodal and lineage B being bimodal. The TMRCA of lineage A and lineage B analyzed using BEAST dated to 0.4305 million years ago (effective sampling size, 5,335.203; 95% credible interval, 0.3049–0.5687 million years ago). The average K2P genetic distance (only COI) among the three *Terapon* species was 0.3156,

TABLE 1. Sampling locations and diversity indices based on concatenated sequences (1,753 bp) in 11 populations of *Terapon jarbua* from Taiwanese waters. Abbreviations are as follows: N = sample size, h = haplotype diversity, and π = nucleotide diversity.

Code	Location	N	Cyt b	COI	h (mean ± SD)	π (mean ± SD)
TP	New Taipei City	10	10	10	0.8889 ± 0.0754	0.0049 ± 0.0020
YL	Yilan County	7	7	7	0.8571 ± 0.1371	0.0013 ± 0.0009
HL	Hua-lian City	8	8	8	1.0000 ± 0.0625	0.0037 ± 0.0022
HFB	Hua-lian County	10	10	10	1.0000 ± 0.0447	0.0065 ± 0.0037
TT	Tai-Tung County	10	10	10	1.0000 ± 0.0447	0.0065 ± 0.0036
KH	Kaohsiung County Taiwan	9	9	9	1.0000 ± 0.0524	0.0079 ± 0.0045
CI	Chiayi County	8	8	8	0.9286 ± 0.0844	0.0051 ± 0.0030
TC	Taichung County	5	5	5	0.9000 ± 0.1610	0.0063 ± 0.0040
HC	Hsinchu County	10	10	10	0.9778 ± 0.0540	0.0055 ± 0.0031
PH	Penghu County	10	10	10	1.0000 ± 0.0447	0.0056 ± 0.0032
RP	Guangdong, China	9	9	9	0.9167 ± 0.0920	0.0046 ± 0.0027
Total		96	96	96		

and the distance between lineage A and lineage B within *T. jarbua* was 0.007 (Table 4).

DISCUSSION

Genetic Diversity

High haplotype diversity (h = 0.86–1) and low nucleotide diversity (π = 0.0013–1) were found in the 11 populations of *T. jarbua*. This phenomenon is commonly observed in marine fishes and has been interpreted as a sign of historical population expansion (Grant and Bowen 1998). The results of the mismatch distribution tests for lineages A and B also support the expansion model. Such results have been inferred to be closely related to population expansion after glaciation in several marine fishes (Tzeng 2007; Horne et al. 2008; Reece et al. 2010). In addition, lineage A showed a typical unimodal

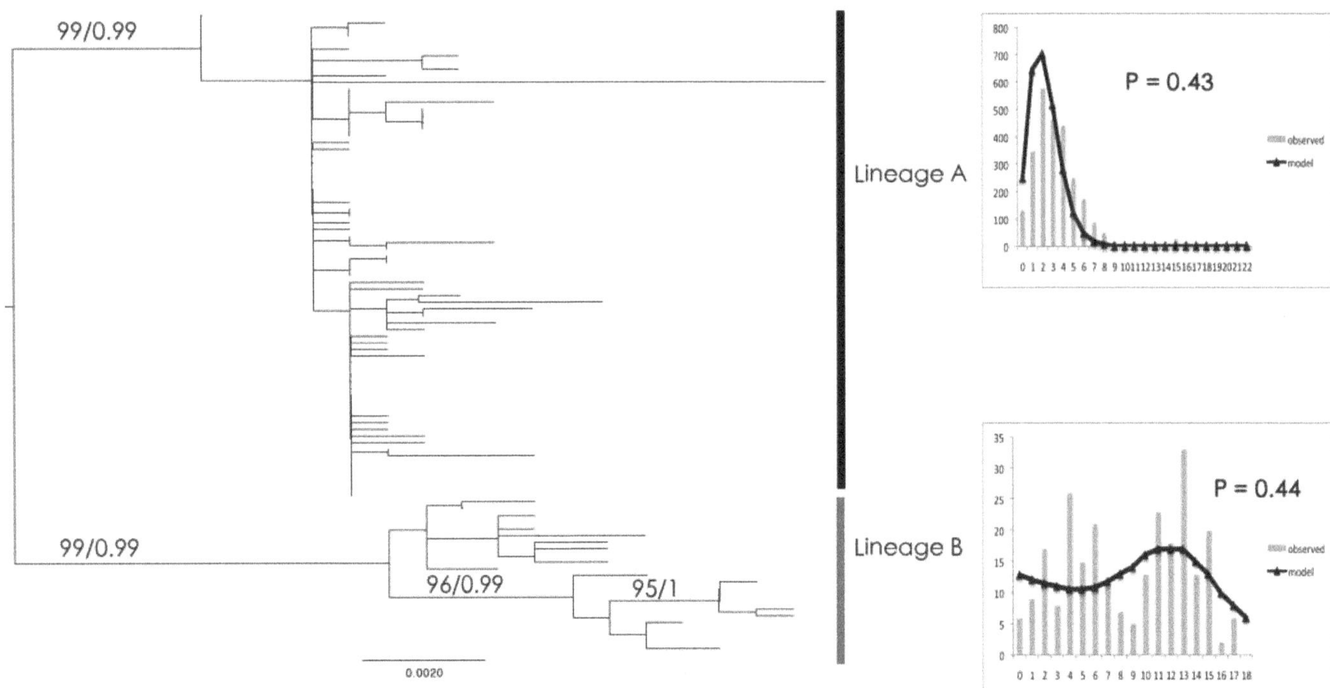

FIGURE 2. The left-hand panel shows the maximum likelihood phylogenetic trees for lineages A and B of *Terapon jarbua* based on 96 concatenated mitochondrial DNA sequences. Only nodes with bootstrap scores >55% majority rule (for maximum likelihood) and >95% majority probabilities (for Bayesian probability values) are presented. The right-hand panels show the mismatch distributions for the two lineages based on the same data set. In each panel, the black line is the expected curve under a constant population size and the gray bars indicate the observed frequencies of haplotype pairs with the number of pairwise differences.

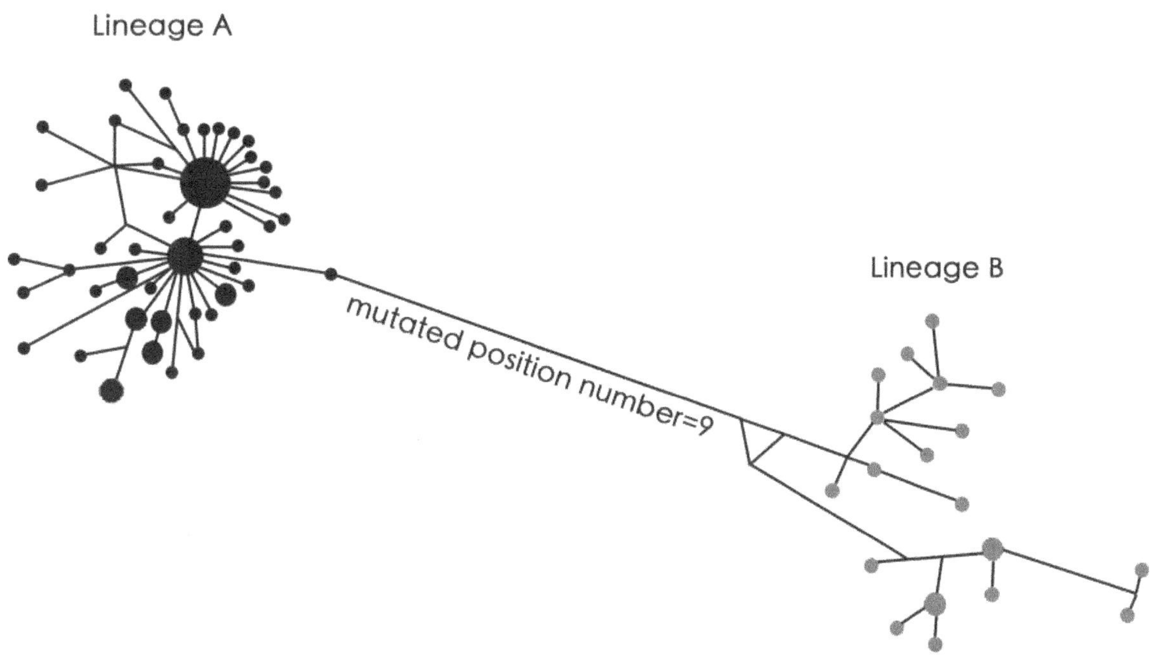

FIGURE 3. Median-joining network (Bandelt et al. 1999) for 96 concatenated sequences from *Terapon jarbua*. The size of each circle is proportional to the number of individuals with that haplotype; black circles represent lineage A fish and gray circles lineage B fish (see Figure 1).

distribution while lineage B showed a bimodal distribution, which can be interpreted as the populations' having gone through expanding and diminishing sizes and structured sizes, respectively (Excoffier and Schneider 1999). Historical population expansion is also supported by the median-joining network, which shows a bush-like pattern of haplotypes for lineage A.

Population Genetic Structure

During the Pleistocene Epoch, several land bridges formed that enclosed the Sea of Japan, partially enclosed the South China Sea, and partially or fully exposed the areas now occupied by the East China and Yellow Seas as a result of lower sea levels that had a great influence on surface currents (Ujiié et al. 1991; Wang 1999). In addition, current hydrodynamic regimes (such as deep-sea gaps along Taiwan and the Japanese Archipelago) may have had profound effects on the genetic divergence of marine populations (Kojima et al. 2006; Kawane et al. 2008). These changes in environmental and geographical regimes are suggested as the mechanism that influenced the population genetic structures of the fishes in the northwest Pacific (Liu et al. 2007; Hu et al. 2011; Shen et al. 2011). Owing to the influence of historical events, the

TABLE 2. Pairwise θ estimates among *Terapon jarbua* populations based on concatenated mtDNA haplotype frequencies. The location codes are given in Table 1. Values in bold italics indicate $P < 0.05$.

Code	TP	YL	HL	HFB	TT	KH	CI	TC	HC	PH	RP
TP											
YL	−0.06026										
HL	−0.07653	−0.05435									
HFB	−0.04984	−0.03621	−0.04203								
TT	−0.08188	−0.03968	−0.06053	−0.09184							
KH	−0.05835	−0.0174	−0.03763	−0.05802	−0.0781						
CI	−0.04342	−0.01801	−0.02048	−0.01379	−0.01193	−0.03135					
TC	0.02836	0.02656	0.02565	***0.09856***	0.06344	0.08484	0.12017				
HC	−0.04959	−0.04346	−0.04044	−0.06323	−0.03142	0.0076	−0.02038	0.01428			
PH	−0.0787	−0.04351	−0.05018	−0.05778	−0.06578	−0.05868	−0.06978	0.05208	−0.04246		
RP	0.03506	−0.01888	−0.02105	−0.01745	−0.02602	−0.04141	−0.02345	***0.17288***	0.03284	−0.05845	

TABLE 3. Analysis of molecular variance (Schneider et al. 2000) based on concatenated mitochondrial DNA haplotype sequences according to two distinct models for the geographic partitioning of populations.

Comparison	df	% Total variance	Φ statistic
East versus west versus China			
Among groups	2	−1.89	$\Phi_{ct} = -0.01864$
Among populations within groups	8	−1.64	$\Phi_{sc} = -0.01613$
Within populations	85	103.51	$\Phi_{st} = -0.03508$
East versus west + China			
Among groups	1	−0.75	$\Phi_{ct} = -0.00754$
Among populations within groups	9	−2.41	$\Phi_{sc} = -0.02393$
Within populations	85	103.16	$\Phi_{st} = -0.03164$

population genetic structure of *T. jarbua* would be expected to exhibit distinct genetic lineages derived from sympatric individuals. Owing to the influence of hydrodynamic barriers, it should be closely associated with the distribution of populations. The present study shows that in Taiwanese waters *T. jarbua* is composed of two genetically divergent lineages. Based on haplotype distribution, we found no geographical gradient or separation across the 11 locations around Taiwan, including one from the coast of China. This result is also supported by AMOVA, which reveals no geographic structure among the 11 locations. Because the distributions of the two haplotype groups overlap, no geographic structure was observed. The genetic partition that we discovered in *T. jarbua* is unlikely to have been caused by current hydrodynamic regimes such as deep-sea gaps, more likely being due to historical events. A similar result was found in a genetic barcoding study of marine fishes in the South China Sea, with two genetically divergent clades (Zhang and Hanner 2012).

The population genetic structure of *T. jarbua* fits the historical scenario. Therefore, we suggest that these two distinct lineages formed in the marginal seas of the northwest Pacific (the Sea of Japan and the South China Sea) during periods of low sea level (Voris 2000), with the estimated TMRCA of lineages A and B being dated to 0.4305 million years ago. After the end of glaciation, the two lineages may have recolonized the same waters around Taiwan and now coexist. However, given the broad distribution of this species, the historical events may have occurred elsewhere. Further studies with large-scale sampling of *T. jarbua* may provide more information regarding the history of genetic divergence.

TABLE 4. *Terapon jarbua* K2P interspecies genetic distance matrix.

Species	*T. theraps*	*T. puta*	*T. jarbua* (lineage A)	*T. jarbua* (lineage B)
T. theraps				
T. puta	0.326			
T. jarbua (lineage A)	0.308	0.32		
T. jarbua (lineage B)	0.304	0.32	0.007	

Species Boundary and Connectivity

The existence of cryptic lineages is common in marine environments (Knowlton 1993, 2000; Rocha and Bowen 2008). Such lineages often represent previously undiscernible species, and several new marine fish species have been described based on cryptic lineages discovered in phylogenetic studies (Drew et al. 2010; Liu et al. 2013). Intraspecies divergence is commonly observed across different taxa in the northwest Pacific, including fishes, crustaceans, and mollusks (review in Shen et al. 2011). However, the threshold for determining cryptic species varies depending on the taxon and genetic marker used. Although morphological characters are the major criteria for defining species, genetic distance matrices are commonly used to infer the level of differentiation (Baldwin et al. 2011; Liu et al. 2013). The genus *Terapon* is comprised of three valid species: *T. jarbua*, *T. puta,* and *T. theraps*. The genetic distance between lineages A and B did not exceed the average interspecies genetic distance for the genus. Therefore, we suggest that these two lineages diverged relatively recently and can be considered two genetic stocks of the same species. Recently, Lavergne et al. (2014) revealed the complex genetic structure of *T. jarbua* in the wider Gulf of Aden (in the range of tens of kilometers even without a geographic barrier between adjacent locations). In contrast to what we found in our study, connectivity is low there. The high genetic connectivity that we found around Taiwan is possibly due to the shared haplotype between localities as well as the mixed haplotypes of the two lineages at most locations. As it happens, two samples collected from the Gulf of Aden shared a unique COI haplotype that can also be found in the South China Sea (Lavergne et al. 2014). This suggests that the range of population expansion after glacial retreat was not restricted to the northwest Pacific but also extended into the Indian Ocean. Broad-scale sampling is needed to test this hypothesis and to understand the historical and contemporary population connectivity within the Indo-Pacific region.

Stock Reassessment

During the past few decades, mtDNA has become the most effective molecule to use for assessing intraspecific genetic variation and genealogy. A well-known case is the European Anchovy *Engraulis encrasicolus*, in which extensive genetic variation was discovered among Mediterranean populations by means of mtDNA analysis (Magoulas et al. 1996, 2006). Based on these studies, the Mediterranean Sea, Bay of Biscay, and Senegal coastal waters (southeastern Atlantic) are characterized by the coexistence of two highly divergent mtDNA clades. This mixed-stock phenomenon is common in fishes across their distribution, which creates some difficulties in fisheries management (Cao et al. 2014). Genetic approaches allow the contributions of different populations to mixed-population fisheries to be inferred (Ruzzante et al. 2006). Even though these approaches have been heavily applied in the past decade to resolve intraspecies issues such as the genetic structure and phylogeography of marine organisms, genetic information has only been used to reassess life history traits between genetic stocks in a few cases (von der Heyden et al. 2014).

Recent studies have indicated that late in the year the population size of California coastal Chinook Salmon *Oncorhynchus tshawytscha* is larger than that of the Klamath River population (Satterthwaite et al. 2014). This indicates that there are two genetic stocks that act differently during their life span. The *Terapon jarbua* populations in Taiwanese waters have previously been considered to be a single stock, and several biological studies, including those of reproductive biology (Miu et al. 1990; Chang 2006), feeding behavior (Lin and Lee 1991), and growth (Wei 1995), have been conducted under this assumption. Therefore, the biological parameters of *T. jarbua* (such as growth rate, age at maturity, and fecundity) that have been based on the single-stock assumption should now be reexamined based on genetic stock information to eliminate the effects of mixing stocks.

CONCLUSIONS

To summarize the present study, we found (1) that two genetically distinct stocks of *T. jarbua* (based on a concatenated sequence of COI and cyt *b*) occur sympatrically in Taiwanese waters, (2) that the average genetic distance between the two lineages is much lower than the average interspecies genetic distance for the genus *Terapon* (which indicates that the divergence between lineages was relatively recent), and (3) that reassessment of biological traits should be conducted based on the genetic information provided in this study. Further genetic studies with rangewide sampling using both mitochondrial and nuclear DNA markers may provide more information on the population genetic structure of this species.

ACKNOWLEDGMENTS

We are grateful to the anonymous referees for their constructive comments. We thank Ming-Tai Chou, Zen-Wei Lin, Tzu-Wei Liu, and Chih-Hao Cheng for their assistance in sampling and Chien-Yao Lu for providing the specimens from Penghu, and we specially thank Buford Pruitt Jr. for English editing. This study was supported by grants to T.Y.L. from the Ministry of Science and Technology (NSC 102-2621-B-110-001) and Asia-Pacific Ocean Research Center, National Sun Yat-Sen University. Shang-Yin Vanson Liu and I-Hsiang Huang contributed equally to this article.

REFERENCES

Allen, G. R., and M. V. Erdmann. 2012. Reef fishes of the East Indies, volumes I–III. Tropical Reef Research, Perth, Australia.

Baldwin, C. C., C. I. Castillo, L. A. Weigt, and B. C. Victor. 2011. Seven new species within western Atlantic: *Starksia atlantica*, *S. lepicoelia*, and *S. sluiteri* (Teleostei, Labrisomidae), with comments on congruence of DNA barcodes and species. ZooKeys 79:21–72.

Bandelt, H. J., P. Forster, and A. Röhl. 1999. Median-joining networks for inferring intraspecific phylogenies. Molecular Biology and Evolution 16:37–48.

Begg, G. A., K. D. Friedland, and J. B. Pearce. 1999. Stock identification and its role in stock assessment and fisheries management: an overview. Fisheries Research 43:1–8.

Begg, G. A., and J. R. Waldman. 1999. An holistic approach to fish stock identification. Fisheries Research 43:35–44.

Cano, J. M., T. Shikano, A. Kuparinen, and J. Merila. 2008. Genetic differentiation, effective population size, and gene flow in marine fishes: implications for stock management. Journal of Integrated Field Science 5:1–10.

Cao, J., S. B. Truesdell, and Y. Chen. 2014. Impacts of seasonal stock mixing on the assessment of Atlantic Cod in the Gulf of Maine. ICES Journal of Marine Science 71:1443–1457.

Chang, C. W., Y. T. Wang, and W. N. Tzeng. 2010. Morphological study on vertebral deformity of the Thornfish *Terapon jarbua* in the thermal effluent outlet of a nuclear power plant in Taiwan. Journal of the Fisheries Society of Taiwan 37:1–11.

Chang, L. M. 2006. Reproductive biology of Thornfish, *Terapon jarbua*, from the surrounding waters off southwestern Taiwan. Master's thesis. National Taiwan Ocean University, Keelung, Taiwan.

Cutarelli, A., M. G. Amoroso, A. De Roma, S. Girardi, G. Galiero, A. Guarino, and F. Corrado. 2014. Italian market fish species identification and commercial frauds revealing by DNA sequencing. Food Control 37:46–50.

Davis, A. M., P. J. Unmack, B. J. Pusey, J. B. Johnson, and R. G. Pearson. 2012. Marine–freshwater transitions are associated with the evolution of dietary diversification in terapontid grunters (Teleostei: Terapontidae). Journal of Evolutionary Biology 25:1163–1179.

Drew, J. A., G. R. Allen, and M. V. Erdmann. 2010. Congruence between mitochondrial genes and color morphs in a coral reef fish: population variability in the Indo-Pacific damselfish *Chrysiptera rex* (Snyder, 1909). Coral Reefs 29:439–444.

Drummond, A. J., and A. Rambaut. 2007. BEAST: Bayesian evolutionary analysis by sampling trees. BMC Evolutionary Biology [online serial] 7:214.

Excoffier, L., and H. L. Lischer. 2010. Arlequin suite ver 3.5: a new series of programs to perform population genetics analyses under Linux and Windows. Molecular Ecology Resources 10:564–567.

Excoffier, L., and S. Schneider. 1999. Why hunter–gatherer populations do not show signs of Pleistocene demographic expansions. Proceedings of National Academy of Sciences of the USA 96:10597–10602.

Excoffier, L., P. E. Smouse, and J. M. Quattro. 1992. Analysis of molecular variance inferred from metric distance among DNA haplotypes: application to human mitochondrial DNA restriction data. Genetics 131:479–491.

Grant, W., and B. Bowen. 1998. Shallow population histories in deep evolutionary lineages of marine fishes: insights from sardines and anchovies and lessons for conservation. Journal of Heredity 89:415–426.

Horne, J. B., L. van Herwerden, J. H. Choat, and D. R. Robertson. 2008. High population connectivity across the Indo-Pacific: congruent lack of phylogeographic structure in three reef fish congeners. Molecular Phylogenetics and Evolution 49:629–638.

Horne, J. B., P. Momigliano, D. J. Welch, S. J. Newman, and L. van Herwerden. 2012. Searching for common threads in threadfins: phylogeography of Australian polynemids in space and time. Marine Ecology Progress Series 449:263–276.

Hu, Z. M., S. Uwai, S. H. Yu, T. Komatsu, T. Ajisaka, and D. L. Duan. 2011. Phylogeographic heterogeneity of the brown macroalga *Sargassum horneri* (Fucaceae) in the northwestern Pacific in relation to late Pleistocene glaciation and tectonic configurations. Molecular Ecology 20:3894–3909.

Jackson, B. C. J., M. X. Kirby, W. H. Berger, K. A. Bjorndal, L. W. Botsford, B. J. Bourque, R. H. Bradbury, R. Cooke, J. Erlandson, J. A. Estes, T. P. Hughes, S. Kidwell, C. B. Lange, H. S. Lenihan, J. M. Pandolfi, C. H. Peterson, R. S. Steneck, M. J. Tegner, and R. R. Warner. 2001. Historical overfishing and the recent collapse of coastal ecosystems. Science 293: 629–638.

Kawane, M., K. Wada, and K. Watanabe. 2008. Comparisons of genetic population structures in four intertidal brachyuran species of contrasting habitat characteristics. Marine Biology 156:193–203.

Knowlton, N. 1993. Sibling species in the sea. Annual Review of Ecology and Systematics 24:189–216.

Knowlton, N. 2000. Molecular genetic analyses of species boundaries in the sea. Hydrobiologia 420:73–90.

Kojima, S., S. Kamimura, A. Iijima, T. Kimura, T. Kurozumi, and T. Furota. 2006. Molecular phylogeny and population structure of tideland snails in the genus *Cerithidea* around Japan. Marine Biology 149:525–535.

Lavergne, E. 2012. Estuarine fish biodiversity of Socotra Island (northwestern Indian Ocean): from the community to the functioning of *Terapon jarbua* populations. Doctoral dissertation. University of Western Brittany, Brest, France.

Lavergne, E., I. Calvès, A. L. Meistertzheim, G. Charrier, U. Zajonz, and J. Laroche. 2014. Complex genetic structure of a euryhaline marine fish in temporarily open/closed estuaries from the wider Gulf of Aden. Marine Biology 161:1113–1126.

Lim, H. C., A. T. Ahmad, A. A. Nuruddin, and S. A. Mohd Nor. In press. Cytochrome *b* gene reveals panmixia among Japanese Threadfin Bream, *Nemipterus japonicus* (Bloch, 1791), populations along the coasts of peninsular Malaysia and provides evidence of a cryptic species. Mitochondrial DNA. DOI: 10.3109/19401736.2014.883602.

Lin, K., and S. C. Lee. 1991. Feeding habits of the Thornfish, *Terapon Jarbua* (Forskal) from the estuary of Tanshui River and its adjacent waters. Journal of the Fisheries Society of Taiwan 18:257–264.

Liu, J. X., T. X. Gao, S. F. Wu, and Y. P. Zhang. 2007. Pleistocene isolation in the northwestern Pacific marginal seas and limited dispersal in a marine fish, *Chelon haematocheilus* (Temminck and Schlegel, 1845). Molecular Ecology 16:275–288.

Liu, S. Y. V., C. L. C. Chan, O. Lin, C. S. Hu, and C. A. Chen. 2013. DNA Barcoding of shark meats identify species composition and CITES-listed species from the markets in Taiwan. PLoS (Public Library of Science) ONE [online serial] 8:e79373.

Liu, S. Y. V., H. C. H. Ho, and C. F. Dai. 2013. A new species of *Pomacentrus* (Actinopterygii: Pomacentridae) from Micronesia, with comments on its phylogenetic relationships. Zoological Studies 52:1–8.

Magoulas, A., R. Castilho, S. Caetano, S. Marcato, and T. Patamello. 2006. Mitochondrial DNA reveals a mosaic pattern of phylogeographical structure in Atlantic and Mediterranean populations of anchovy (*Engraulis encrasicolus*). Molecular Phylogenetics and Evolution 39:734–746.

Magoulas, A., N. Tsimenides, and E. Zouros. 1996. Mitochondrial DNA phylogeny and the reconstruction of the population history of a species: the case of the European Anchovy (*Engraulis encrasicolus*). Molecular Biology and Evolution 13:178–190.

Miu, T. C., S. C. Lee, and W. N. Tzeng. 1990. Reproductive biology of *Terapon jarbua* from the estuary of Tamshui River. Journal of the Fisheries Society of Taiwan 17:9–20.

Rambaut, A., M. A. Suchard, D. Xie, and A. J. Drummond. 2014. Tracer version 1.6. Available: http://beast.bio.editor.ac.uk/Tracer. (July 2014).

Reece, J. S., B. W. Bowen, K. Joshi, V. Goz, and A. Larson. 2010. Phylogeography of two moray eels indicates high dispersal throughout the Indo-Pacific. Journal of Heredity 101:391–402.

Rocha, L. A., and B. W. Bowen. 2008. Speciation in coral reef fishes. Journal of Fish Biology 72:1101–1121.

Ronquist, F., and J. P. Huelsenbeck. 2003. MrBayes 3: Bayesian phylogenetic inference under mixed models. Bioinformatics 19:1572–1574.

Russell, E. S. 1931. Some theoretical considerations on the "overfishing" problem. Journal de Conseil International pour l'Exploration de la Mer 6:3–20.

Ruzzante, D. E., S. Mariani, D. Bekkevold, C. André, H. Mosegarrd, L. A. W. Clausen, T. G. Dahlgren, W. F. Hutchinson, E. M. C. Hatfield, E. Torstensen, J. Brigham, E. J. Simmonds, L. Laikre, L. C. Larsson, R. J. M. Stet, N. Ryman, and G. R. Carvalho. 2006. Biocomplexity in a highly migratory pelagic marine fish, Atlantic Herring. Proceedings of the Royal Society of London B 273:1459–1464.

Satterthwaite, W. H., M. S. Mohr, M. R. O'Farrell, E. C. Anderson, M. A. Banks, S. J. Bates, M. R. Bellinger, L. A. Borgeron, E. D. Crandall, J. C. Garza, B. J. Kormos, P. W. Lawson, and M. J. Palmer-Zwahlen. 2014. Use of genetic stock identification data for comparison of the ocean spatial distribution, size at age, and fishery exposure of an untagged stock and its indicator: California coastal versus Klamath River Chinook Salmon. Transactions of American Fisheries Society 143:117–133.

Schneider, S., D. Roessli, and L. Excoffier. 2000. Arlequin: a software for population genetics data analysis user manual ver 2.000. University of Geneva, Geneva, Switzerland.

Shen, K. N., B. W. Jamandre, C. C. Hsu, W. N. Tzeng, and J. D. Durand. 2011. Plio–Pleistocene sea level and temperature fluctuations in the northwestern Pacific promoted speciation in the globally distributed Flathead Mullet *Mugil cephalus*. BMC Evolutionary Biology [online serial] 11:83.

Singh, N. S., B. K. Behera, S. P. Kunal, P. Das, P. Paria, and A. P. Sharma. In press. Genetic stock structure of *Osteobrama belangeri* (Valenciennes, 1844) in the Indian region. Mitochondrial DNA. DOI: 10.3109/19401736.2014.883602.

Slatkin, M., and R. R. Hudson. 1991. Pairwise comparisons of mitochondrial DNA sequences in stable and exponentially growing populations. Genetics 129:555–562.

Storfer, A., M. A. Murphy, J. S. Evans, C. S. Goldberg, S. Robinson, S. F. Spear, R. Dezzani, E. Delmelle, L. Vierling, and L. P. Waits. 2007. Putting the "landscape" in landscape genetics. Heredity 98:128–142.

Swearer, S. E., J. S. Shima, M. E. Hellberg, S. R. Thorrold, G. P. Jones, D. R. Robertson, S. G. Morgan, K. A. Selkoe, G. M. Ruiz, and R. R. Warner. 2002. Evidence of self-recruitment in demersal marine populations. Bulletin of Marine Science 70:251–271.

Taillebois, L., M. Castelin, J. R. Ovenden, C. Bonillo, and P. Keith. 2013. Contrasting genetic structure among populations of two amphidromous fish species (Sicydiinae) in the central West Pacific. PLoS (Public Library of Science) One [online serial] 8:e75465.

Tamura, K., G. Stecher, D. Peterson, A. Filipski, and S. Kumar. 2013. MEGA6: molecular evolutionary genetics analysis, version 6.0. Molecular Biology and Evolution 30:2725–2729.

Thompson, J. D., D. G. Higgins, and T. J. Gibson. 1994. CLUSTAL W: improving the sensitivity of progressive multiple sequence alignments through sequence weighting, position-specific gap penalties, and weight matrix choice. Nucleic Acids Research 22:4679–4680.

Tzeng, T. D. 2007. Population structure and historical demography of the Spotted Mackerel (*Scomber australasicus*) off Taiwan inferred from mitochondrial control region sequencing. Zoological Studies 46:656–663.

Ujiié, H., Y. Tanaka, and T. Ono. 1991. Late Quaternary paleoceanographic record from the middle Ryukyu Trench slope, northwest Pacific. Marine Micropaleontology 18:115–128.

Vari, R. P. 2001. Teraponidae. Pages 3305–3316 *in* K. E. Carpenter and V. H. Niem, editors. FAO species identification guide for fisheries purposes: the living marine resources of the western central Pacific, volume 5. Food and Agriculture Organization of the United Nations, Rome.

Viñas, J., N. Sanz, L. Peñarrubia, R.-M. Araguas, J.-L. García-Marín, M.-I. Roldán, and C. Pla. 2014. Genetic population structure of European Anchovy in the Mediterranean Sea and the northeast Atlantic Ocean using sequence analysis of the mitochondrial DNA control region. ICES Journal of Marine Science 71:391–397.

von der Heyden, S., M. Beger, R. J. Toonen, L. van Herwerden, M. A. Juinio-Meñez, R. Ravago-Gotanco, C. Fauvelot, and G. Bernardi. 2014. The application of genetics to marine management and conservation: examples from the Indo-Pacific. Bulletin of Marine Science 90:123–158.

Voris, H. K. 2000. Maps of Pleistocene sea levels in Southeast Asia: shorelines, river systems, and time durations. Journal of Biogeography 27:1153–1167.

Wang, D. Y., and B. Q. Huang. 1995. Hematological effects of detergent LAS toxicity in young Thornfish, *Terapon jarbua*. Journal of the Fisheries Society of Taiwan 22:163–168.

Wang, P. X. 1999. Response of western Pacific marginal seas to glacial cycles: paleoceanographic and sedimentological features. Marine Geology 156:5–39.

Ward, R. D., T. S. Zemlak, B. H. Innes, P. R. Last, and P. D. N. Hebert. 2005. DNA barcoding Australia's fish species. Philosophical Transactions of the Royal Society of London B 360:1847–1857.

Wei, X. B. 1995. Daily age and growth of Thornfish (*Terapon jarbua* Forsskal) in the coastal waters of northern Taiwan as revealed from otolith daily growth increment. Master's thesis. National Taiwan University,Taipei.

Worm, B., E. B. Barbier, N. Beaumont, J. Emmett Duffy, C. Folke, B. S. Halpern, J. B. C. Jackson, H. K. Lotze, F. Micheli, S. R. Palumbi, E. Sala, K. A. Selkoe, J. J. Stachowicz, and R. Watson. 2006. Impacts of biodiversity loss on ocean ecosystem services. Science 314:787–790.

Wu, Z. C., Q. H. Xu, J. F. Zhu, X. J. Dai, and L. X. Xu. 2014. Genetic population structure of the Bigeye Tuna *Thunnus obesus* in the central Pacific Ocean based on mtDNA Cyt*b* sequences. Fisheries Science 80:415–426.

Zhang, J., and R. Hanner. 2012. Molecular approach to the identification of fish in the South China Sea. PLoS (Public Library of Science) ONE [online serial] 7:e30621.

10

Feeding Ecology of Juvenile Yellowfin Tuna from Waters Southwest of Taiwan Inferred from Stomach Contents and Stable Isotope Analysis

Jinn-Shing Weng
Executive Yuan, Council of Agriculture, Fisheries Research Institute, Coastal and Offshore Resources Research Center, Kaohsiung 80627, Taiwan; and Department of Environmental Biology and Fisheries Science, National Taiwan Ocean University, Keelung 20224, Taiwan

Ming-An Lee
Department of Environmental Biology and Fisheries Science, National Taiwan Ocean University, Keelung 20224, Taiwan

Kwang-Ming Liu*
Institute of Marine Affairs and Resource Management, National Taiwan Ocean University, Keelung 20224, Taiwan; and George Chen Shark Research Center, National Taiwan Ocean University, Keelung 20224, Taiwan

Ming-Shu Hsu, Mine-Kune Hung, and Long-Jing Wu
Executive Yuan, Council of Agriculture, Fisheries Research Institute, Coastal and Offshore Resources Research Center, Kaohsiung 80627, Taiwan

Abstract

The Yellowfin Tuna *Thunnus albacares* is one of the major fish species caught around subsurface fish aggregation devices (FADs) in the waters southwest of Taiwan. However, how it interacts with other organisms around FADs is poorly known. In this study, the diet and feeding habits of juvenile Yellowfin Tuna were estimated from the analysis of stomach contents from 1,477 specimens with FLs ranging from 24 to 108 cm and stable isotope analysis (202 specimens) collected around FADs in the waters southwest of Taiwan. The analysis of stomach contents indicated that juvenile Yellowfin Tuna with FL < 50 cm mainly feed on larval purpleback flying squid *Sthenoteuthis oualaniensis*, larval shrimps, and zooplanktonic organisms such as amphipods. Yellowfin Tuna with FL of ~50 cm switch their diet to teleost fishes such as Japanese Barracudina *Lestrolepis japonica*, Skinnycheek Lanternfish *Benthosema pterotum*, and fishes in the families Exocoetidae and Scombridae. Stable isotope analysis indicated that the $\delta^{15}N$ values ranged between 6.2‰ and 12.6‰, and the estimated trophic position varied from 3.18 $\pm$ 0.24 for tuna with FL < 30 cm, while it reached 4.59 $\pm$ 0.50 for those with FL > 50 cm and 4.75 $\pm$ 0.06 for those with FL > 90 cm. Based on the distinct diet shift of the juvenile Yellowfin Tuna, demonstrated by both stomach contents and stable isotope analyses, this study concluded that the tuna shift their diet at approximately 50 cm FL.

Subject editor: Suam Kim, Pukyong National University, Busan, South Korea

*Corresponding author: kmliu@mail.ntou.edu.tw

The Yellowfin Tuna *Thunnus albacares* is distributed worldwide in tropical and subtropical waters (Collette and Nauen 1983) and is a common target species of commercial fisheries (Buckley et al. 1989). This Yellowfin Tuna in the western Pacific Ocean grow fast (von Bertalanffy $K = 0.392$ years^{-1}) and reach an asymptotic length of 175 cm FL and live for about 7.7 years (Su et al. 2003); at 50% maturity Wang (2005) measured sizes at 107.8 cm and 112.5 cm FL for females and males, respectively. The catch of Yellowfin Tuna has increased as a result of the use of fish aggregating devices (FADs) in subtropical waters, which attract the Yellowfin Tuna to linger around FADs.

Because juvenile Yellowfin Tuna found around subsurface FADs are a major target species of longline and trolling fisheries in the waters southwest of Taiwan, large amounts of fish are caught by these fisheries. In recent years, the impact of fishing on the Yellowfin Tuna stock was a cause of concern for the local government and environmental groups. Consequently, the government of Taiwan prohibited FAD deployment after 2006.

To understand the interaction between Yellowfin Tuna and other organisms, several attempts have been made to observe their feeding habits, including prey composition (Dragovich 1970; Dragovich and Potthoff 1972), feeding behavior (Bertrand et al. 2002; Potier et al. 2004), and feeding strategies (Rohit et al. 2010). In addition, several studies have documented the feeding habits of Yellowfin Tuna around FADs in different waters, including those around American Samoa (Buckley and Miller 1994) and of the Atlantic Ocean near the equator (Ménard et al. 2000b). Buckley and Miller (1994) indicated that food consumption in Yellowfin Tuna associated with FADs comprised a larger proportion of their body weight than it did in fish that were not FAD associated. However, all of the aforementioned studies focused on adult fish. As juvenile tuna are generally more prevalent around natural or artificial FADs (Ménard et al. 2000a), it is crucial to understand the predator–prey relationship between juvenile tuna and other organisms. Unfortunately, only a few studies have focused on the diet of juvenile Yellowfin Tuna (Brock 1985; Maldeniya 1996; Graham et al. 2007).

Analyzing stomach contents is the most common method of identifying prey items to assess interactions between predators and prey (MacDonald et al. 1982). However, because the metabolic rate of Yellowfin Tuna is fast (Olson and Boggs 1986), some of their quickly digested food is often missed using this approach (MacDonald et al. 1982). Stable isotope analysis can enhance the understanding of a fish's feeding habits as well as the food sources and trophic levels for a particular species (Peterson et al. 1985; Peterson and Fry 1987).

Stable isotope methods have been used in several studies that investigated the trophic interactions (Fry 1988; Hobson and Welch 1992), temporal and spatial variations in food web dynamics (Deegan and Garritt 1997; ÓReilly et al. 2002), migration (Fry et al. 2003), feeding and habitats (Harrigan et al. 1989), and ontogenetic shifts (Renones et al. 2002) of aquatic organisms. Stable isotope analysis can provide further information to that obtained from the analyses of stomach contents (Pauly et al. 1998; Pinnegar et al. 2003) and can be used to determine diet or niche shifts (Grey 2001; Post 2003; Hammerschlag-Peyer et al. 2011) or to acquire accurate data regarding feeding habits (Harvey et al. 2002). Stable isotope analyses are based on measurements of the carbon and nitrogen isotope compositions of an organism's tissues, which contain information about the food or nutrient sources of the organisms consumed. During digestion and absorption, isotopic fractionation of $\delta^{15}N$ and $\delta^{13}C$ relative to prey items occurs at each trophic level (Minagawa and Wada 1984). For example, when trophic-related fractionation causes a mean value change of 3.4‰ (DeNiro and Epstein 1981) or 2.96‰ (Vanderklift and Ponsard 2003) on $\delta^{15}N$ and -1.1‰ (DeNiro and Epstein 1978) or 0.5 ± 0.13‰ (mean $\pm$ SE) (McCutchan et al. 2003) on $\delta^{13}C$, the trophic level of the predator increases by one level. Furthermore, metabolic activity affects the tissue turnover rate, which in turn affects the stable isotope values in different tissues (Fry and Arnold 1982). More metabolically active tissue will reflect more rapid changes in food habits than would less metabolically active tissues (Hobson and Clark 1992). In an organism, spatial and background information can be derived from various tissues. The muscle below the dorsal fin, which exhibits a smaller variance in $\delta^{15}N$ and $\delta^{13}C$ values in stable isotope analysis, has been widely used to evaluate food web structures (Pinnegar and Polunin 1999; Deudero et al. 2004).

Although studies have investigated the diet shifts of juvenile Yellowfin Tuna in Hawaiian waters (Graham et al. 2007) and food webs in the eastern Pacific Ocean (Olson et al. 2010) by using stable isotope analysis, no such study has been conducted on Yellowfin Tuna in Taiwanese waters. Hence, the objectives of this study were to investigate the feeding ecology and examine a possible ontogenetic shift in the diet and trophic level of juvenile Yellowfin Tuna around FADs in the waters southwest of Taiwan using stomach contents analysis and stable isotope analysis. The intent is that the results derived from this study can be used in ecosystem-based fishery management of this species in the future.

METHODS

Data collection.—Yellowfin Tuna were collected during the day near FADs in the waters southwest of Taiwan during January 2010–December 2011 using a 10-m trolling vessel (Figure 1). The specimens were stored on ice immediately after harvest and transported to the laboratory for analysis.

Stomach contents analysis.—Fork length (to 0.1 cm) and body weight (BW; to 0.01 kg) measurements were taken of all specimens. The stomach of each specimen was removed and the contents collected; prey species were identified to the lowest taxonomic level. Then, the prey species were counted and

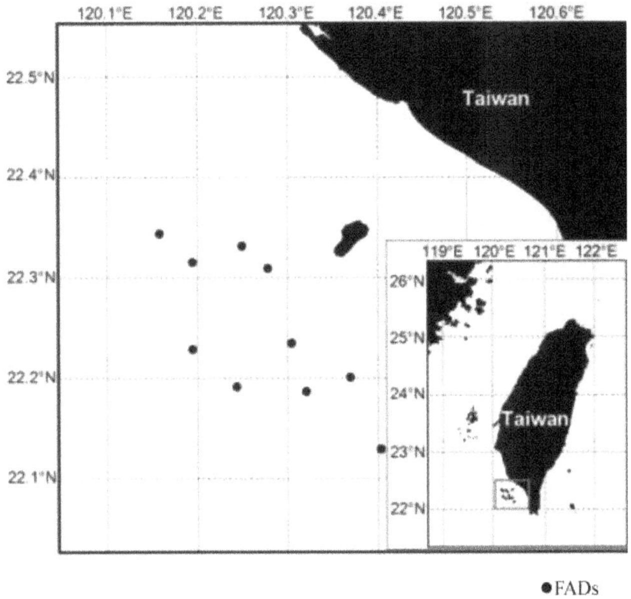

FIGURE 1. Sampling locations for Yellowfin Tuna around the subsurface FADs in southwestern waters of Taiwan.

weighed to within 0.01 g according to a previously described method (Hyslop 1980). The number of empty stomachs (<0.01 g prey/kg BW) was also counted. The feeding activity for size-classes was evaluated by the trend of the repletion index, which was expressed as grams of stomach contents per kilogram BW (Graham et al. 2007). These indices were calculated for each size-class of tuna and were used to discern differences in foraging success based on predator size. The Shapiro–Wilk test indicated that the repletion indices were not normally distributed; therefore, the Kruskal–Wallis test (Zar 2010) was used to examine the median of repletion indices among size-classes.

The importance of various prey species to the diet of the Yellowfin Tuna was assessed by calculating the following dietary indices:

$$\%N_i = \frac{N_i}{\sum_k N} \quad \text{and} \quad \%W_i = \frac{W_i}{\sum_k W},$$

where $\%N_i$ and $\%W_i$ are the percent abundance and percent weight of the ith prey species, respectively, k is the total number of species, N_i is number of individuals of the ith species, and W_i is the total weight of the ith species. These indices have been used extensively to analyze stomach contents data (Hyslop 1980). In addition, the mean percent abundance ($\%MN$) and the mean percent weight ($\%MW$) of the prey species were calculated following a previously described method (Graham et al. 2007).

To examine the feeding overlap in diet among different size-classes of Yellowfin Tuna, we applied the method described by Graham et al. (2007) to divide the specimen sizes into five classes (classes I through V according to FL; i.e., I: <30 cm, II: 30–50 cm, III: 50–70 cm, IV: 70–90 cm, and V: 90–110 cm), and Morisita's original index (Horn 1966) was used to estimate the overlap by using the following equation:

$$C_{mh} = \frac{2\sum_{i=1}^{S} p_{A,\,i} \times p_{B,\,i}}{\sum_{i=1}^{S} p_{A,\,i}^2 + \sum_{i=1}^{S} p_{B,\,i}^2},$$

where C_{mh} is the Morisita–Horn index of overlap between Yellowfin Tuna size-classes A and B. The values $P_{A,\,i}$ and $P_{B,\,i}$ represent the contribution of the ith prey species to A and B in terms of the species dietary indices (i.e., $\%N$, $\%W$, $\%MN$, and $\%MW$), and S represents the total number of identified prey species in the feeding regime of the predator. A value for $C_{mh} \geq 0.6$ indicates a significant feeding overlap in diet for the fish between the two size-classes (Zaret and Rand 1971).

Stable isotope analysis.—The liver and white muscle tissues under the second dorsal fin of each specimen of the subsample randomly chosen from each size-class of Yellowfin Tuna as well as tissues of the prey fishes, the muscle of cephalopods, and the entire bodies of other prey were collected for stable isotope analysis. Theses tissues and prey were cleaned twice with distilled water immediately after dissection and stored at −40°C. They were then thawed and dried at 60°C for 27 h and homogenized to a fine powder. Aliquots of the homogenized samples (0.7–1.2 mg) were weighed using a Sartorius R200D digital analytical balance and packed into 8 × 5-mm tin cups to analyze the $\delta^{15}N$ and $\delta^{13}C$ values. The accuracy of the analysis was 0.2% for nitrogen and carbon, as estimated from standards analyzeds with the samples. The samples were combusted in an elemental analyzer (Flash EA-1112; Thermo Fisher Scientific, Bremen, Germany) to produce CO_2 and N_2, and the gas was analyzed using an isotope-ratio mass spectrometer (IMRS Delta V Advantage, Thermo Fisher Scientific) to determine its isotopic composition. The isotopic values were expressed as the difference in parts per thousand (‰) from the respective standards (Pee Dee belemnite limestone for $\delta^{13}C$ and N_2 in air for $\delta^{15}N$):

$$\delta X = \left[\left(R_{\text{sample}}/R_{\text{standard}}\right)\right] \times 10^3,$$

where X represents ^{15}N or ^{13}C, and R represents the ratio of ^{15}N:^{14}N or ^{13}C:^{12}C of the sample or the standard material (Peterson and Fry 1987). The mean and SD of reference materials are as measured for USGS40 and L-glutamic acid. Nitrogen and carbon isotopic composition analyzed by the USGS40 was used to correct the basic sample. The USGS40 $\delta^{15}N = -4.52 \pm 0.1$‰ (air N_2) and $\delta^{13}C = -26.39 \pm 0.1$‰ (Vienna Pee Dee Belemnite).

An ANOVA was used to examine the homogeneity of the mean values of $\delta^{13}C$ among size-classes. Then, Tukey's multiple comparison method was used for pairwise comparison. As the Shapiro–Wilk test indicated that $\delta^{15}N$ values were not normally distributed, the Kruskal–Wallis test (Zar 2010) was used to examine the mean values of $\delta^{15}N$ among size-classes and Dunn's test was used for posterior comparison.

Because lipid content in fish tissue varies depending on species, it is usually estimated using an organic solvent extraction of lipids following Folch et al. (1957). Lipid extraction may result in the loss of some nonlipid compounds and thus alter $\delta^{15}N$ (Sotiropoulos et al. 2004; Sweeting et al. 2006; Logan et al. 2008; Elsdon et al. 2010; Varela et al. 2011). A mathematical lipid correction of $\delta^{13}C$ values provides a simple alternative to chemical lipid extraction. The normalization model of McConnaughey and McRoy (1979) is based on empirical equations describing the relationship between the two equations. The proportion of the lipid content (L) of the sample is calculated from the sample's C:N ratio:

$$L = 93/(0.246\mathrm{C{:}N} - 0.775) - 1.$$

Lipid-normalized $\delta^{13}C'$ is calculated from L and the measured $\delta^{13}C$ of the sample:

$$\delta^{13}C' = \delta^{13}C + D(I + 3.9) / (1 + 287L),$$

where D is the isotopic difference between lipids and proteins (D is assumed to be 6%; McConnaughey and McRoy 1979) and I is a constant (= −0.207).

Food web studies generally explore the trophic structure and feeding relationships of aquatic organisms (Aberle and Malzahn 2007). Therefore, the following equation was used to estimate the trophic position (TP) of Yellowfin Tuna (Vander Zanden and Rasmussen 1999; Olson et al. 2010):

$$TP_{YFi} = \left(\frac{\delta^{15}N_{YFi} - \delta^{15}N_{baseline}}{TEF}\right) + 2,$$

where TP_{YFi} is the estimated trophic position for Yellowfin Tuna at the site of sample i, $\delta^{15}N_{YFi}$ is the $\delta^{15}N$ value of Yellowfin Tuna at the site of sample i, $\delta^{15}N_{baseline}$ is the corrected isotope signature of the Yellowfin Tuna, and TEF is the trophic enrichment factor. The $\delta^{15}N$ value of consumers cannot be considered as an absolute measure of trophic position. Therefore, it is necessary to correct fish $\delta^{15}N$ signatures to account for variation in the $\delta^{15}N$ values of primary consumers (e.g., zooplankton, chironomids, and amphipods: Vander Zanden and Rasmussen 1999). The relationship can then be used to calculate baseline conditions ($\delta^{15}N_{baseline}$) that are used to correct $\delta^{15}N$ values for the secondary consumers (Vander Zanden and Rasmussen 1999). We used zooplankton as the isotopic baseline for the primary consumers when calculating the trophic position of the Yellowfin Tuna. The mean $\delta^{15}N$ value was 5.1 ± 0.3‰ (n = 5), and primary producers are at trophic level 1; primary consumers, such as herbivorous fishes and zooplankton, are at trophic level 2 or slightly higher (Post 2002). The TEF is a suitable indicator for estimating the isotopic relationships between Yellowfin Tuna and their prey. As no stable isotope analysis has been performed for other tuna species or sharks in Taiwanese waters, the average TEF value of marine fishes (2.4‰) has been used (Vanderklift and Ponsard 2003) when evaluating Yellowfin Tuna feeding habits (Olson et al. 2010).

RESULTS

Stomach Contents Analysis

In total, 1,477 Yellowfin Tuna (24–108 cm FL) were caught by means of trolling during the day. Of this total, 349 (23.6%) had empty stomachs.

The analysis of 1,128 Yellowfin Tuna stomachs indicated that the major contributors of %N were unidentified zooplankton (23.9%), amphipods (16.2%), crab larvae (8.0%), and purpleback flying squid (7.6%). The major %W contributors were Exocoetidae species (30.0%), unidentified fishes (17.8%), Bullet Tuna *Auxis rochei* (6.5%), purpleback flying squid (5.5%), and Japanese Barracudina (4.05%). The repletion indices of the Yellowfin Tuna ranged from 1.3 to 59.3 g/kg, and the Kruskal–Wallis test indicated that these values differed significantly by size (H = 34.01, P < 0.01; Table 1). Dunn's posterior test indicated that the repletion indices of Yellowfin

TABLE 1. Stomach contents of juvenile Yellowfin Tuna by size-class.

Size-class (cm FL)		Stomach contents						
				Repletion index (g/kg)				
		n	% with prey	Range	Mean	±SD	Median
I	<30	41	53.9	1.3–47.5	13.0	18.2	3.8
II	30–50	987	77.8	1.3–46.1	7.0	12.2	2.4
III	50–70	68	76.1	1.4–42.8	5.3	8.9	1.4
IV	70–90	19	76.3	1.3–29.5	5.0	7.6	5.7
V	90–110	13	92.9	1.5–59.3	23.1	15.1	23.7

Tuna in classes III and II were significantly different from those in classes I, IV, and V, but no significant difference was found for those between classes I and II.

The prey species were categorized into 17 groups to examine the major prey species of Yellowfin Tuna of different sizes. The major *%MN* contributors were mantis shrimp *Faughnia* spp. (49.5%) and purpleback flying squid (26.6%) for class I; unidentified zooplankton (25.7%), Amphipoda species (15.5%), and purpleback flying squid (9.4%) for class II; Scombridae species (25.1%), crab larvae (11.7%), and mantis shrimp (10.9%) for class III; Skinnycheek Lanternfish (74.2%), crab larvae (5.7%), and mantis shrimp (4.9%) for class IV; and Exocoetidae species (80.6%), and Scombridae species (11.1%) for class V (Table 2).

The major *%MW* contributors were purpleback flying squid (42.9%) and mantis shrimp (13.5%) for class I; unidentified fishes (30.1%), purpleback flying squid (11.4%), Scombridae species (10.0%), and Japanese Barracudina (6.3%) for classs II; Scombridae species (43.5%), Amphipoda species (6.1%), Carangidae species (5.6%), and purpleback flying squid (4.1%) for class III; Skinnycheek Lanternfish (39.1%), Carangidae species (29.6%), Scombridae species (10.1%) for class IV; and Exocoetidae species (70.4%) for class V (Table 2).

Using the four metrics (*%MN*, *%N*, *%MW*, and *%W*) Morisita's original index indicated there was a significant overlap of the diets of fish in size-classes II and III (Table 3). No significant overlap existed between size-classes III and IV or between size-classes IV and V. These results suggested that between size-classes II and III, at an FL of approximately 50 cm, Yellowfin Tuna shifted their diets. Both classes II and III overlapped at FLs of approximately 50 cm; thus, their feeding habits were similar and exhibited significant overlap. Tuna in other size-classes ate different prey and had very low values for the Morisita index.

Stable Isotope Analysis

A total of 154 samples of Yellowfin Tuna white muscle and liver tissues plus 48 prey species were examined. Averaged $\delta^{13}C$ and $\delta^{13}C'$ values were -17.1 ± 0.50‰ (mean ± SD) and -17.0 ± 0.50‰ for the white muscle tissue and -18.6 ± 0.6‰ and -18.5 ± 0.6‰ for liver tissue, respectively. A significant difference in $\delta^{13}C'$ values was found between the white muscle and liver tissues (paired *t*-test: $P < 0.01$). The $\delta^{15}N$ values were estimated to range from 6.2‰ to 12.6‰ for the white muscle tissue and from 7.1‰ to 12.2‰ for the liver tissue. The $\delta^{15}N$ values of the white muscle and liver tissues of the juvenile Yellowfin Tuna differed significantly among the size classes (Kruskal–Wallis test: $P < 0.01$). The mean ± SD $\delta^{15}N$ of the white muscle tissue was 7.6 ± 0.7‰ for size-class I. The difference in $\delta^{15}N$ values between class I and class III was 3.7‰ (Table 4). The $\delta^{15}N$ values of the muscle and

TABLE 2. Comparision of the 17 prey groups in the diets of juvenile Yellowfin Tuna measured as (*%MN*) and (*%MW*), by size-class (cm FL).

	Size-class									
	<30 cm		30–50 cm		50–70 cm		70–90 cm		90–110 cm	
Prey item	*%MN*	*%MW*	*%MN*	*%MW*	*%MN*	*%MW*	*%MN*	*%MW*	*%MN*	*%MW*
Japanese Barracudina *Lestrolepis japonica*			1.6	6.3	0.3	0.3				
Skinnycheek Lanternfish *Benthosema pterotum*			0.5	0.3	0.4	0.1	74.2	39.1		
Moonfish *Mene maculata*			0.4	0.2	0.3	0.5				
Flying fishes (Exocoetidae)			0.1	6.5	0.3	0.8			80.6	70.4
Butterfly fishes (Chaetodontidae)	0.9	0.3	2.7	1.3	0.4	0.1				
Bullet Tuna *Auxis rochei*			1.6	10.0	25.1	43.5	1.4	10.1	11.1	28.4
Puffers (Tetraodontidae)			0.5	0.4	0.3	1.9				
Blue Jack Mackerel *Trachurus picturatus* and scads (Carangidae)			0.2	0.9	0.3	5.6	0.8	29.6		
Unidentified fishes	3.7	9.5	5.8	30.1	5.2	21.8	0.8	9.2	8.3	1.2
Mantis shrimp *Faughnia* spp.	49.5	13.5	7.4	5.2	10.9	1.9	4.9	0.1		
Other shrimps	0.9	0.1	6.2	2.8	4.0	0.5				
Crab larvae	4.6	0.8	7.0	1.0	11.7	0.5	5.7	0.1		
Purpleback flying squid *Sthenoteuthis oualaniens*	26.6	42.9	9.3	11.5	0.3	4.1	2.2	2.5		
Unidentified squids	3.7	28.3	10.2	9.1	2.2	6.5	0.7	0.4		
Amphipods			15.5	4.2	10.1	6.1				
Unidentified zooplankton	4.6	1.7	25.7	6.5	16.3	2.2	1.6	0.5		
Other prey	5.5	2.9	5.3	3.8	11.9	3.6	7.7	8.4		

TABLE 3. Moriseta's original index of the dietary overlap of juvenile Yellowfin Tuna by size-class. An asterisk indicates significant overlap ($C_{mh} \geq 0.6$). Values in bold italics below the diagonal are *%MN* and *%MW*.

	Size-class and FL (cm)				
	I	II	III	IV	V
Size-class	<30	30–50	50–70	70–90	90–110
	Overlap of prey taxa in terms of *%N* and *%MN*				
I		0.32	0.40	0.08	0.01
II	***0.38***		0.85*	0.06	0.09
III	***0.33***	***0.67****		0.09	0.07
IV	***0.08***	***0.06***	***0.08***		0.005
V	***0.006***	***0.08***	***0.08***	***0.002***	
	Overlap of prey taxa in terms of *%W* and *%MW*				
I		0.47	0.18	0.08	0.001
II	***0.52***		0.76*	0.23	0.016
III	***0.22***	***0.65****		0.39	0.13
IV	***0.08***	***0.24***	***0.33***		0.03
V	***0.002***	***0.22***	***0.32***	***0.07***	

liver tissues showed a distinct positive shift at a FL of 50 cm (Figure 2). The $\delta^{15}N$ values of the white muscle (8.8 ± 1.3‰) and liver tissues (8.9 ± 1.3‰) of the tuna with FL < 50 cm (classes I and II) were significantly lower than those of the muscle (11.5 ± 0.9‰) and liver tissues (10.8 ± 0.9‰) of the tuna with FL > 50 cm (classes III, IV, and V) (Dunn's test: $P < 0.01$). Yellowfin Tuna > 50 cm FL exhibited a wider range of $\delta^{15}N$ values than those < 50 cm FL (Figure 2). The $\delta^{15}N$ values of the white muscle and liver tissues increased with the weight of the fish, reaching an asymptote for weights > 5.5 kg. This relationship can be expressed as: $\delta^{15}N = 7.03 + 3.97[1 - \exp(-0.63W)]$ ($r^2 = 0.62$, $P < 0.01$) for white muscle and $\delta^{15}N = 6.77 + 4.98[1 - \exp(-0.48W)]$ ($r^2 = 0.72$, $P < 0.01$) for liver (Figure 2). The $\delta^{15}N$ values of the white muscle and liver tissues increased 3.3‰ and 1.9‰ for the size-classes of <50 cm FL and >50 cm FL, respectively.

The mean $\delta^{13}C'$ values of the white muscle and liver tissues were −17.4 ± 0.4‰ and −18.2 ± 0.8‰, respectively, for size-class I. These mean $\delta^{13}C'$values were −16.7 ± 0.4‰ and −18.2 ± 0.4‰ for size-class III, and the differences in $\delta^{13}C'$ values from class I to class III were 0.7‰ and 0‰ in white muscle and liver tissues, respectively (Table 4). The $\delta^{13}C'$ values of the white muscle and liver tissues of Yellowfin Tuna < 50 cm FL (−17.3 ± 0.4‰ and −18.5 ± 0.6‰, respectively) were not significantly different from those of tuna > 50 cm FL (−16.8 ± 0.5‰ and −18.3 ± 0.4‰). The $\delta^{13}C'$ values of the white muscle tissue were negatively related to FL and BW ($P < 0.05$), but these relationships were not significant for the liver tissue ($P > 0.05$; Figure 3a, b). Carbon isotopic concentration was negatively correlated with FL and BW (Figure 3a, b), but this relationship was very weak. The $\delta^{13}C'$ values of the white muscle and liver tissues did not have a significant relationship with FL or BW.

TABLE 4. Mean $\delta^{15}N$ (‰) and $\delta^{13}C'$ (‰) (±SD) of juvenile Yellowfin Tuna by size-class.

	$\delta^{15}N$ (±SD)		$\delta^{13}C'$ (±SD)	
Size-class (*n*)	Muscle	Liver	Muscle	Liver
I (12)	7.6 (0.6)	7.9 (0.5)	−17.5 (1.3)	−18.3 (0.8)
II (29)	9.0 (1.4)	9.2 (1.3)	−17.2 (0.41)	−18.5 (0.5)
III (20)	11.3 (1.3)	10.8 (1.3)	−16.7 (0.4)	−18.2 (0.4)
IV (9)	11.3 (1.0)	10.6(1.3)	−17.0 (0.8)	−18.4 (0.4)
V (7)	11.7 (0.1)	10.4 (0.9)	−16.9 (0.2)	−18.5 (0.3)

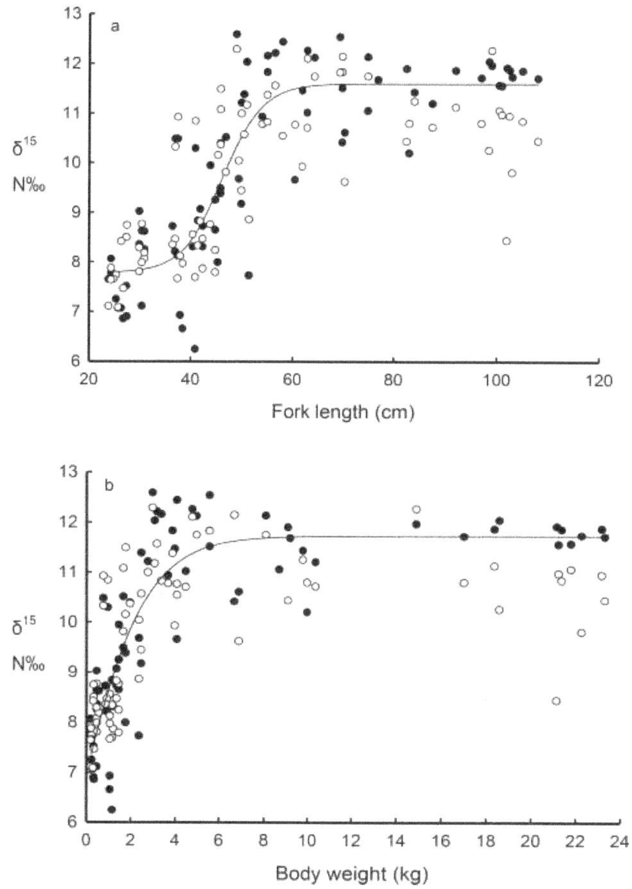

FIGURE 2. Relationships between $\delta^{15}N$ values and both (**a**) fork length and (**b**) body weight of Yellowfin Tuna. White muscle is represented by a solid line and filled circles, and liver is represented by a dashed line and open circles. In panel (a) the four-parameter, sigmoid models were estimated as: $\delta^{15}N = 7.78 + 3.818/\{1 + \exp[-(FL - 46.2)/3.91]\}$ ($r^2 = 0.856$) for white muscle and $\delta^{15}N = 8.03 + 2.887/\{1 + \exp[-(FL - 44.5)/3.34]\}$ ($r^2 = 0.797$) for livers. In panel (b) the three-parameter exponential growth models were estimated as: $\delta^{15}N = 7.03 + 3.97[1 - \exp(-0.63BW)]$ ($r^2 = 0.62$) for white muscle and $\delta^{15}N = 6.77 + 4.98[1 - \exp(-0.48BW)]$ ($r^2 = 0.723$) for livers.

The $\delta^{15}N$ values ranged from 4.2‰ to 11.3‰ for the 48 prey specimens (Table 5). The trophic niche of the juvenile

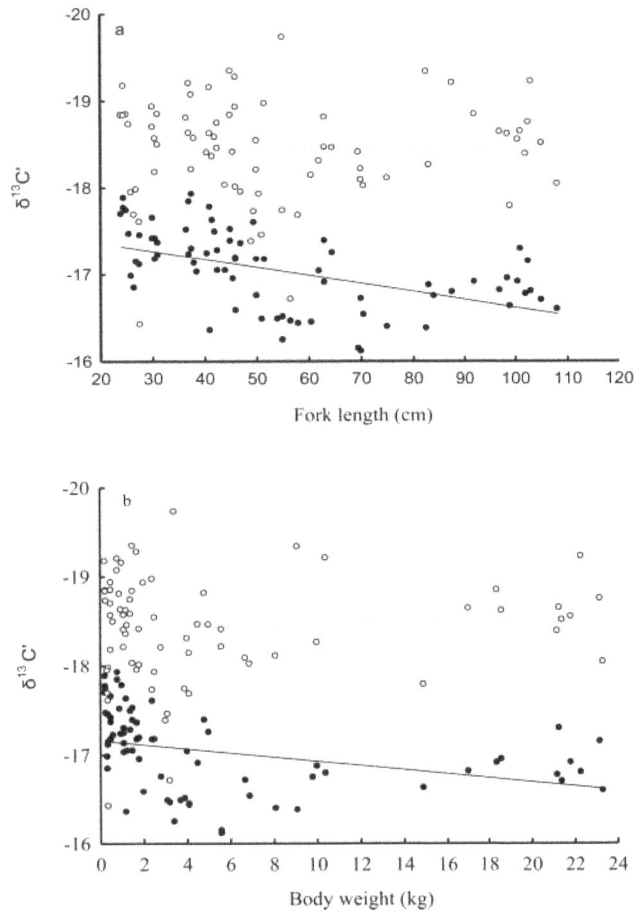

FIGURE 3. Relationship between $\delta^{13}C'$ values and both (**a**) fork length and (**b**) body weight of Yellowfin Tuna. White muscle is represented by the a solid line and filled circles, and liver is represented by a dashed line and open circles. In panel (a) $\delta^{13}C' = -17.546 - 0.0093FL$ for white muscle and $\delta^{13}C' = -18.367 + 0.0012FL$ for liver. In panel (b) $\delta^{13}C' = -17.158 + 0.023BW$ for white muscle and $\delta^{13}C' = -18.397 + 0.0069BW$ for liver.

TABLE 5. Mean $\delta^{15}N$ (‰) and $\delta^{13}C'$ (‰) (±SD) of juvenile Yellowfin Tuna by prey items.

Prey item (*n*)	$\delta^{15}N$ (±SD)	$\delta^{13}C'$ (±SD)
Mackerel *Rastrelliger* spp. (2)	8.6 (0.2)	−17.8 (0.6)
Bullet Tuna *Auxis rochei* (4)	10.8 (1.1)	−16.9 (0. 6)
Japanese Barracudina *Lestrolepis japonica* (3)	11.3 (0.9)	−17.7 (0.3)
Skinnycheek Lanternfish *Benthosema pterotum* (3)	11.0 (0.8)	−17.0 (0.7)
Mackerel Scad *Decapterus macaerllus* (2)	9.2 (0.2)	−17.1 (0.1)
Whitetip Scad *Decapterus maruadsi* (2)	9.2 (0. 5)	−19.0 (0.2)
Moonfish *Mene maculata* (2)	9.3 (0.21)	−18.6 (0.2)
Flying fishes (Exocoetidae) (5)	10.4 (0.1)	−17.7 (0.2)
Butterfly fishes (larval Chaetodontidae) (2)	10.4 (1.1)	−20.2 (0.1)
Pacific Pomfret *Brama japonica* (1)	9.3	−19.6
Purpleback flying squid *Sthenoteuthis oualaniensis* (4)	8.4 (2.4)	−18.27 (0.5)
Octopus *Octopus* sp. (1)	8.4	−17.74
Mantis shrimp *Faughnia* spp. (5)	4.8 (1.5)	−17.87 (1.1)
Shrimp postlarva (4)	4.2 (1.0)	−17.87 (1.1)
Amphipoda (5)	7.4 (2.1)	−19.7 (1.6)
Crab megalopa (3)	5.1 (0.3)	−19.7 (1.2)

Yellowfin Tuna from the different size-classes and of the prey species are shown in Figure 4a, b. The postlarval shrimp, larval crabs, and mantis shrimp, the major components of size-class I, showed an average $\delta^{15}N$ value that was 3.4‰ less than that of the prey tissue. The $\delta^{13}C'$ values of the prey varied from −16.9‰ to −20.2‰, but this information could not be used to identify the diet inputs because there was almost no difference in the $\delta^{13}C'$ values of the white muscle between the size-classes < 50 cm FL (−17.3 ± 0.4‰) and >50 cm FL (−16.8 ± 0.5‰).

Trophic Position

The trophic structure of the Yellowfin Tuna by size-class is shown in Figure 4a and Table 6. The TP of the Yellowfin Tuna ranged from 3.2 to 4.8 depending on size, with a mean ± SD of 4.5 ± 0.6 ($r^2 = 0.75$, $P < 0.01$) (Figure 5). The TP of Yellowfin Tuna with FL < 30 cm (class I) was 3.2 ± 0.2, which represents a probable baseline level for this species at an age < 4 months (Hampton and Fournier 2001). The TP of the tuna with FL > 50 cm (class III) was 4.6 ± 0.50, and the TP reached a peak level of 4.8 ± 0.1 at FL > 90 cm.

DISCUSSION

Stomach Contents

The juvenile Yellowfin Tuna associated with FADs in the waters southwest of Taiwan feed on different prey. The stomach contents analysis showed that a diet shift occurs around the size of 50 cm FL when they are still young of the year (age 0) according to Su et al. (2003). Yellowfin Tuna < 50 cm FL mainly feed on crustaceans and squid larvae; the tuna then switch at > 50 cm FL to a more varied diet from higher trophic levels, including teleost fishes. Maldeniya (1996) documented similar results showing that Yellowfin Tuna < 40 cm FL fed on zooplankton, crustaceans, and cephalopods, whereas those with FL ≥ 50 cm shifted their diet to feed on fish. The ontogenetic stages of a fish involve physical, structural, and physiological changes that result in behavioral changes (Noakes and Godin 1988). The time when fish shift to piscivory varies among individuals for the same cohort (Post 2003). Early hatched individuals with higher growth rates can make an

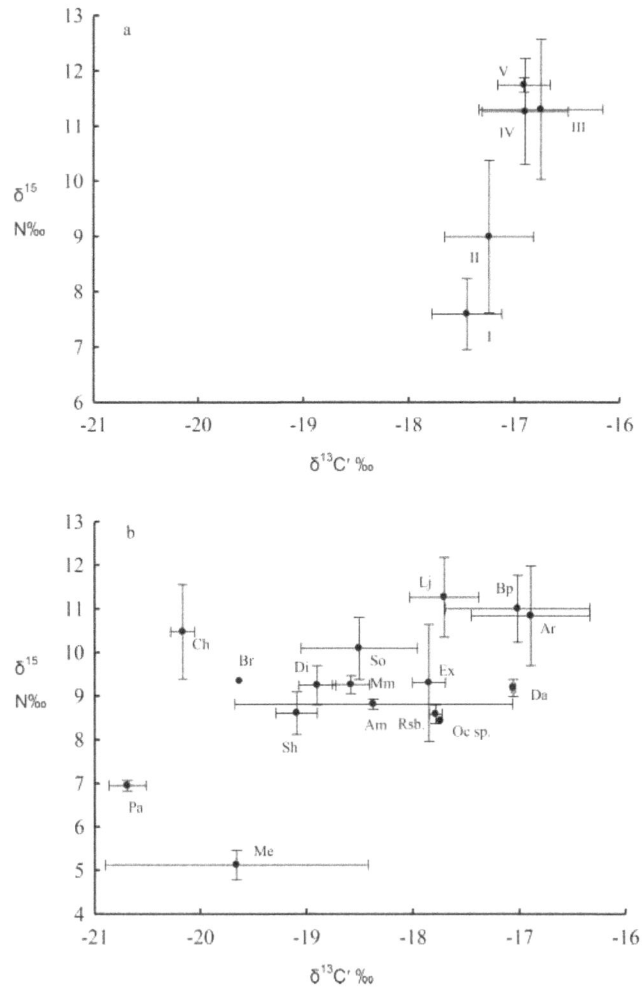

FIGURE 4. Mean $\delta^{13}C'$ and $\delta^{15}N$ values of (**a**) the white muscle of Yellowfin Tuna size-classes I–V and (**b**) prey species. R sp. = *Rastrelliger* sp.; Ar = *Auxis rochei*; Lj = *Lestrolepis japonica*; Oc sp. = *Octopus* sp.; Ex = Exocoetidae; Am = Amphipoda; Ch = larval Chaetodontidae; Sh = postlarval shrimps; Bp = *Benthosema pterotum*; So = *Sthenoteuthis oualaniensis*; Pa = *Faughina* spp.; Me = larval crabs; Br = *Brama* sp. Vertical and horizontal bars represent SD values.

early transition to piscivory, which may result in an increase in growth and a decrease in mortality (Post 2003). Such phenomena can have critical implications for population dynamics, community structure, and ecosystem function (Olson 1996; Hammerschlag-Peyer et al. 2011). Similar behavior was observed for juvenile Yellowfin Tuna in this study. For the same age-0 cohort, some fast-growing individuals switched to piscivory, but the remainder continued to feed on crustaceans and squid larvae.

Grubbs (2010) documented that larger fish consume a wider range of prey than do smaller ones. On the other hand, McCormick (1998) concluded that juveniles of various fish species typically consume a wider variety of prey than adults do because of the differences in their size (Herrel and Gibb 2006). We found that juvenile Yellowfin Tuna in classes II and III consumed a wider variety of prey species than those in classes IV and V. This is likely because some individuals in classes II and III had not yet shifted to piscivory, but all Yellowfin Tuna in classes IV and V fed mainly fish. The ontogenetic diet shifts can reduce the intraspecific competition and increase the survival rate of juvenile Yellowfin Tuna.

The relationship between the sizes of predators and prey is perhaps the most crucial factor that influences the predator–prey relationships (Miller et al. 1988; Fuiman and Magurran 1994). Stronger predators have a greater ability to chase prey. In feeding strategies of fish, mouth size and gape size limit feeding on prey larger than the fish's own mouth (Renones et al. 2002; Arim et al. 2007). To match increased energy requirements, a quantitative or qualitative change in diet is expected (Cooper et al. 2007). Olson and Boggs (1986) mentioned that the daily ration of Yellowfin Tuna is 3.9–6.7% of their body mass and their energy requirement is a function of swimming speed. As the fish grow larger and have faster swimming speeds, the need for more energy can be achieved by switching to a prey species with high lipid and energy content (Olson and Boggs 1986). The mortality in the age-0 stage is very high and those individuals that shift to pisicivory can grow faster by gaining more energy from their prey and thus have a better chance to survive (Olson 1996). Similar findings observed in this study suggest that Yellowfin Tuna acquire more energy by shifting their diet to fish when their FLs > 50 cm.

Sheldon et al. (1977) and Smetacek (1999) concluded that the size of a fish is a crucial factor that affects the trophic interactions between various marine organisms because most predators are larger than their prey. Therefore, predators can feed on the largest prey possible to maximize energy intake (Schoener 1971; Charnov 1976). In this study, many of the stomachs of the Yellowfin Tuna in the smallest size-class contained prey

TABLE 6. Statistics of isotopic $\delta^{15}N$ and $\delta^{13}C'$ values in the muscle tissue of juvenile Yellowfin Tuna by size-class and estimated trophic position.

Size-class	Mean FL (cm) ± SD	BW range (kg)	$\delta^{15}N$ range (‰)	$\delta^{13}C'$ range (‰)	Mean TP ± SD
I (<30 cm)	26.2 ± 2.6	0.2 to 0.6	6.9 to 9.0	−16.8 to −17.9	3.2 ± 0.2
II (30–50 cm)	41.2 ± 5.9	0.6 to 2.8	6.2 to 12.6	−16.6 to −17.8	3.6 ± 0.6
III (50–70 cm)	61.8 ± 8.1	2.8 to 8.7	7.7 to 12.5	−16.1 to −17.4	4.6 ± 0.5
IV (70–90 cm)	88.9 ± 8.0	9.1 to 14.6	11.1 to 12.0	−16.4 to −18.5	4.7 ± 0.2
V (90–110 cm)	103.1 ± 2.6	14.6 to 23.4	11.5 to 11.9	−16.6 to −17.3	4.8 ± 0.1

species that were relatively large. For example, the Yellowfin Tuna with a FL of 40 cm preyed on Japanese Barracudina with a FL of 20 cm. These results are similar to those reported in the aforementioned studies, in which fishes shifted their diet to consume prey that would provide maximum energy.

Yellowfin Tuna in size-classes II and III (*%MN*) exhibited feeding overlap at a FL of approximately 50 cm (Table 3); this result was consistent with the diet shift and findings reported by Graham et al. (2007) regarding Yellowfin Tuna with FLs ranging from 45 to 50 cm. Yellowfin Tuna of size-classes IV and V (2 years old and older: Su et al. 2003) primarily fed on Skinnycheek Lanternfish and flying fish and exhibited no feeding overlap with those of other size-classes. The samples from classes IV and V used in this study were all collected in April. During the period of April–July, exocoetid fishes are prevalent in Taiwanese waters and these fishes are also the main prey of tunas (Oxenford and Hunte 1999) and Dolphinfish *Coryphaena hippurus* (Wu et al. 2006). Therefore, the stomachs of Yellowfin Tuna $\geq$ 100 cm FL mainly contained these prey (Table 2).

Stable Isotope Analysis

Sarà and Sarà (2007) and Varela et al. (2013) indicated that larger predators often exhibit higher δ^{15}N values than smaller predators do. Bearhop et al. (2004) reported significant variations in δ^{15}N values and increased diversification of prey among tuna at FL > 45 cm. Graham et al. (2007) and Olson et al. (2010) reported that the stable isotope values of Yellowfin Tuna tend to increase with an increase in body size. In this study, δ^{15}N and $\delta^{13}C'$ values of the muscle tissues of Yellowfin Tuna < 30 cm FL were 7.6 ± 0.6‰ and −17.5 ± 1.3‰,

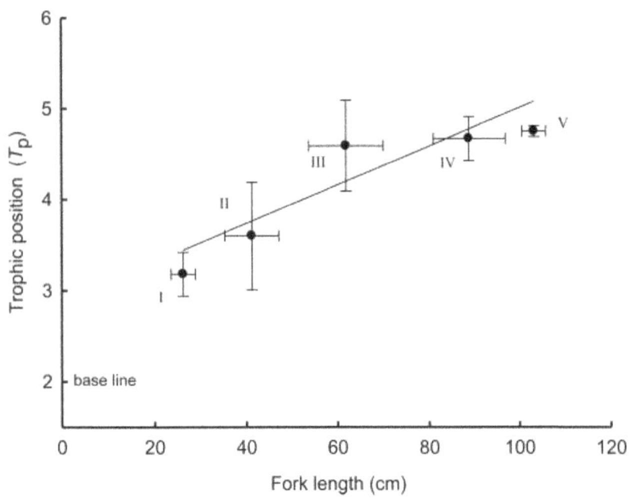

FIGURE 5. Relationship between trophic position (TP) and fork length of Yellowfin Tuna using δ^{15}N concentration of the zooplankton (5.13 ± 0.26‰) as a baseline. The TP–FL relationship of size-classes I–V was fitted by TP = 4.4827 + 0.0043FL (r^2 = 0.75). Vertical and horizontal bars represent SD values.

while for those > 50 cm FL, the δ^{15}N and $\delta^{13}C'$ values were 11.5 ± 0.9‰ and −16.8 ± 0.5‰, respectively (Table 4). A significant difference in δ^{15}N values was observed between Yellowfin Tuna < 30 cm FL and those > 50 cm (*t*-test: $P < 0.01$); these results are consistent with those of the aforementioned studies (Graham et al. 2007; Olson et al. 2010).

The δ^{13}C is derived from organic carbon in the ecosystem, and it has been used to track the movement of animals between areas with different food sources (Kurle and Worthy 2001). In this study, very few differences in δ^{13}C values were found among the size-classes (Tables 4, 5), suggesting that the prey and predators were likely from the same ecosystem. The stable isotope $\delta^{13}C'$ value of the muscle tissue of the juvenile Yellowfin Tuna varied with size ($P < 0.05$), but such variation could not be found for the liver tissue ($P > 0.05$) (Figure 3a, b). Graham (2008) estimated the half-life turnover rate of the liver to be shorter than that of muscle tissue (12 versus 63 d), because the metabolic rate of the liver is higher than that of muscle tissue (Suzuki et al. 2005; Guelinckx et al. 2007). Weng et al. (2013) reported that juvenile Yellowfin Tuna that lingered around a single FAD in the waters southwest of Taiwan for >31 d fed on various types of prey during this period and their muscle tissue exhibited varied stable isotope N and C values. The movement of juvenile Yellowfin Tuna from other waters to Taiwanese waters may alter their food intake and increase their nutrient supply. Future archival tag studies may facilitate clarification of this phenomenon.

Although the stable isotope analysis did not provide information regarding prey diversity, variability in the δ^{15}N values of predators can provide information that can be used in trophic level assessment (Bearhop et al. 2004). A trophic level change of 3.4‰ in δ^{15}N and −1.1‰ in δ^{13}C values indicates the existence of a predator–prey relationship (DeNiro and Epstein 1978, 1981). Thus, the difference of 4.1‰ in δ^{15}N values between the Yellowfin Tuna of size-class II and the postlarval shrimp, larval crabs, and mantis shrimp implies a predator–prey relationship (Tables 4, 5). However, nutrient sources affect the N values of food webs due to spatial and temporal variations (O'Reilly et al. 2002). Even identical prey collected in different waters may exhibit different stable isotope values (Graham et al. 2007; Sarà and Sarà 2007; Varela et al. 2013). According to a survey conducted by Dore et al. (2002), surface mixed waters are primarily affected by N_2 fixation, which changes with the seasons, potentially causing the aforementioned differences in the stable isotope values of the prey.

Epply and Peterson (1979) indicated that in the open sea ecosystem, the nutrients of surface waters are limited between the flux and the mixed layer. Therefore, the nitrogen content below the mixed layer that varies among seasons will be higher than that of the surface waters. Significant seasonal changes have been observed in the waters southwest of Taiwan. During autumn and winter, the Kuroshio Current branches northward (high temperature and salinity) and the China coastal currents (low temperature and salinity) flow southward into west Taiwan, whereas during spring and summer, the South China Sea surface current (high temperature and low salinity) flows northward into west Taiwan waters (Jan et al. 2002, 2010). The China coastal currents and Kuroshio Current system bring rich nutrients to this important fishing ground, and juvenile Yellowfin Tuna feed on several juvenile fishes, Parasquillidae species, Amphipoda species, and postlarval shrimp. Prey at this stage float on surface waters, and the stable isotope $\delta^{15}N$ values of these prey are lower than those of mesopelagic prey (Rau et al. 1989). For the Yellowfin Tuna with an average (±SD) FL of 53.9 cm (±2.1), the $\delta^{15}N$ value was 11.3‰ (±1.5) and the TP was 4.6 (±0.6); the fish shifted their diet to scombroid fishes and Japanese Barracudina.

Because the size range of the specimens did not cover all of the age-classes, the results obtained in this study can only be applied to juvenile Yellowfin Tuna. To provide more information on the food web in the waters southwest of Taiwan future studies should focus on collecting large-sized specimens and investigating the carbon and nitrogen composition of environmental nutrients.

Conclusions

Taken together, the results of the stomach contents and stable isotope analyses provide important evidence regarding ontogenetic shifts in the diet, feeding strategies, and stable isotope variations of juvenile Yellowfin Tuna. The stomach contents and stable isotope values differed significantly among tuna at different trophic positions. This ontogenetic shift in the diet of juvenile Yellowfin Tuna occurs at a FL of 50 cm.

ACKNOWLEDGMENTS

This study was financially supported by the Council of Agriculture, Executive Yuan, Taiwan, ROC on contracts 99AS-10.2.1-AI-A1(3) and 100AS-10.2.1-AI-A1. We thank captains J. J. Lee and F. M. Tsay for their help with sample collection, and C. Y. Chen for laboratory preparation. Appreciation extends to W.Y. Kao, and T. C. Wu, Institute of Biology and Evolutionary Biology, National Taiwan University, for their assistance with stable isotope analysis.

REFERENCES

Arim, M., F. Bozinovic, and P. A. Marquet. 2007. On the relationship between trophic position, body mass and temperature: reformulating the energy limitation hypothesis. Oikos 116:1524–1530.

Bearhop, S., C. E. Adams, S. Waldron, R. A, Fuller, and H. MacLeod. 2004. Determining trophic niche width: a novel approach using stable isotope analysis. Journal of Animal Ecology 73:1007–1012.

Bertrand, A., F. X. Bard, and E. Josse. 2002. Tuna food habits related to the micronekton distribution in French Polynesia. Marine Biology 140:1023–1037.

Brock, R. E. 1985. Preliminary study of the feeding habits of pelagic fish around Hawaiian fish aggregation devices or can fish aggregation devices enhance local fisheries productivity? Bulletin of Marine Science 37:40–49.

Buckley, R. M., D. G. Itano, and T. W. Buckley. 1989. Fish aggregation device (FAD) enhancement of offshore fisheries in American Samoa. Bulletin of Marine Science 44:942–949.

Buckley, T. W., and B. S. Miller. 1994. Feeding habits of Yellowfin Tuna associated with fish aggregation devices in American Samoa. Bulletin of Marine Science 55:445–459.

Charnov, E. L. 1976. Optimal foraging, the marginal value theorem. Theoretical Population Biology 9:129–136.

Collette, B. B., and C. E. Nauen. 1983. Scombrids of the world. An annotated and illustrated catalogue of tunas, mackerels, bonitos and related species known to date, volume 2. FAO (Food and Agriculture Organization of the United Nations) Fisheries Synopsis 125.

Cooper, A. B., N. Pettorelli, and S. M. Durant. 2007. Large carnivore menus: factors affecting hunting decisions by cheetahs in the Serengeti. Animal Behavior 73:651–659.

Deegan, L. A., and R. H. Garritt. 1997. Evidence for spatial variability in estuarine food webs. Marine Ecology Progress Series 147:31–47.

DeNiro, M. J., and S. Epstein. 1978. Influence of diet on the distribution of carbon isotopes in animals. Geochimica et Cosmochimica Acta 42:495–506.

DeNiro, M. J., and S. Epstein. 1981. Influence of diet on the distribution of nitrogen isotopes in animals. Geochimica et Cosmochimica Acta 45:341–351.

Deudero, S., J. K. Pinnegar, N. V. C. Polunin, G. Morey, and B. Morales-Nin. 2004. Spatial variation and ontogenic shifts in the isotopic composition of Mediterranean littoral fishes. Marine Biology 145:971–981.

Dore, J. E., J. R. Brum, L. M. Tupas, and D. M. Karl. 2002. Seasonal and interannual variability in sources of nitrogen supporting export in the oligotrophic subtropical North Pacific Ocean. Limnology and Oceanography 47:1595–1607.

Dragovich, A. 1970. The food of skipjack and Yellowfin Tunas in the Atlantic Ocean. U.S. National Marine Fisheries Service Fishery Bulletin 68:445–460.

Dragovich, A., and T. Potthoff. 1972. Comparative study of food of skipjack and Yellowfin Tunas off the coast of west Africa. U.S. National Marine Fisheries Service Fishery Bulletin 70:1087–1110.

Elsdon, T. S., S. Ayvazian, K. W. McMahon, and S. R. Thorrold. 2010. Experimental evaluation of stable isotope fractionation in fish muscle and otoliths. Marine Ecology Progress Series 408:195–205.

Epply, R. W., and B. J. Peterson. 1979. Particulate organic matter flux and planktonic new production in the deep ocean. Nature 282:677–680.

Folch, J., M. Lees, and G. H. S. Stanley. 1957. A simple method for the isolation and purification of total lipids from animal tissues. Journal of Biological Chemistry 226:497–509.

Fry, B. 1988. Food web structure on Georges Bank from stable C, N, and S isotopic compositions. Limnology and Oceanography 33:1182–1190.

Fry, B., and C. Arnold. 1982. Rapid $^{13}C/^{12}C$ turnover during growth of brown shrimp (*Penaeus aztecus*). Oecologia 54:200–204.

Fry, B., D. M. Baltz, M. C. Benfield, J. W. Fleeger, A. Gace, H. L. Haas, and Z. J. Quiñones-Rivera. 2003. Stable isotope indicators of movement and residency for brown shrimp (*Farfantepenaeus aztecus*) in coastal Louisiana marshscapes. Estuaries 26:82–97.

Fuiman, L. A., and A. E. Magurran. 1994. Development of predator defenses in fishes. Reviews in Fish Biology and Fisheries 4:145–183.

Graham, B. S. 2008. Trophic dynamic and movements of tuna in the tropical Pacific Ocean inferred from stable isotope analysis. Doctoral dissertation. University of Hawaii, Manoa.

Graham, B. S., D. Grbbus, K. Holland, and B. N. Popp. 2007. A rapid ontogenetic shift in the diet of juvenile Yellowfin Tuna from Hawaii. Marine Biology 150:647–658.

Grey, J. 2001. Ontogeny and dietary specialization in Brown Trout (*Salmo trutta* L.) from Loch Ness, Scotland, examined using stable isotopes of carbon and nitrogen. Ecology of Freshwater Fish 10:168–176.

Grubbs, R. D. 2010. Ontogenetic shifts in movements and habitat use. Pages 319–350 *in* J. C. Carrier, J. A. Musick, M. R. Heithaus, editors. Sharks and their relatives II: biodiversity, adaptive physiology, and conservation. CRC Press, Boca Raton, Florida.

Guelinckx, J., J. Maez, P. Van Den Driessche, B. Geysen, F. Dehairs, and F. Ollevier. 2007. Changes in δ^{13} C and δ^{15} N in different tissues of juvenile and goby *Pomatoschistus minutus*: a laboratory diet-switch experiment. Marine Ecology Progress Series 341:205–215.

Hammerschlag-Peyer, C. M., L. A. Yeager, M. S. Araújo, and C. A. Layman. 2011. A hypothesis-testing framework for studies investigating ontogenetic niche shifts using stable isotope ratios. PLoS (Public Library of Science) ONE [online serial] 6(11):e27104.

Hampton, J., and D. A. Fournier. 2001. A spatially disaggregated, length-based, age-structured population model of Yellowfin Tuna (*Thunnus albacares*) in the western and central Pacific Ocean. Marine and Freshwater Research 52:937–963.

Harrigan, P., J. C. Zieman, and S. A. Macko. 1989. The base of nutritional support for the Gray Snapper (*Lutjanus griseus*): An evaluation based on a combined stomach content and stable isotope analysis. Bulletin of Marine Science 44:65–77.

Harvey, C. J., P. C. Hanson, T. E. Essington, P. B. Brown, and J. F. Kitchell. 2002. Using bioenergetics models to predict stable isotope ratios in fishes. Canadian Journal of Fisheries and Aquatic Sciences 59:115–124.

Herrel, A., and A. C. Gibb. 2006. Ontogeny of performance in vertebrates. Physiological and Biochemical Zoology 79:1–6.

Hobson, K., and H. Welch. 1992. Determination of trophic relationships within a high Arctic marine food web using δ^{13} C and δ^{15} N analysis. Marine Ecology Progress Series 84:9–18.

Hobson, K. A., and R. G. Clark. 1992. Assessing avian diets using stable isotopes I: turnover of ^{13}C in tissues. Condor 94:181–188.

Horn, H. S. 1966. Measurement of "overlap" in comparative ecological studies. American Naturalist 100:419–424.

Hyslop, E. J. 1980. Stomach contents analysis – a review of methods and their application. Journal of Fish Biology 17:411–429.

Jan, S., Y. H. Tseng, and D. E. Dietrich. 2010. Sources of water in the Taiwan Strait. Journal of Oceanography 66:211–221.

Jan, S., J. Wang, C. S. Chern, and S. Y. Chao. 2002. Seasonal variation of the circulation in the Taiwan Strait. Journal of Marine Systems 35:249–268.

Kurle, C. M., and G. A. J. Worthy. 2001. Stable isotope assessment of temporal and geographic differences in feeding ecology of northern fur seals (*Callorhinus ursinus*) and their prey. Oecologia 126:254–265.

Logan, J. M., T. D. Jardine, T. J. Miller, S. E. Bunn, R. A. Cunjak, and M. E. Lutcavage. 2008. Lipid corrections in carbon and nitrogen stable isotope analyses: comparison of chemical extraction and modelling methods. Journal of Animal Ecology 77:838–846.

MacDonald, J. S., K. G. Waiwood, and R. H. Green. 1982. Rates of digestion of different prey in Atlantic Cod (*Gadus morhua*), Ocean Pout (*Macrozoarces americanus*), Winter Flounder (*Pseudopleuronectes americanus*), and American Plaice (*Hippoglossoides platessoides*). Canadian Journal of Fisheries and Aquatic Sciences 39:651–659.

Maldeniya, R. 1996. Food consumption of Yellowfin Tuna, *Thunnus albacares*, in Sri Lankan waters. Environmental Biology of Fishes 47:101–107.

McConnaughey, T., and C. P. McRoy. 1979. Food-web structure and the fractionation of carbon isotopes in the Bering Sea. Marine Biology 53:257–262.

McCormick, M. I. 1998. Ontogeny of diet shifts by a microcarnivorous fish, *Cheilodactylus spectabilis*: relationship between feeding mechanics, microhabitat selection and growth. Marine Biology 132:9–20.

McCutchan, J. H., W. M. Lewis, C. Kendall, and C. C. McGrath. 2003. Variation in trophic shift for stable isotope ratios of carbon, nitrogen, and sulfur. Oikos 102:378–390.

Ménard, F., A. Fonteneau, D. Gaertner, V. Nordstrom, B. Stéquert, and E. Marchal. 2000a. Exploitation of small tunas by a purse-seine fishery with fish aggregating devices and their feeding ecology in an eastern tropical Atlantic ecosystem. ICES Journal of Marine Science 57:525–530.

Ménard, F., B. Stéquert, A. Rubin, M. Herrera, and E. Marchal. 2000b. Food consumption of tuna in the Equatorial Atlantic Ocean: FAD-associated versus unassociated schools. Aquatic Living Resources 13:233–240.

Miller, T. J., L. B. Crowder, J. A. Rice, and E. A. Marschall. 1988. Larval size and recruitment mechanisms in fishes: toward a conceptual framework. Canadian Journal of Fisheries and Aquatic Sciences 45:1657–1670.

Minagawa, M., and E. Wada. 1984. Stepwise enrichment of ^{15}N along food chains: further evidence and the relation between δ^{15}N and animal age. Geochimica et Cosmochimica Acta 48:1135–1140.

Noakes, D. L. G., and J-G. J. Godin. 1988. Ontogeny of behavior and concurrent developmental changes in sensory systems in teleost fishes. Pages 345–395 *in* W. S. Hoar and D. J. Randall, editors. Fish physiology, volume 11. Academic Press, New York.

O'Reilly, C. M., R. E. Hecky, A. S. Cohen, and P. D. Plisnier. 2002. Interpreting stable isotopes in food webs: recognizing the role of time averaging at different trophic levels. Limnology and Oceanography 47:306–309.

Olson, M. H. 1996. Ontogenetic niche shifts in Largemouth Bass: variability and consequences for first-year growth. Ecology 77:179–190.

Olson, R. J., and C. H. Boggs. 1986. Apex predation by Yellowfin Tuna (*Thunnus albacares*): independent estimates from gastric evacuation and stomach contents, bioenergetics, and cesium concentrations. Canadian Journal of Fisheries and Aquatic Sciences 43:1760–1775.

Olson, R. J., B. N. Popp, B. S. Graham, G. A. López-Ibarra, F. Galván-Magaña, C. E. Lennert-Cody, N. Bocanegra-Castillo, N. J. Wallsgrove, E. Gier, V. Alatorre-Ramirez, L. T. Balance, and B. Fry. 2010. Food-web inferences of stable isotope spatial patterns in copepods and Yellowfin Tuna in the pelagic eastern Pacific Ocean. Progress in Oceanography 86:124–138.

Oxenford, H. A., and W. Hunte. 1999. Feeding habits of the Dolphinfish (*Coryphaena hippurus*) in the eastern Caribbean. Scientia Marina 63:303–315.

Pauly, D., A. W. Trites, E. Capuli, and V. Christensen. 1998. Diet composition and trophic levels of marine mammals. ICES Journal of Marine Science 55:467–481.

Peterson, B. J., and B. Fry. 1987. Stable isotopes in ecosystem studies. Annual Review of Ecology and Systematics 18:293–320.

Peterson, B. J., R. W. Howarth, and R. H. Garritt. 1985. Multiple stable isotopes used to trace the flow of organic matter in estuarine food webs. Science 227:1361–1363.

Pinnegar, J. K., and N. V. C. Polunin. 1999. Differential fractionation of δ^{13}C and δ^{15}N among fish tissues: implications for the study of trophic interactions. Functional Ecology 13:225–231.

Pinnegar, J. K., N. V. Polunin, and F. Badalamenti. 2003. Long-term changes in the trophic level of western Mediterranean fishery and aquaculture landings. Canadian Journal of Fisheries and Aquatic Sciences 60:222–235.

Post, D. M. 2002. Using stable isotopes to estimate trophic position: models, methods, and assumptions. Ecology 83:703–718.

Post, D. M. 2003. Individual variation in the timing of ontogenetic niche shifts in Largemouth Bass. Ecology 84:1298–1310.

Potier, M., F. Marsic, V. Lucas, R. Sabatié, J. P. Hallier, and F. Ménard. 2004. Feeding partitioning among tuna taken in surface and mid-water layer: the case of yellowfin (*Thunnus albacares*) and bigeye (*T. obserus*) in the western tropical Indian Ocean. Western Indian Ocean Journal of Marine Science 3:51–62.

Rau, G. H., M. Heyraud, and R. D. Cherry. 1989. ^{15}N/ ^{14}N and ^{13}C/ ^{12}C in mesopelagic shrimp from the Northeast Atlantic Ocean: evidence for differences in diet. Deep-Sea Research Part A Oceanographic Research Papers 36:1103–1110.

Renones, O., N. V. C. Polunin, and R. Goni. 2002. Size related dietary shifts of *Epinephelus marginatus* in a western Mediterranean littoral ecosystem: an isotope and stomach content analysis. Journal of Fish Biology 61:122–137.

Rohit, P., G. S. Rao, and K. Ram Mohan. 2010. Feeding strategies and diet composition of Yellowfin Tuna *Thunnus albacares* (Bonnaterre, 1788) caught along Andhra Pradesh, east coast of India. Indian Journal of Fisheries 57:13–19.

Sarà, G., and R. Sarà. 2007. Feeding habits and trophic levels of Bluefin Tuna *Thunnus thynnus* of different size classes in the Mediterranean Sea. Journal of Applied Ichthyology 23:122–127.

Schoener, T. W. 1971. Theory of feeding strategies. Annual Review of Ecology and Systematics 2:369–404.

Sheldon, R. W., W. H. Sutcliffe Jr., and M. A. Paranjape. 1977. Structure of pelagic food chain and relationship between plankton and fish production. Journal of the Fisheries Research Board of Canada 34:2344–2353.

Smetacekm, V. 1999. Revolution in the ocean. Nature 401:647.

Sotiropoulos, M. A., W. M. Tonn, and L. I. Wassenaar. 2004. Effects of lipid extraction on stable carbon and nitrogen isotope analyses of fish tissues: potential consequences for food web studies. Ecology of Freshwater Fish 13:155–160.

Su, N. J., C. L. Sun, and S. Z. Yeh. 2003. Estimation of growth parameters and age composition for Yellowfin Tuna, *Thunnus albacares*, in the western Pacific using the length-based MULTIFAN method. Journal of Fisheries Society of Taiwan. 30:171–184.

Suzuki, K. W., A. Kasai, K. Nakayama, and M. Tanaka. 2005. Differential isotopic enrichment and half-life among tissues in Japanese temperate bass (*Lateolabrax japonicus*) juveniles: implications for analyzing migration. Canadian Journal of Fisheries and Aquatic Sciences 62: 671–678.

Sweeting, C. J., N. V. C. Polunin, and S. Jennings. 2006. Effects of chemical lipid extraction and arithmetic lipid correction on stable isotope ratios of fish tissues. Rapid Communications in Mass Spectrometry 20:595–601.

Vander Zanden, M. J., and J. B. Rasmussen. 1999. Primary consumer $\delta^{15}N$ and $\delta^{13}C$ and the trophic position of aquatic consumers. Ecology 80:1395–1404.

Vanderklift, M. A., and S. Ponsard. 2003. Sources of variation in consumer-diet $\delta^{15}N$ enrichment: a meta-analysis. Oecologia 136:169–182.

Varela, J. L., A. Larrañaga, and A. Medina. 2011. Prey-muscle carbon and nitrogen stable isotope discrimination factors in Atlantic Bluefin Tuna (*Thunnus thynnus*). Journal of Experimental Marine Biology and Ecology 406:21–28.

Varela, J. L., E. Rodríguez-Marín, and A. Medina. 2013. Estimating diets of pre-spawning Atlantic Bluefin Tuna from stomach content and stable isotope analyses. Journal of Sea Research 76:187–192.

Wang, W. R. 2005. Reproductive biology of Yellowfin Tuna *Thunnus albacares* in the western Pacific Ocean. Master's thesis. National Taiwan University, Taipei City.

Weng, J. S., M. K. Hung, C. C. Lai, L. J. Wu, M. A. Lee, and K. M. Liu. 2013. Fine-scale vertical and horizontal movements of juvenile Yellowfin Tuna (*Thunnus albacares*) associated with a subsurface fish aggregating device (FAD) off southwestern Taiwan. Journal of Applied Ichthyology 29:990–1000.

Wu, C. C., J. C Lin., and W. C. Su. 2006. Diet and feeding habits of Dolphin Fish (*Coryhaena hippurus*) in the waters off eastern Taiwan. Journal of Taiwan Fisheries Research 14:13–27.

Zar, J. H. 2010. Biostatistical analysis, 5th edition. Pearson Education, Upper Saddle River, New Jersey.

Zaret, T., and A. S. Rand. 1971. Completion in tropical stream fishes support for the competitive exclusion principle. Ecology 52:336–342.

11

Estimating Sturgeon Abundance in the Carolinas Using Side-Scan Sonar

H. Jared Flowers*
North Carolina Cooperative Fish and Wildlife Research Unit, Department of Applied Ecology, North Carolina State University, 127 David Clark Labs, Campus Box 7617, Raleigh, North Carolina 27695-7617, USA

Joseph E. Hightower
U.S. Geological Survey, North Carolina Cooperative Fish and Wildlife Research Unit, Department of Applied Ecology, North Carolina State University, 127 David Clark Labs, Campus Box 7617, Raleigh, North Carolina 27695-7617, USA

Abstract

Sturgeons (Acipenseridae) are one of the most threatened taxa worldwide, including species in North Carolina and South Carolina. Populations of Atlantic Sturgeon *Acipenser oxyrinchus* in the Carolinas have been significantly reduced from historical levels by a combination of intense fishing and habitat loss. There is a need for estimates of current abundance, to describe status, and for estimates of historical abundance in order to provide realistic recovery goals. In this study we used *N*-mixture and distance models with data acquired from side-scan sonar surveys to estimate abundance of sturgeon in six major sturgeon rivers in North Carolina and South Carolina. Estimated abundances of sturgeon greater than 1 m TL in the Carolina distinct population segment (DPS) were 2,031 using the count model and 1,912 via the distance model. The Pee Dee River had the highest overall abundance of any river at 1,944 (count model) or 1,823 (distance model). These estimates do not account for sturgeon less than 1 m TL or occurring in riverine reaches not surveyed or in marine waters. Comparing the two models, the *N*-mixture model produced similar estimates using less data than the distance model with only a slight reduction of estimated precision.

Sturgeon (Acipenseridae) populations worldwide have declined from historical levels as a result of a combination of factors, including overharvest and habitat alteration (Secor 2002; Pitkitch et al. 2005; Nelson et al. 2013). In the eastern USA, Atlantic Sturgeon *Acipenser oxyrinchus oxyrinchus* was listed as "endangered" under the U.S. Endangered Species Act (ESA) in 2012. Under the listing, five separate Atlantic Sturgeon distinct population segments (DPS) were established. One DPS was categorized as threatened, but the other four were listed as endangered, including the Carolina and South Atlantic DPSs (Figure 1). The Carolina DPS includes riverine populations ranging from the Roanoke River in North Carolina to the Santee–Cooper system in South Carolina. The South Atlantic DPS includes rivers from the Ashepoo–Combahee–Edisto basin in South Carolina to the St. Johns River, Florida. Given the status of sturgeons, it is important to assess and monitor the state of populations. Unfortunately, sturgeons have not been well studied in many Southeastern systems since the closure of commercial U.S. fisheries in 1998 (ASSRT 2007; Peterson et al. 2008). Reliable estimates of abundance

Subject editor: Donald Noakes, Vancouver Island University, Nanaimo, British Columbia, Canada

*Corresponding author: hjflower@ncsu.edu

are necessary for making management decisions and developing recovery strategies.

A common method for estimating absolute abundance of fish species is mark–recapture; however, there are drawbacks to these studies, including cost and potential harm to the study species. The endangered status of sturgeon limits the sampling methods researchers can employ because extra care must be taken when sampling and handling specimens (Damon-Randall et al. 2010; Kahn and Mohead 2010). Hydroacoustics provides a nonintrusive method for sampling endangered species. Sturgeon and related Paddlefish *Polyodon spathula* have been identified as potentially suitable targets for hydroacoustic studies (Nealson and Brundage 2007; Bergman 2011; Nelson et al. 2013). Side-scan sonar is a relatively old hydroacoustic technology that has been used increasingly in fisheries studies in recent years (Kaeser and Litts 2008; Foote 2009). In a previous paper, we described a procedure to rapidly survey sturgeon populations using side-scan sonar and occupancy modeling approaches (Flowers and Hightower 2013). We reexamine those data here, with a goal of estimating Atlantic Sturgeon abundance in the Carolina DPS. Large body size, including a maximum length of over 3 m (Gross et al. 2002) and distinctive shape make this species an ideal hydroacoustic target. This anadromous species, spends large amounts of time feeding and migrating in marine environments, using freshwater rivers for spawning and a summer dormant period (ASSRT 2007). During summer, Atlantic Sturgeon may aggregate in lower portions of rivers, often in deep holes near the freshwater–saltwater interface, although locations vary with river characteristics and flow conditions (Moser and Ross 1995; Collins et al. 2000).

Sturgeon density can be obtained from counts based on side-scan images and the length and width of the survey transect. These counts can be analyzed with an *N*-mixture model (Royle 2004) that uses a distribution, such as a Poisson, to simulate abundance and binomial distribution to simulate detection. Count data can also be analyzed using distance sampling methods (Buckland et al. 2001; Royle et al. 2004) based on the distance on side-scan images from sturgeon to the survey transect. We compare results from the various data types and models and provide abundance estimates for Atlantic Sturgeon recovery planning.

METHODS

Study sites.—Field work took place in six river systems in North Carolina and South Carolina (Figure 1). Systems were chosen based on available information about the status of

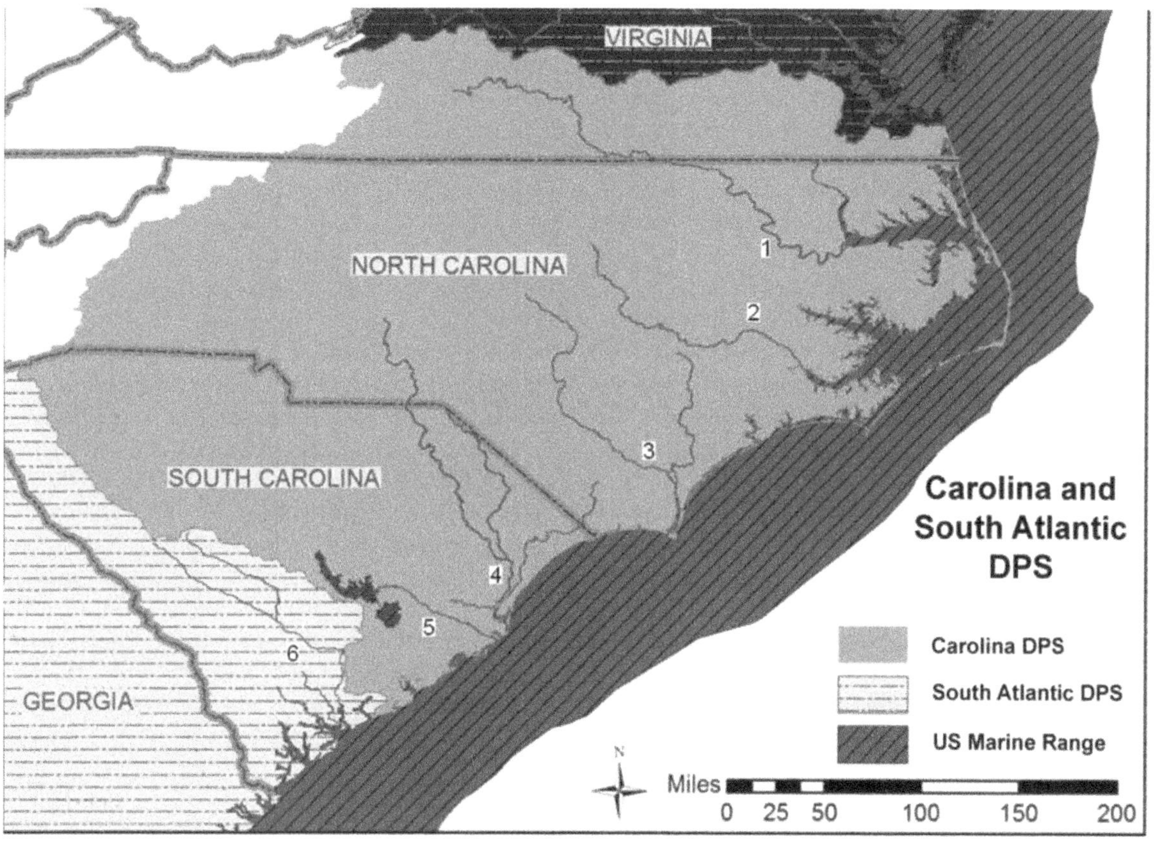

FIGURE 1. Lower portions of six river systems in the Carolina and South Atlantic distinct population segments (DPS) surveyed using side-scan sonar; from north to south they are the Roanoke (1), Neuse (2), and Cape Fear (3) rivers in North Carolina and the Pee Dee/Waccamaw (4), Santee (5), and Edisto (6) rivers in South Carolina.

sturgeon populations or the potential for presence of sturgeon in that system. Five of the rivers were located in the Carolina DPS (Roanoke, Neuse, Cape Fear, Pee Dee, and Santee) while the sixth was located in the South Atlantic DPS (Edisto). The Tar River was the only Carolina DPS river excluded from the survey, based on smaller size (Oakley and Hightower 2007) and lack of recent anecdotal evidence of sturgeon. Mark–recapture studies or other types of sturgeon population studies have not been performed on these systems.

Side-scan surveys.—Surveys were conducted in coastal rivers near the freshwater–saltwater interface where sturgeon have been observed (Moser and Ross 1995; Collins et al. 2000; Peterson et al. 2008). Repeated side-scan sonar surveys were performed over 3 d at each of the six rivers. The three sampling days were as close as possible, often consecutive, in order to reduce the chance of movement between sites, as needed to minimize violations of the population closure assumptions of these methods. Riverine survey reaches began downstream in tidal portions of elevated salinity (>15 mg/L) and ranged upstream a distance of at least 30 km inland on the main stem or to a point where river depths were too shallow to effectively operate the side-scan sonar. The total distance surveyed in each river varied, ranging from 40 to 80 km because of braids and side channels. Full details regarding survey methods were provided in Flowers and Hightower (2013).

We used an Edgetech 4125-P side-scan sonar unit operated in high-frequency (1,250 kHz) mode, with a total swath width of 50 m. The unit was deployed at a constant depth of approximately 1 m below the water's surface for all systems and surveys. Side-scan sonar data were collected and processed using Chesapeake Technology's SonarWiz.Map software. When a potential sturgeon was observed on side-scan images, the target was marked, its GPS coordinates were taken, and the target's body length and distance from the transect were measured.

Each river was an individual study area subdivided into 2-km sites, for a total of 179 sites. Sites of 1 km and 4 km were previously evaluated for their effect on occupancy, with results bracketing the 2-km sites (Flowers and Hightower 2013). The large amount of data collected during the surveys required extended time for analysis. Side-scan sonar file processing took 1–3 d/survey day, depending on how many targets needed to be recorded.

Abundance estimation.—Only targets clearly identified as sturgeon (Flowers and Hightower 2013) were used in our analyses. For the *N*-mixture model, side-scan observer data were compiled into a data matrix containing river, site, and the three daily counts of individual sturgeon. Sites were arranged in rows ($N = 179$), while river index and daily counts were listed in columns ($N = 4$). A separate matrix for distance sampling data incorporated corrected horizontal distance from a side-scan transect's centerline to each sturgeon identified in a given site for each survey day. The number of rows in this matrix was equal to the number of distance observations by site ($N =$ 1,436) and had columns representing river, site, sample day, and distance ($N = 4$). A "slant range correction" was used to account for positioning errors because the cross-track coordinates of a side-scan image are a function of range to the sonar rather than horizontal distance on the bottom (Cobra et al. 1992). Although our side-scan swath was 25 m/side, based on slant-range, the maximum horizontal range was actually 24 m.

The counts and distance data sets were analyzed using procedures in package unmarked (Fiske and Chandler 2011) in program R (R Development Core Team 2009). For both analytical methods, we assumed that we correctly identified targets classified as sturgeon, although additional sturgeon were assumed to be present but not identifiable from a particular side-scan image. Count data were analyzed in unmarked using the pcount procedure, based on the *N*-mixture framework proposed by Royle (2004). The general form of the *N*-mixture model for site abundance is

$$N_i \sim f(\lambda, \theta) \text{ for } i = 1, 2, ..., M \tag{1}$$

and for the detection process is

$$y_{ij}|N_i \sim \text{Binomial}(N_i, p) \text{ for } j = 1, 2, ..., J_i \tag{2}$$

where λ is the abundance per site *i* and *p* is the detection probability. A discrete distribution, such as the Poisson or negative-binomial, is used for *f*, with support restricted to $N_i \geq 0$ (Fiske and Chandler 2011). The θ are additional parameters of *f* other than the abundance rate for distributions such as the negative binomial (Fiske and Chandler 2011).

This method uses temporally repeated counts of individuals at a site to estimate the abundance of a closed population. The framework uses a combination of two different processes to model detection and abundance. Detection probability is modeled using a binomial distribution, while abundance is modeled using a Poisson, negative binomial, or zero-inflated Poisson distribution. Sites were assumed to be closed for the duration of the experiment, and probability of detecting an individual animal was assumed to be constant (Royle 2004).

Distance sampling data were analyzed using the gdistsamp function based on the method proposed by Royle et al. (2004) and extended by Chandler et al. (2011). This model is similar to the *N*-mixture model, although it adds a component to the likelihood to address variability in detection probability as a function of range from the survey transect. Here transect-level abundance is modeled the same as the *N*-mixture model, but the detection process is modeled as

$$y_{ij}|N_i \sim \text{Multinomial}(N_i, \pi_{ij}) \text{ for } i = 1, \ldots, M \quad j = 1, \ldots., J, \tag{3}$$

where N_i is the latent abundance at site *i*, as with the *N*-mixture model, and π_{ij} is the multinomial cell probability for transect *i* in distance-class *j*, computed by integrating a detection function such as the half-normal (with scale parameter σ) over each distance interval (Fiske and Chandler 2011). Abundance is modeled using a Poisson or negative-binomial distribution. Four functions (uniform, half-normal, exponential, hazard), described in further depth by Buckland et al. (2001), were used to model detection, with distance data formatted in 4-m bins over the 24 m/side width of the side-scan swath. As above, sites were assumed to be closed over sampling occasions. Additional assumptions related to distance sampling were (1) sturgeon were distributed according to some stochastic process based on an underlying density, (2) sturgeon were detected at their initial locations, prior to any movement (to avoid double-counting), and (3) distances were correctly grouped by intervals (Buckland et al. 2001). It is generally assumed for distance surveys that survey lines are randomly placed, but here, we used a single mid-river survey line. Thus, we further assumed that sturgeon were randomly distributed so that our estimated abundances were representative of the entire surveyed sections of rivers.

For all models, scenarios were run using river as a site-level covariate to model the variability in abundance across rivers. For each method (count and distance), the most appropriate model was selected using Akaike's information criterion (AIC). We evaluated model selection using ΔAIC, i.e., the difference between the best model and each model, and AIC weight (AICwt), representing the relative likelihood of each model. Goodness of fit for the best models was evaluated using the parametric bootstrapping function in unmarked, parboot. One hundred bootstraps were run for each model, and the Freeman–Tukey test was used to assess fit. Estimates of abundance and 95% confidence intervals were calculated for each of the best models using the predict function in unmarked.

We estimated the proportion of the river scanned based on the total surface area and the area surveyed by side-scan sonar (48 m × 179 2-km sites). River area measurements were performed in Google Earth Pro (Google, Inc.). Our side-scan surveys only covered 17.2 km^2, which was 0.13 of the area of the survey reach for all rivers combined. Proportions for individual rivers were 0.13 (Roanoke), 0.08 (Neuse), 0.26 (Cape Fear), 0.11 (Pee Dee), 0.29 (Santee), and 0.11 (Edisto).

RESULTS

Counts of sturgeon during surveys ranged from 0 to 109 across sites (Figure 2). Of the 179 sites, 113 had no sturgeon detected, 55 had between 1 and 10 sturgeon, and only 11 had more than 10 sturgeon. Clearly, sturgeon were not uniformly distributed across sites or systems. When detected, they were predominantly alone or in small numbers and not in large aggregations, except in a few instances.

The best performing *N*-mixture model used a negative binomial distribution for abundance and included river as a covariate (Table 1). There was negligible support for a Poisson distribution for abundance (with or without zero inflation) or

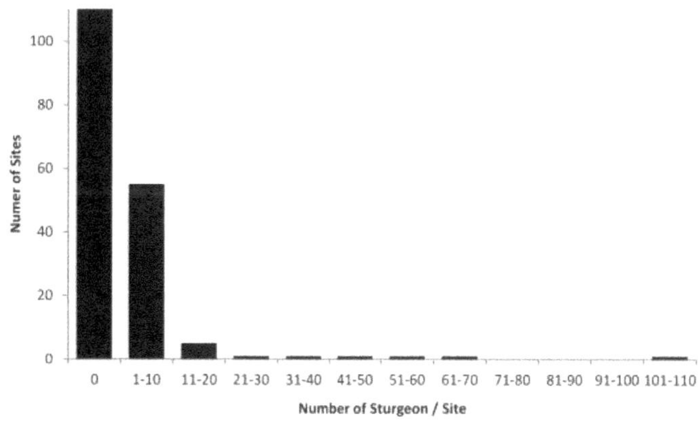

FIGURE 2. The distribution of Atlantic Sturgeon counts per site across all rivers (179 sites) sampled in the Carolina and South Atlantic distinct population segments. Most sites had detections of either no sturgeon or 1–10. The maximum number of sturgeon counted at a single site was 109.

models without river-specific estimates. Estimated abundance varied substantially among sites and rivers, the highest observed and estimated counts being for the Pee Dee and Edisto rivers (Table 2; Figure 3). The Pee Dee River had the highest estimated abundance for a single river (1,944) and comprised a large fraction of estimated total abundance for the Carolina DPS (Table 2). The selected model appeared to be adequate based on a goodness-of-fit test (Figure 4). The data do appear to be overdispersed, with estimated dispersion $\alpha > 0$, at 0.13.

The best-performing distance model also utilized a negative binomial distribution for abundance, with river as a covariate (Table 3). As in the *N*-mixture models, there was negligible support for a Poisson distribution for abundance. The preferred model used a half-normal distribution for describing detection probability. The pattern of detections by range bins (Figure 5)

TABLE 1. Comparison of *N*-mixture models as applied to Atlantic Sturgeon in the Carolina and South Atlantic distinct population segments with and without river as a covariate for abundance and using alternative distributions for abundance. Model fit is compared using Akaike's information criterion (AIC); a lower value indicates a better performing model. The number of parameters in each model is denoted by nPars, ΔAIC is the difference between AIC value and the best model, and AICwt represents the relative likelihood of each model.

Model	nPars	AIC value	ΔAIC	AICwt
Negative binomial, with river	8	1,208.48	0.00	1.00
Negative binomial	3	1,292.65	84.17	<0.001
Zero-inflated Poisson, with river	8	2,147.44	938.96	<0.001
Zero-inflated Poisson	3	2,663.59	1,455.11	<0.001
Poisson, with river	7	2,735.87	1,527.39	<0.001
Poisson	2	3,909.23	2,700.75	<0.001

TABLE 2. Abundance estimates and 95% confidence intervals in parentheses for riverine Atlantic Sturgeon >1 m TL within six North Carolina and South Carolina rivers (see Figure 1). Results for each analytical approach are from the model with lowest Akaike information criterion. Maximum counts (per survey day) and number of survey sites for each river system are listed for comparison. The Carolina distinct population segment (DPS) estimate is a total for all rivers except the Edisto.

River	*N* sites	Maximum counts	*N*-mixture survey area	Distance survey area
Roanoke	30	4	10.9 (3–36)	10.3 (3–34)
Neuse	22	1	2.7 (0–23)	2.7 (0–21)
Cape Fear	38	23	73.1 (35–152)	75.8 (37–156)
Pee Dee	37	419	1,943.8 (1,036–3,646)	1,823.3 (976–3,406)
Santee	30	0	0.0 (0–0)	0.0 (0–0)
Edisto	22	104	343.5 (150–788)	326.3 (143–744)
Carolina DPS	157	447	2,030.5(1,075–3,858)	1,912.0 (1,016–3,616)

was not well described by the half-normal distribution (or any other alternative), but the half-normal did account for the low detections at ranges greater than 20 m. Detections probably decreased at the far edge of the swath because acoustic shadows would have been outside the sonar image field-of-view. Detections increased gradually at distances of 8–20 m off the transect, perhaps because the more gradual angle relative to the sonar provided a clearer acoustic shadow for identification. Abundance estimates from the distance model were quite similar to those of the *N*-mixture model, the Carolina DPS estimate being 1,823 (Table 2; Figure 3). The preferred distance model was supported by the goodness-of-fit test (Figure 4). The data do appear to be overdispersed, with estimated dispersion $\alpha > 0$, at 0.274.

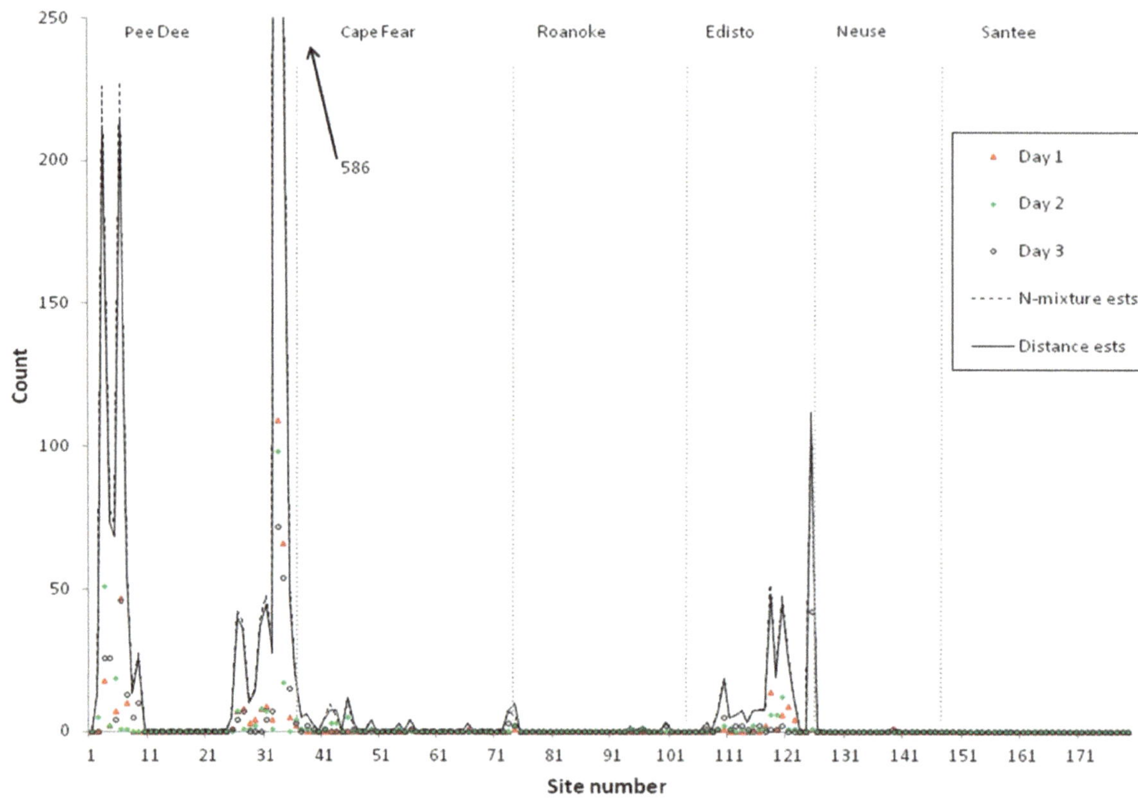

FIGURE 3. Site-specific estimates of Atlantic Sturgeon abundance in the Carolina and South Atlantic distinct population segments derived by the distance (solid) and *N*-mixture (dashed) models and overlaid onto counts of sturgeon per site for each survey day. Model estimates followed trends in site counts, and there was little difference in estimates between models.

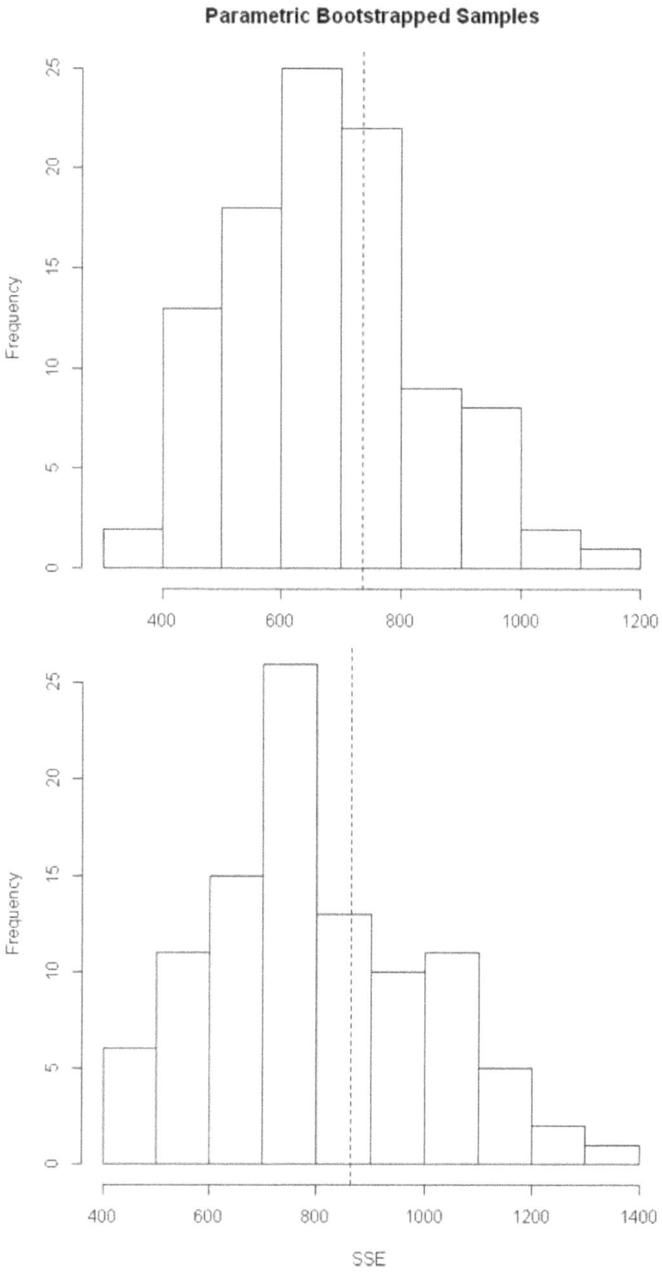

FIGURE 4. Results of parametric bootstrap goodness-of-fit testing as related to Atlantic Sturgeon in the Carolina and South Atlantic distinct population segments. The upper panel is for the *N*-mixture model; the lower panel is for the distance model. The dotted line is the Freeman–Tukey test statistic. Both models adequately explain the observed data.

DISCUSSION

Our abundance estimates are the first of any kind on these rivers and only the fifth estimate of Atlantic Sturgeon abundance rangewide (Kocik et al. 2013; Kahn et al. 2014). Sturgeon have been observed in these rivers and sampled in various studies (Moser and Ross 1995; Collins and Smith 1997; Collins et al. 2000; Armstrong and Hightower 2002; Oakley and Hightower 2007; authors' unpublished data), but no assessments have been undertaken. A recent range-wide study estimates that there may be 6,615–29,784 ocean-going Atlantic Sturgeon in the Carolina DPS (Kocik et al. 2013).

Our Atlantic Sturgeon population estimates are not all-inclusive. Estimates are limited to sturgeon >1 m because smaller sturgeon are difficult to identify confidently using side-scan images. Thus, all identified sturgeon were likely Atlantic Sturgeon due to their greater abundance and larger overall size. Only the largest Shortnose Sturgeon *A. brevirostrum* would be close to the 1 m size threshold for identification. Estimates are also limited to the reaches of rivers surveyed. This zone, the freshwater–saltwater interface, probably contained most of a river's available Atlantic Sturgeon encompassing ≥1 m, although side-scan detections of sturgeon sometimes occurred to the upper extent of survey reaches (Flowers and Hightower 2013, Supplement A). We did not observe sturgeon in our surveyed portion of the Santee River, resulting in a 0 population estimate, but sturgeon have been observed there in the past (Collins et al. 2000). A different survey design would be required to produce estimates that would apply to each entire river.

Both models require that sites be closed with respect to mortality, recruitment, and movement. Our survey passes were typically done on successive days to minimize these events, and telemetry data showed limited movement between sampling occasions (Flowers and Hightower 2013). Prior studies have shown that Atlantic Sturgeon occupy deep areas and show little or no movement for extended periods during summer (Moser and Ross 1995). Distance sampling assumes that transects are located randomly and independent of animals' locations (Thomas et al. 2010). Our transects extended through the entire study reach so were therefore independent of sites containing sturgeon. However, we do recommend that future side-scan surveys of sturgeon be done with transects at varying locations cross-channel.

Based on telemetry studies (Erickson et al. 2011; Parauka et al. 2011; Oliver et al. 2013), sturgeon are known to move between river systems. Therefore, a sturgeon detected in a given river system may not belong to that river's reproductive population. We also do not know what proportion of sturgeon remain in river systems during summer versus move to marine habitats. Generally it is thought that juvenile Atlantic Sturgeon stay in riverine areas year-round, while adults move into rivers during spring for spawning but move out into marine environments during other times (Smith 1985; Bain 1997). Our study and others (Moser and Ross 1995; Collins et al. 2000) have observed adult and subadult Atlantic Sturgeon in riverine areas during summer. In an ongoing study in the Roanoke River–Albemarle Sound system, most sonic-tagged adult Atlantic Sturgeon were present during the summer months (two of three in 2011, five of five in 2012, and three of six in 2013), not in offshore marine waters (authors' unpublished data). This summering behavior may be similar to observations for

TABLE 3. Comparison of distance models as applied to Atlantic Sturgeon in the Carolina and South Atlantic distinct population segments with and without river as a covariate for abundance and using alternative distributions for abundance and detection. Model fit is compared using Akaike's information criterion (AIC); a lower value indicates a better performing model. For all but the top performing model, AIC weights were <0.001. The number of parameters in each model is denoted by nPars, ΔAIC is the difference in a model's AIC value and the best model, and AICwt represents the relative likelihood of each model.

Model	nPars	AIC	ΔAIC	AICwt
Negative binomial, half-normal, with river	9	443.54	0.00	1.00
Negative binomial, exponential, with river	9	462.09	18.55	<0.001
Negative binomial, uniform, with river	8	480.34	36.81	<0.001
Negative binomial, half-normal	4	528.58	85.04	<0.001
Negative binomial, exponential	4	547.13	103.59	<0.001
Negative binomial, uniform	3	565.38	121.84	<0.001
Negative binomial, hazard, with river	10	980.19	536.65	<0.001
Negative binomial, hazard	5	1,061.96	618.43	<0.001
Poisson, hazard, with river	9	1,864.23	1,420.69	<0.001
Poisson, half-normal, with river	8	1,980.57	1,537.03	<0.001
Poisson, exponential, with river	8	1,999.12	1,555.58	<0.001
Poisson, uniform, with river	7	2,017.37	1,573.83	<0.001
Poisson, hazard	4	3,031.24	2,587.70	<0.001
Poisson, half-normal	3	3,147.59	2,704.05	<0.001
Poisson, exponential	3	3,166.14	2,722.60	<0.001
Poisson, uniform	2	3,184.39	2,740.85	<0.001

Gulf Sturgeon *A. oxyrinchus desotoi* (Wooley and Crateau 1985; Clugston et al. 1995). Additionally, presence of adult sturgeon in rivers during summer may be related to fall spawning (Smith et al. 1984; Balazik et al. 2012; Smith et al. 2015).

The best individual model for each method used a negative binomial distribution because the data appeared to be overdispersed, based on estimated values of the overdisperion parameter α. Precision was similar for both models, although the distance model had slightly narrower confidence intervals. The distance model used replicate counts to provide information about the probability of being available for detection (Chandler et al. 2011). This could represent temporary emigration or in this case fish that are present but oriented in a way that does not allow for detection. The replicate counts allow the model to relax the assumption that detection probability is 1 on the transect.

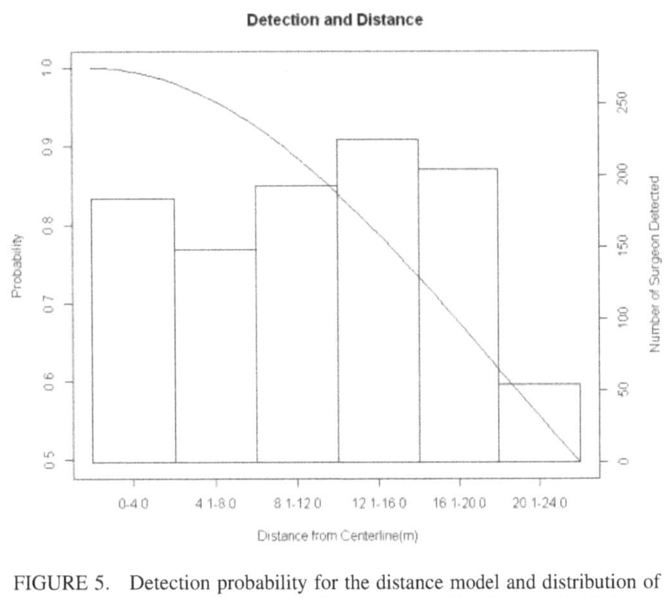

FIGURE 5. Detection probability for the distance model and distribution of side-scan sonar targets (i.e., Atlantic Sturgeon in the Carolina and South Atlantic distinct population segments) across the scan swath for all surveyed rivers (see Figure1). Target distance is measured from centerline of side-scan image to body of sturgeon target. The detection probability function shown is the half-normal for the top AIC selected distance model, σ = 20.3.

For both models, estimated detection probability was a product of true detection probability and availability of an animal within the site to be detected by the survey. These two components can sometimes be estimated separately, such as in occupancy modeling with multiple gears (Hines et al. 2010; Flowers and Hightower 2013). However, there is insufficient information available in this study to parse the effects of detection probability and availability using only one survey gear. While our assumption of site closure was probably not violated for the 2-km sites, it was likely that all sturgeon did not remain within our side-scan swaths during our surveys, based on the variability of sturgeon counted in individual sites over the course of our surveys. If sturgeon were moving randomly in and out of our survey swaths, our abundance estimates should reflect the numbers of sturgeon in our surveyed rivers. If sturgeon movement was more limited, abundance estimates may need to be scaled up to account for areas not surveyed by the sonar.

The distance model incorporates information about the spatial distribution of animals within the survey swath to make inference about changes in detection probability with distance, while the *N*-mixture model assumes an average detection probability throughout the swath. Even though the half-normal function was selected as the best to model detection across the sampling swath, it did not appear to fit the distance observations well (Figure 5). We evaluated the results of other poorer fitting distributions, but abundance estimates were comparable to the half-normal. Another option was to truncate our data observations to a maximum distance of 20 m. In this case the uniform distribution fit best and estimates were more similar to those of the *N*-mixture model. Distance-sampling methods may not be well suited for use with side-scan sonar because the survey width is arbitrarily truncated by the width of the sonar swath, which explains why estimates were so similar for each model. Apparent changes in detection probability across the swath may be more of an artifact of interpreting side-scan images rather than of detection itself. Distance sampling may be more effective over wider swath widths, where there is greater contrast between image quality of near and far targets.

CONCLUSIONS

Side-scan sonar can be used to survey sturgeons and potentially other large fishes and various analytical approaches, such as *N*-mixture and distance models, are well suited for side-scan data. It is interesting that both abundance estimation models provided such similar estimates despite using different data sets and assumptions about detection. The *N*-mixture model requires less data (counts only, no distances off transect) than the distance-sampling model and only had slightly more uncertainty around abundance estimates. Further studies are needed to determine whether the pattern in our distance data are a result of the distribution of sturgeon within our study, characteristics of side-scan data, or both. If patterns are related to side-scan characteristics, then new detection functions and analysis techniques may be needed to properly analyze data.

Our abundance modeling provided useful information about the status of sturgeon within our sampled rivers systems. We were able to produce estimates of sturgeon populations in river systems using a fraction of the effort of traditional netting programs and without having to handle our target species. Covariates that could influence abundance can also be incorporated into all models (Kéry and Schaub 2012). Abundance estimates from side-scan surveys can be used in conjunction with other data sources, such as genetic abundance estimates or traditional mark–recapture tagging studies to improve population abundance estimates. Side-scan sonar can provide absolute abundance estimates in discrete areas while tagging (especially with sonic tags) could provide the information needed for expanding to systemwide estimates. Netting would also allow for information to be collected on smaller individuals and species composition. Side-scan sonar can also provide habitat information about sturgeon locations and identify potential areas where netting operations could be performed safely.

ACKNOWLEDGMENTS

Funding for this project was provided by National Oceanographic and Atmospheric Administration. We thank T. Jackson for his work in the field and in the laboratory reviewing side-scan and video data. We appreciate C. Collier and M. Loeffler (North Carolina Division of Marine Fisheries), and W. Post (South Carolina Department of Natural Resources) allowing us the use of their sonic-tagged sturgeon during the course of the study. We also thank N. Cole, J. Hughes, and A. Prince for assistance during preliminary phases of our side-scan sonar work. B. Gardner and K. Pollock provided comments on an earlier version of the manuscript. The Cooperative Fish and Wildlife Research Unit is jointly supported by North Carolina State University, North Carolina Wildlife Resources Commission, U.S. Geological Survey, U.S. Fish and Wildlife Service, and Wildlife Management Institute. Any use of trade, product, or firm names is for descriptive purposes only and does not imply endorsement by the U.S. Government.

REFERENCES

Armstrong, J. L., and J. E. Hightower. 2002 Potential for restoration of the Roanoke River population of Atlantic Sturgeon. Journal of Applied Ichthyology 18:475–480.

ASSRT (Atlantic Sturgeon Status Review Team). 2007. Status review of Atlantic Sturgeon (*Acipenser oxyrinchus oxyrinchus*). Report to National Marine Fisheries Service, Northeast Regional Office, Gloucester, Massachusetts.

Bain, M. B. 1997. Atlantic and Shortnose sturgeons of the Hudson River: common and divergent life history attributes. Environmental Biology of Fishes 48:347–358.

Balazik, M. T., G. C. Garman, J. P. Van Eenennaam, J. Mohler, and L. C. Woods III. 2012. Empirical evidence of fall spawning by Atlantic Sturgeon in the James River, Virginia. Transactions of the American Fisheries Society 141:1465–1471.

Bergman, P. 2011. Videography monitoring of adult sturgeon in the Feather River basin, CA. Report to Anadromous Fish Restoration Program, Cramer Fish Sciences, Gresham, Oregon.

Buckland, S. T., D. R. Anderson, K. P. Burnham, J. L. Laake, D. L. Borchers, and L. Thomas. 2001. Introduction to distance sampling. Oxford University Press, Oxford, UK.

Chandler, R. B., J. A. Royle, and D. I. King. 2011. Inference about density and temporary emigration in unmarked populations. Ecology 92:1429–1435.

Clugston, J. P., A. M. Foster, and S. H. Carr. 1995. Gulf Sturgeon, *Acipenser oxyrinchus desotoi*, in the Suwannee River, Florida, USA. *In* A. D. Gershanovich and T. I. J. Smith, editors. Proceedings international symposium on sturgeons. Russian Federal Research Institute of Fisheries and Oceanography Publications, Moscow.

Cobra, D. T., A. V. Oppenheim, and J. S. Jaffe. 1992. Geometric distortions in side-scan sonar images: a procedure for their estimation and correction. IEEE (Institute of Electrical and Electronics Engineers) Journal of Oceanic Engineering 17:252–268.

Collins, M. R., and T. I. J. Smith. 1997. Distributions of Shortnose and Atlantic sturgeons in South Carolina. North American Journal of Fisheries Management 17:995–1000.

Collins, M. R., T. I. J. Smith, W. C. Post, and O. Pashuk. 2000. Habitat utilization and biological characteristics of adult Atlantic Sturgeon in two South Carolina rivers. Transactions of the American Fisheries Society 129:982–988.

Damon-Randall, K., R. Bohl, S. Bolden, D. Fox, C. Hager, B. Hickson, E. Hilton, J. Mohler, E. Robbins, T. Savoy, and A. Spells. 2010. Atlantic Sturgeon research techniques. NOAA Technical Memorandum NMFS-NE-215.

Erickson, D. L., A. Kahnle, M. J. Millard, E. A. Mora, M. Bryja, A. Higgs, J. Mohler, M. DuFour, G. Kenney, J. Sweka, and E. Pikitch. 2011. Use of pop-up satellite archival tags to identify oceanic-migratory patterns for adult Atlantic Sturgeon, *Acipenser oxyrinchus oxyrinchus* Mitchell, 1815. Journal of Applied Ichthyology 27:356–365.

Fiske, I., and R. B. Chandler. 2011. unmarked: an R package for fitting hierarchical models of wildlife occurrence and abundance. Journal of Statistical Software 43:1–23.

Flowers, H. J., and J. E. Hightower. 2013. A novel approach to surveying sturgeon using side-scan sonar and occupancy modeling. Marine and Coastal Fisheries: Dynamics, Management, and Ecosystem Science [online serial] 5:211–223.

Foote, K. G. 2009. Acoustic methods: brief review and prospects for advancing fisheries research. Fish and Fisheries Series 31:313–343.

Gross, M. R., J. Repka, C. T. Robertson, D. H. Secor, and W. Van Winkle. 2002. Sturgeon conservation: insights from elasticity analysis. Pages 13–30 *in* W. Van Winkle, P. Anders, D. Dixon, and D. Secor, editors. Biology, management and protection of North American sturgeons. American Fisheries Society, Symposium 28, Bethesda, Maryland.

Hines, J. E., J. D. Nichols, J. A. Royle, D. I. MacKenzie, A. M. Gopalaswamy, N. Samba Kumar, and K. U. Karanth. 2010. Tigers on trails: occupancy modeling for cluster sampling. Ecological Applications 20:1456–1466.

Kaeser, A. J., and T. L. Litts. 2008. An assessment of deadhead logs and large woody debris using side scan sonar and field surveys in streams of southwest Georgia. Fisheries 33:589–597.

Kahn, J., C. Hager, J. C. Watterson, J. Russo, K. Moore, and K. Hartman. 2014. Atlantic Sturgeon annual spawning run estimate in the Pamunkey River, Virginia. Transactions of the American Fisheries Society 143:1508–1514.

Kahn, J., and M. Mohead. 2010. A protocol for use of Shortnose, Atlantic, Gulf, and Green sturgeons. NOAA Technical Memorandum NMFS-OPR-45.

Kéry, M, and M. Schaub 2012. Bayesian population analysis using WinBUGS: a hierarchical perspective. Academic Press, Waltham, Massachusetts.

Kocik, J., C. Lipsky, T. Miller, P. Rago, and G. Shepherd. 2013. An Atlantic Sturgeon population index for ESA management analysis. Northeast Fisheries Science Center, Reference Document 13–06, Woods Hole, Massachusetts.

Moser, M. L., and S. W. Ross. 1995. Habitat use and movements of Shortnose and Atlantic sturgeons in the lower Cape Fear River, North Carolina. Transactions of the American Fisheries Society 124:225–234.

Nealson, P. A., and H. M. Brundage III. 2007. Feasibility assessment of split-beam hydroacoustic techniques for monitoring adult Shortnose Sturgeon in the Delaware River. Pages 405–415 *in* J. Munro, D. Hatin, J. E. Hightower, K. McKown, K. J. Sulak, A. W. Kahnle, and F. Caron, editors. Anadromous sturgeons: habitats, threats, and management. American Fisheries Society, Symposium 56, Bethesda, Maryland.

Nelson, T. C., P. Doukakis, S. T. Lindley, A. D. Schreier, J. E. Hightower, L. R. Hildebrand, R. E. Whitlock, and M. A. H. Webb. 2013. Research tools to investigate movements, migrations, and life history of sturgeons (Acipenseridae), with an emphasis on marine-oriented populations. PLoS ONE [online serial] 8:e71552.

Oakley, N. C., and J. E. Hightower. 2007. Status of Shortnose Sturgeon in the Neuse River, North Carolina. Pages 273–284 *in* J. Munro, D. Hatin, J. E. Hightower, K. McKown, K. J. Sulak, A. W. Kahnle, and F. Caron, editors. Anadromous sturgeons: habitats, threats, and management. American Fisheries Society, Symposium 56, Bethesda, Maryland.

Oliver, M. J., M. W. Breece, D. A Fox, D. E. Haulsee, J. T. Kohut, J. Manderson, and T. Savoy. 2013. Shrinking the haystack: using an AUV in an integrated ocean observatory to map Atlantic Sturgeon in the coastal ocean. Fisheries 38:210–216.

Parauka, F. M., M. S. Duncan, and P. A. Lang. 2011. Winter coastal movement of Gulf of Mexico sturgeon throughout northwest Florida and southeast Alabama. Journal of Applied Ichthyology 27:343–350.

Peterson, D. L., P. Schueller, R. DeVries, J. Fleming, C. Grunwald, and I. Wirgin. 2008. Annual run size and genetic characteristics of Atlantic Sturgeon in the Altamaha River, Georgia. Transactions of the American Fisheries Society 137:393–401.

Pitkitch, E. K., P. Doukakis, L. Lauck, P. Chakrabarty, and D. L. Erickson. 2005. Status, trends, and management of sturgeon and Paddlefish fisheries. Fish and Fisheries 6:233–265.

R Development Core Team. 2009. R: A language and environment for statistical computing. R Foundation for Statistical Computing, Vienna. Available: www.R-project.org. (April 2014).

Royle, J., D. K. Dawson, and S. Bates. 2004. Modeling abundance effects in distance sampling. Ecology 85:1591–1597.

Royle, J. A. 2004. N-mixture models for estimating population size from spatially replicated counts. Biometrics 60:108–115.

Secor, D. H. 2002. Atlantic Sturgeon fisheries and abundances during the late nineteenth century. Pages 89–101 *in* W. Van Winkle, P. Anders, D. Dixon, and D. Secor, editors. Biology, management and protection of North American sturgeons. American Fisheries Society, Symposium 28, Bethesda, Maryland.

Smith, J. A., H. J. Flowers, and J. E. Hightower. 2015. Fall spawning of Atlantic Sturgeon in the Roanoke River, North Carolina. Transactions of the American Fisheries Society 144:48–54.

Smith, T. I. J. 1985. The fishery, biology, and management of Atlantic Sturgeon, *Acipenser oxyrinchus*, in North America. Environmental Biology of Fishes 14:61–72.

Smith, T. I. J., D. E. Marchette, and G. F. Ulrich. 1984. The Atlantic Sturgeon fishery in South Carolina. North American Journal of Fisheries Management 4:164–176.

Thomas, L., S. T. Buckland, E. A. Rexstad, J. L. Laake, S. Strindberg, S. L. Hedley, J. R. B. Bishop, T. A. Marques, and K. P. Burnham. 2010. Distance software: design and analysis of distance sampling surveys for estimating population size. Journal of Applied Ecology 47:5–14.

Wooley, C. M., and E. J. Crateau. 1985. Movement, microhabitat, exploitation, and management of Gulf of Mexico sturgeon, Apalachicola River, Florida. North American Journal of Fisheries Management 5:590–605.

12

Marine Distribution and Habitat Use of Atlantic Sturgeon in New York Lead to Fisheries Interactions and Bycatch

Keith J. Dunton*
School of Marine and Atmospheric Sciences, Stony Brook University, Stony Brook, New York 11794-5000, USA

Adrian Jordaan
Department of Environmental Conservation, University of Massachusetts Amherst, Holdsworth Hall, 160 Holdsworth Way, Amherst, Massachusetts 01003, USA

David O. Conover
School of Marine and Atmospheric Sciences, Stony Brook University, Stony Brook, New York 11794-5000, USA

Kim A. McKown and Lisa A. Bonacci
New York State Department of Environmental Conservation, Division of Fish, Wildlife and Marine Resources, Bureau of Marine Resources, 205 North Belle Mead Road, Suite 1, East Setauket, New York 11733, USA

Michael G. Frisk
School of Marine and Atmospheric Sciences, Stony Brook University, Stony Brook, New York 11794-5000, USA

Abstract

Population declines of Atlantic Sturgeon *Acipenser oxyrinchus oxyrinchus* prompted initial fisheries closures and an eventual endangered or threatened species listing across the U.S. portion of their range in 2012. Atlantic Sturgeon aggregations and migration routes along the coast of Long Island overlap with commercial fishing activities that may lead to incidental take in nondirected fisheries. Thus, understanding the distribution and movement of Atlantic Sturgeon in relation to commercial fisheries can help management agencies determine impacts and develop bycatch mitigation measures. Stratified random sampling and targeted bottom trawl surveys were used to identify the temporal and spatial use of marine habitat in New York waters. The majority of survey captures were restricted to depths of less than 15 m and known aggregation areas. During the aggregation periods (May, June, September, and October) in known aggregation areas, catches were an order of magnitude higher than in other areas and months of the year. Northeast Fisheries Observer Program bycatch data from 1989 to 2013 was analyzed for the New York region and suggested that bycatch occurs within two main gear types: otter bottom trawls and sink gill nets. Trawling bycatch contained primarily subadult Atlantic Sturgeon and is highest during the Summer Flounder *Paralichthys dentatus* fishery in New York State waters. Trawling overlaps spatially and temporally with identified Atlantic Sturgeon aggregation areas, while bycatch in gill nets targeted adult fish farther offshore in federal waters. Bycatch in these fisheries may be a regional threat to recovery, and spatial and temporal closures, gear modifications, or other bycatch reduction techniques are suggested to protect aggregating and migrating fish.

Subject editor: Carl Walters, University of British Columbia, Canada

*Corresponding author: keith.dunton@stonybrook.edu

For Atlantic Sturgeon *Acipenser oxyrinchus oxyrinchus*, a long history of commercial exploitation led to a century of population declines and a recent listing of the New York Bight (NYB) population as endangered (USOFR 2012). In the late 19th century, New York coastal gill-net fisheries targeted adults for caviar and flesh and juveniles for flesh (Murawski and Pacheco 1977; Smith and Clugston 1997). The New York fishery experienced peak landings of 231,000 kg in 1898, but quantities had declined to 30,000 kg by 1901 (Murawski and Pacheco 1977). This "boom and bust" pattern in fisheries landings was also experienced by other Atlantic Sturgeon populations as local fisheries collapsed throughout the eastern United States (Smith and Clugston 1997). In the mid-1970s a resurgent Atlantic Sturgeon fishery developed with southeastern stocks initially representing 80% of the landings, a proportion that persisted through the early 1980s (Smith and Clugston 1997). During the late 1980s until the closure of the fishery in 1996, the fishery overwhelmingly focused in the NYB (Smith and Clugston 1997; Kahnle et al. 2007). Increased fishing mortality from 1977 to 1995 was coincident with an 80% decline in the Hudson River population (Peterson et al. 2000). In order to protect the remaining Atlantic Sturgeon, a 40-year moratorium was enacted in 1998 (ASMFC 1998), followed by a National Oceanic and Atmospheric Administration (NOAA) listing as an endangered species in 2012. The Hudson River stock remains the largest Atlantic Sturgeon population, although still considered at risk of extinction (ASSRT 2007), with an estimated 870 spawning individuals, of which 270 are female (Kahnle et al. 2007).

Assessing the current health of Hudson River Atlantic Sturgeon has been problematic. Surveys that target different areas and life history stages indicate opposing abundance trends. In the Hudson River, surveys that target premigrant juveniles have shown an increasing population trend since the 1990s, while marine-based surveys targeting migrants have continued to show declining numbers (Kahnle et al. 2007). Marine-based surveys continue to show low abundance estimates of mature adults, raising the concern that an increase in premigrant juveniles is not resulting in an increased abundance of late juvenile and adult Atlantic Sturgeon (ASMFC 2007). At the current estimated population abundance, levels of bycatch mortality greater than 6% in the marine environment could hinder recovery (Boreman 1997; ASMFC 2007; Kahnle et al. 2007). Atlantic Sturgeon are particularly vulnerable to incidental bycatch in coastal fisheries. Mortality during marine migrations may explain why surveys show no increase in the abundance of older individuals (Stein et al. 2004a).

Understanding Atlantic Sturgeon use of marine habitat, where the majority of their life cycle is spent, is required for the development of effective habitat-related Endangered Species Act rules or area-based management solutions (Nemeth 2005). A relatively large amount of research has been conducted on the freshwater habitat use of Atlantic Sturgeon; however, there is a paucity of information characterizing marine habitat use. Available marine research suggests that Atlantic Sturgeon occur in shallow inshore areas of the continental shelf and form marine aggregations near the mouths of river systems (Stein et al. 2004a, 2004b; Laney et al. 2007; Dunton et al. 2010; Erickson et al. 2011). Along the coast of Long Island, New York, fisheries and other anthropogenic stressors commonly co-occur in locations identified as Atlantic Sturgeon aggregation sites and migration pathways, potentially causing an increased risk of bycatch mortality or other negative interactions (Dunton et al. 2010). Indeed, bycatch documented along the eastern seaboard (Stein et al. 2004a) and recent anecdotal information reported to the New York Department of Environmental Conservation suggest that Atlantic Sturgeon are caught in commercial trawl and gill-net fisheries along the coast of Long Island (K. A. McKown, unpublished data).

The objectives of this study were to describe Atlantic Sturgeon habitat use along the coast of Long Island and assess the potential for fisheries interactions and bycatch mortality. Specifically, we provide spatial distribution and abundance estimates from a 3-year trawl survey conducted in the coastal waters of New York and commercial bycatch and fishery interactions are examined using data collected by the Northeast Fisheries Observer Program (Karp et al. 2011). Additionally, targeted trawling was employed to measure the potential impacts of fisheries directed on aggregation areas.

METHODS

Sampling was conducted utilizing a depth-stratified random design based on 10-m depth intervals using the inshore strata and depth zones designed for the Northeast Fisheries Science Center's inshore trawl survey (Sosebee and Cadrin 2006). Strata 1, 2, 12, 13, and 14 were truncated to exclusively New York waters (Figure 1). Cruises consisted of two different surveys, the New York young-of-the-year Bluefish *Pomatomus saltatrix* survey and the New York ocean trawl survey for Atlantic Sturgeon that occurs throughout the year. A full description of the surveys can be found in Dunton et al. (2010). To briefly summarize, surveys used a three-to-one two-seam trawl (headrope 25 m, footrope 30.5 m) with forward netting beginning at 12-cm mesh and tapering down to 8-cm stretched mesh lined with a 6.0-mm-mesh cod end. Surveys were conducted with the RV *Seawolf* and included the waters inshore of 30 m from the easternmost point of Long Island to the entrance of New York Harbor. Tows were standardized to 20 min duration and 3–3.5 knots. Not all depth strata were sampled each survey.

In addition to stratified random sampling, directed research tows specifically targeting Atlantic Sturgeon aggregations occurred opportunistically in April, May, and June, and in October, November, and December of 2007–2013, largely in strata 12 and 13. Tow duration was reduced to 5–15 min during targeted sampling of aggregations in order to be in

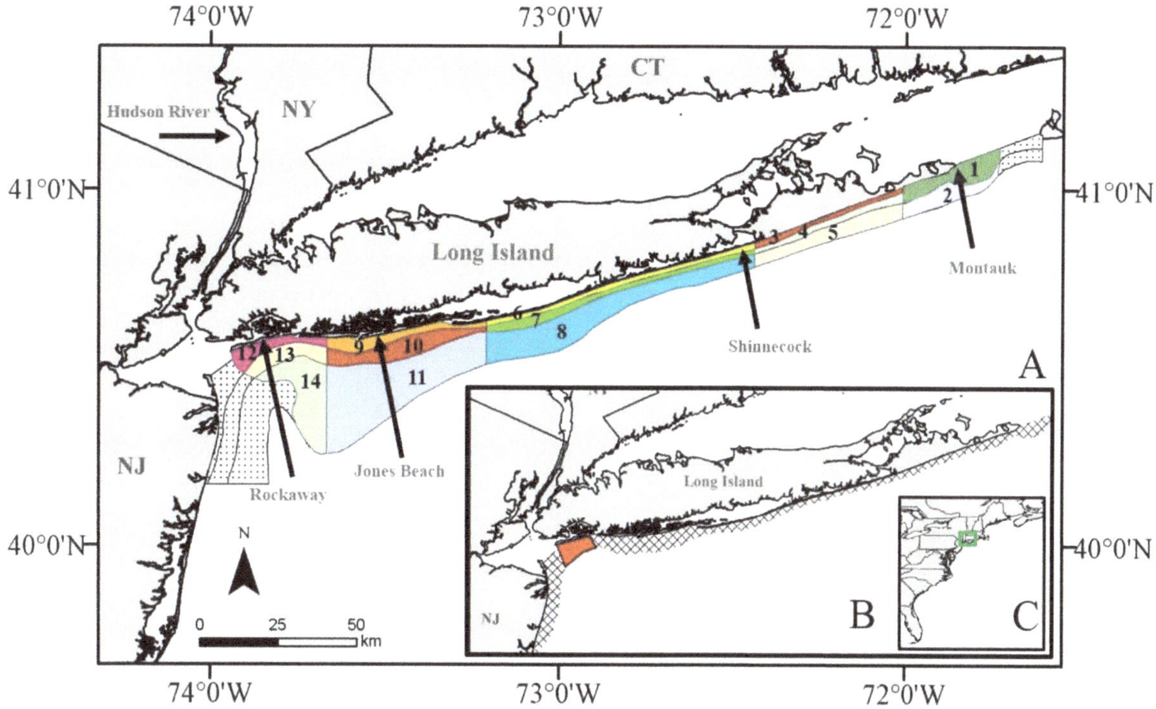

FIGURE 1. Map of the study site **(A)** showing strata used in the stratified random trawl survey along the south shore of Long Island. Strata 1, 2, 12, 13, and 14 were truncated from the Northeast Fisheries Science Center's inshore trawl survey to include New York waters only (dotted areas represent full strata). Targeted sampling was restricted to strata 12 and 13. Strata 12, 9, 6, and 3 have a depth of 0–10 m, strata 13, 10, 7, 4, and 1 have a depth 10–20 m, and strata 14, 11, 8, 5, and 2 have a depth of 20–30 m. Locations important to this study are noted. The inset maps show **(B)** the Atlantic Sturgeon aggregation area (red area) and their migration corridors (hatched) and **(C)** the regional location of the study site. The state abbreviations are as follows: NY = New York, CT = Connecticut, and NJ = New Jersey.

compliance with the Endangered Species Permit 16422 issued to Stony Brook University.

All Atlantic Sturgeon captured were measured to the nearest centimeter (fork and total length), weighed (kg) using a platform scale, and examined for prior tags. If no tags were detected, the Atlantic Sturgeon were double-tagged with an external United States Fish and Wildlife Service (USFWS) carlin or dart tag and an internally implanted 125- or 134.2-kHz passive integrated transponder (PIT). External tags, supplied by the USFWS, had reporting information printed directly on the tag to enable fisherman and the general public to report information on encountered Atlantic Sturgeon. Internally implanted PIT tags are long-term tags utilized by the scientific community to identify individuals. All Atlantic Sturgeon captured after April 2012 were collected under Endangered Species Permit 16422 issued to Stony Brook University.

Spatial distribution of catch.—Atlantic Sturgeon captures were mapped using ArcGIS 9.2 (ESRI; Redlands, California). Map base layers were obtained from the United States Geological Survey Coastal and Marine Geology Program GIS catalog. Atlantic Sturgeon captures were plotted using graduated symbols in the following categories: 1, 2, 3, 4, 6, and 10 Atlantic Sturgeon per tow.

Catch per unit effort.—Catch per unit effort (CPUE) was calculated as fish per minute to allow comparison of the varying tow times during targeted trawling. The CPUEs for the random trawl survey were weighted, taking into account the stratified design using the modified equation from Perry and Smith (1994) and Dunton et al. (2010):

$$\text{Weighted CPUE} = \frac{W_h}{n_h}\frac{y_{hi}}{y_{st}},$$

where W_h = the proportion of the survey area in stratum h; n_h = the number of tows in stratum h; y_{hi} = the CPUE (number of fish per minute) in tow i and stratum h; and y_{st} = the stratified mean abundance (CPUE × W_h).

Targeted sampling CPUE was not weighted due to its nonstratified design. The CPUE estimates were compared using 95% confidence intervals for random versus targeted sampling, western Long Island versus central-east Long Island, and peak versus nonpeak periods, using bootstrap resampling (n = 10,000). Confidence intervals were bias-corrected using the methods described by Efron (1987).

Fishery interactions.—Patterns in fishery interactions and regional bycatch of Atlantic Sturgeon off Long Island were examined using two methods: (1) collecting data on Atlantic Sturgeon tagged in this study and reported as bycatch from commercial fisheries using the external USFWS dart or carlin tags and (2) examining data collected by the Northeast Fisheries Observer Program (NEFOP) from 1989 to 2013. The NEFOP collects catch and species data, including geographic location and gear information, for each tow as well as biological data for bycatch species of interest, including number caught, length, weight, and condition (dead, alive, injured). There are some caveats to the observer data, including annual variability and relatively low coverage (Warden and Orphanides 2008). Additionally, only vessels with state and federal permits have observers through NEFOP, and no observer data were recorded for vessels that held only state permits. The NEFOP data for catch and total length was examined for all gear types for vessels originating from New York and New Jersey ports for the Northwest Atlantic Fisheries Organization's statistical areas 611, 612, and 613. Statistical area 611 included trips confined to marine waters south of Long Island in the Atlantic Ocean and excluded trips in Long Island Sound, Block Island Sound, and Gardiners Bay. Trips and bycatch were summarized by state, gear type, and season. Analysis of variance (ANOVA) was used to compare the size distributions of Atlantic Sturgeon captured in gill-net and trawl fisheries.

RESULTS

Stratified Random Survey

A total of 149 Atlantic Sturgeon were captured in 10,380 min (n = 519 tows) of stratified random trawling (Table 1). Atlantic Sturgeon distributions varied by strata throughout the season (Figure 2; Table 1). The weighted survey average CPUE was 0.023 sturgeon/min (SD, 0.159), with the highest average weighted CPUE in May, followed by October, November, September, and June (Table 1). The lowest weighted CPUEs were observed in January, followed by March and August (Table 1), with no fish captured in April. The CPUE was highest along western Long Island in strata 12, 9, and 10 (Table 1; Figure 2). Few Atlantic Sturgeon were captured off eastern Long Island (<8%). No Atlantic Sturgeon were captured at depths of 20–30 m (strata 2, 5, 8, 11, and 14) (Table 1; Figure 2).

Targeted Survey

Targeted trawling along western Long Island during spring and fall captured an additional 825 Atlantic Sturgeon in 4,144 min (n = 312 bottom tows) for a mean CPUE of 0.226 sturgeon/min (SD, 0.470) (Table 2). The highest CPUEs were

TABLE 1. Weighted CPUE ± SD (sturgeon/min) of Atlantic Sturgeon by month and stratum for random trawl surveys in 2005–2007.

Month, stratum, and average	2005	2006	2007	Average	Total sturgeon	Total minutes
January		0.005 ± 0.025	0.011 ± 0.046	0.008 ± 0.037	4	1,180
March		0.009 ± 0.031		0.009 ± 0.031	2	540
April		0.000 ± 0.000	0.000 ± 0.000	0.000 ± 0.000	0	600
May	0.029 ± 0.058	0.003 ± 0.010	0.117 ± 0.594	0.058 ± 0.407	34	1,280
June	0.021 ± 0.062	0.010 ± 0.030	0.017 ± 0.024	0.016 ± 0.042	37	1,980
August	0.005 ± 0.024	0.027 ± 0.122		0.015 ± 0.083	5	900
September	0.000 ± 0.000	0.056 ± 0.165	0.019 ± 0.049	0.020 ± 0.090	21	1,620
October		0.034 ± 0.097	0.034 ± 0.070	0.034 ± 0.085	36	1,080
November	0.016 ± 0.086	0.044 ± 0.157		0.030 ± 0.127	10	1,200
1	0.000 ± 0.000	0.000 ± 0.000	0.000 ± 0.000	0.000 ± 0.000	0	620
2	0.000 ± 0.000	0.000 ± 0.000	0.000 ± 0.000	0.000 ± 0.000	0	400
3	0.036 ± 0.096	0.009 ± 0.025	0.002 ± 0.006	0.012 ± 0.048	11	1,000
4	0.000 ± 0.000	0.000 ± 0.000	0.000 ± 0.000	0.000 ± 0.000	0	640
5	0.000 ± 0.000	0.000 ± 0.000	0.000 ± 0.000	0.000 ± 0.000	0	400
6	0.010 ± 0.029	0.003 ± 0.008	0.003 ± 0.007	0.004 ± 0.015	9	1,440
7	0.007 ± 0.029	0.006 ± 0.026	0.000 ± 0.000	0.005 ± 0.025	2	880
8	0.000 ± 0.000	0.000 ± 0.000	0.000 ± 0.000	0.000 ± 0.000	0	580
9	0.011 ± 0.039	0.118 ± 0.210	0.044 ± 0.070	0.059 ± 0.130	41	1,060
10	0.000 ± 0.000	0.083 ± 0.211	0.000 ± 0.000	0.040 ± 0.151	6	700
11	0.000 ± 0.000	0.000 ± 0.000	0.000 ± 0.000	0.000 ± 0.000	0	540
12	0.036 ± 0.118	0.048 ± 0.115	0.187 ± 0.657	0.101 ± 0.426	78	1,200
13	0.000 ± 0.000	0.023 ± 0.053	0.000 ± 0.000	0.011 ± 0.037	2	520
14	0.000 ± 0.000	0.000 ± 0.000	0.000 ± 0.000	0.000 ± 0.000	0	400
Yearly average	0.011 ± 0.054	0.022 ± 0.095	0.034 ± 0.252	0.023 ± 0.159	149	10,380

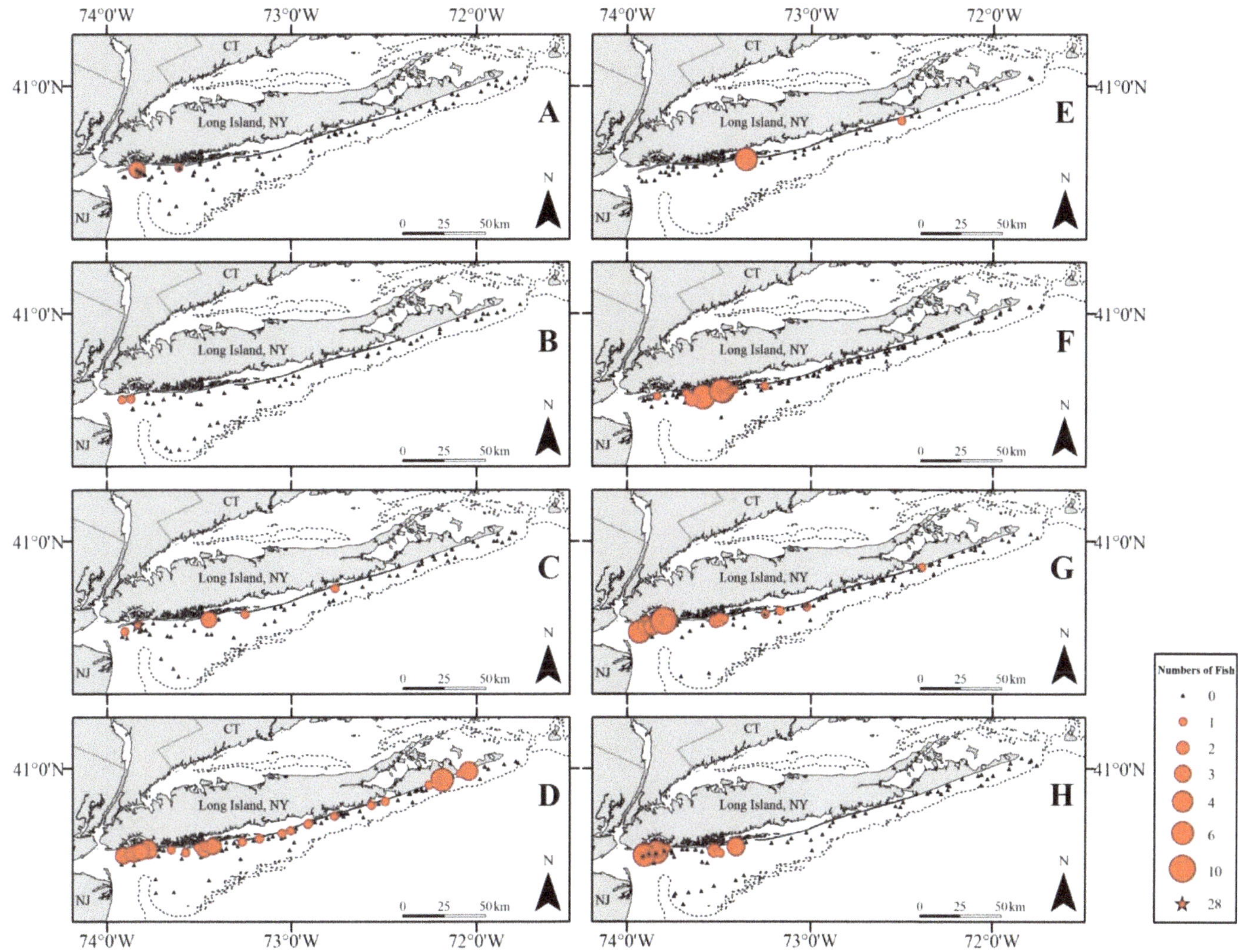

FIGURE 2. Abundance and distribution of Atlantic Sturgeon captured in random trawl surveys for **(A)** January, **(B)** April, **(C)** May, **(D)** June, **(E)** August, **(F)** September, **(G)** October, and **(H)** November. The dashed black line represents the 30-m depth contour (the farthest extent of the survey), and the black triangles represent the locations of tows in which no Atlantic Sturgeon were captured.

observed during May 2011, when 82 fish were captured in 61 min (n = 8 tows) of trawling for an average CPUE of 1.533 sturgeon/min (SD, 1.301), and May 2007, when 141 Atlantic Sturgeon were captured in 164 min (n = 8 trawls) for a trip CPUE of 0.848 sturgeon/min (SD, 0.493) (Table 2). The highest observed individual tow CPUEs were 3.8 sturgeon/min (May 2011) and 3.25 sturgeon/min (May 2012). Two targeted trawling trips yielded no Atlantic Sturgeon captured (December 2010, October 2012).

Bootstrap Values

Bootstrap resampling was conducted to compare random versus targeted sampling, western Long Island versus central-east Long Island locations, and specific time periods (May, June, September, and October versus November, January, March, and April) (Table 3). Targeted trawling had an order of magnitude higher weighted CPUE than did random trawling, with bootstrap-generated means of 0.226 sturgeon/min (SD, 0.027) and 0.023 sturgeon/min (SD, 0.007), respectively. Significantly higher catches occurred in western Long Island at 0.048 sturgeon/min (SD, 0.016) during the peak months (Table 3).

Fishery Interactions

Tagging data.—A combined total of 974 Atlantic Sturgeon were tagged through random and targeted efforts. Atlantic Sturgeon ranged in size from 64 to 195 cm fork length, with a mean of 101 cm fork length (SD, 19.8). A total of 13 tagged fish were captured in commercial and recreational fisheries: 10 were reported by commercial trawl fisheries, 2 in commercial gill-net fisheries, and 1 in recreational nontarget hook-and-line fisheries. All commercial recaptures were reported to the

TABLE 2. Catch per unit effort ± SD (sturgeon/min) for Atlantic Sturgeon in targeted trawls for 2006–2013. Values were not weighted for targeted trawling.

Year and average	April	May	June	October	November	December	Average CPUE	Total sturgeon	Total minutes
2006					0.221 ± 0.170		0.221 ± 0.170	31	140
2007	0.017 ± 0.026	0.848 ± 0.493	0.020 ± 0.045		0.145 ± 0.441		0.256 ± 0.490	200	764
2008					0.041 ± 0.057		0.041 ± 0.057	14	340
2010		0.026 ± 0.032	0.043 ± 0.068			0.000 ± 0.000	0.026 ± 0.046	28	1,070
2011		1.533 ± 1.301		0.247 ± 0.487	0.428 ± 0.319		0.504 ± 0.721	272	642
2012		0.369 ± 0.560		0.000 ± 0.000			0.287 ± 0.516	196	645
2013		0.088 ± 0.106		0.418 ± 0.533			0.149 ± 0.271	84	543
Average	0.017 ± 0.026	0.291 ± 0.580	0.039 ± 0.064	0.215 ± 0.443	0.237 ± 0.344	0.000 ± 0.000	0.226 ± 0.470	825	4,144

TABLE 3. Bootstrap (n = 10,000) results for determining differences among CPUE (sturgeon/min) of Atlantic Sturgeon for random versus targeted trawling, for region, and for periods of higher versus lower abundance for the stratified random survey.

Comparisons	Factors	Mean ± SD	Max	Bias corrected 95% CI
Random and targeted trawling	Random	0.023 ± 0.007	0.060	0.013–0.041
	Target	0.226 ± 0.027	0.339	0.178–0.281
Region (random trawling)	West	0.048 ± 0.016	0.133	0.027–0.092
	Central	0.004 ± 0.001	0.010	0.001–0.007
	East	0.004 ± 0.001	0.010	0.001–0.007
Time (random trawling)	May-Jun-Sep-Oct	0.024 ± 0.006	0.053	0.011–0.028
	Nov-Jan-Mar-Apr	0.005 ± 0.002	0.013	0.002–0.009

USFWS directly by commercial fishermen with no observers on board. Trawl fisheries targeting Winter Flounder *Pseudopleuronectes americanus* or Summer Flounder *Paralichthys dentatus* accounted for 50% (n = 6) of the commercial recaptures, with other recaptures reported in trawl fisheries for Atlantic horseshoe crab *Limulus polyphemus* (n = 1), Red Hake *Urophycis chuss* or longfin inshore squid *Loligo pealeii* (n = 1), and skate *Leucoraja* spp. (n = 2) (Figure 3). Recaptures also occurred in gill-net fisheries targeting Goosefish *Lophius americanus* (n = 1) and dogfish (genus *Squalus* or *Mustelus*) (n = 1) (Figure 3). Commercial recaptures of Atlantic Sturgeon tagged in New York were largely concentrated in two regions: off of Highlands, New Jersey, and Jones Beach, New York (Figure 3).

Northeast Fisheries Observer Program data.—Captures in observed gill-net fishing trips occurred more frequently off the coast of New Jersey, while captures during bottom trawling trips were more frequent off the coast of New York (Figure 4). In the area studied, there were 24,674 observed trips for all gear types, and a total of 413 Atlantic Sturgeon were observed as bycatch off the south shore of Long Island from 1989 to 2013 (Dunton 2014; Figure 4). Bycatch of Atlantic Sturgeon occurred in the fixed or anchor sink gill net (n = 200 sturgeon; CPUE = 0.04 sturgeon/observed trip), otter bottom trawl fish (n = 182 sturgeon; CPUE = 0.03 sturgeon/observed trip), drift-sink gill net (n = 19 sturgeon; CPUE = 0.02 sturgeon/observed trip), drift-floating gill net (n = 6 sturgeon; CPUE < 0.01 sturgeon/observed trip), and twin otter bottom trawl (n = 6 sturgeon; CPUE = 0.50 sturgeon/observed trip) (Dunton 2014; Figure 4). Atlantic Sturgeon were not observed in anchored floating gill nets, hydraulic clam dredges, sea scallop dredges, bottom longlines, fish traps or pots, lobster pots or

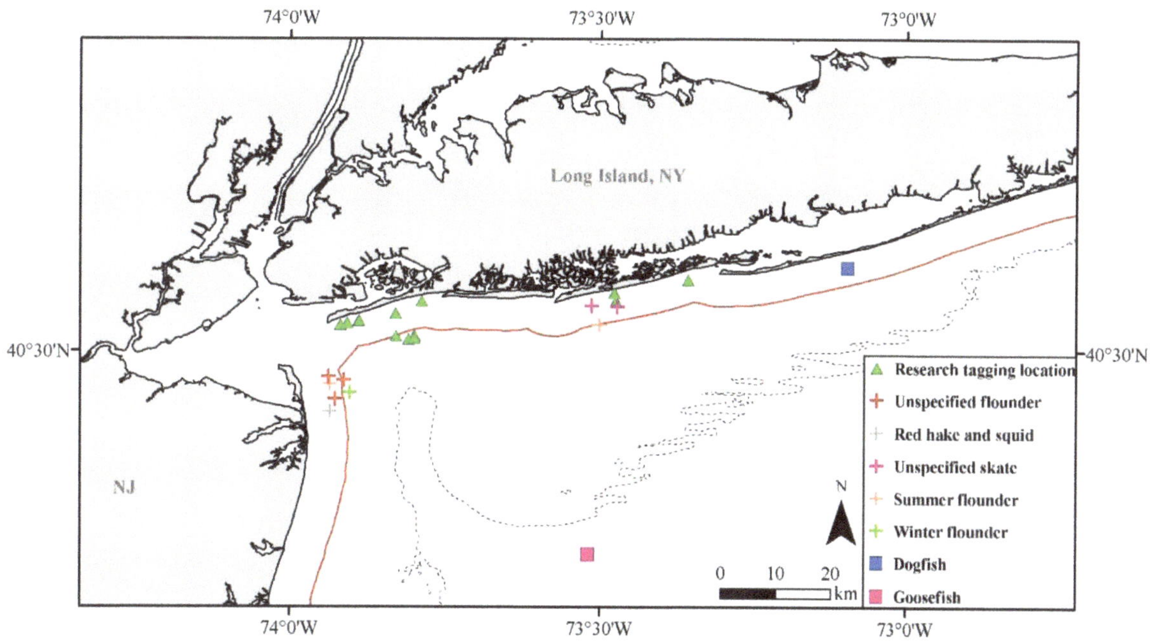

FIGURE 3. Atlantic Sturgeon captured and tagged in New York research trawls (green triangles) and recaptured in commercial fisheries for target species that used bottom trawls (crosses) and gill nets (squares). One commercial recapture off Maryland is not shown. The dashed black line represents the 30-m depth contour, and the red line indicates the Federal–State boundary (4.83 km).

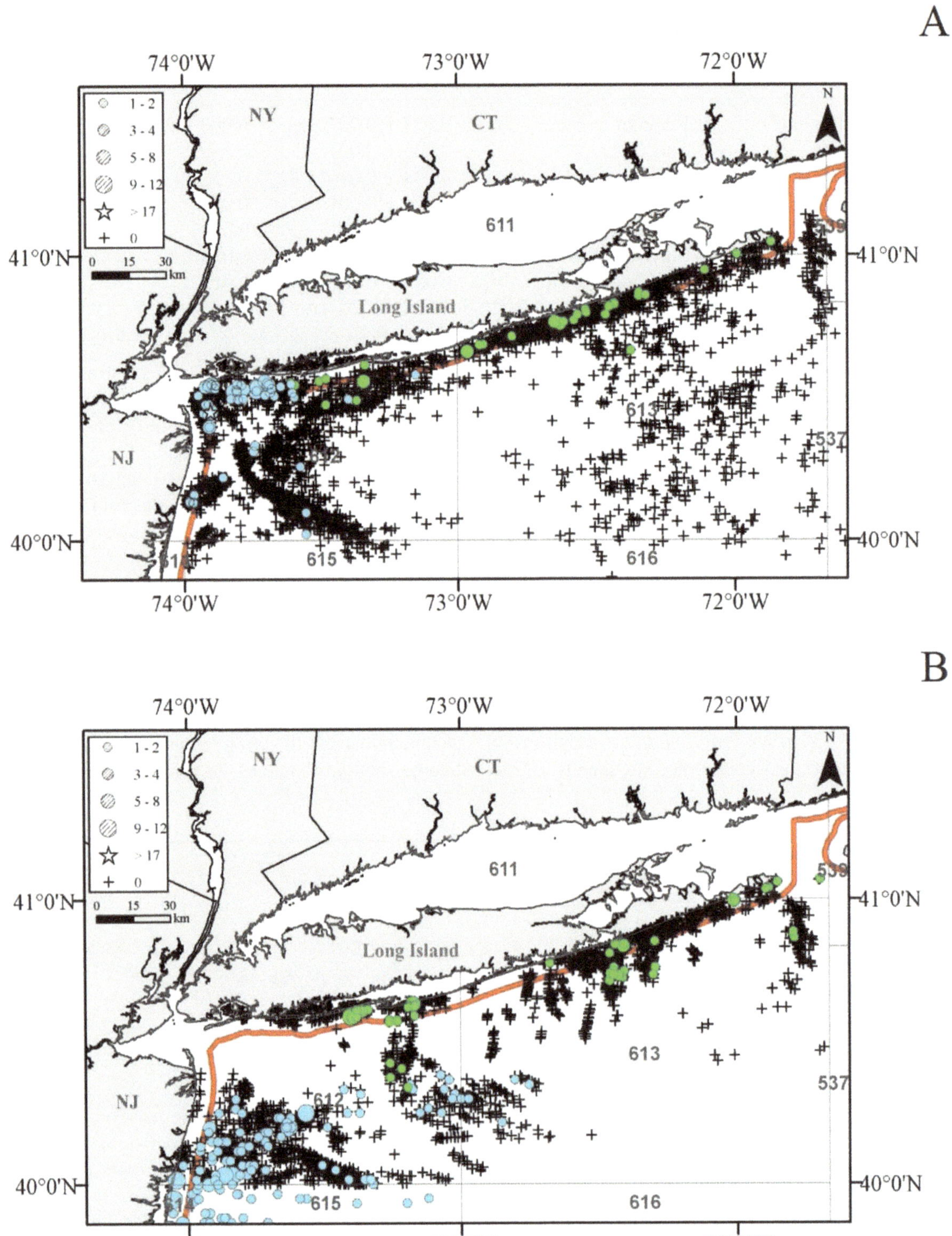

FIGURE 4. Atlantic Sturgeon bycatch observed by the Northeast Fisheries Observer Program from **(A)** bottom otter trawls (fish and twin) and **(B)** gill nets (floating sink-drift and sink fixed-anchored combined). Green circles represent bycatch from vessels that landed in New York ports and blue circles are from vessels that landed in New Jersey ports within statistical areas 611 (modified, only tows south of Long Island and Block Island are included), 612, and 613 (gray numbers and gray lines). The Federal–State boundary (4.83 km) is indicated by the red line.

traps, scallop bottom trawls, midwater otter trawls, midwater paired otter trawls, and troll lines. For the two main gear types, gill nets and trawls, bycatch and observer coverage varied spatially and seasonally (Figure 5).

Atlantic Sturgeon captured in gill nets (all types combined) were significantly (ANOVA: $F_{1,\,371} = 6.39$, $P = 0.012$) larger (mean length ± SD, 128 ± 29 cm FL) than those captured in otter trawls (120 ± 36 cm FL; Figure 6). Of the 413 observed Atlantic Sturgeon bycatch events in the NYB, direct mortality was highest in gill nets, with 58.22% of captured Atlantic Sturgeon released alive, 38.22% dead, and 3.56% with unknown status. Trawls released 94.15% alive, 3.72% dead, and 2.13% with unknown status.

While otter bottom trawls were distributed coastwide, 64% of the observed trawl bycatch occurred less than 4.83 km from shore within state waters (Figure 4; Figure 5). Observed fisheries targeted 30 different species, but Atlantic Sturgeon were captured in only 13 of these fisheries (fisheries for Summer Flounder, longfin inshore squid, Winter Flounder, Striped Bass *Morone saxatilis*, Winter Skate *Leucoraja ocellata*, Silver Hake *Merluccius bilinearis*, Weakfish *Cynoscion regalis*, Bluefish, Red Hake *Urophycis chuss*, Scup *Stenotomus chrysops*, Little Skate *Leucoraja erinacea*, Tautog *Tautoga onitis*, and one unidentified species). The Summer Flounder fishery generated 73% of Atlantic Sturgeon trawl bycatch (Figure 7A). Much of the observed bycatch was from vessels engaged in the Summer Flounder fishery landing catch in New Jersey (Figure 4). Atlantic Sturgeon captures and observer coverage varied seasonally (Figure 5), with peak catches in April (0.07 sturgeon/observed tow), May (0.04 sturgeon/observed tow), June (0.04 sturgeon/observed tow), and September (0.04 sturgeon/observed tow). Low bycatch was observed in February (0.01 sturgeon/observed tow), July (0.02 sturgeon/observed tow), October (0.01 sturgeon/observed tow), and November (0.01 sturgeon/observed tow). No bycatch was observed in January, March, August, and December. Bycatch in trawls was also depth dependent, with 90% of all Atlantic Sturgeon captured in water less than 20 m deep (Figure 8).

Gill nets were concentrated off the coast of New Jersey, with 30% of total observed bycatch occurring within state limits (Figure 4; Figure 5). Gill-net fisheries off Long Island were focused in three regions: Jones Beach, Shinnecock, and Montauk, New York (Figure 4; Figure 5). Atlantic Sturgeon captures and observer coverage varied seasonally (Figure 5). The highest observed Atlantic Sturgeon captures occurred in November, April, December, and May (0.11, 0.09, 0.08 and 0.07 sturgeon/observed set, respectively), with intermediate catches observed in June, August, and October (0.04, 0.04, and 0.03 sturgeon/observed set, respectively). The lowest observed bycatch occurred in January, February, March, July (0.02, 0.01, 0.02, and 0.01 sturgeon/observed set, respectively). Bycatch of Atlantic Sturgeon was observed to occur in all nine target-species categories (Goosefish, Summer Flounder, Striped Bass, Winter Skate, Weakfish, Bluefish, Spiny Dogfish *Squalus acanthias*, and an unknown), with 62% in the Goosefish gill-net fishery concentrated within area 612 in federal waters (Figures 4, 7B). No trends were seen in Atlantic Sturgeon captures by depth (Figure 8).

DISCUSSION

Shallow marine distributions during migration and the formation of aggregations are placing Atlantic Sturgeon at risk for bycatch in coastal trawl and gill-net fisheries based in New York and New Jersey. Research survey and commercial fishery observer data suggest that interactions between Atlantic Sturgeon and fisheries are most likely to occur during seasonal migrations along coastal New York during April to June and October to November. The highest bycatch of Atlantic Sturgeon was observed in the Summer Flounder bottom trawl fishery and the Goosefish gill-net fishery. Gill nets primarily appear to intercept migrating adult Atlantic Sturgeon, while trawls encounter juveniles during periods of aggregation around the mouth of the Hudson River. The seasonally predictable timing of migration and aggregation formation provides an opportunity to develop limited spatial and temporal management to reduce bycatch (Armstrong et al. 2013; Dunton 2014).

Atlantic Sturgeon occur in large numbers in the federal and state waters outside of the Hudson River, one of the previously identified aggregation areas along the East Coast for both subadult (Dunton et al. 2010) and adult Atlantic Sturgeon (Erickson et al. 2011). Atlantic Sturgeon aggregations (Dunton et al. 2010; Erickson et al. 2011) were generally restricted to shallow depths (<20 m) in New York waters, following a seasonal pattern with peak abundance during the spring and fall. Dunton et al. (2010) also found that subadult Atlantic Sturgeon had a significant habitat preference for shallow depths (<20 m) in the region. The aggregation area located in the Rockaway region accounted for the majority of Atlantic Sturgeon captured. When this area was specifically targeted, CPUE rose to a maximum of 61 times the average of random surveys, reaching around 3.8 sturgeon/min.

The Rockaway aggregation area experienced the highest bycatch from otter trawling during months of peak Atlantic Sturgeon abundance. Bycatch was primarily the result of the Summer Flounder fishery (73% of the total trawl bycatch) operating out of New Jersey. Commercial trawlers also reported 83% of tag returns from commercial fisheries in the Rockaway area, with all but one capture occurring in the NYB. Coastwide estimates of all recaptured tagged Atlantic Sturgeon within the USA (including all participating researchers and reports to the USFWS Atlantic Sturgeon database) ranged between 8% and 14% for trawl fisheries. When reported bycatch in the USFWS database for the New Jersey–New York coastal subregion was examined separately, bycatch increased to 30% (S. Eyler, USFWS Maryland Fishery Resources Office, personal communication; Eyler et al. 2009).

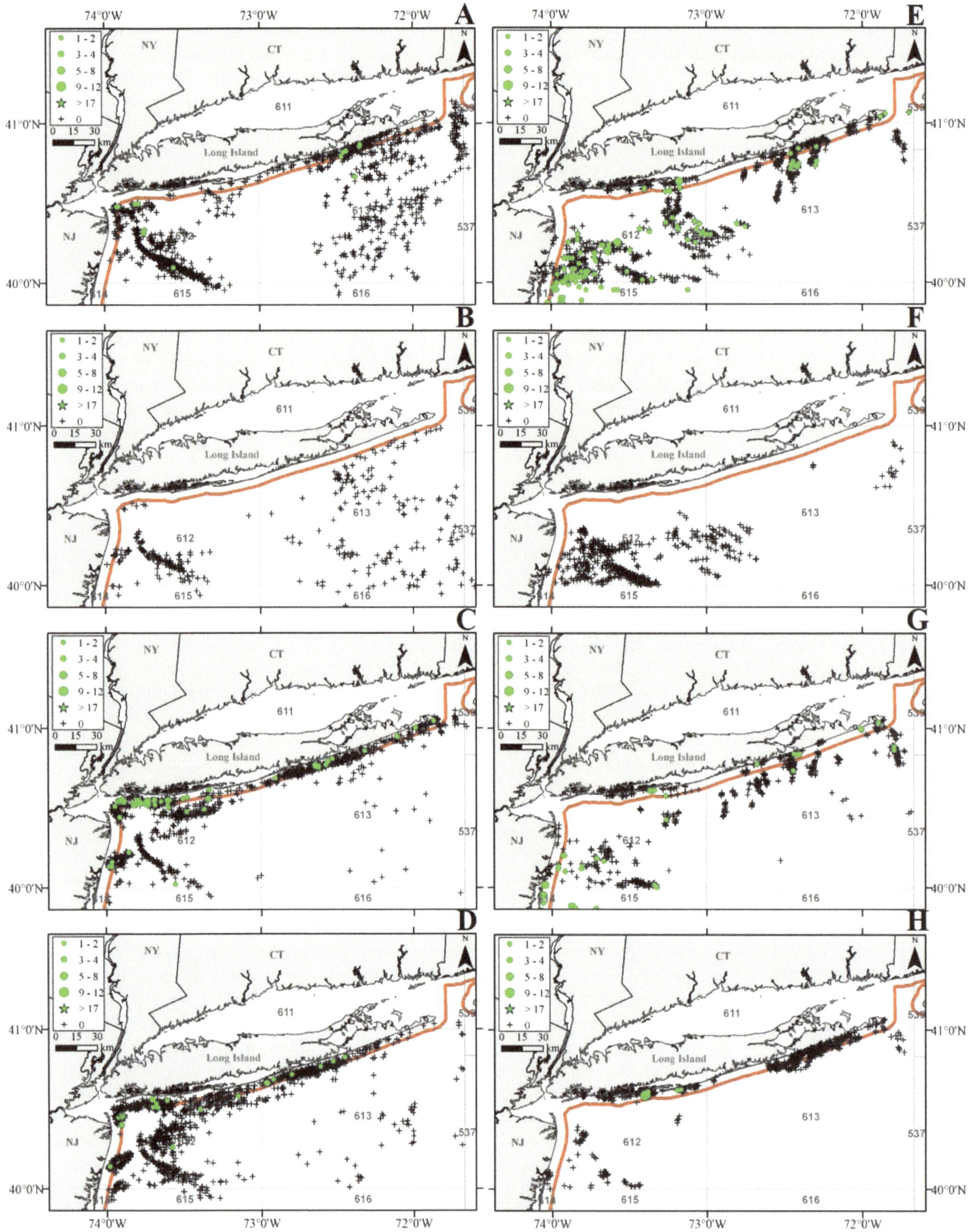

FIGURE 5. Commercial bycatch of Atlantic Sturgeon observed by the Northeast Fisheries Observer Program by season for otter trawls in (A) fall, (B) winter, (C) spring, and (D) summer and for gill nets in (E) fall, (F) winter, (G) spring, and (H) summer. The Federal–State boundary (4.83 km) is indicated by the red line.

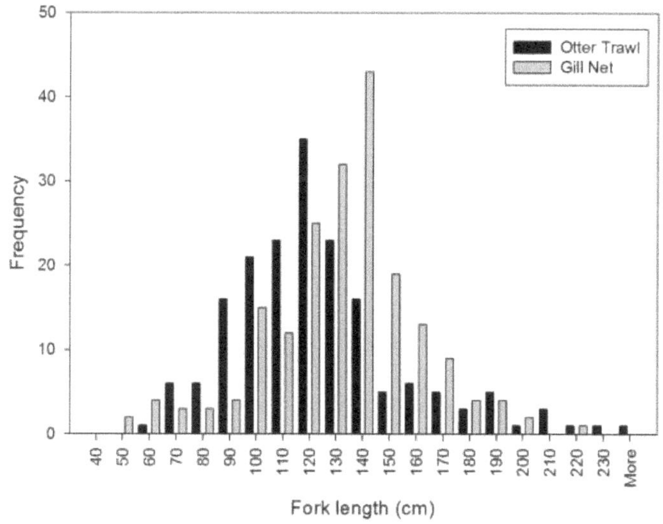

FIGURE 6. Size distribution of Atlantic Sturgeon captured as bycatch in commercial fisheries that use bottom otter trawls (twin and fish) and gill nets (all types). This information was recorded through the Northeast Fisheries Observer Program.

Based on tag reporting and analysis of NEFOP data, trawling within the NYB region appears to represent a bigger threat to Atlantic Sturgeon recovery compared to other locations.

Coastwide bycatch of Atlantic Sturgeon occurs in a wide range of fisheries, with the largest bycatch reported for gill-net (sink and drift) fisheries (Stein et al. 2004a; ASMFC 2007; Eyler et al. 2009). Subadults and adults are presumed to have similar seasonal migration patterns (Bain et al. 2000; Stein et al. 2004a; Erickson et al. 2011; Dunton et al. 2012) that place them in the path of both gill-net and trawl fisheries. Trawls passing through aggregation areas when Atlantic Sturgeon are present could result in large bycatch, as could gill nets set in the narrow migration corridor along the shoreline in depths < 15 m during Atlantic Sturgeon migrations. In response to high Atlantic Sturgeon mortality associated with gill nets in the Goosefish fishery, the fishing industry has been experimenting with gear modifications. Currently, evidence suggests that these experimental nets can decrease Atlantic Sturgeon bycatch (Fox et al. 2013; Pingguo and Jones 2013). However, similar modifications to trawling gear have not been developed or widely tested. The mortality rate for Atlantic Sturgeon captured in commercial trawling is unknown; however, direct observations of deceased juvenile Atlantic Sturgeon are frequently made on beaches close to the Rockaway aggregation region during the spring and fall and adjacent to coastal areas that experience high-frequency bottom trawling (K. J. Dunton, unpublished data; T. Lomschumbo, Gateway National Recreation Area, personal communication). Although these observations suggest that mortality is occurring in local commercial trawl fisheries, the exact cause of mortality for these fish is unknown.

While bycatch-related mortality is believed to be lower in trawls than in gill nets, the potential for large catches in aggregation areas and repetitive habitat use may lead to a significant impact on the recovering population. An accurate estimate of population mortality of Atlantic Sturgeon due to bycatch is not possible at this time because of unknown postrelease survival (Stein et al. 2004a). Relatively little direct mortality is observed for fish captured by otter trawls, with 94% of otter trawl captures observed by NEFOP being released alive. However, the delayed effects of stress and injuries may occur for weeks after the initial capture (Davis 2002; Broadhurst et al. 2006). Beardsall et al. (2013) showed that 60-min experimental trawls are a moderate stressor leading to survival rates of around 94% for acoustically tagged fish tracked for an average of 56 d in the Minas Basin, Canada. However, the small sample size ($n = 34$) and the short duration for some tracked fish (3 d), limit inference to coastwide fisheries, particularly in more southerly (warmer) regions. The estimate of postrelease survival by Beardsall et al. (2013) coupled with NEFOP observations of captured Atlantic Sturgeon released alive, indicates that fisheries mortality is certainly higher when postrelease mortality is considered for trawls. Further, multiple possibilities for capture during the 7 year (or longer) juvenile marine stage and as adults in gill-net fisheries increase the cumulative likelihood of mortality associated with fishery interactions.

Many factors, such as catch size, species composition, water and air temperatures, tow duration, and handling time, are likely to influence survival (Beardsall et al. 2013). Tow durations in commercial fisheries in the New York region last between 60 and 180 min (NEFOP). The total number of Atlantic Sturgeon captured in NEFOP-observed tows was typically 1 to 4 animals, although bycatch events as high as 60 individuals have been reported to the USFWS (Dunton 2014). A large catch of Atlantic Sturgeon may increase mortality due to increased handling time and injuries during capture. Since a 6% mortality rate would put recovering populations at risk (ASMFC 2007), a detailed study of handling practices once Atlantic Sturgeon are on deck is warranted, as well as how handling practices are influenced by the total catch size. Understanding bycatch mortality is essential to developing best-practices during fishing operations, thereby increasing survival rates.

The high potential for Atlantic Sturgeon bycatch in trawl fisheries that overlap aggregation areas in coastal Long Island emphasizes the need for bycatch reduction measures. Most states in the region offer Atlantic Sturgeon de facto protection from bycatch by limiting or excluding trawling in state waters (Dunton et al. 2010); however, New York limits trawling only within a 2.41-km arc of coastline around navigable inlets. The NEFOP data shows that 64% of Atlantic Sturgeon bycatch in otter trawling occurs in state waters. A majority of incidents occur in known aggregation areas along western Long Island. Aggregation areas in the NYB have been recommended as closed areas and essential marine habitat for subadults in order to mitigate fisheries impacts (Dunton et al. 2010; Dunton 2014). Given the nature of Atlantic Sturgeon habitat preference and movements, New York state could adopt restricted

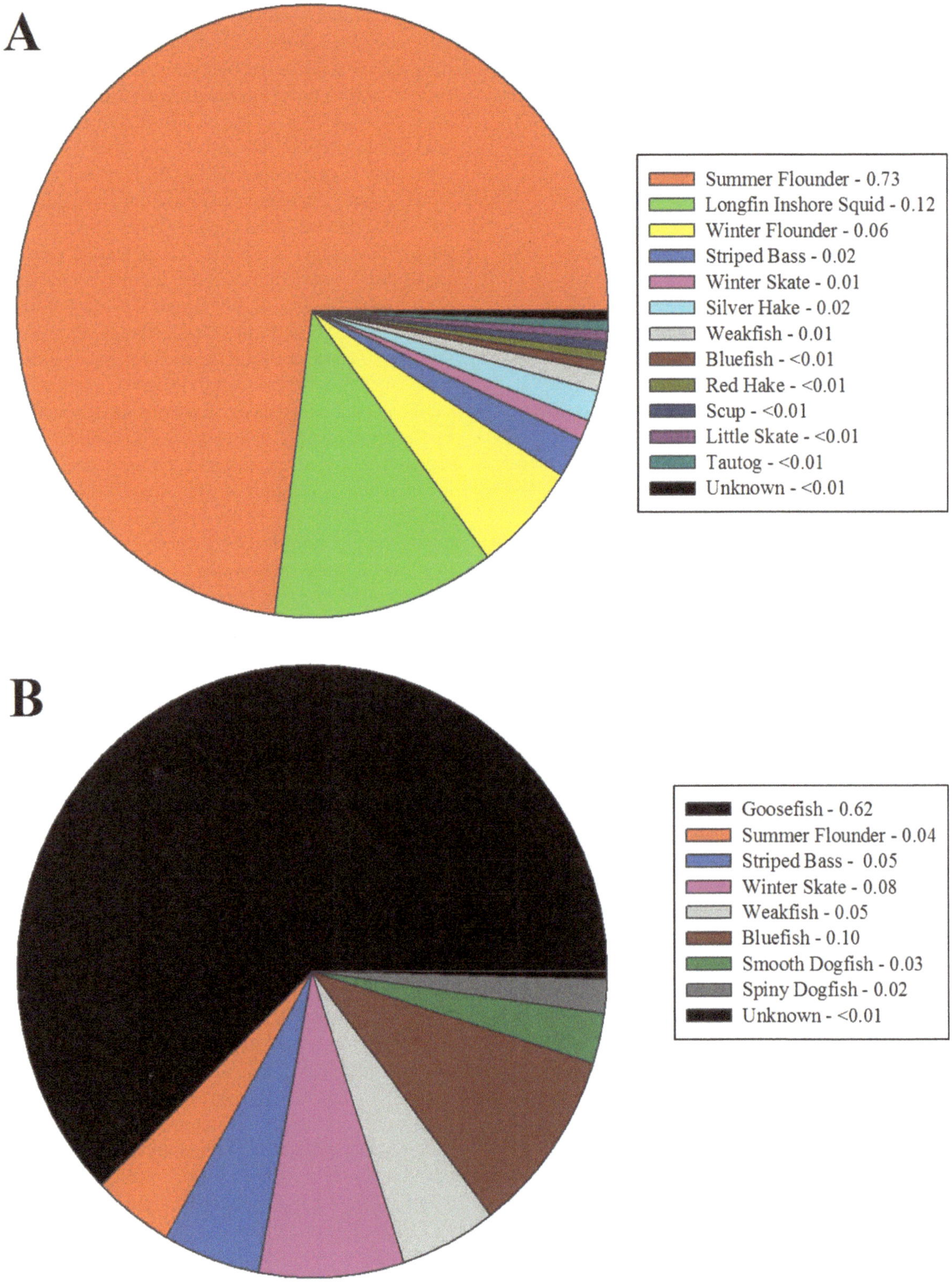

FIGURE 7. Proportion of Atlantic Sturgeon bycatch observed through the Northeast Fisheries Observer Program organized by commercial target fish species for **(A)** bottom and twin otter trawls (fish only) and **(B)** gill nets (all types). The proportion of Atlantic Sturgeon bycatch by target species is shown as the numbers in the legend.

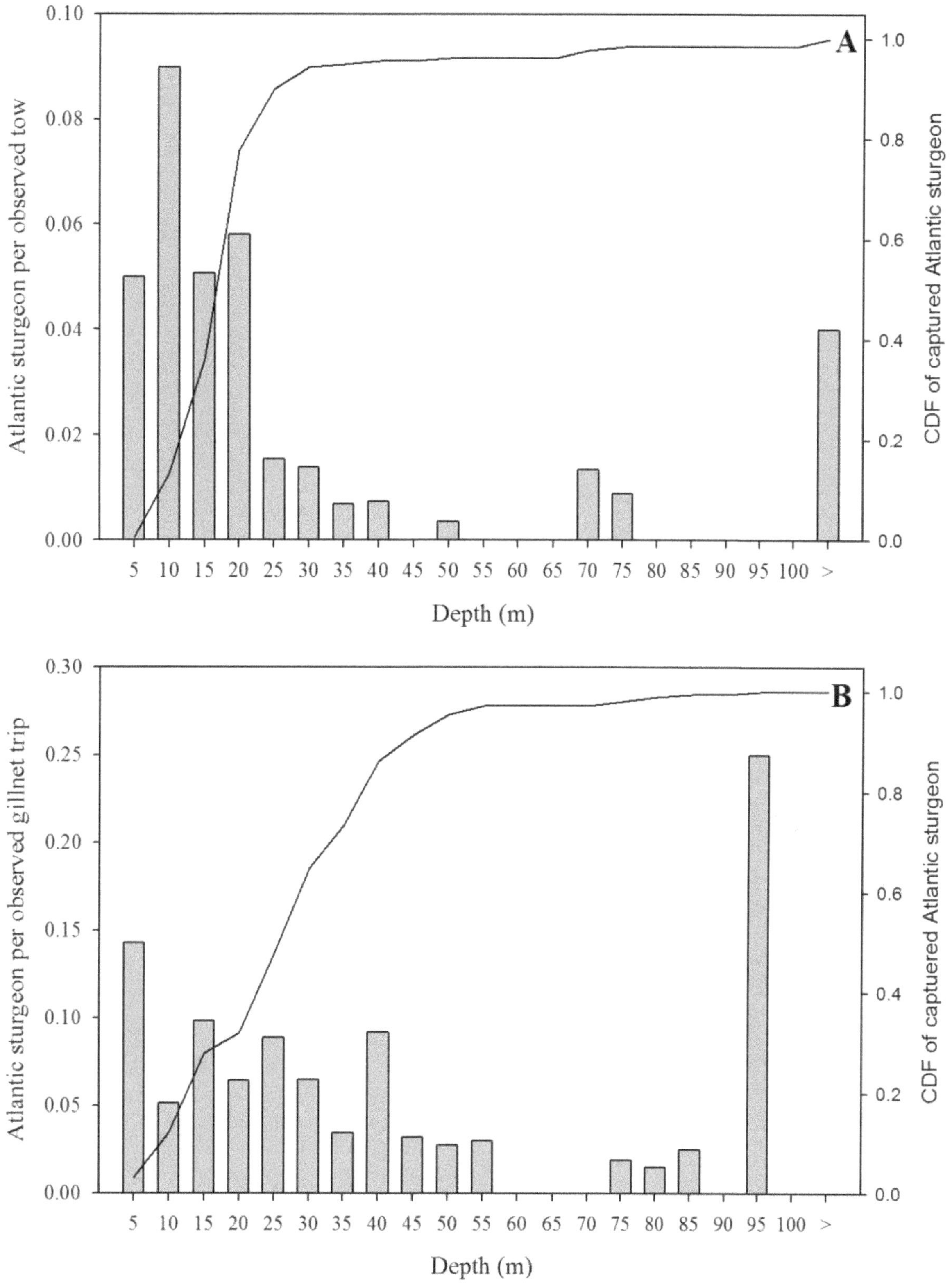

FIGURE 8. Observed Atlantic Sturgeon bycatch (gray bars; primary *y*-axis) and cumulative distribution function (CDF; black line; secondary *y*-axis) of captured Atlantic Sturgeon within commercial fisheries by depth for **(A)** bottom otter trawls (fish and twin) and **(B)** gill nets (all types combined).

TABLE 4. Regional spatial restrictions for bottom trawling.

State	Bottom trawl limits	Regulatory code
Massachusetts	Spatial and temporal trawl closure areas to protect spawning fish	Code of Massachusetts Regulations, Title 322, section 8.09
New Hampshire	Complete ban on trawls in state waters	New Hampshire Revised Statutes Annotated, section 211.49
New Jersey	No trawling within 3.22 km of coast (limited exception for shrimp trawls)	New Jersey Administrative Code, Title 7:25, section 18.14(b)
New York	2.41–3.22-km arc seaward of inlets with Atlantic Ocean; various spatial restrictions within Atlantic Ocean, Long Island Sound, and inland bays	New York Environmental Law, section 13-0341
Delaware	Bottom trawling prohibited in state waters, except for scientific purposes	Delaware Code Annotated, Title 7, section 927
Maryland	Bottom trawling prohibited within 1.61 km of coastal shore or in Chesapeake Bay or bays behind the Atlantic barrier islands	Code of Maryland Regulations, Title 08.02, section 05.03

trawl zones similar to those of neighboring states to protect important Atlantic Sturgeon aggregation areas (Table 4) or employ less restrictive spatial and temporal closures to protect migrating fish (Dunton et al. 2010). Bycatch of Atlantic Sturgeon continues to be a chronic source of mortality, and the fact that impacts may not be evenly distributed across distinct population segments (Stein et al. 2004a; Dunton et al. 2012) supports the need for ongoing research. Regardless, protecting aggregation areas is vital. Since they may contain a large proportion of the population(s) during the spring and fall, sufficient levels of bycatch during those periods may impede recovery (ASSRT 2007; Jaric and Gessner 2013).

The ultimate goal of Atlantic Sturgeon management is to rebuild the population to a level that can sustain harvest (ASMFC 1998). In order to do so, threats to recovery, such as incidental take in fisheries, must be considered and addressed. Our results provide information that can be used for understanding the future impacts of bycatch as well as the potential targeted harvest of Atlantic Sturgeon. If a future fishery targeted Atlantic Sturgeon aggregations, high catch rates could be maintained in the short term, even though declines within the aggregation area would likely go undetected (Frisk et al. 2011). Metrics derived from commercial fisheries would need to be used with great caution out of concerns about hyperstability in stock estimates (Erisman et al. 2011; Jaric and Gessner 2013). Potential impacts can come from dredging, sand mining, pipeline construction, and wind farm development, as well as from commercial fishing. The improved coordination of research and planning among state agencies is required to develop integrated policies for managing all impacts to this endangered fish. Determining marine habitat use throughout the species range will be essential for effective management. In addition to protection in the Hudson River, concentrations of Atlantic Sturgeon, combined with the high incidence of bycatch during the spring and fall off western Long Island, indicate the need for spatial and temporal marine fisheries closures to reduce bycatch and allow population recovery. Because several distinct and endangered population segments are inadvertently caught in the NYB (Dunton et al. 2012), protecting aggregation areas off Long Island will impact the recovery of significant segments of the Atlantic Sturgeon population.

ACKNOWLEDGMENTS

Funding for this research was provided by a Rutgers University Bluefish Recruitment Dynamics Grant, two separate USFWS State Wildlife Grant awards, and a NOAA Species Recovery Grant Program NA07NMF4550320 under Memorandums of Understanding with the New York Department of Environmental Conservation. Keith Dunton was also supported through the Hudson River Foundation Graduate Fellowship program (Award 5539) and a Steven Berkeley Fellowship award from the American Fisheries Society. Thanks to Gina Shield, NOAA Northeast Fisheries Science Center, for providing assistance with the Northeast Fisheries Observer Program, and Shelia Eyler, USFWS, for managing the Atlantic Sturgeon tagging database. We would like to thank M. Wiggins for assistance in survey design and technical support and Captain S. Cluett and the crew of the RV *Seawolf* for help with collecting.

REFERENCES

Armstrong, M. P., M. J. Dean, W. S. Hoffman, D. R. Zemeckis, T. A. Nies, D. E. Pierce, P. J. Diodati, and D. J. McKiernan. 2013. The application of small scale fishery closures to protect Atlantic Cod spawning aggregations in the inshore Gulf of Maine. Fisheries Research 141:62–69.

ASMFC (Atlantic States Marine Fisheries Commission). 1998. Amendment 1 to the interstate fishery management plan for Atlantic Sturgeon. ASMFC, Fishery Management Report 31, Washington D.C.

ASMFC (Atlantic States Marine Fisheries Commission). 2007. Estimation of Atlantic Sturgeon bycatch in coastal Atlantic commercial fisheries of New England and the mid-Atlantic. ASMFC, Washington, D.C.

ASSRT (Atlantic Sturgeon Status Review Team). 2007. Atlantic Sturgeon status review team status review of Atlantic Sturgeon (*Acipenser oxyrinchus oxyrinchus*). Report to National Marine Fisheries Service, Northeast Regional Office, Gloucester, Massachusetts.

Bain, M., N. Haley, D. Peterson, J. Waldman, and K. Arend. 2000. Harvest and habitats of Atlantic Sturgeon *Acipenser oxyrinchus* Mitchill, 1815 in the Hudson River estuary: lessons for sturgeon conservation. Boletin Instituo Espanol de Oceanografia 16:43–54.

Beardsall, J. W., M. F. McLean, S. J. Cooke, B. C. Wilson, M. J. D adswell, A. M. Redden, and M. J. Stokesbury. 2013. Consequences of incidental otter trawl capture on survival and physiological condition of threatened Atlantic Sturgeon. Transactions of the American Fisheries Society 142:1202–1214.

Boreman, J. 1997. Sensitivity of North American sturgeons and Paddlefish to fishing mortality. Environmental Biology of Fishes 48:399–405.

Broadhurst, M. K., P. Suuronen, and A. Hulme. 2006. Estimating collateral mortality from towed fishing gear. Fish and Fisheries 7:180–218.

Davis, M. W. 2002. Key principles for understanding fish bycatch discard mortality. Canadian Journal of Fisheries and Aquatic Sciences 59:1834–1843.

Dunton, K. J. 2014. Population dynamics of juvenile Atlantic Sturgeon, *Acipenser oxyrinchus oxyrinchus*, within the northwest Atlantic Ocean. Doctoral dissertation. Stony Brook University, Stony Brook, New York.

Dunton, K. J., D. D. Chapman, A. Jordaan, K. A. Feldheim, S. J. O'Leary, K. A. McKown, and M. G. Frisk. 2012. Genetic mixed-stock analysis of Atlantic Sturgeon *Acipenser oxyrinchus oxyrinchus* in a heavily exploited marine habitat indicates the need for routine genetic monitoring. Journal of Fish Biology 80:207–217.

Dunton, K. J., A. Jordaan, K. A. McKown, D. O. Conover, and M. G. Frisk. 2010. Abundance and distribution of Atlantic Sturgeon (*Acipenser oxyrinchus*) within the northwest Atlantic Ocean, determined from five fishery-independent surveys. U.S. National Marine Fisheries Service Fishery Bulletin 108:450–464.

Efron, B. 1987. Better bootstrap confidence intervals. Journal of the American Statistical Association 82:171–185.

Erickson, D., A. Kahnle, M. Millard, E. Mora, M. Bryja, A. Higgs, J. Mohler, M. DuFour, G. Kenney, and J. Sweka. 2011. Use of pop-up satellite archival tags to identify oceanic-migratory patterns for adult Atlantic Sturgeon, *Acipenser oxyrinchus oxyrinchus* Mitchell, 1815. Journal of Applied Ichthyology 27:356–365.

Erisman, B. E., L. G. Allen, J. T. Claisse, D. J. Pondella, E. F. Miller, J. H. Murray, and C. Walters. 2011. The illusion of plenty: hyperstability masks collapses in two recreational fisheries that target fish spawning aggregations. Canadian Journal of Fisheries and Aquatic Sciences 68:1705–1716.

Eyler, S., M. Mangold, and S. Minkkien. 2009. Atlantic coast sturgeon tagging database. U.S. Fish and Wildlife Service, Maryland Fishery Resources Office, Summary Report, Annapolis, Maryland.

Fox, D. A., J. L. Armstrong, L. M. Brown, and K. Wark. 2013. Year three, the influence of sink gillnet profile on bycatch of Atlantic Sturgeon in the mid-Atlantic monkfish fishery. Northeast Fisheries Science Center, Sturgeon Gillnet Study EA-133F-12-RQ-0697, Completion Report, Woods Hole, Massachusetts.

Frisk, M. G., D. E. Duplisea, and V. M. Trenkel. 2011. Exploring the abundance-occupancy relationships for the Georges Bank finfish and shellfish community from 1963 to 2006. Ecological Applications 21:227–240.

Jarić, I., and J. Gessner. 2013. A life-stage population model of the European Sturgeon (*Acipenser sturio*) in the Elbe River. Part I: general model outline and potential applications. Journal of Applied Ichthyology 29:483–493.

Kahnle, A. W., K. A. Hattala, and K. McKown. 2007. Status of Atlantic Sturgeon of the Hudson River estuary, New York, USA. Page 347–363 *in* J. Munro, D. Hatin, J. E. Hightower, K. McKown, K. J. Sulak, A. W. Kahnle, and F. Caron, editors. Anadromous sturgeons: habitats, threats, and management. American Fisheries Society, Symposium 56, Bethesda, Maryland.

Karp, W. A., L. L. Desfosse, and S. G. Brooke. 2011. U.S. National bycatch report. NOAA Technical Memorandum NMFS-F/SPO.

Laney, R. W., J. E. Hightower, B. R. Versak, M. F. Mangold, W. Cole, and S. E. Winslow. 2007. Distribution, habitat use, and size of Atlantic Sturgeon captured during cooperative winter tagging cruises, 1988–2006. Pages 167–182 *in* J. Munro, editor. Anadromous sturgeons: habitats, threats, and management. American Fisheries Society, Symposium 56, Bethesda, Maryland.

Murawski, S. A., and A. L. Pacheco. 1977. Biological and fisheries data on Atlantic Sturgeon, *Acipenser oxyrhynchus* (Mitchill). National Marine Fisheries Service, Sandy Hook Laboratoy Technical Series Report 10, Highlands, New Jersey.

Nemeth, R. S. 2005. Population characteristics of a recovering US Virgin Islands Red Hind spawning aggregation following protection. Marine Ecology Progress Series 286:81–97.

Perry, R. I., and S. J. Smith. 1994. Identifying habitat associations of marine fishes using survey data: an application to the northwest Atlantic. Canadian Journal of Fisheries and Aquatic Sciences 51:589–602.

Peterson, D. L., M. B. Bain, and N. Haley. 2000. Evidence of declining recruitment of Atlantic sturgeon in the Hudson River. North American Journal of Fisheries Management 20:231–238.

Pingguo, H., and N. Jones. 2013. Design and test of a low profile gillnet to reduce Atlantic Sturgeon and sea turtle by-catch in mid-Atlantic monkfish fishery. National Oceanic and Atmospheric Administration Fisheries, Completion Report EA133F-12-SE-2094, Silver Spring, Maryland.

Smith, T. I., and J. P. Clugston. 1997. Status and management of Atlantic Sturgeon, *Acipenser oxyrinchus*, in North America. Environmental Biology of Fishes 48:335–346.

Sosebee, K. A., and S. X. Cadrin. 2006. A historical perspective on the abundance and biomass of northeast complex stocks from NMFS and Massachusetts inshore bottom trawl surveys, 1963–2002. Northeast Fisheries Science Center, Reference Document 06-05, Woods Hole, Massachusetts.

Stein, A. B., K. D. Friedland, and M. Sutherland. 2004a. Atlantic Sturgeon marine bycatch and mortality on the continental shelf of the Northeast United States. North American Journal of Fisheries Management 24:171–183.

Stein, A. B., K. D. Friedland, and M. Sutherland. 2004b. Atlantic Sturgeon marine distribution and habitat use along the northeastern coast of the United States. Transactions of the American Fisheries Society 133:527–537.

USOFR (U.S. Office of the Federal Register). 2012. Endangered and threatened wildlife and plants; threatened and endangered status for distinct population segments of Atlantic Sturgeon in the northeast region. Federal Register 77:22(6 February 2012):5880–5912.

Warden, M. L., and C. D. Orphanides. 2008. Preparation of the Northeast Fisheries Observer Program gillnet data for use in bycatch analyses of protected species. Northeast Fish Science Center, Reference Document 08-17, Woods Hole, Massachusetts.

13

Harvesting the Beach Clam *Tivela mactroides*: Short- and Long-Term Dynamics

Márcia Regina Denadai
Centro Universitário Módulo, Rua Frei Pacífico Wagner 653, 11660-903 Caraguatatuba, São Paulo, Brazil; and Departamento de Oceanografia Biológica, Instituto Oceanográfico da Universidade de São Paulo, Praça do Oceanográfico 191, 05508-120 São Paulo, São Paulo, Brazil

Maíra Pombo
Departamento de Oceanografia Biológica, Instituto Oceanográfico da Universidade de São Paulo, Praça do Oceanográfico 191, 05508-120 São Paulo, São Paulo, Brazil

Ligia Coletti Bernadochi
Centro Universitário Módulo, Rua Frei Pacífico Wagner 653, 11660-903 Caraguatatuba, São Paulo, Brazil

Alexander Turra*
Departamento de Oceanografia Biológica, Instituto Oceanográfico da Universidade de São Paulo, Praça do Oceanográfico 191, 05508-120 São Paulo, São Paulo, Brazil

Abstract

Small-scale fisheries are frequently overlooked for research and management, and their social and environmental impacts are often overlooked as well, preventing the implementation of appropriate actions for their sustainability. Additionally, the dynamics of beach clam fisheries and their importance for local communities are not well understood. A study on the population biology of the clam *Tivela mactroides* in Caraguatatuba Bay, southeastern Brazil, revealed intense harvesting of this resource by both residents and tourists. To assess the extent and dynamics of clamming, the number of harvesters was recorded during the course of the day in vacation and nonvacation periods throughout 2003–2005 and 2007–2008; the number of other beach users, weather conditions, and tide height were also recorded. The overall amount of clams harvested was estimated based on censuses of clammers and interviews to calculate the amount of clams harvested per collecting event. The intensity and dynamics of the harvesting activity varied on all the temporal scales evaluated. The estimated amount of clams harvested per year decreased from the first (24.6 kg/year) to the second (8.8 kg/year) group of sampling years, presumably due to clam mass-mortality events and smaller shell sizes in 2007–2008, although clam abundance increased enormously. Vacation periods (presence of tourists) influenced the number of harvesters and the daily dynamics of clamming activity, although this relationship was only evident during 2003–2004. The number of harvesters increased with the number of tourists, except in periods of very high tourist activity, when harvesting decreased. Clamming was more widespread during the day under high tourist activity but during nonvacation times was concentrated in morning low-tide periods. Weather had a partial influence on clamming, with harvesters absent only during intense rain occasions. The understanding of the dynamics of this *Tivela mactroides* fishery highlights key points for planning and implementing management measures, which will involve continuous monitoring of stocks, harvesting, and food safety.

Subject editor: Silvia Salas, Cinvestav del IPN, Mérida Yucatán, Mexico

*Corresponding author: turra@usp.br

Beach clams are among the most important components of the intertidal macrofauna and are harvested throughout the world. This kind of harvesting may be conducted on a large commercial scale, especially in temperate regions, where clams tend to be larger and more commercially valuable. But most commonly, clamming occurs as a part of recreational activities and artisanal fisheries (MacLachlan et al. 1996; Dadon 2005; Defeo and MacLachlan 2005).

Consequently, most data on clam fisheries are derived from the temperate zones, mainly because landing statistics are more commonly reported and can be easily tracked for species of high commercial interest, whose yields are systematically computed in markets, fishing warehouses, or co-operatives (Caddy 1989). On the other hand, in contrast to the significant number of studies on most temperate sandy-beach bivalves, there is a lack of information on clam-harvesting issues, including the social aspects of clammers, for most clam species worldwide and this is even more pronounced for small-sized tropical species (Narchi 1976; Schaeffer-Novelli 1980; Viegas 1981; Paes 1989; Lima et al. 2000; Laudien et al. 2003; Hartill et al. 2005; Herrmann et al. 2009; Abrahão et al. 2010). This situation is often an important factor preventing the establishment of appropriate management strategies to allow sustainable use of these resources. The lack of data for some locations, however, should not be understood as reflecting lower abundance or importance of this resource, especially for local communities. Similarly, the need for appropriate management of clamming should not be underestimated, either to assure the maintenance of the clam populations or to assure food security.

A prominent example is *Tivela mactroides*, a common bivalve in moderately exposed sandy beaches, especially near river mouths, from São Paulo, Brazil, on the east coast of South America to the Caribbean Sea (McLachlan et al. 1996; Denadai et al. 2005). Little information on this species has been published, especially about harvesting. The only available estimates for *T. mactroides* consumption are from Venezuela, where the species was reported as highly abundant and was harvested for both recreational and commercial use (Prieto 1983; McLachlan et al. 1996). In 1992, the Venezuelan agency Servicio Autónomo de los Recursos Pesqueros y Acuícolas (SARPA) estimated that 354 tons of *T. mactroides* were collected in that country, which is likely an underestimate (McLachlan et al. 1996). No systematically collected temporal data set yet exists to allow for the continuous evaluation of a stock of this species or the extent of its harvesting.

During research on the biology of a tropical–subtropical population of *T. mactroides* in Caraguatatuba Bay, São Paulo State, southeastern Brazil (Denadai et al. 2005), intense harvesting was observed. The total amount of these clams harvested and consumed has not yet been estimated, so the extent of this informal activity is unknown. In the case of *T. mactroides* in Caraguatatuba Bay, this estimate is of interest for public health since the water quality in Caraguatatuba Bay is occasionally poor (CETESB 2014) and the consequent contamination of the clams may render them unfit for human consumption. Mass-mortality events have also been reported for this clam in the area (Turra et al. 2014). Although natural mass-mortality events are often reported for other clam populations (Dadon 2005; Aburto and Stotz 2013), the possible consequences for consumers of a large die-off cannot be overlooked.

Therefore, this area was chosen to evaluate the harvesting dynamics, over different temporal scales, based on censuses conducted year-round during two biennia (2003–2004 and 2007–2008). The different temporal scales encompassed (1) daylong assessments, (2) comparisons between vacation and nonvacation periods (which alter the human activity on sandy beaches), and (3) comparisons between the two biennia. Data for environmental factors (weather and tide height) and the number of beach users were included in order to evaluate their possible influence on the dynamics of this activity. The data from interviews of harvesters on the amount of clams collected allowed us to estimate the total amount of *T. mactroides* collected during the two periods. Since information on the population biology of *T. mactroides* is available for the same biennia (Turra et al. 2014), it was possible to assess the relationship between the stock and its harvesting.

METHODS

Study Area

Caraguatatuba Bay is located on the northern coast of the state of São Paulo and is limited to the south by Arpoar Point (23°43′25.3″S, 45°24′07.1″W) and to the north by Camaroeiro Point (23°37′41.1″S, 45°24′02.4″W) (Figure 1). The bay is bordered by a long sandy beach (about 16 km long) consisting of several named beaches (Enseada, Flecheiras, Porto Novo, Romance, Palmeiras, Pan-Brasil, Indaiá, Centro, and Camaroeiros) that are not separated by physical barriers. The beaches are subject to moderate wave energy, being sheltered by São Sebastião Island. Three main rivers (Juqueriquerê, Lagoa, and Santo Antônio) flow into the bay, transporting continental nutrients as well as organic pollution.

A 4-km stretch of beach between two rivers in the central-southern part of the bay (23°42′073″S, 45°25′447″W; 23°40′046″S, 45°25′490″W; Figure 1) was chosen as the study area because harvesting was concentrated there. No clamming activity was seen south of this area, possibly because that area is a tidal flat that requires walking longer distances and great effort to screen the sediment far from the water during low tides. North of the sampling area, the slope steepens and the short intertidal flat is less influenced by the rivers and is rapidly covered by rising tides.

FIGURE 1. Map of Caraguatatuba Bay showing the area where the censuses were conducted.

Sampling Procedures

Census and dynamics of harvesters.—The area was divided into 14 inline stations, each including central buildings or kiosks on the beach, which were used as observation points. At each of these, the number of people harvesting (harvesters or clammers) and other beach users were counted at six fixed daily times, 2 h apart (0700, 0900, 1100, 1300, 1500, and 1700 hours and occasionally, during the summer, also at 1900 hours), from an observation point to the limits of each counting station (midway between observation points). To exclude biases due to groups of clammers, whenever it was realized that a group of people was harvesting clams, the number of people composing each group was recorded and used in the subsequent analyses. This procedure was performed in two different biennia: on 40 d between May 2003 and January 2005 and on 56 d between March 2007 and April 2008. The sampling days were scheduled to cover a variety of periods, such as vacation or nonvacation, weekends and weekdays, and all seasons.

Weather (classified as clear, few clouds, partly cloudy, overcast, light rain, or rain) was recorded for each sampling time and observation point, and flood and ebb tide peaks (tide height, in meters) were recorded daily. Also, nonvacation periods (weekdays) were distinguished from vacation times (weekends, holidays, and summer holidays). These data were used in analyses of the temporal dynamics of harvesting.

Amount of clams harvested.—Interviews were conducted with harvesters working in the study area, who were approached and asked to answer the questions while collecting clams. The question of concern here was the amount of clams harvested per harvesting event, so whenever collectors were working in groups, only one person was interviewed. The interview was semistructured; because it may be difficult for collectors to state a precise amount per event, there was also an option with predetermined ranges of amounts (up to 1, over 1 up to 2, over 2 up to 5, over 5 up to 10, over 10 up to 20, and over 20 kg) for the interviewee to select. Other relevant questions, concerning socio-economic aspects of the harvesters and the fishery, were included and are being addressed in another context (A. Turra and colleagues, unpublished). The sampling intervals also covered different periods of the year (summer holidays, public holidays, weekends, and weekdays) and different daylight-hour intervals. The interviews were conducted from October 2003 through January 2005 (first biennium) and March 2007 through March 2008 (second biennium). From the information about the amount of clams harvested per harvesting event and the number of harvesters, the amount of clams harvested in the area was estimated.

Data Analysis

Census and dynamics of harvesters.—To assess the daylong panorama of the harvesting and the potential factors grouping harvesters on the beach, descriptive plots were constructed showing the percentage of the daily number of harvesters and other beach users in each 2-h interval. For this purpose, the data from all sampling points for the same 2-h interval and day were combined and converted to percentages. Raw daily values, and the respective tide and weather conditions, were added to the plots.

To assess how the two variables (number of harvesters and number of other beach users) were related to each other, the daily counts of harvesters were plotted and regressed as a function of the number of people on the beach. This was performed separately for each biennium. Since a clamming trip generally did not exceed 2 h (Turra and colleagues, unpublished) and recreational activity usually extended for a longer period, different units were used for each category in the regression test: the daily censuses of harvesters were based on the sum of all subareas and periods of a day, while the daily censuses of other beach users were based on a mean value across sampling points and periods of the day. This was obtained by calculating hourly mean values based on the sum of all sampling points and then obtaining a final daily mean value based on the sum of all time intervals.

To assess the influence of the clam stock on harvesting activity, the daily number of harvesters was also compared with the respective characterization of the clam population. For this purpose, data from the present study were plotted together with data on population dynamics and secondary production of *T. mactroides* (Turra et al. 2014). This produced a graph of the number and size of individual clams over time during similar study periods (2003–2004 and 2007–2008).

Since all analyses were performed separately for each biennium, comparisons between them were allowed at every step.

Estimate of the amount of T. mactroides *harvested.*—The next step attempted to estimate the total amount of *T.*

mactroides harvested per year over the sampling periods. The censuses of harvesters were combined with data from the interviews, which provided the mean amount of *T. mactroides* collected per harvesting event.

All reports on the amounts of clams collected per event (in kg) were classified into the predetermined options, retaining raw values when the amount reported exceeded 20 kg. A weighted average was then calculated: the number of times that a category was chosen (i.e., *n* people who gave that answer) was multiplied by its mean value (for each category, respectively: 0.5, 1.5, 3.5, 7.5, 15.0, and "mean raw value") and the sums of these values of all categories were added together and divided by the overall number of answers. The procedure was performed separately for each biennium.

Estimates were then made using the following steps:

1. The mean number of harvesters per day: based on the daily censuses, the mean number of harvesters per day was calculated for vacation and nonvacation periods for each sampling biennium. A factorial ANOVA or the equivalent nonparametric test was used to compare sampling years and vacation and nonvacation periods, followed by the respective post hoc test. Whenever no distinction between or among groups was detected, the group was treated as an overall mean for subsequent calculations.
2. The number of days within a period by year: the total numbers of vacation and nonvacation days within 1 year were calculated based on annual calendars. Occasional differences between the biennia were accounted for.
3. The number of harvesters within periods of a year (step 1 × step 2): the means obtained by step 1 (mean number of harvesters per day) were multiplied by the respective yearly number of days in each category (step 2) to estimate the overall number of harvesters in a year and in each period (vacation or nonvacation) of a year.
4. The mean number of harvesters per group: since only one person responded on behalf of a group during interviews, whenever the clammers were working in groups, the estimate of the amount of clams collected was based on harvesting events, i.e., on the groups rather than on individuals who were collecting clams, although some "groups" were composed of only a single individual. As described in step 1, the mean numbers of harvesters within groups were calculated and compared using a factorial ANOVA between sampling biennia and vacation or nonvacation periods. Whenever they differed significantly, different mean values were used in the subsequent procedures, and an overall mean value was used whenever they did not differ.
5. The number of harvesting events within the periods of a year (step 3/step 4): to convert the number of harvesters to the number of harvesting events, the number of harvesters within the periods of a year (step 3) divided by the mean number of harvesters per group (step 4) gave the number of harvesting events within the different periods of a year (vacation or nonvacation), again for each biennium separately.
6. The amount (kg) of clams collected per harvesting event: data from interviews were used to estimate the mean amount of clams collected per harvesting event for each biennium separately.
7. The amount (kg) of clams collected per period of the year (step 5 × step 6): finally, the extrapolation of data from the censuses of harvesters to the amount of clams collected per year consisted of multiplying the values from step 5 (number of harvesting events within a period) by the respective mean amount of clams collected per harvesting event (step 6) for each biennium separately.

RESULTS

Harvester Censuses and Dynamics

The daylong plots (Figures 2, 3) showed that when the number of other beach users increased, harvesting tended to be more equally distributed throughout the day, and at these times the harvesting activity tended to follow the daily peaks and troughs of the tourist concentration. In contrast, in nonvacation periods the time of day (early morning and late afternoon) and low tides were the main concentration factors. In these cases, most of the periods for which harvesting was not associated with ebb tides were during neap tides. Weather (except for heavy rain) had little influence on clamming activity whether tourists were present or not. These conditions were similar in both biennia, even from late April 2007 on, when there was a clear decrease in the number of harvesters (Figure 3). These results indicate that clam harvesting in Caraguatatuba Bay was influenced mainly by a social component (tourism), but when this component was not present, tide stage became important.

When the daily number of harvesters was tested as a function of the mean daily number of other users of the beach, dependence between the variables was observed in the first biennium, resulting in a negative quadratic function. Three main stages can be recognized in Figure 4 (left panel). When there were less than 400 people on the beach, a strong positive relationship between the variables was apparent, with small differences between the minimum and maximum number of harvesters. At around 400–850 beach users, the number of harvesters reached its peak but also very low numbers were observed, i.e., the number of harvesters varied widely. When the highest number of people were at the beach (over 850), the number of harvesters and the variation declined further. The numbers of harvesters and other users were square root transformed and applied to a linear regression, resulting in a significant ($P < 0.01$) linear function ($r^2 = 0.524$) of the number of harvesters according to the number of tourists, described as $y = 2.234 + 0.183x$.

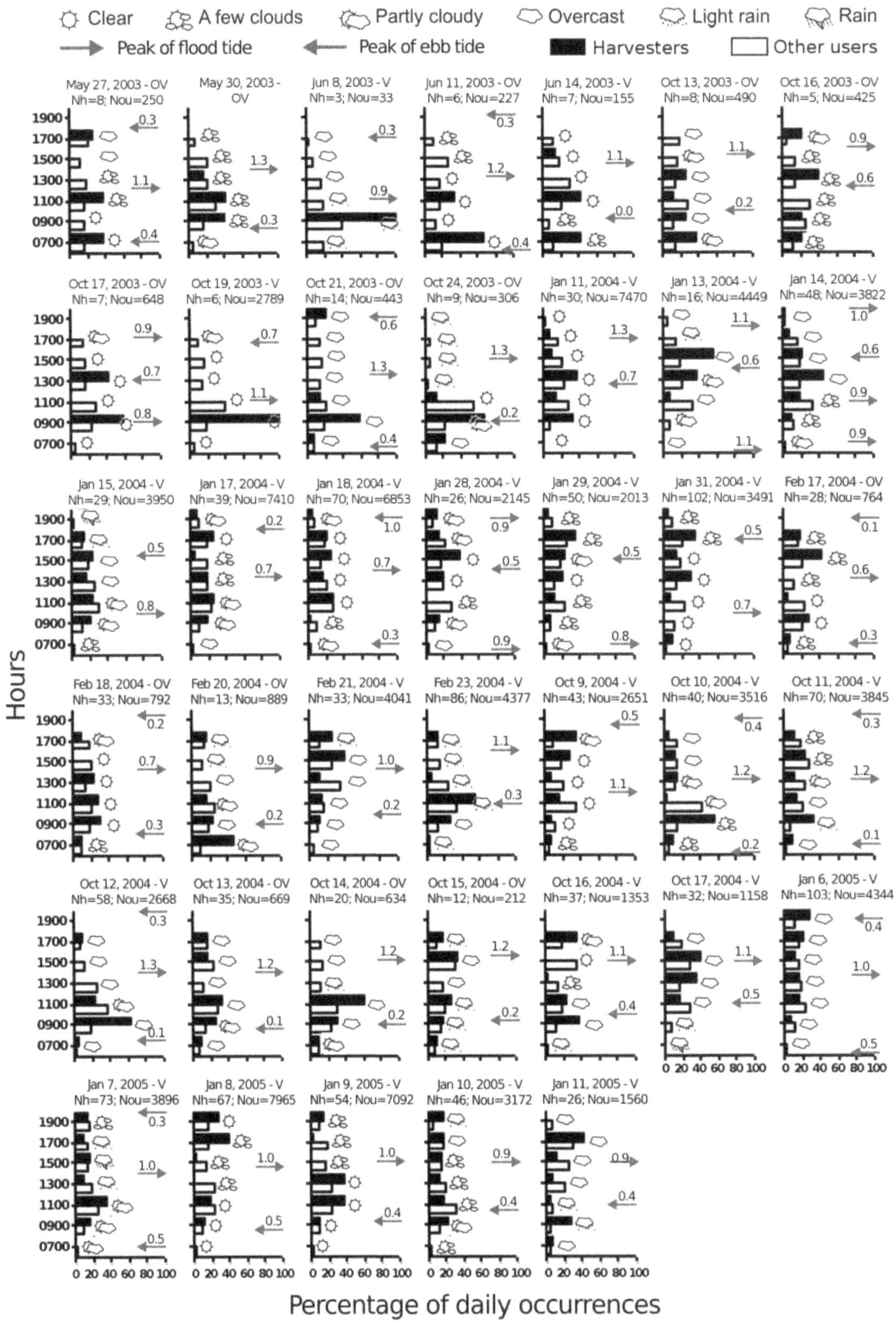

FIGURE 2. Percentage of the daily number of clam harvesters (Nh) and other beach users (Nou) in each 2-h interval from 0700–1700 hours (occasionally in summer to 1900 hours). Samples were taken in Caraguatatuba Bay on 40 d over the period from May 2003 through January 2005. Vacation days (V) were distinguished from nonvacation days (OV), and weather conditions (drawings) and tidal peaks (arrows) were recorded.

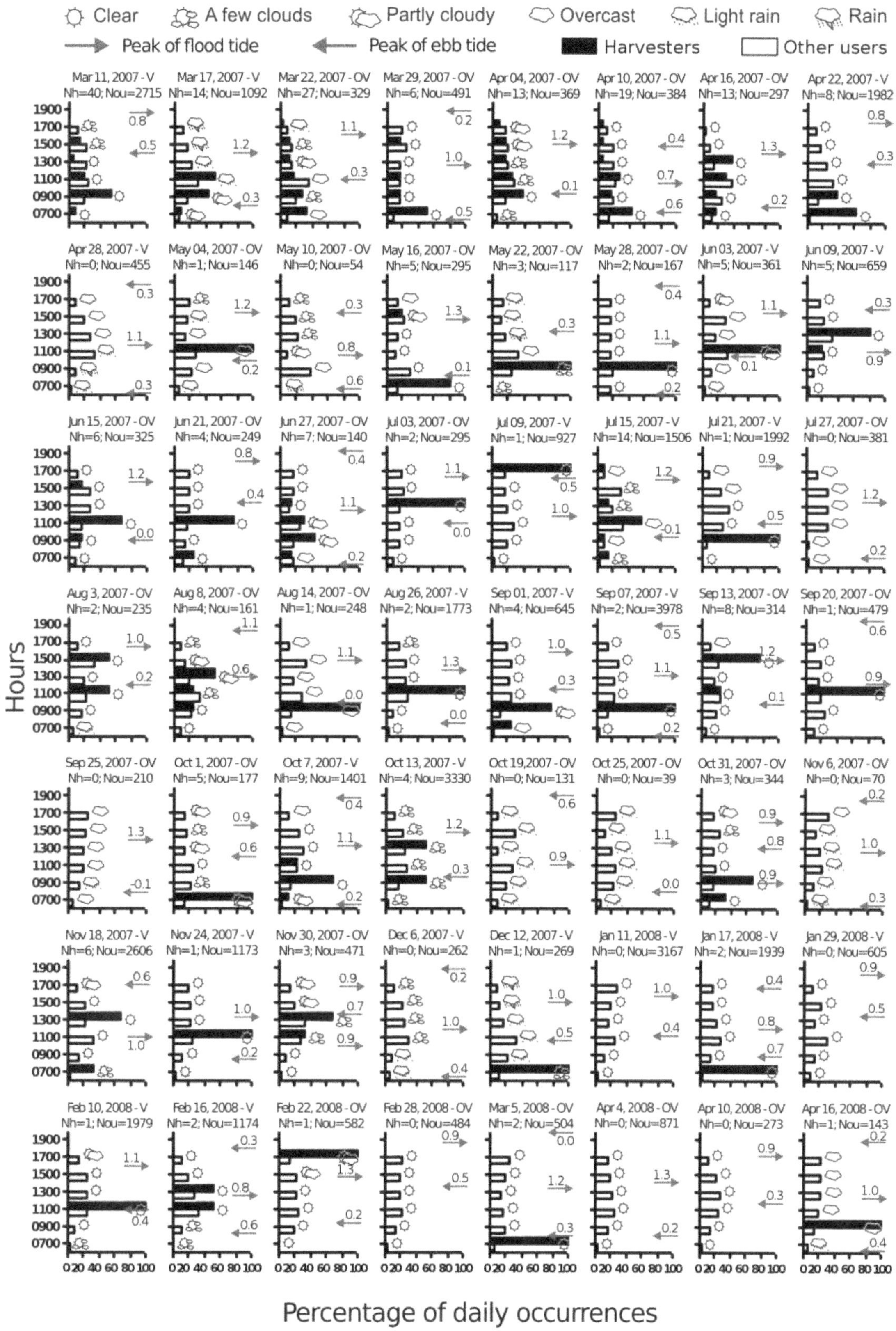

FIGURE 3. Percentage of the daily number of clam harvesters (Nh) and other beach users (Nou) in each 2-h interval from 0700–1700 hours. Samples were taken in Caraguatatuba Bay on 56 d over the period from March 2007 through April 2008. Vacation days (V) were distinguished from nonvacation days (OV), and weather conditions (drawings) and tidal peaks (arrows) were recorded.

For the second biennium, these variables showed no significant relationship, in spite of the model tested and although one outlier (corresponding to the first sampling day of the second biennium) was disregarded (Figure 4, right panel, point 1). Notably, due to the high concentration of dots on the left side of the graph (Figure 4, right panel), the number of harvesters was considerably lower, as was the number of other users in general. The numbers of harvesters and other users were log transformed [$\log_{10}(n + 1)$] and applied to a linear regression, and still no significant relationship between variables was identified ($P = 0.068$, $r^2 = 0.060$).

Another essential factor to consider is the characteristics of the *T. mactroides* population itself, here provided by the plot of the censuses together with data from Turra et al. (2014), with information on the means for both the number and size of the clams from this population (Figure 5). During the first biennium, the first census period (May–June 2003) began when the density of the clam population was among the lowest recorded, while the mean size of individual clams was intermediate (around 20 mm). The following census period of the first biennium (October 2003) was similar to the first with respect to the harvesting pattern, i.e., a relatively low number of harvesters and small differences between vacation and non-vacation periods. The abundance of clams, on the other hand, was about twice as high and their mean size about half that in the first period. The third census period (January–February 2004) was during the peak tourist season, including summer and Carnival holidays. The number of harvesters was extremely variable; the differences were associated with variations in tourism intensity, as described in the above analysis. By this time (January–February 2004), both the abundance and size of the clams had increased. These conditions remained quite similar in subsequent census periods of the first biennium, with respect to both harvesting intensity and clam population structure. The clam population structure may have had some influence on the harvesting intensity, since the abundance, and mostly size, were higher in periods when more harvesting events occurred. However, in this first period, clamming activity seemed to be more closely related to the concentration of people on the beach rather than to fluctuations in the clam population itself. In summation, the number of harvesters was positively associated with vacation periods, but these periods also showed wide variability in the number of harvesters during the first biennium.

In contrast, the characteristics of the clam population during the second biennium provided evidence that its features influenced the harvesting dynamics. During the beginning of the second biennium, the abundance was higher than in any period in the first biennium (Figure 5; note the scale differences in the middle panel between the left and right *y*-axes, which represent the two biennia, respectively), and the mean size of individual clams was also among the largest in the entire study. Notwithstanding, the related harvester censuses showed relatively high numbers and, as expected, the highest number of harvesters on a vacation day. These numbers are especially high, since these first censuses (March–April 2007) did not include peak tourist seasons. However, from May 2007 on, the abundance of clams increased dramatically, together with a sharp decrease in mean size. Subsequent censuses showed fewer harvesters, a situation that persisted through the end of the period. Even lower numbers of harvesters were recorded in the last censuses, which were conducted

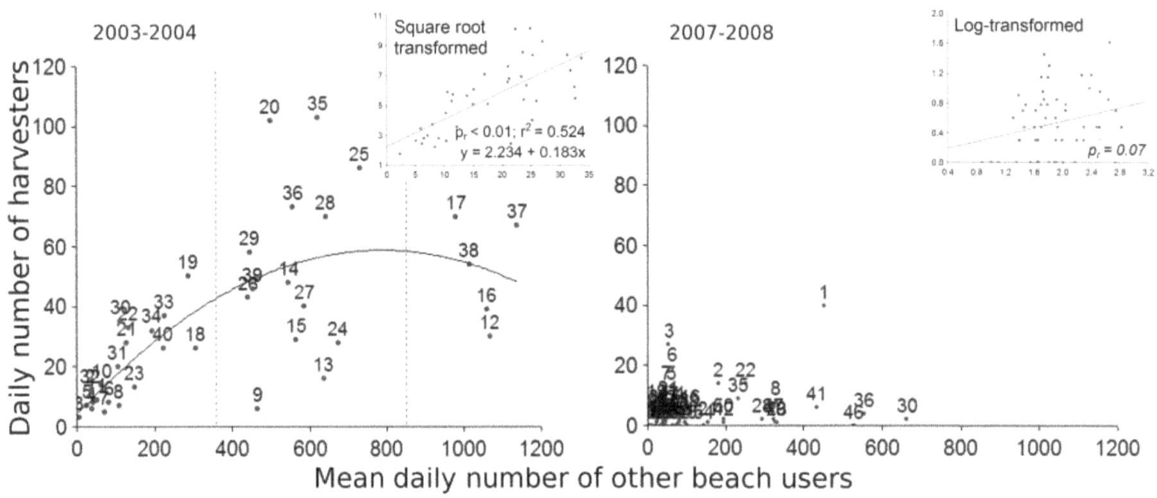

FIGURE 4. Relationship between the total daily number of clam harvesters of *Tivela mactroides* and the mean daily number of other people on the beach, based on counts at 2-h intervals. Plots are provided separately for the first (2003–2004; left panel) and the second (2007–2008; right panel) biennia assessed. The respective transformed data and the regression results are displayed in each panel at the upper right. The numbered replicates (numbered black dots) are equivalent to the sequence shown in Figures 2 (biennium 1) and 3 (biennium 2). The dashed lines in the left panel divide the graph into the three main stages identified.

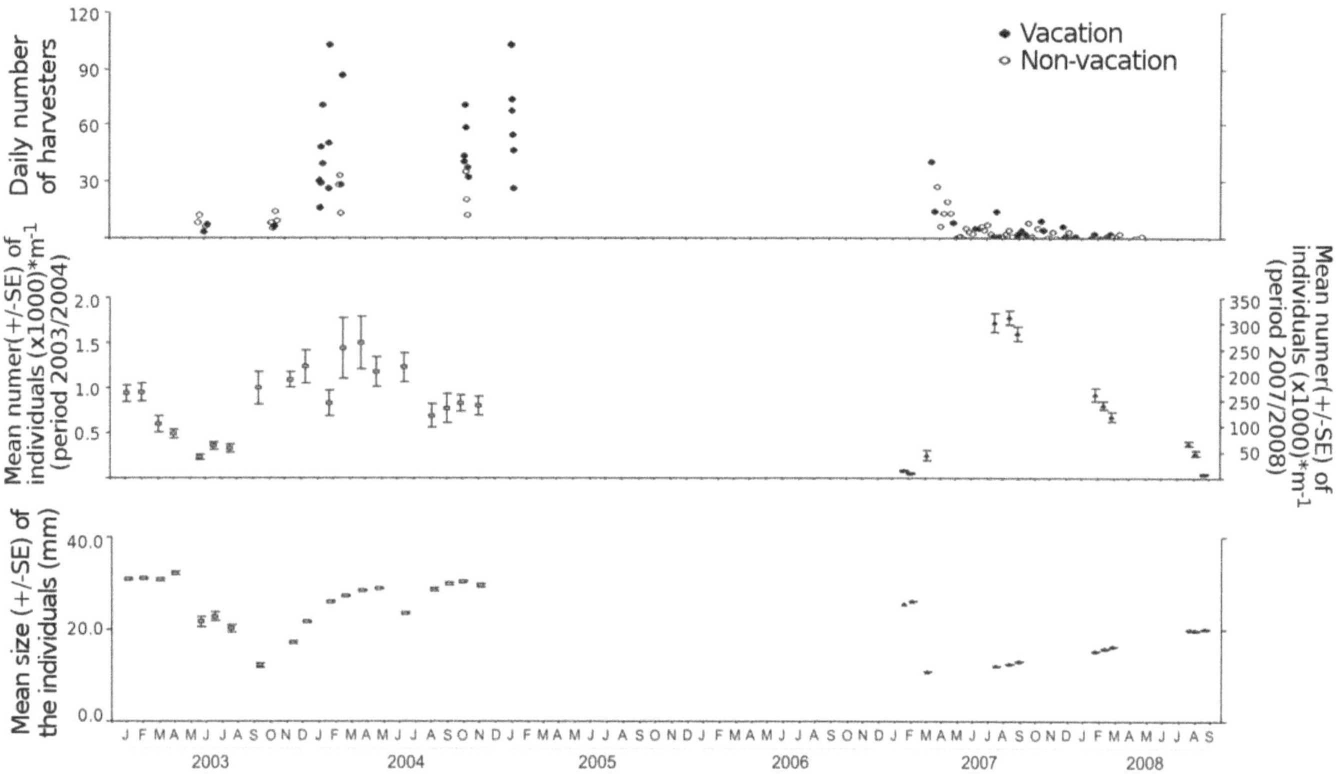

FIGURE 5. Temporal variation of the daily number of harvesters during vacation and nonvacation periods. Plots were obtained, together with data on temporal variation of *Tivela mactroides* density (individuals/m; unit obtained from the conversion of the number of individuals per strip transect of known length, 56 m, and width, 1 m) and size (mm), from Turra et al. (2014).

during vacation periods. During this period, the number of tourists was also considerably lower than the number in the first-biennium peak season (Figures 2, 3), possibly minimizing the wide variations in the number of harvesters observed during the first biennium. Therefore, the extreme conditions of the clam population (high abundance of small-sized individuals), and also the decrease in tourism activity (secondarily), seemed to strongly affect harvesting activity in the 2007–2008 biennium.

Estimate of the Amount of Clams Harvested

The mean number of harvesters differed significantly, both between the biennia and between vacation and nonvacation periods ($F_b = 80.83$, df $= 1$, $P < 0.01$; $F_s = 17.12$, df $= 1$, $P < 0.01$, respectively), with a significant interaction between factors ($F_i = 14.54$, df $= 1$, $P < 0.01$). The vacation period during the first biennium had the highest mean ± SD values (48.30 ± 26.95 harvesters), followed by the nonvacation condition (21.60 ± 19.60 harvesters), while the second period showed a significantly lower mean number of harvesters and no difference between the touristic conditions (5.30 ± 8.60 and 4.21 ± 6.01 harvesters, respectively; $P = 0.82$).

The pattern that was seen with the mean number of harvesters also occurred for the mean number of harvesters per group, with a significant difference between the two factors ($F_b = 58.23$, df $= 1$, $P < 0.01$; $F_s = 4.84$, df $= 1$, $P = 0.03$) and a significant interaction between factors ($F_i = 12.48$, df $= 1$, $P < 0.01$). Vacation periods in the first biennium showed the highest mean ± SD value (2.62 ± 0.62 harvesters/group), followed by nonvacation periods (1.91 ± 0.64 harvesters/group); the second period showed significantly lower values, with no difference between touristic conditions (1.41 ± 0.56 and 1.27 ± 0.41 harvesters/group, respectively; $P = 0.42$).

The responses to the interviews about the amount of clams (in kg) collected per harvesting event (Figure 6) were processed to determine a weighted average for each biennium. These averages were used to estimate the annual amount of clams collected in the study area. For the first biennium the average obtained was 4.74 kg, rising to 6.89 kg of clams collected in the second biennium (see Table 1 for a summary of the values used and the results).

There was a sharp difference between the estimates for the total amount of clams harvested in the study area in the two biennia. For the first biennium, this estimate was about 24.6 tons/year. Even with the higher estimated weight of clams

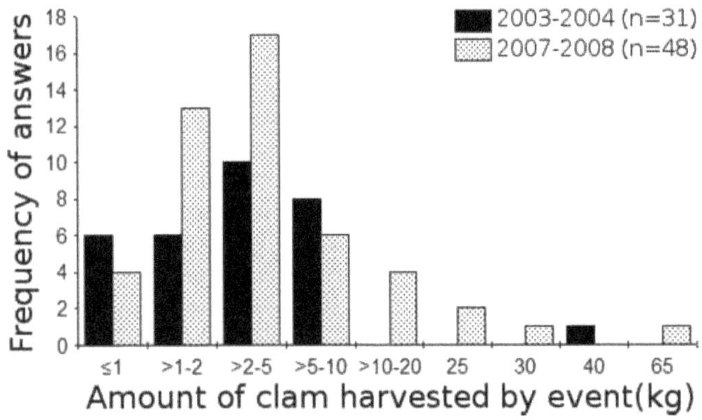

FIGURE 6. Frequency of responses to semistructured interviews of *Tivela mactroides* harvesters at Caraguatatuba Bay about the amount (in kg) of clams collected per harvesting event. Interviews were conducted from October 2003 through January 2005 (first biennium) and March 2007 through March 2008 (second biennium).

collected per harvesting event for the second biennium, the total amount of clams harvested for this period decreased by nearly one-third, to about 8.8 tons/year (Table 1).

DISCUSSION

Harvester Censuses and Dynamics

Collecting clams has a strong recreational appeal (MacLachlan et al. 1996; Griffiths et al. 2006; Dyson and Huppert 2010), and indeed, tourism considerably influenced the present results for harvesting intensity. However, as the routine of tourists differs substantially from that of local harvesters, the harvesting activity during peaks of tourism was perceptibly more diffused during the day, regardless of tide stage and weather conditions. Also, the reasons why tourists collect clams are very likely to differ because they are more likely to do it as a leisure activity and rarely depend on this resource or customarily consume clams. Local harvesters may collect clams for nearly opposite reasons, i.e., subsistence or economic reasons.

Taking into account the periods of low tourism activity, tide was a main component driving more intense harvesting, which tended to be concentrated during low tides. The concentration of clammers was uncorrelated with tide height only during neap-tide periods, when differences between low and high tides are minimal and apparently were not sufficiently different to affect the harvesting activity. Because the harvesters in this situation are much more likely to be local residents and frequent clammers, the results reflect more familiarity with the factors influencing the activity. Weather, except for heavy rain, did not seem to have a strong influence on harvesting and was the environmental condition that least affected this activity. This condition seems quite natural: for tourists it would be expected that only heavy rain would inhibit their leisure activities and, similarly, resident harvesters would prefer to wait for more favorable conditions. They might also be expected to avoid periods of more intense sunshine, but as these conditions may last for long periods, waiting may not be feasible.

The close relationship between harvesting and tourism was evident, showing not only that harvesting activity increases with increasing tourism but also that this concomitant increase has a limit. Clamming activity might have a snowball effect with respect to the presence of tourists: even tourists who have never collected a clam before are attracted by the activity when they see it being done and feel comfortable eating the clams because the local people do it. The more people collect clams, the more the activity is noticed by other tourists, and so on. After a given point, however, the number of tourists comes to be negatively correlated with the number of harvesters. Following the previous argument, if local harvesters cease their activity during periods of overcrowding, tourists would be less likely to collect clams, leading to the reverse effect. That is likely to happen because local harvesters will exploit the touristic activity as an informal source of revenue, a more profitable activity than clam harvesting during vacation periods. Another contributing factor may be that peak-season tourists, who are mainly day-trippers, usually have a different profile than those who visit the beach more frequently. The latter group would have more time available for different activities and more facilities in which to prepare their own meals.

The different results found in the second biennium can be attributed to the low number of harvesters and the considerably fewer tourists observed but also to the smaller size of the clams. It would be interesting to have data on the fluctuations in the number of tourists over time for the municipality, but unfortunately this information is not available for the study period. This could indicate whether the decrease in visitation was limited to the bay or was representative for the municipality as a whole. In any event, the extremely high abundance of clams in this period and location led to mass-mortality events of *T. mactroides*, most likely due to density-dependent processes (Turra et al. 2014), which may have been the main factor preventing visitation. In addition to the consequent foul odor and unpleasant sight, it is practically unavoidable to question why the mortality occurred and to be afraid to eat the clams (Figure 7). Even though high variability and density-dependent processes are normal features of clam populations (Dadon 2005; Aburto and Stotz 2013; Turra et al. 2014), Caraguatatuba Bay is an urban area that is subject to oil spills and discharges of industrial, domestic, and agricultural effluents. Factors that could cause or contribute to mass mortalities include ocean storms, excessive algal mats, parasite proliferation, and also contamination from oil spills (Araújo 2001; Jorge 2003; Dunham et al. 2007).

In summary, the mass-mortality events had a strong effect on the harvesting dynamics. During the first biennium, the relatively constant and moderate abundance and size of individual clams provided suitable conditions for harvesting. Wide fluctuations in the number of harvesters were hypothesized to

TABLE 1. Compiled data and schematic calculation of the estimated amount of *Tivela mactroides* harvested per year in Caraguatatuba Bay. Censuses of harvesters were conducted in two biennia, 2003–2004 and 2007–2008, distinguishing vacation (V) from nonvacation periods (N). For each period, the following variables were measured or calculated: the mean number of harvesters per day (A), total number of vacation or nonvacation days in a year (B), mean number of people per group (C), mean number of harvesting events per day (D), and mean amount of clams collected per harvesting event (E). A specific mean was calculated within each period (Specific), but a common mean value was used for periods that showed no statistical differences, so the values that were actually used are also shown (Used). The variables that were used in the final estimates are shown in bold italics. The steps of the calculations are shown with the column title, where applicable.

		(A) Mean number of harvesters/day			(C) Mean number of people/group		(D) Number of harvesting events/day (= A/C)			Estimate of clams harvested (kg/year) (= B·D·E)	
Biennium	Period	Specific	Used	(B) Number of days in a year	Specific	Used	Specific	Used	(E) Mean amount collected (kg)	Partial	Annual
1 (2003–2004)	V	48.30	***48.30***	***150***	2.62	***2.62***	18.43	***18.43***	***4.74***	13,107	24,632
	N	21.60	***21.60***	***215***	1.91	***1.91***	11.31	***11.31***		11,525	
2 (2007–2008)	V	5.30	***4.66***	***152***	1.41	***1.33***	3.98	***3.50***	***6.89***	3,669	8,811
	N	4.21	***4.66***	***213***	1.27	***1.33***	3.42	***3.50***		5,142	

FIGURE 7. *Tivela mactroides* individuals covering the beach in Caraguatatuba Bay during a series of mass-mortality events in 2007 and 2008. On the day of the photograph, this abundance was consistent alongshore and across shore (detail).

be due to changes in tourism intensity, but probably the increasing harvesting activity was related to the increased abundance and size of the clams, characterizing it as an opportunistic activity. At the beginning of sampling in the second biennium, the abundance and size of clams were still high. The number of harvesters was correspondingly high, even for the equivalent period in the first biennium. However, despite the subsequent astonishing increase in the abundance of clams, the number of harvesters decreased considerably. This change may be related to the noticeable decrease in the size of the clams, making them unattractive for harvesting. Furthermore, these factors (unusually high density and small-sized individuals) led to mass-mortality events (Turra et al. 2014), which probably deterred tourism and further harvesting.

Estimate of the Amount of Clams Harvested

The demand for a resource is a key factor for environmental management. In the case of clams, if these filter feeders are exposed to pollutants they are very likely to transmit the contamination to their consumers. Closures of clam fisheries are common after blooms of toxic algae (Dyson and Huppert 2010; Chadsey et al. 2012), and Chen (1997) reported, following an instance of arsenic poisoning, that even consuming small amounts of clams exceeded the allowed limits. Data on *T. mactroides* revealed contamination by *Escherichia coli, Staphylococcus aureus, Vibrio cholerae* and *Salmonella* sp. in Caraguatatuba Bay, which raises concerns regarding clam consumption (M. R. Denadai and colleagues, unpublished).

In Caraguatatuba Bay, the historical lack of reports about the exploitation of *T. mactroides* leads to underestimation of this activity. For example, according to some informal reports, restaurants from the city of São Paulo occasionally order large amounts of *T. mactroides*. The present estimated amount of clams harvested reached almost 25 tons/year, which is not negligible. Although the larger clams in temperate zones tend to have much greater economic value (MacLachlan et al. 1996), some examples from tropical regions illustrate the significance of the present data. Murray-Jones and Steffe (2000) estimated that for a commercially exploited species on a 32-km-long beach in Australia (Caraguatatuba is 16 km long and the study area where the activity is concentrated was 4 km long), 46.5 tons/year was taken by recreational harvesters. On the Eastern Cape of South Africa, the sustainable amount of the clam *Donax serra* harvested for commercial exploitation was estimated at 100 tons/year (Sims-Castley and Hosking 2003).

Given the high abundance of *T. mactroides* in Caraguatatuba Bay, the current level of artisanal or recreational harvesting does not seem to be harming the clam population. This harvesting activity has probably been conducted for decades without damaging the resource, but changes in harvesting intensity and demand over time, as well as other causes of mortality (mass mortality and predation), may have affected clam abundance. An increase in recreational clammers, who tend to have few or no concerns about a minimum size for harvesting (Murray-Jones and Steffe 2000), may compromise not only the clam population but also the entire intertidal community (Dadon 2005).

The disturbance in the clam population that occurred in the second biennium of the study provided an opportunity to assess how the harvesters behaved in these circumstances. The enormous number of clams present did not encourage more people to harvest them; on the contrary, the number of harvesters decreased nearly tenfold. However, people who were still collecting the clams were possibly encouraged by their availability given the increase in the average amount collected per harvesting event mentioned in the interviews, an amount that increased from 4.74 to 6.89 kg.

Management Considerations

The enormous variability in the clam population abundance, the lack of concern and assessment about harvesting numbers, and underestimates of clam consumption due to lack of commercial sales in small-scale fisheries have led to a dearth of management efforts and of integration between policy makers and resource users (Murray-Jones and Steffe 2000; Castilla and Defeo 2001; Hartill et al. 2005; Wiber et al. 2010; Aburto and Stotz 2013). Lack of information and control measures may lead to dramatic consequences for the clam stocks, intertidal benthic community, organisms that depend on clams as a food source, and even social, economic, and cultural aspects of local and traditional communities.

Exploited beach clams generally burrow near the sediment surface in the intertidal zone of sandy beaches. They are easily harvested without the need of boats or special equipment, using only hands, feet, or in some cases, very simple tools. Access to sandy beaches is normally unrestricted, and there are no physical or economic barriers to harvesting resources from these environments. In this situation, one possible way to control indiscriminate harvesting would be based on a comanagement strategy supported by biological and social data on the activity. Castilla and Defeo (2001) recommended cooperative management efforts among fishermen, scientists, and managers to regulate the use of beach shellfisheries. Systematic data collection on stocks and harvesting activity is important to provide for sustainable use by local fishing communities. Studies in South America have already identified some local populations of *T. mactroides* that are close to extinction due to uncontrolled harvesting and that have not recovered after long-term closures (Ortega et al. 2012), which exemplifies the risks of neglecting the management of harvesting activity.

Since coastal environments are threatened by a variety of factors, from waste dumping to climate change, that require complex and long-term measures, macrofauna structure and stocks of beach shellfish may also be affected to some extent. However, overexploitation of this resource is one of the factors that can be effectively minimized with the aid of some simple but strategic tools, such as actions to increase awareness about its use. This study provides information on the dynamics of clam harvesting and the influence of time of day, tide, vacation–nonvacation period, stock (number and size), and mass-mortality events on the number of harvesters. The dynamics and intensity of clamming proved to vary over time (between the two biennia). In addition to information about the socioeconomic profile of the clammers, participatory planning strategies can be designed to allow communication between users, managers, and scientists to discuss best management practices for sustaining the *T. mactroides* fishery in Caraguatatuba Bay. Knowledge of the temporal dynamics of the activity may also inform proposals for education and communication strategies to reach most of the users.

For example, the mass-mortality events may have been caused by unsuitable sanitation or pollution conditions. On the other hand, if these mortality events occur naturally, due to a density-dependent phenomenon (e.g., Turra et al. 2014), people should be advised to exploit the clams as much as they want to, otherwise the resource will be lost. The clammers could also be informed about food safety if the quality of the clams is monitored. This study highlights the need for long-term monitoring of clam harvesting as an important part of managing this traditional and informal activity, to assess the possible risks of overexploitation, and to assure the safety of clam consumers and also provides important information to guide the adaptive management of this resource and this socioeconomic activity.

ACKNOWLEDGMENTS

We are grateful to the Fundação de Amparo à Pesquisa do Estado de São Paulo (São Paulo Research Foundation) for the scholarship and grants that made this study possible. We also thank the Project AWARE Foundation and the PADI Foundation for the grants in support of this research. Our special thanks to Professor Cecilia Amaral (UNICAMP) for her valuable contributions during the development of this study.

REFERENCES

Abrahão, J. R., R. S. Cardoso, L. Q. Yokoyama, and A. C. Z. Amaral. 2010. Population biology and secondary production of the stout razor clam *Tagelus plebeius* (Bivalvia, Solecurtidae) on a sandflat in southeastern Brazil. Zoologia (Curitiba) 27:54–64.

Aburto, J., and W. Stotz. 2013. Learning about TURFs and natural variability: failure of surf clam management in Chile. Ocean and Coastal Management 71:88–98.

Araújo, C. M. Y. 2001. Biologia reprodutiva do berbigão *Anomalocardia brasiliana* (Mollusca, Bivalvia, Veneridae) na reserva extrativista marinha do pirajubaé. [Reproductive biology of the clam *Anomalocardia brasiliana* (Mollusca, Bivalvia, Veneridae) in the Marine Extractive Reserve of Pirajubaé.] Doctoral dissertation. University of São Paulo, São Paulo, Brazil.

Caddy, J. F. 1989. Marine invertebrate fisheries: their assessment and management. Wiley, New York.

Castilla, J. C., and O. Defeo. 2001. Latin American benthic shellfisheries: emphasis on co-management and experimental practices. Reviews in Fish Biology and Fisheries 11:1–30.

CETESB (Companhia Ambiental do Estado de São Paulo). 2013. Relatório de qualidade das praias litorâneas no estado de São Paulo [Report of quality of coastal beaches of the state of São Paulo.] CETESB, São Paulo, Brazil.

Chadsey, M., V. L. Trainer, and T. M. Leschine. 2012. Cooperation of science and management for harmful algal blooms: domoic acid and the Washington coast razor clam fishery. Coastal Management 40:33–54.

Chen, S. H. 1997. An investigation on the effect of arsenic poisoning on humans, agriculture, and fisheries. Journal of the Chinese Chemical Society 44:115–121.

Dadon, J. R. 2005. Changes in the intertidal community structure after mortality event in sandy beaches of Argentina. Contributions to Zoology 74:27–39.

Defeo, O., and A. MacLachlan. 2005. Coupling between macrofauna community structure and beach type: a deconstructive meta-analysis. Marine Ecology Progress Series 433:29–41.

Denadai, M. R., A. C. Z. Amaral, and A. Turra. 2005. Along-and across-shore components of the spatial distribution of the clam *Tivela mactroides* (Born 1778) (Bivalvia, Veneridae). Journal of Natural History 39:3275–3295.

Dunham, J. S., B. Koke, G. E. Gillespieand, and G. Meyer. 2007. An exploratory survey for littleneck clams (*Protothaca staminea*) in the Broughton Archipelago, British Columbia—2006. Canadian Manuscript Report of Fisheries and Aquatic Sciences 2787.

Dyson, K., and D. D. Huppert. 2010. Regional economic impacts of razor clam beach closures due to harmful algal blooms (HABs) on the Pacific coast of Washington. Harmful Algae 9:264–271.

Griffiths, J., M. N. Dethier, A. Newsom, J. E. Byers, J. J. Meyer, F. Oyarzun, and H. Lenihan. 2006. Invertebrate community responses to recreational clam digging. Marine Biology 149:1489–1497.

Hartill, B. W., M. Cryer, and M. A. Morrison. 2005. Estimates of biomass, sustainable yield and harvest: neither necessary nor sufficient for the management of non-commercial urban intertidal shellfish fisheries. Fisheries Research 71:209–222.

Herrmann, M., D. Carstensen, S. Fisher, J. Laudien, P. E. Penchaszadeh, and W. E. Arntz. 2009. Population structure, growth, and production of the wedge clam *Donax hanleyanus* (Bivalvia: Donacidae) from northern Argentinean beaches. Journal of Shellfish Research 28:511–526.

Jorge, R. A. D. L. V. C. 2003. Avaliação de impacto ambiental sobre o ecossistema marinho utilizando larvas de mexilhões (*Perna perna*) (Linnaeus, 1758) (Mollusca:Bivalvia) como bioindicadores, através de técnicas ecotoxicológicas. [Assessment of the environmental impact on the marine ecosystem using mussels larvae (*Perna perna*) (Linnaeus, 1758) (Mollusca: Bivalvia) as bioindicators, through ecotoxicological techniques.] Doctoral dissertation. University of São Paulo, São Paulo, Brazil.

Laudien, J., T. Brey, and W. E. Arntz. 2003. Population structure, growth and production of the surf clam *Donax serra* (Bivalvia, Donacidae) on Namibian two sandy beaches. Estuarine, Coastal and Shelf Science 58(Supplement):105–115.

Lima, M., A. Brazeiro, and O. Defeo. 2000. Population dynamics of the yellow clam (*Mesodesma mactroides*): recruitment variability, density-dependence and stochastic processes. Marine Ecology Progress Series 207:97–108.

McLachlan, A., J. E. Dugan, O. Defeo, A. D. Ansell, D. M. Hubbard, E. Jaramillo, and P. Penchaszadeh. 1996. Beach clam fisheries. Oceanography and Marine Biology: an Annual Review 34:163–232.

Murray-Jones, S., and S. Steffe. 2000. A comparison between the commercial and recreational fisheries of the surf clam, *Donax deltoids*. Fisheries Research 44:219–233.

Narchi, W. 1976. Ciclo anual da gametogenese de *Anomalocardia brasiliana* (Gmelin, 1791) (Mollusca Bivalvia). [Annual cycle of the gametogenesis of *Anomalocardia brasiliana* (Gmelin, 1791) (Mollusca Bivalvia).] Boletim de Zoologia da Universidade de São Paulo 1:331–350.

Ortega, L., J. C. Castilla, M. Espino, C. Yamashiro, and O. Defeo. 2012. Effects of fishing market price, and climate on two South American clam species. Marine Ecology Progress Series 469:71–85.

Paes, E. T. 1989. Biologia e ecologia de *Donax gemmula* (Bivalvia, Donacidae) na zona de arrebentação da Praia do Cassino, Rio Grande, RS; e observações sobre a fauna acompanhante. [Biology and ecology of *Donax gemmula* (Bivalvia, Donacidae) in the surf zone of Cassino Beach, Rio Grande, RS; with comments on the accompanying fauna.] Master's thesis. Fundação Universidade do Rio Grande, Rio Grande, Brazil.

Prieto, A. S. 1983. Ecologia de *Tivela mactroides* (Born, 1778) (Mollusca, Bivalvia) en Playa Güiria (Sucre, Venezuela). [Ecology of *Tivela mactroides* (Born, 1778) (Mollusca, Bivalvia) at Güiria Beach (Sucre, Venezuela).] Boletín del Instituto Oceanográfico, Universidad de Oriente 22:7–19.

Schaeffer-Novelli, Y. 1980. Análise populacional de *Anomalocardia brasiliana* (Gmelin, 1791), na Praia do Saco da Ribeira, Ubatuba, Estado de São Paulo. [Population analysis of *Anomalocardia brasiliana* (Gmelin, 1791) at Saco da Ribeira Beach, Ubatuba, State of Sao Paulo.] Boletim do Instituto Oceanográfico 29:351–355.

Sims-Castley, R., and S. G. Hosking. 2003. A social cost-benefit analysis of a small-scale clam fishery in the Eastern Cape, South Africa. African Journal of Marine Science 25:159–168.

Turra, A., M. Petracco, A. C. Z. Amaral, and M. R. Denadai. 2014. Temporal variation in life-history traits of the clam *Tivela mactroides* (Bivalvia:Veranidae): density-dependent process in sandy beaches. Estuarine, Coastal and Shelf Science 150(Part A):157–164.

Viegas, O. 1981. Dinâmica populacional e produção de *Tagelus plebeius* (Solcurtidae: Bivalvia) no Canal do Calunga-Macéio-Alagoas. [Population dynamics and production *Tagelus plebeius* (Solcurtidae: Bivalvia) in the Channel Calunga-Maceió-Alagoas.] Doctoral dissertation. University of Brasília, Federal District, Brazil.

Wiber, M. G., M. A. Rudd, E. Pinkerton, A. T. Charles, and A. Bull. 2010. Coastal management challenges from a community perspective: the problem of "stealth privatization" in a Canadian fishery. Marine Policy 34:598–605.

14

Growth and Mortality in Coastal Populations of Winter Flounder: Implications for Recovery of a Depleted Population

Melissa A. Yencho
School of Marine and Atmospheric Sciences, Stony Brook University, Stony Brook, New York 11794-5000, USA

Adrian Jordaan
Department of Environmental Conservation, University of Massachusetts Amherst, Holdsworth Hall, Amherst, Massachusetts 01003, USA

Robert M. Cerrato
School of Marine and Atmospheric Sciences, Stony Brook University, Stony Brook, New York 11794-5000, USA

Hannes Baumann
Department of Marine Sciences, University of Connecticut, 1080 Shennecossett Road, Groton, Connecticut 06340-6048, USA

Michael G. Frisk*
School of Marine and Atmospheric Sciences, Stony Brook University, Stony Brook, New York 11794-5000, USA

Abstract

We studied growth, mortality, and settlement distributions of juvenile Winter Flounder *Pseudopleuronectes americanus* in two bays of Long Island, New York, to better understand localized population dynamics of a species experiencing a protracted population decline. Juvenile mortality in Long Island bays ranged between 0.02 and 0.04 per day and was as high as or higher than values reported for other systems. Settlement distributions had multiple peaks (cohorts) occurring between March and late July in 2007 and between February and May in 2008. The presence of multiple cohorts limited the usefulness of field-derived, length-based estimates of growth, resulting in unrealistic values compared with otolith-based measures (field based: −0.05 to 0.25 cm/d; otolith based: 0.05–0.06 cm/d). Thus, we recommend the use of otolith methods or the repeated measurement of individuals to estimate growth of juvenile Winter Flounder. Otolith-based growth rate was significantly higher for Port Jefferson Harbor during 2007 than for all other year × location combinations. Together with previous research showing genetic differentiation and migratory diversity, our finding of multiple spawning cohorts in Long Island Winter Flounder suggests a degree of isolation, and local management will be needed to support healthy populations. Future research

Subject editor: Debra J. Murie, University of Florida, Gainesville

*Corresponding author: mfrisk@notes.cc.sunysb.edu

to determine adult spawning, migratory behavior, stock structure, duration of the larval period, and settlement timing is required to unravel the complex behavior of Winter Flounder.

The Winter Flounder *Pseudopleuronectes americanus* is a right-eyed flounder that is found in coastal waters from Labrador, Canada, to Georgia, USA, but it is most common between Nova Scotia, Canada, and Delaware, USA (Klein-MacPhee 2002). Winter Flounder populations in the northeastern USA have a history of supporting recreational and commercial fisheries in federal waters (Brown and Gabriel 1998) and in nonfederal waters (e.g., Long Island, New York, bays) since the early 1900s (Lobell 1939). However, a 2008 stock assessment placed the southern New England–Mid-Atlantic Bight population at 9% of the management target, and a closure was enacted in federal waters south of Cape Cod to support rebuilding of the stock (NEFSC 2008). Coincidentally, fishery catches and abundance indices of Long Island bay populations have declined by more than an order of magnitude since the 1980s (Figure 1), and Winter Flounder are nearly absent from areas that once supported active fisheries (Socrates and Colvin 2006; Sagarese 2009; O'Leary et al. 2013).

In addition to the documented overfishing, other causes of Winter Flounder population declines have been hypothesized, including targeted fishing of spawning adults, environmental change, and increased predation by crustaceans and finfishes (Fairchild and Howell 2000; Keller and Klein-MacPhee 2000; Manderson et al. 2000; Taylor 2004; Scharf et al. 2006). Despite declines in the adult populations targeted by fisheries, abundance and length frequency data from a seining survey conducted by the New York State Department of Environmental Conservation indicated that production of newly settled young-of-the-year (age-0) Winter Flounder in Long Island estuaries remained relatively stable during the initial population decline in the 1990s (Socrates and Colvin 2006; Yencho 2009). It appears that postsettlement mortality constrained recruitment to the adult stock to near zero and produced a bottleneck during the first year of life. Therefore, high postsettlement mortality within Long Island bays may be an important factor limiting population recovery.

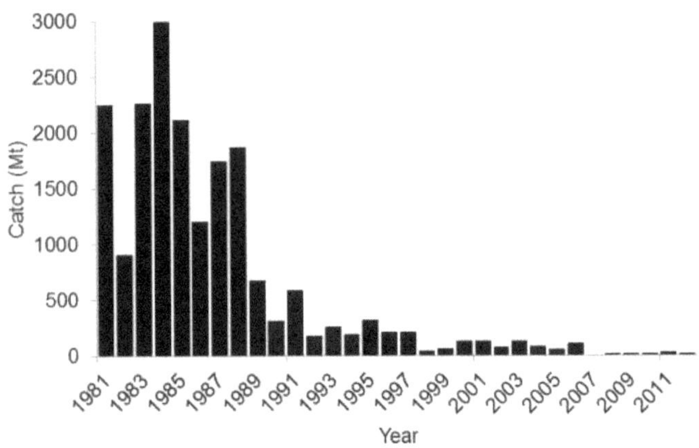

FIGURE 1. Estimates of Winter Flounder recreational catch (metric tons [mt]) in New York's inland waters (NOAA 2012).

According to the traditional life history model of Winter Flounder, adults migrate from deeper offshore waters into inshore estuarine habitats for spawning in winter and then move back offshore during the spring (Lobell 1939; Perlmutter 1947; Van Guelpen and Davis 1979; Collette and Klein-MacPhee 2002). However, historical evidence suggests a more complex population structure in Long Island bays, consisting of distinct behavioral groups termed resident "bay fish" and migratory "ocean fish" (Lobell 1939; Perlmutter 1947). Homing and fine-scale population structuring in Winter Flounder have been observed before (Saila 1961; Buckley et al. 2008; O'Leary et al. 2013), and recent tagging evidence suggests that some fish exhibit persistent residency in inshore estuarine habitats (DeCelles and Cadrin 2010; Sagarese and Frisk 2011).

Our overall goal was to better understand the growth, mortality, and settlement behavior of Winter Flounder in Long Island bays. Specifically, we used relative abundance and otolith microstructure in age-0 Winter Flounder to estimate (1) growth and mortality and (2) settlement date distributions. We discuss our results in the context of recovering depleted populations, and we explore the implications of population structuring and field-based approaches for estimating growth.

METHODS

Study locations.—Age-0 Winter Flounder were obtained from two Long Island bays: Port Jefferson Harbor (PJ) and Shinnecock Bay (SB; Figure 2). The two areas differ in habitat type, temperature regime, and depth profile. Port Jefferson Harbor is connected to Long Island Sound and is a deep, well-stratified bay with a coolwater refuge during the summer months. The PJ coastline is primarily rocky and gravelly, with patches of soft bottom and vegetation. In contrast, SB is directly connected to the Atlantic Ocean and is a shallow, vertically mixed bay with little temperature refuge during the warm summer months.

Sampling regime.—Field sampling was conducted between March and November 2007 and between May and November 2008 (Figure 2). In 2007, a beach seine (61 × 3 m; 6-mm mesh) was used in PJ and SB. In addition, a smaller beach seine (30 × 3 m; 6 mm mesh) was occasionally used and was assumed to sample half the area relative to the larger net. The small net was not used to describe

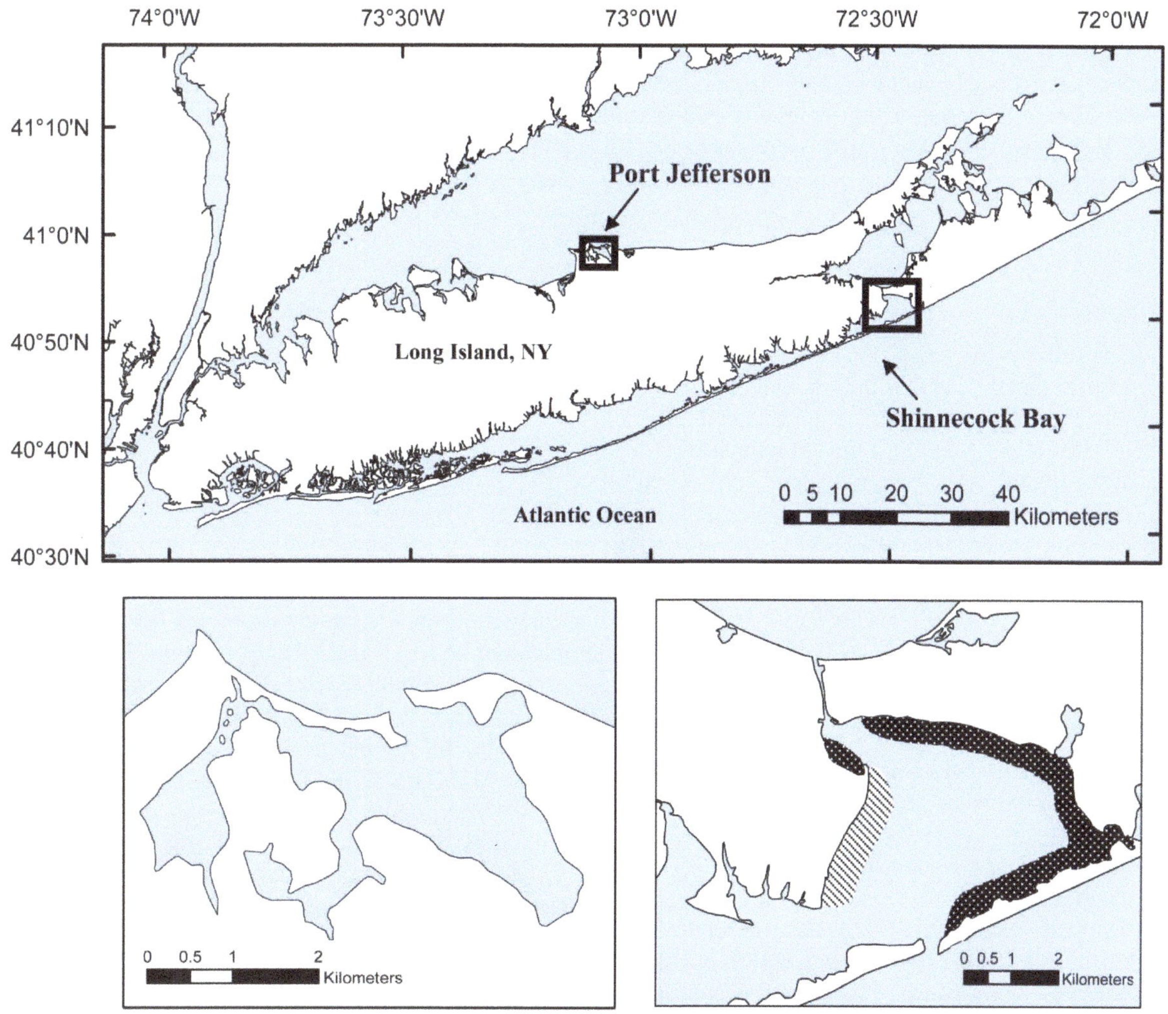

FIGURE 2. Winter Flounder sampling locations in Port Jefferson Harbor (bottom left panel) and Shinnecock Bay (bottom right panel), Long Island, New York. Beam-trawl sampling covered the entire area of Port Jefferson Harbor. For Shinnecock Bay, the gray hatched portion represents the sampling area used in both 2007 and 2008; the dark stippled area indicates the additional area sampled in 2008.

the "descending segment" of relative abundance used in mortality estimation (see description of catch-curve analysis below). However, it was used for results pertaining to beach seining conducted in SB on April 26, May 22, June 5, and June 19, 2007. The smaller seine had a design identical to that of the large net and was assumed to have similar selectivity. It is possible that the shorter offshore reach of the smaller net resulted in the sampling of a smaller proportion of the population; however, results were very similar for the two nets. Beach seining in PJ and SB was conducted every other week during 2007 and monthly during 2008. A single pull of the net was completed at each sampling station; the stations were randomly selected based on 500-m intervals along the perimeter of each bay. All captured Winter Flounder were enumerated, and the TL of each fish was measured to the nearest millimeter.

In 2007, a 1-m beam trawl (6-mm mesh) was utilized to sample known Winter Flounder habitat within SB (Figure 2). Three-minute tows were conducted at stations that were randomly selected from a grid of equal-sized cells. Preliminary analysis of the 2007 SB beam-trawling data suggested that this method provided better survey estimates than beach seining; thus, beam trawling was expanded in 2008 to cover the entire area of PJ and to include nearshore habitat throughout SB (Figure 2). Consequently, seining was reduced to once per month for both bays in 2008. The CPUE was calculated to estimate relative abundance. Effort was defined as one pull of the large seine and as minutes towed for beam trawling. The objective was to successfully sample 7–12 stations/d via seine or beam trawl, but weather and other issues sometimes prevented completion of sampling at the desired number of stations.

Catch-curve analysis was employed to estimate mortality (Quinn and Deriso 1999) of Winter Flounder for each year, bay, and gear type. Values of relative abundance ($\log_e$CPUE) were plotted over time, resulting in a piece-wise linear relationship with two segments. The ascending segment represents increasing selectivity to the gear until full selectivity is reached at the peak. The descending segment beyond the peak represents losses resulting from natural mortality (Quinn and Deriso 1999). Relative abundance data were corrected for selectivity by using

$$s_x = \frac{1}{1+\exp^{-g(l-lh)}},$$

where s_x is selectivity for length interval x; g is the shape parameter; l is fish length; and lh is the length at 50% selectivity. The descending segment of relative abundance (corrected for selectivity) was regressed against time to estimate the slope (instantaneous mortality M). The parameters were estimated by minimizing the sum of squares for all catch data using Solver in Microsoft Excel 2007 (Microsoft Corporation). The selectivity correction was applied to the number of fish in each length-class and was used to estimate relative abundance. Several steps were taken to ensure that spawning and settlement had ended at the time indicated by the peak of the catch-curve (i.e., that no newly recruited age-0 fish were entering the sampling areas after selectivity was reached). First, length frequency data obtained on dates before and after the peak were compared to determine when settlement had ended. Second, fish larger than 10 cm were excluded from early dates (since that size indicates an age greater than 1 year for fish in the Long Island region).

Two growth rate estimates were calculated: one derived from length data and another derived from otolith-based age–length relationships. Growth rates from length data were calculated by taking the difference between average lengths obtained from sampling dates over time,

$$G = \frac{L_2 - L_1}{t_2 - t_1},$$

where L_2 is the average length at time t_2; and L_1 is the average length at time t_1. Growth rates were estimated for intervals between sampling events and were expressed in centimeters per day. Otolith-derived growth rate estimates were obtained from linear regressions of TL against the daily age (days postsettlement) estimated from daily increment counts (Sogard and Able 1992). Slope estimates yielded the growth rate in centimeters per day.

Otolith preparation.—The total numbers of sagittal otoliths processed for aging were 87 and 88 from SB Winter Flounder (2007 and 2008 samples, respectively) and 76 and 82 from PJ fish. Otoliths were dissected and the remaining tissue was removed by using a Nikon Eclipse 80i dissection microscope; the otoliths were then dried for 24 h on glass slides. After drying, cyanoacrylate glue was used to affix each otolith to a clean slide. Otoliths were polished in the sagittal plane with 30- and 3-micron lapping film (3M) until the micro-increments between the secondary primordia (a proxy for settlement; Sogard 1991) and the otolith edge were sufficiently discernible. Postsettlement ages were determined from calibrated photographs of the polished otoliths (Nikon Model DX1200C digital camera attached to a compound microscope) by marking and counting the micro-increments that had formed outward of the secondary primordia (Image Pro-Plus version 6.0; Media Cybernetics). Micro-increments were assumed to form daily (Sogard 1991). Each otolith was read three times, and each reading was given a qualitative measure of reliability ranging from 1 (least confidence in reading) to 5 (most confidence in reading). Within- and between-reader aging precision was estimated using average percent error (APE; Campana and Jones 1992),

$$\text{APE}_j = 100 \cdot \frac{1}{R}\sum_{i=1}^{R}\frac{|X_{ij}-\overline{X}_j|}{\overline{X}_j},$$

where X_{ij} is the ith reading for the jth fish; R is the number of readings; and $\overline{X}_j$ is the mean of readings for the jth fish (Campana 2001). Within-reader APE was calculated based on all otoliths read, and a random subsample of 17 otoliths was examined by an independent reader to estimate between-reader APE (Frisk and Miller 2006).

Statistical analysis of growth and mortality.—Likelihood ratio tests were used to examine the coincidence of linear regressions for mortality and growth models. The tests compared an overall relationship (i.e., with a common slope, intercept, and variance) to a relationship based on individual models for each location and year. The procedure tested the null hypothesis that linear relationships were derived from a common data population, and the test statistic was compared with a chi-square distribution. Significant differences were further evaluated by using Tukey's honestly significant difference (HSD) test. Additionally, mortality and growth rate estimates were compared with previously published values for Winter Flounder.

Settlement date analysis.—Estimates of individual age and daily instantaneous mortality were used to evaluate the strength of settlement throughout the season. The analysis assumes that mortality is constant from the date of settlement until the date of capture; the approximate timing and magnitude of settlement are determined. Abundance at settlement (N_S) was calculated as

$$N_s = N_c \cdot \exp^M,$$

where N_c is the relative abundance at capture; and M is the cumulative instantaneous mortality from settlement to capture. Relative abundance at settlement was estimated by using selectivity-corrected catch values. A modal analysis was performed to determine the modes of cohorts or spawning cycles. The analysis utilized maximum likelihood to fit a multinomial distribution of expected frequencies to the observed settlement frequencies by fitting the mean ordinal date (x), variance (σ^2), and height (h) of the distributions (Hastings and Peacock 1975; Haddon 2001). Modal analysis assumes that each detected mode is normally distributed and corresponds to a separate discrete cohort or settlement group. Both statistical interpretation (Akaike's information criterion [AIC]) and biological interpretation of trends were used as criteria to identify cohorts.

RESULTS

Sampling

During 2007, seining was conducted on a total of 7 d in PJ and on 10 d in SB (Table 1); the number of seining stations per sampling day ranged between 5 and 12 (mean = 8 stations) for PJ and ranged between 2 and 12 (mean = 7 stations) for SB. During 2008, seining was conducted on 5 d in PJ and on 4 d in SB. The number of stations seined per sampling day ranged between 8 and 11 (mean = 10 stations) for PJ and ranged between 6 and 10 (mean = 8 stations) for SB. Length frequencies showed one or more cohorts that were first captured by beach seining at approximately 2–3 cm TL and that reached 7–9 cm TL by autumn (Figure 3).

Beam trawling was conducted in SB during 2007 and in both PJ and SB during 2008. In SB during 2007, beam-trawl samples were collected on 12 d (Table 1); the number of locations per sampling day ranged between 7 and 19 (mean = 16 trawl locations). In 2008, beam trawling was conducted on 9 d in PJ and 10 d in SB. The number of locations per sampling day ranged between 7 and 19 (mean = 16 trawl locations) for PJ and ranged between 6 and 21 (mean = 16 trawl locations) for SB. Length frequencies showed one or more cohorts that were first captured by beam trawling at approximately 2.5–4.5 cm TL and that reached 7–9 cm TL by autumn (Figure 4).

Mortality

Gear selectivity parameters were estimated for beach seining (lh = 4.97; g = 0.95) and beam trawling (lh = 5.05; g = 1.30) and were used to correct abundance estimates. In 2007, new settlement of age-0 fish had stopped and full selectivity had occurred on July 5 and July 3 for seining in PJ and SB, respectively, and on July 2 for beam trawling in SB. In 2008, new settlement of age-0 individuals had ended and full selection to the gear took place on July 8 and July 21 for beam trawling in PJ and SB, respectively. Relative abundance declined at all locations after selection to the gear (Figure 5). Values of M ranged from 0.02 to 0.04 per day (Table 2). Calculation of M was not possible for 2008, as too few age-0 Winter Flounder were caught by beach seines in either PJ or SB. A likelihood ratio test for coincident regressions indicated that mortality was not statistically different between locations and years ($P > 0.459$). Raw and selectivity-corrected relative abundance data are provided in the Supplement (available online).

Growth

The APE for within-reader precision in estimating daily increments was 5.6% ($n = 163$) for 2007 and 4.7% ($n = 170$) for 2008. Between-reader precision was 5.1% ($n = 17$). Significant length–age relationships were found for both locations in both years, and otolith-derived growth rates ranged from 0.05 to 0.06 cm/d (Figure 6; Table 3). A likelihood ratio test of coincident regressions was significant ($P < 0.001$), and multiple comparisons indicated that the otolith-derived growth rate for PJ during 2007 was significantly higher than those for all other year × location combinations (Tukey's HSD test: $P < 0.001$; Table 4). No other differences in otolith growth rate were observed between years or between locations (however, the difference between PJ and SB in 2008 was nearly

TABLE 1. Summary of catch and sampling effort for age-0 Winter Flounder collected from Port Jefferson Harbor (PJ) and Shinnecock Bay (SB), Long Island, New York. The CPUE is expressed as number of individuals per seine haul for beach seine data and number of individuals per minute towed for beam-trawl data; CPUE estimates are based on days when age-0 fish were captured (not on the total number of sampling days).

Gear type	Location	Year	Number of sampling days	Number of tows	Time towed (min)	Total number of age-0 fish	CPUE
Beach seine	PJ	2007	7	89		679	17.42
	PJ	2008	5	48		48	1.92
	SB	2007	10	52.5		202	7.7
	SB	2008	4	32		114	5.88
Beam trawl	PJ	2008	9	163	494	63	0.23
	SB	2007	12	98	352	358	1.53
	SB	2008	10	174	527	68	2.13
Total						1,532	

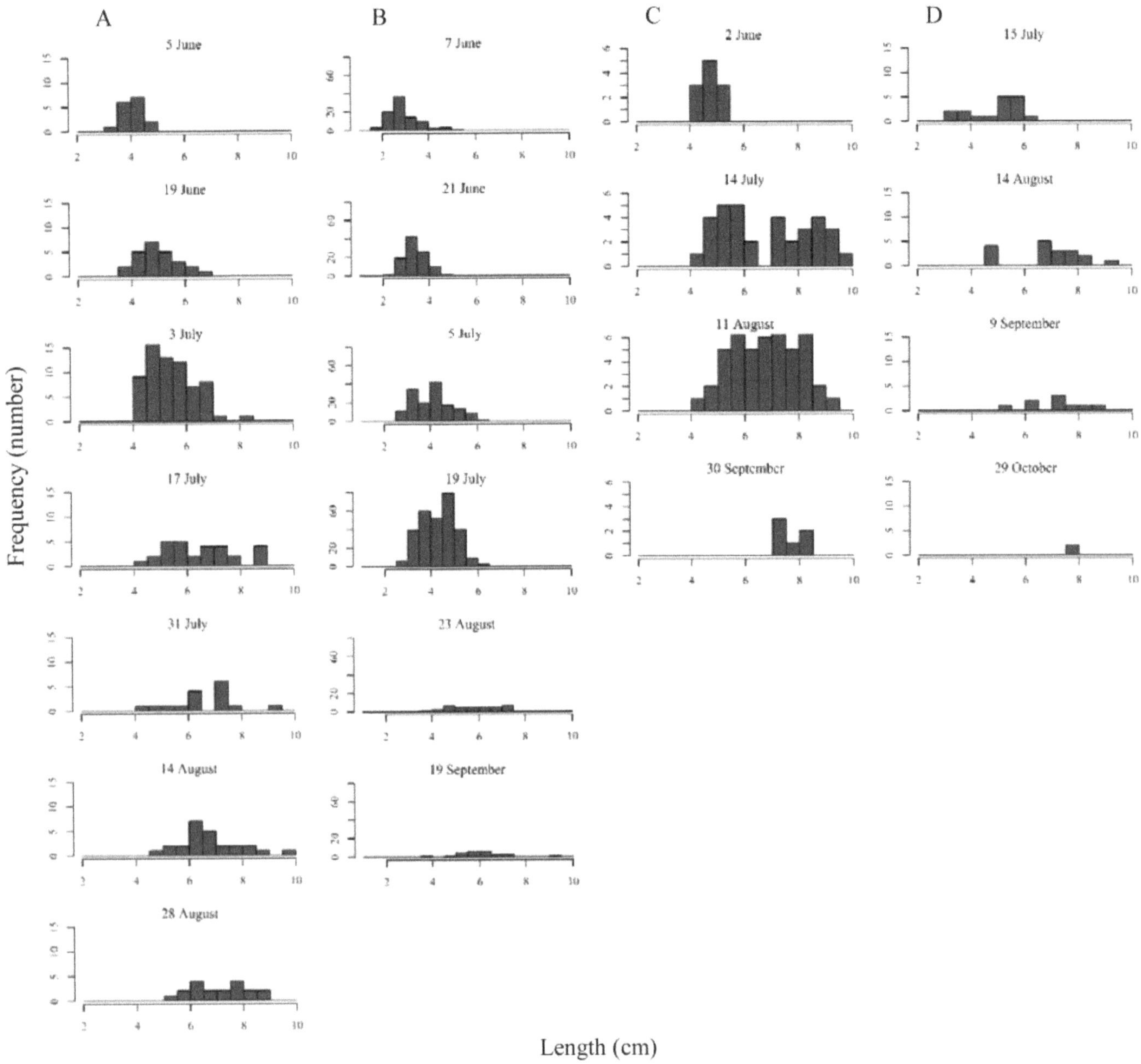

FIGURE 3. Length frequency histograms (cm TL) for age-0 Winter Flounder sampled by beach seining in **(A)** Shinnecock Bay during 2007, **(B)** Port Jefferson Harbor during 2007, **(C)** Shinnecock Bay during 2008, and **(D)** Port Jefferson Harbor during 2008. Note that individuals greater than 10 cm were excluded, as were dates when no Winter Flounder were captured (Port Jefferson Harbor: May 10, 2007, and July 3, 2008; Shinnecock Bay: April 24, April 26, and May 22, 2007).

significant: $P = 0.054$). Field-derived growth rates were highly variable and in some cases unrealistic, ranging from -0.25 to 0.12 cm/d throughout the season (Table 5). As a result, no statistical comparisons were made using field-derived growth estimates.

Settlement Date

Modal analyses of PJ data indicated that mean peak settlement occurred during late April and early May in 2007, with smaller peaks occurring at the end of May and at the end of July (Figure 7A). The three-mode model had the lowest AIC value (i.e., best performance) and matched observed trends more consistently than the two-mode and one-mode models. For PJ in 2008, only a single-mode model was fitted after visual inspection, and settlement was estimated to occur between March and May (Table 6; Figure 8A).

Modal analyses for SB in 2007 indicated that mean peak settlement occurred in late March and April, a smaller peak was present during May, and settlement also took place during

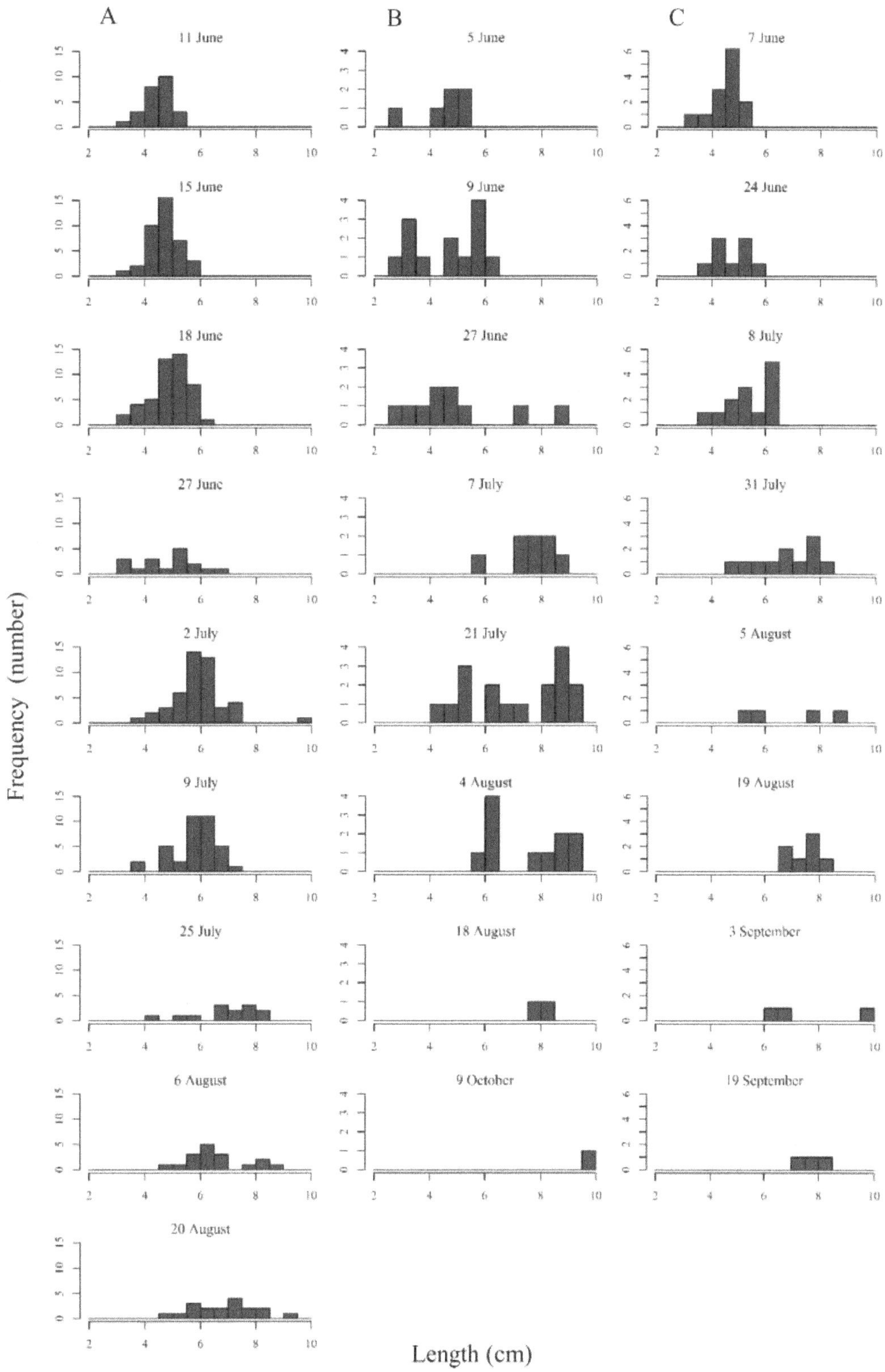

FIGURE 4. Length frequency histograms (cm TL) for age-0 Winter Flounder sampled by beam trawling in **(A)** Shinnecock Bay during 2007, **(B)** Shinnecock Bay during 2008, and **(C)** Port Jefferson Harbor during 2008. Note that individuals greater than 10 cm were excluded, as were dates when no Winter Flounder were captured (Port Jefferson Harbor: May 24, 2008; Shinnecock Bay: June 6, 2007; September 28 and October 25, 2007 [when only two fish were captured]; and April 29 and May 13, 2008).

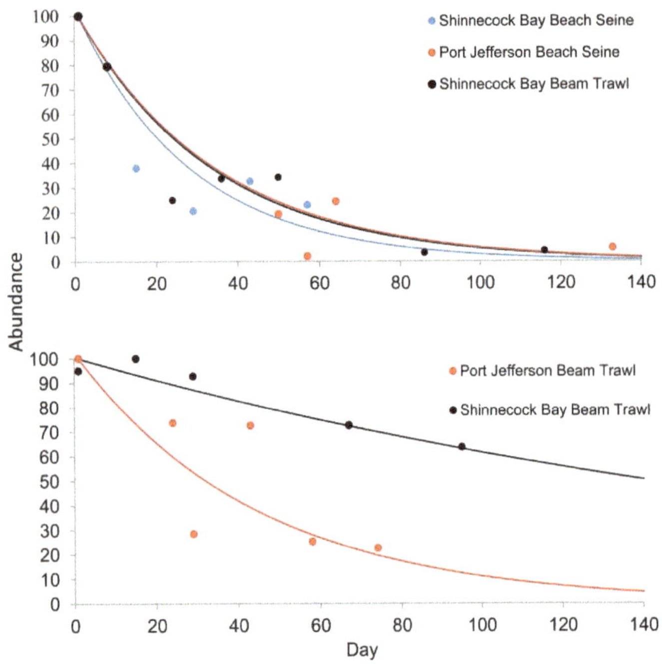

FIGURE 5. Changes in estimated relative abundance during the first 140 d postsettlement for age-0 Winter Flounder in Long Island bays: (upper panel) 2007 and (lower panel) 2008. Using the projection equation $N_t = N_0e^{-Mt}$, the number of fish on day t (N_t) was estimated relative to an initial abundance of 100 individuals (N_0 = day of settlement; M = mortality rate estimated from catch-curves).

July and August (Figure 7B; Table 6). The three-mode model had the lowest AIC in comparison with the two-mode and one-mode models (Table 6). For SB data from 2008, visual inspection suggested that only one mode was likely, so the two- and three-mode models were not attempted. Settlement in 2008 occurred between late February and early May (Table 6; Figure 8B).

Comparison with Published Growth and Mortality Rates

Estimates of Winter Flounder mortality were similar to or within the higher range of estimates from research previously conducted in Connecticut and Rhode Island (Table 5). Growth rate estimates in the present study were within the range of or lower than field- and otolith-derived measures from other studies (Table 5).

DISCUSSION

The population dynamics of age-0 Winter Flounder were investigated in response to the species' dramatic decline in abundance along the coast of Long Island. Mortality rates in the present study were within the range estimated for Winter Flounder in the Connecticut River and New Haven Harbor during the late 1990s (Meise et al. 1999) and also coincided with the upper range of estimated M for Rhode Island

TABLE 2. Statistical analysis of catch-curve results for Winter Flounder sampled from Shinnecock Bay and Port Jefferson Harbor, Long Island (n = number of sampling days; M = daily instantaneous mortality rate; L95% = lower 95% confidence limit of M; U95% = upper 95% confidence limit of M; y = y-intercept).

Gear type	Year	n	M	L95%	U95%	y	r^2	F	P
				Shinnecock Bay					
Beam trawl	2007	7	−0.03	−0.04	−0.02	1,169.95	0.89	36.06	0.002
Beach seine	2007	5	−0.04	−0.06	−0.01	1,406.57	0.90	21.92	0.018
Beam trawl	2008	4	−0.03	−0.07	0.01	1,372.15	0.90	17.19	0.053
				Port Jefferson Harbor					
Beach seine	2007	6	−0.03	−0.07	0.01	1,141.09	0.52	4.26	0.108
Beam trawl	2008	6	−0.02	−0.04	−0.01	884.98	0.68	8.77	0.041

TABLE 3. Otolith-derived growth rates (G, cm/d; i.e., slope value) for Winter Flounder sampled from Shinnecock Bay and Port Jefferson Harbor, Long Island (n = number of observations; L95% = lower 95% confidence limit of G; U95% = upper 95% confidence limit of G; y = y-intercept).

Year	n	G	L95%	U95%	y	r^2	F	P
				Shinnecock Bay				
2007	87	0.05	0.05	0.06	1.2	0.64	152.33	<0.001
2008	88	0.05	0.046	0.06	1.1	0.65	162.64	<0.001
				Port Jefferson Harbor				
2007	76	0.06	0.05	0.07	−0.03	0.67	149.57	<0.001
2008	82	0.05	0.04	0.06	1.6	0.69	178.81	<0.001

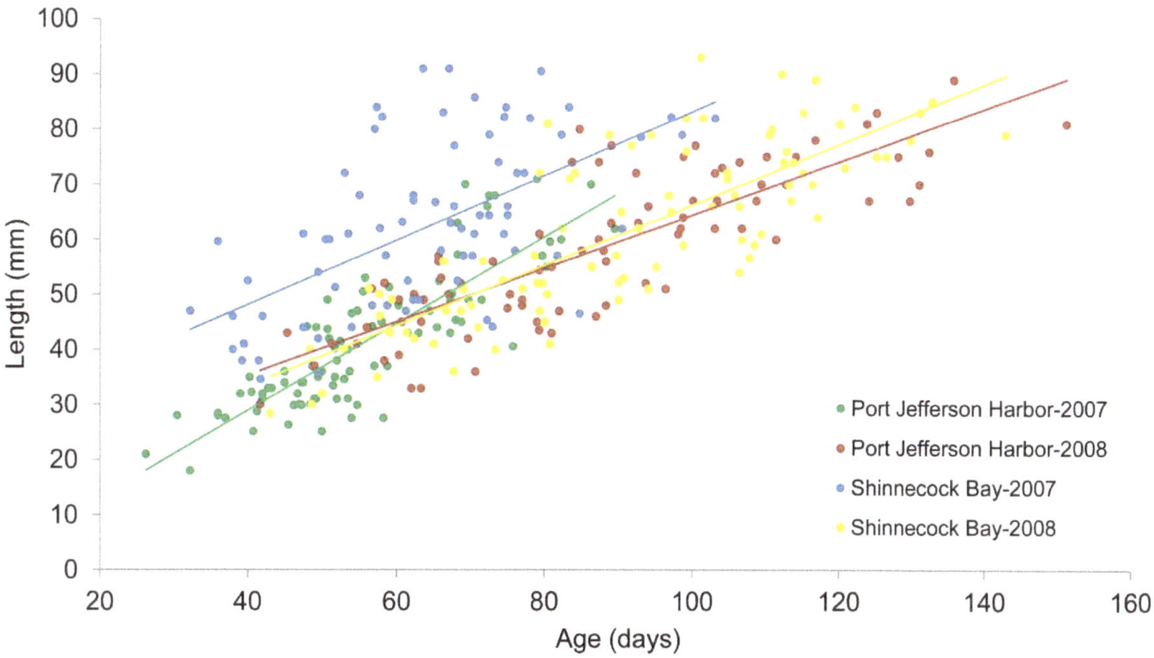

FIGURE 6. Otolith-derived growth rates of age-0 Winter Flounder sampled from Port Jefferson Harbor and Shinnecock Bay, Long Island, during 2007 and 2008. The only significant difference in growth rates was a higher value for Port Jefferson Harbor in 2007 than for all other year × location combinations ($P < 0.001$; see Table 4).

(DeLong et al. 2001) but were much higher than estimates for the Niantic River, Connecticut, and the Mystic River estuary, Connecticut (Pearcy 1962; Rose et al. 1996). In contrast, growth rates of Long Island Winter Flounder were within the range of or lower than estimates reported for other populations. Although our estimates of vital rates were within the range of previous estimates, it is difficult to gauge population health without the availability of historical baselines for Long Island Winter Flounder. Productivity of the Long Island coastal populations may not be high enough to promote recovery; therefore, research addressing a wider range of life history stages is needed.

There is little doubt that Winter Flounder inhabiting Long Island bays are in a depleted state, and a recent genetic analysis has detected inbreeding (O'Leary et al. 2013). For depleted populations, it is particularly pressing to understand which factors might suppress recruitment and therefore which management options are potentially most conducive to stock recovery. Mortality of age-0 Winter Flounder is affected by a suite of environmental variables, including prey availability, density of intraspecific and interspecific competitors, temperature, salinity, dissolved oxygen, pollution, and the presence of physical structures in nursery grounds (Sogard 1991; Bejda et al. 1992; Able et al. 1999; DeLong et al. 2001; Meise et al. 2003). Although many potential impediments to recovery exist, predation on age-0 Winter Flounder is a likely cause of the higher M observed in our study relative to studies in the Niantic River and the Mystic River estuary. Migratory predators originating from other systems can act as subsidized consumers in that population productivity is decoupled from local drivers (sensu Polis et al. 1997). For example, the Striped Bass *Morone saxatilis* is a migratory generalist predator with consumption rates that can exceed local prey resources (Hartman and Margraf 2003). Even low individual predation rates by predators at high abundance can lead to substantial mortality of depleted Winter Flounder populations. Summer Flounder *Paralichthys dentatus* and Bluefish *Pomatomus saltatrix* have been shown to consume age-0 Winter Flounder (Manderson et al. 2000, 2004; Sagarese et al. 2011) and may represent a substantial source of mortality for age-0 fish (Nuttall et al. 2011). Thus, recovered populations of piscivorous predators are potentially impeding the recovery of resident Winter Flounder populations in the Long Island region. Additional research is needed to estimate age-0 Winter Flounder consumption by finfish species

TABLE 4. Results (P-values) from Tukey's honestly significant difference tests comparing otolith-derived growth rates of age-0 Winter Flounder sampled from Port Jefferson Harbor (PJ) and Shinnecock Bay (SB), Long Island, during 2007 and 2008.

Location and year	PJ 2007	SB 2007	PJ 2008
PJ 2007			
SB 2007	<0.001		
PJ 2008	<0.001	0.054	
SB 2008	<0.001	0.963	0.145

TABLE 5. Comparison of daily mortality (*M*) and growth rate (*G*) estimates for age-0 Winter Flounder. To facilitate comparison across studies, values of *M* were converted to daily rates, and values of *G* were converted to centimeters per day.

Study	Location	M (d^{-1})	Field G (cm/d)	Otolith G (cm/d)
Present study, 2007	Shinnecock Bay, New York	0.03–0.04	−0.2	0.05
Present study, 2007	Port Jefferson Harbor, New York	0.03	−0.06	0.06
Present study, 2008	Shinnecock Bay	0.04	−0.25	0.05
Present study, 2008	Port Jefferson Harbor	0.02	0.01–0.12	0.05
Pearcy 1962	Mystic River estuary, Connecticut	0.01		
Meise et al. 1999	Connecticut River and New Haven, Connecticut	0.01–0.03	0.02–0.05	
Meise et al. 2003	Navesink River–Sandy Hook Bay, New Jersey		0.02–0.03	0.1
Rose et al. 1996	Niantic River, Connecticut	0.01		
Delong et al. 2001	Rhode Island	0.01–0.02		
Sogard and Able 1992	New Jersey			0.06–0.12
Sogard et al. 2001	New Jersey			0.03–0.14

in order to confirm or refute predation as the mechanism behind the species' decline and to identify the predators with the greatest impact.

In addition to predation by piscivorous fishes, crustaceans have been documented to prey on age-0 Winter Flounder. Crustacean predators include the sand shrimp *Crangon septemspinosa*, green crab *Carcinus maenas*, and blue crab *Callinectes sapidus* (Taylor 2004, 2005; Collier et al. 2014), the latter of which have experienced an increase in abundance throughout the region (Nuttall et al. 2011). Winter Flounder eggs and small juveniles were detected in the stomachs of sand shrimp sampled from May to early June (Taylor 2004). Taylor (2005) found evidence that green crabs begin to prey on age-0 Winter Flounder in late spring and early summer, possibly accounting for 0.4–7.7% of daily mortality in some systems. Taylor's (2005) results also indicated that 1.1–32.3% of the Winter Flounder year-class could potentially be consumed and that variation in the abundance and size of green crabs coincided with variability in age-0 survival. Thus, any future work should examine predation from both vertebrates and invertebrates.

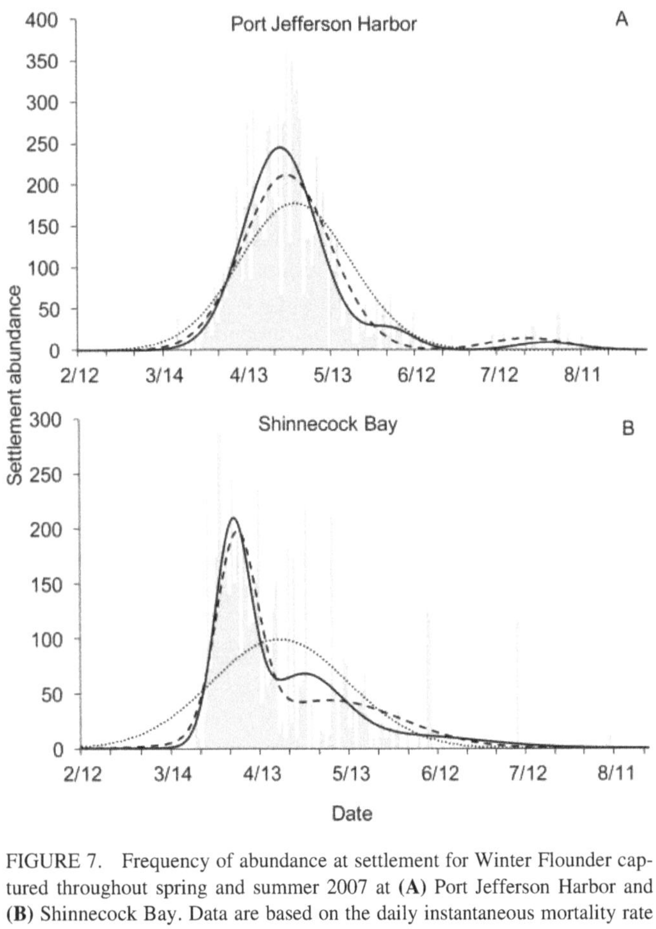

FIGURE 7. Frequency of abundance at settlement for Winter Flounder captured throughout spring and summer 2007 at **(A)** Port Jefferson Harbor and **(B)** Shinnecock Bay. Data are based on the daily instantaneous mortality rate and age–length relationship estimated for each region (solid line = three-mode fit; dashed line = two-mode fit; dotted line = single-mode fit).

Growth rates previously estimated based on length frequency analysis or otolith increment analysis of juvenile Winter Flounder have shown high variation, making it difficult to decipher the relative influence of causal factors, such as habitat quality and temperature (Phelan et al. 2000). Calculations based on length frequency, the pooling of individuals, or both may introduce bias to estimates because they (1) ignore small fluctuations that occur over time or (2) obscure individual variations that may prove important (Miller et al. 1988; Pepin 1993). Periods of starvation or high metabolic costs can lead to losses in length and weight, and increased size-based mortality could generate bias in length-based estimators (Jordaan and Brown 2003). In the case of Winter Flounder, our results indicate a bias due to multiple cohorts of newly settled individuals of a smaller size and younger age, which likely explains the high variability in length-based growth estimates. If discrete cohorts are not easily identifiable or separable in the analyses, otolith approaches and/or repeated measurements of individuals should be used when estimating growth in age-0 Winter Flounder so as to avoid unrealistic results.

TABLE 6. Modal analyses for age-0 Winter Flounder collected from Port Jefferson Harbor and Shinnecock Bay, Long Island, in 2007 and 2008. The mean ordinal date (x), variance (σ^2), and height parameter (h) of the normal distribution are provided along with Akaike's information criterion (AIC) values. Note that two-mode and three-mode models were not fitted for 2008 data.

			Mode		
Model	AIC	Parameter	1	2	3
		Port Jefferson Harbor, 2007			
Three-mode	76,175	x	114.74	153.88	210.45
		σ^2	13.16	8.70	11.10
		h	8,068.26	559.31	262.72
Two-mode	76,923	x	117.01	203.44	
		σ^2	15.77	14.00	
		h	8,378.88	512.53	
One-mode	80,792	x	120.00		
		σ^2	20.00		
		h	8,890.72		
		Port Jefferson Harbor, 2008			
One-mode	6,711	x	111.00		
		σ^2	28.76		
		h	772.26		
		Shinnecock Bay, 2007			
Three-mode	49,280	x	93.48	117.00	152.00
		σ^2	6.02	13.15	25.53
		h	2,948.52	2,089.32	719.38
Two-mode	49,819	x	94.82	126.01	
		σ^2	6.95	24.18	
		h	3,115.69	2,640.70	
One-mode	52,535	x	109.16		
		σ^2	23.12		
		h	5,752.23		
		Shinnecock Bay, 2008			
One-mode	12,903	x	84.14		
		s	21.45		
		h	2,887.04		

Given the assumptions that the survival rate is constant and that settlement occurs at approximately 2–3 months posthatch, the time frame of settlement can be used as a proxy for spawning activity, thereby providing insight into the dynamics of age-0 year-classes (Sogard 1991; Pereira et al. 1999). Thus, Winter Flounder spawning in SB during 2007 likely began in late January and continued until mid-March, exhibiting two (and possibly three) peaks. Similarly, in PJ during 2007, three peak spawning periods were likely. The period from late February to early April comprises the normal spawning window for Winter Flounder in this region (Collette and Klein-MacPhee 2002). However, the potential spawning in late May to early June falls outside the commonly observed time frame for this species (Collette and Klein-MacPhee 2002; Klein-MacPhee 2002). Our results are similar to observations of July settlement (i.e., from an April spawning event) in New Jersey (Meise et al. 2003).

Does the combination of multiple spawning cohorts, the presence of migratory and resident fish, and weak genetic differentiation among Winter Flounder populations support the observation of metapopulations with spatial or temporal overlap? Previous research on Winter Flounder has identified spawning periods lasting from January to April, with the presence of peaks in spawning activity (Stoner et al. 1999), which could explain the observed trends in settlement distributions. However, as additional evidence builds on observations that Winter Flounder exhibit natal spawning groups (Lobell 1939; Perlmutter 1947; Saila 1961) and local genetic structure (Buckley et al. 2008; O'Leary et al. 2013), the underlying mechanisms maintaining the population structure will have to be considered. Future research encompassing adult spawning, migratory behavior, stock structure, duration of the larval period, and settlement timing will be required to unravel the complex behavior of Winter Flounder (Frisk et al. 2014).

The regional stock structure of Winter Flounder has been studied (McBride 2014), but there remains a large amount of uncertainty regarding local life history diversity, including

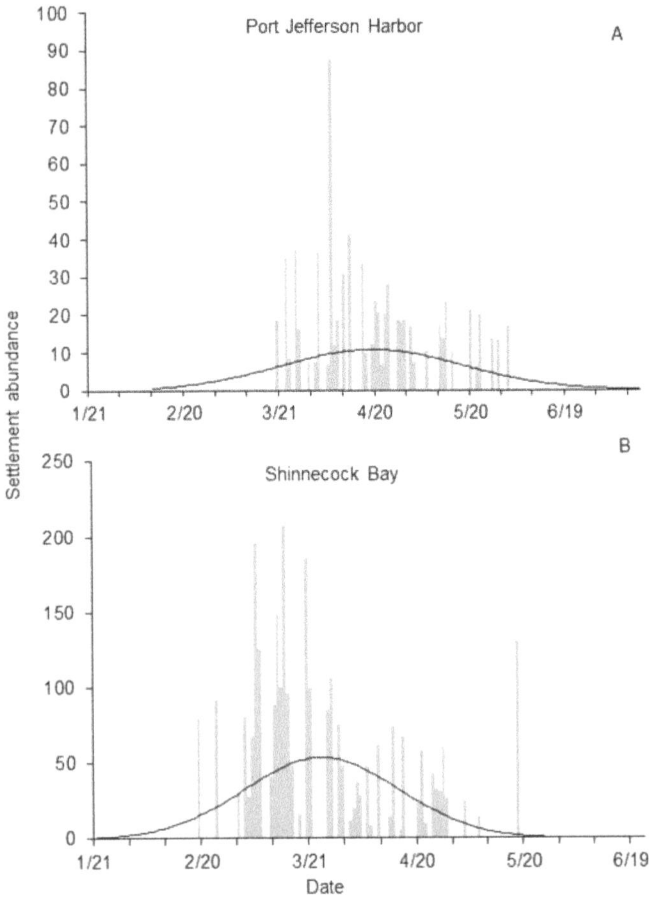

FIGURE 8. Frequency of abundance at settlement for Winter Flounder captured throughout spring and summer 2008 at **(A)** Port Jefferson Harbor and **(B)** Shinnecock Bay. Data are based on the daily instantaneous mortality rate and age–length relationship estimated for each region.

population, demographic, and genetic connectivity (Frisk et al. 2014). This paucity of information means that the extent to which Winter Flounder in Long Island bays are connected to ocean stocks is unknown. The difference in population status between the offshore stocks (at 9% of target levels; NEFSC 2008) and the Long Island populations that have experienced inbreeding suggests a degree of isolation between these two contingents (Secor 1999). Research is needed to link local productivity to competing drivers of density-dependent compensation and predation-based mortality in Winter Flounder. Finally, the conflicting genetic evidence for local population structure (Crivello et al. 2004; Buckley et al. 2008; O'Leary et al. 2013) and against such structure (Wirgin et al. 2014) in coastal Winter Flounder must be reconciled.

ACKNOWLEDGMENTS

This project was made possible by funding from the New York State Department of Environmental Conservation. We also thank Skyler Sagarese, Matthew Nuttall, Amanda Steen, Adam Uhlich, Christopher Martinez, Carolyn Hall, and Keith Dunton for assistance in the extensive field sampling required for this study. We are grateful to Stephanie Vos, Rowena Eng, and Kristin Oppito for help in dissecting age-0 Winter Flounder for otolith collection and to Kestrel Perez for technical assistance and otolith methodology development. We appreciate the assistance of Captain Mark Wiggins and Captain Dave Bowman as well as Richard McIntyre, Brian Gagliardi, Don Getz, and Melanie Meade at the Southampton Marine Station.

REFERENCES

Able, K. W., J. P. Manderson, and A. L. Studholme. 1999. Habitat quality for shallow-water fishes in an urban estuary: the effects of man-made structures on growth. Marine Ecology Progress Series 187:227–235.

Bejda, A. J., B. A. Phelan, and A. L. Studholme. 1992. The effect of dissolved oxygen on the growth of young-of-the-year Winter Flounder, *Pseudopleuronectes americanus*. Environmental Biology of Fishes 34:321–327.

Brown, R., and W. Gabriel. 1998. Winter Flounder. NOAA Technical Memorandum NMFS-NE-115:81–84.

Buckley, L., J. S. Collie, L. E. A. Kaplan, and J. Crivello. 2008. Winter Flounder larval genetic population structure in Narragansett Bay, R.I.: recruitment to juvenile young-of-the-year. Estuaries and Coasts 31:745–754.

Campana, S. E. 2001. Accuracy, precision and quality control in age determination, including a review of the use and abuse of age validation methods. Journal of Fish Biology 59:197–242.

Campana, S. E., and C. M. Jones. 1992. Analysis of otolith microstructure date. Canadian Special Publication of Fisheries and Aquatic Sciences 117:73–100.

Collette, B. B., and G. Klein-MacPhee. 2002. Bigelow and Schroeder's fishes of the Gulf of Maine, 3rd edition. Smithsonian Institution Press, Washington, D.C.

Collier, J. L., S. P. Fitzgerald, L. A. Hice, M. G. Frisk, and A. E. McElroy. 2014. A new PCR-based method shows that blue crabs (*Callinectes sapidus* (Rathbun)) consume Winter Flounder (*Pseudopleuronectes americanus* (Walbaum)). PLoS (Public Library of Science) One [online serial] 9(1): e85101.

Crivello, J. F., D. J. Danila, E. Lorda, M. Keser, and E. F. Roseman. 2004. The genetic stock structure of larval and juvenile Winter Flounder larvae in Connecticut waters of eastern Long Island Sound and estimations of larval entrainment. Journal of Fish Biology 65:62–76.

DeCelles, G. R., and S. X. Cadrin. 2010. Movement patterns of Winter Flounder (*Pseudopleuronectes americanus*) in the southern Gulf of Maine: observations with the use of passive acoustic telemetry. U.S. National Fisheries Service Fishery Bulletin 108:408–419.

DeLong, A. K., J. S. Collie, C. J. Meise, and J. C. Powell. 2001. Estimating growth and mortality of juvenile Winter Flounder, *Pseudopleuronectes americanus*, with a length-based model. Canadian Journal of Fisheries and Aquatic Sciences 58:2233–2246.

Fairchild, E. A., and W. H. Howell. 2000. Predator–prey size relationship between *Pseudopleuronectes americanus* and *Carcinus maenas*. Journal of Sea Research 44:81–90.

Frisk, M. G., A. Jordaan, and T. J. Miller. 2014. Moving beyond the current paradigm in marine population connectivity: are adults the missing link? Fish and Fisheries 15:242–254.

Frisk, M. G., and T. J. Miller. 2006. Age, growth, and latitudinal patterns of two Rajidae species in the northwestern Atlantic: Little Skate (*Leucoraja erinacea*) and Winter Skate (*Leucoraja ocellata*). Canadian Journal of Fisheries and Aquatic Sciences 63:1078–1091.

Haddon, M. 2001. Modelling and quantitative methods in fisheries. Chapman and Hall/CRC Press, New York.

Hartman, K. J., and F. J. Margraf. 2003. U.S. Atlantic coast Striped Bass: issues with a recovered population. Fisheries Management and Ecology 10:309–312.

Hastings, N. A. J., and J. B. Peacock. 1975. Statistical distributions. Butterworth, London.

Jordaan, A., and J. A. Brown. 2003. The risk of running on empty: the influence of age on starvation and gut fullness in larval Atlantic Cod (*Gadus morhua*). Canadian Journal of Fisheries and Aquatic Sciences 60:1289–1298.

Keller, A. A., and G. Klein-MacPhee. 2000. Impact of elevated temperature on the growth, survival, and trophic dynamics of Winter Flounder larvae: a mesocosm study. Canadian Journal of Fisheries and Aquatic Sciences 57:2382–2392.

Klein-MacPhee, G. 2002. Righteye flounders: Family Pleuronectidae. Pages 560–587 *in* B. B. Collette and G. Klein-MacPhee, editors. Bigelow and Schroeder's fishes of the Gulf of Maine, 3rd edition. Smithsonian Institution Press, Washington, D.C.

Lobell, M. J. 1939. A biological survey of the salt waters of Long Island, 1938. Report on certain fishes: Winter Flounder (*Pseudopleuronectes americanus*). *In* Supplement to the 28th annual report of the New York State Conservation Department, part 1. New York State Conservation Department, Albany.

Manderson, J. P., J. Pessutti, J. G. Hilbert, and F. Juanes 2004. Shallow-water predation risk for a juvenile flatfish (Winter Flounder; *Pseudopleuronectes americanus*, Walbaum) in a northwest Atlantic estuary. Journal of Experimental Marine Biology and Ecology 30:137–157.

Manderson, J. P., B. A. Phelan, A. W. Stoner, and J. Hilbert. 2000. Predator–prey relations between age-1+ Summer Flounder (*Paralichthys dentatus*, Linnaeus) and age-0 Winter Flounder (*Pseudopleuronectes americanus*, Walbaum): predator diets, prey selection, and effects of sediments and macrophytes. Journal of Experimental Marine Biology and Ecology 251:17–39.

McBride, R. S. 2014. Managing a marine stock portfolio: stock identification, structure, and management of 25 fishery species along the Atlantic coast of the United States. North American Journal of Fisheries Management 34:710–734.

Meise, C., J. S. Collie, J. Widman, and P. Howell. 1999. Growth and mortality of juvenile Winter Flounder in two New England estuaries. Estuaries 22:297–303.

Meise, C. J., D. L. Johnson, L. L. Stehlik, J. Manderson, and P. Shaheen. 2003. Growth rates of juvenile Winter Flounder under varying environmental conditions. Transactions of the American Fisheries Society 132:335–345.

Miller, T. J., L. B. Crowder, J. A. Rice, and E. A. Marschall. 1988. Larval size and recruitment mechanisms in fishes: toward a conceptual framework. Canadian Journal of Fisheries and Aquatic Sciences 45:1657–1670.

NEFSC (Northeast Fisheries Science Center). 2008. Assessment of 19 northeast groundfish stocks through 2007. Report of the third groundfish assessment review meeting (GARM III), NEFSC, Reference Document 08-15, Woods Hole, Massachusetts.

NOAA (National Oceanic and Atmospheric Administration). 2012. Recreational fisheries statistics database. Available: http://www.st.nmfs.noaa.gov/recreational-fisheries/access-data/run-a-data-query/index. (June 2012).

Nuttall, M. A., A. Jordaan, R. M. Cerrato, and M. G. Frisk. 2011. Identifying 120 years of decline in ecosystem structure and maturity of Great South Bay, New York, using the Ecopath modelling approach. Ecological Modelling 222:3335–3345.

O'Leary, S. J., L. A. Hice, K. A. Feldheim, M. G. Frisk, A. E. McElroy, M. D. Fast, and D. D. Chapman. 2013. Severe inbreeding and small effective number of breeders in a formerly abundant marine fish. PLoS (Public Library of Science) One [online serial] 8(6):e66126.

Pearcy, W. G. 1962. Ecology of an estuarine population of Winter Flounder, *Pseudopleuronectes americanus* (Walbaum), parts I–IV. Bulletin of the Bingham Oceanographic Collection Yale University 18:5–78.

Pepin, P. 1993. An appraisal of the size-dependent mortality hypothesis for larval fish: comparison of a multispecies study with an empirical review. Canadian Journal of Fisheries and Aquatic Sciences 50:2166–2174.

Pereira, J. J., R. Goldberg, J. J. Ziskowski, P. L. Berrien, W. W. Morse, and D. L. Johnson. 1999. Essential fish habitat source document: Winter Flounder, *Pseudopleuronectes americanus*, life history and habitat characteristics. NOAA Technical Memorandum NMFS-NE-138.

Perlmutter, A. 1947. The Blackback Flounder and its fishery in New England and New York. Bulletin of the Bingham Oceanographic Collection Yale University 11.

Phelan, B. A., R. Goldberg, A. J. Bejda, J. Pereira, S. Hagan, P. Clark, A. L. Studholme, A. Calabrese, and K. W. Able. 2000. Estuarine and habitat-related differences in growth rates of young-of-the-year Winter Flounder (*Pseudopleuronectes americanus*) and Tautog (*Tautoga onitis*) in three northeastern U.S. estuaries. Journal of Experimental Marine Biology and Ecology 247:1–28.

Polis, G. A., W. B. Anderson, and R. D. Holt. 1997. Toward an integration of landscape and food web ecology: the dynamics of spatially subsidized food webs. Annual Review of Ecology and Systematics 28:289–316.

Quinn, T. J., and R. B. Deriso. 1999. Quantitative fish dynamics. Oxford University Press, New York.

Rose, K. A., J. A. Tyler, R. C. Chambers, G. Klein-MacPhee, and D. J. Danila. 1996. Simulating Winter Flounder population dynamics using coupled individual-based young-of-the-year and age-structured adult models. Canadian Journal of Fisheries and Aquatic Sciences 53:1071–1091.

Sagarese, S. R. 2009. Inshore movements, residency, and abundance of adult Winter Flounder, *Pseudopleuronectes americanus*, and piscivorous predation on young-of-the-year Winter Flounder within coastal bays of Long Island and an investigation on the effect of temperature and photoperiod on vertebral band deposition in Little Skate *Raja erinacea*. Master's thesis. Stony Brook University, Stony Brook, New York.

Sagarese, S. R., R. M. Cerrato, and M. G. Frisk. 2011. Diet composition and feeding habits of common fishes in Long Island bays, New York. Northeastern Naturalist 18:291–314.

Sagarese, S. R., and M. G. Frisk. 2011. Movement patterns and residence of adult Winter Flounder within a Long Island estuary. Marine and Coastal Fisheries: Dynamics, Management, and Ecosystem Science [online serial] 3:295–306.

Saila, S. B. 1961. A study of Winter Flounder movements. Limnology and Oceanography 6:292–298.

Scharf, F. S., J. P. Manderson, and M. C. Fabrizio. 2006. The effects of seafloor habitat complexity on survival of juvenile fishes: species-specific interactions with structural refuge. Journal of Experimental Biology and Ecology 335:167–176.

Secor, D. H. 1999. Specifying divergent migrations in the concept of stock: the contingent hypothesis. Fisheries Research 43:13–34.

Socrates, J. B., and G. C. Colvin. 2006. A study of the Striped Bass in the marine district of New York State. New York State Department of Environmental Conservation, Anadromous Fish Conservation Act Project AFC-33, Albany.

Sogard, S. M. 1991. Interpretation of otolith microstructure in juvenile Winter Flounder (*Pseudopleuronectes americanus*): ontogenetic development, daily increment validation, and somatic growth relationships. Canadian Journal of Fisheries and Aquatic Sciences 48:1862–1871.

Sogard, S. M., and K. W. Able. 1992. Growth variation of newly settled Winter Flounder (*Pseudopleuronectes americanus*) in New Jersey estuaries as determined by otolith microstructure. Netherlands Journal of Sea Research 29:163–172.

Stoner, A. W., A. J. Bejda, J. P. Manderson, B. A. Phelan, L. L. Stehlik, and J. P. Pessutti. 1999. Behavior of Winter Flounder, *Pseudopleuronectes americanus*, during the reproductive season: laboratory and field observations on spawning, feeding, and locomotion. U.S. National Marine Fisheries Service Fishery Bulletin 97:999–1016.

Taylor, D. L. 2004. Immunological detection of Winter Flounder (*Pseudopleuronectes americanus*) eggs and juveniles in the stomach content of crustacean predators. Journal of Experimental Marine Biology and Ecology 301:55–73.

Taylor, D. L. 2005. Predatory impact of the green crab (*Carcinus maenas* Linnaeus) on post-settlement Winter Flounder (*Pseudopleuronectes americanus* Walbaum) as revealed by immunological dietary analysis. Journal of Experimental Marine Biology and Ecology 324:112–126.

Van Guelpen, L., and C. C. Davis. 1979. Seasonal movements of the Winter Flounder, *Pseudopleuronectes americanus*, in two contrasting inshore locations in Newfoundland. Transactions of the American Fisheries Society 108:26–37.

Wirgin, I., L. Maceda, C. Grunwald, N. K. Roy, and J. R. Waldman. 2014. Coastwide stock structure of Winter Flounder using nuclear DNA analyses. Transactions of the American Fisheries Society 143:240–251.

Yencho, M. A. 2009. Abundance, mortality, age and growth of young-of-the-year Winter Flounder (*Pseudopleuronectes americanus*) in two locations on Long Island. Master's thesis. Stony Brook University, Stony Brook, New York.

15

Linking Behavior, Physiology, and Survival of Atlantic Salmon Smolts During Estuary Migration

Daniel S. Stich*

Department of Wildlife, Fisheries, and Conservation Biology, University of Maine, 5755 Nutting Hall, Orono, Maine 04469, USA

Gayle B. Zydlewski

School of Marine Sciences, University of Maine, 5706 Aubert Hall, Orono, Maine 04469, USA

John F. Kocik

National Oceanic and Atmospheric Administration, Northeast Fisheries Science Center, 17 Godfrey Drive, Suite 1, Orono, Maine 04473, USA

Joseph D. Zydlewski

U.S. Geological Survey, Maine Cooperative Fish and Wildlife Research Unit, University of Maine, 5755 Nutting Hall, Orono, Maine 04469, USA

Abstract

Decreased marine survival is identified as a component driver of continued declines of Atlantic Salmon *Salmo salar*. However, estimates of marine mortality often incorporate loss incurred during estuary migration that may be mechanistically distinct from factors affecting marine mortality. We examined movements and survival of 941 smolts (141 wild and 800 hatchery-reared fish) released in freshwater during passage through the Penobscot River estuary, Maine, from 2005 to 2013. We related trends in estuary arrival date, movement rate, and survival to fish characteristics, migratory history, and environmental conditions in the estuary. Fish that experienced the warmest thermal history arrived in the estuary 8 d earlier than those experiencing the coolest thermal history during development. Estuary arrival date was 10 d later for fish experiencing high flow than for fish experiencing low flow. Fish released furthest upstream arrived in the estuary 3 d later than those stocked further downstream but moved 0.5 km/h faster through the estuary. Temporally, movement rate and survival in the estuary both peaked in mid-May. Spatially, movement rate and survival both decreased from freshwater to the ocean. Wild smolts arrived in the estuary later than hatchery fish, but we observed no change in movement rate or survival attributable to rearing history. Fish with the highest gill Na^+, K^+-ATPase activity incurred 25% lower mortality through the estuary than fish with the lowest gill Na^+, K^+-ATPase activity. Smolt survival decreased (by up to 40%) with the increasing number of dams passed (ranging from two to nine) during freshwater migration. These results underscore the importance of physiological preparedness on performance and the delayed, indirect effects of dams on survival of Atlantic Salmon smolts during estuary migration, ultimately affecting marine survival estimates.

Subject editor: Carl Walters, University of British Columbia, Vancouver

*Corresponding author: daniel.stich@maine.edu

The links between the freshwater experience of anadromous fishes and marine survival are poorly understood in general (McCormick et al. 2009). Information about these relationships could have timely implications for the management and conservation of fisheries. Elevated marine mortality in recent years, due in large part to changes in ocean climate (Friedland et al. 2003; Mills et al. 2013; Friedland et al. 2014), is thought to be a driver behind the recovery failure of many endangered populations of Atlantic Salmon *Salmo salar* (Chaput et al. 2005; Chaput 2012; Miller et al. 2012; Mills et al. 2013; Lacroix 2014). However, high mortality during migration through freshwater and estuarine corridors (Thorstad et al. 2012b; Hayes and Kocik 2014) also likely contributes to reduced population sizes (Parrish et al. 1998). Estimates of marine survival also often include estuarine mortality due to the difficulty in separating these processes (Friedland 1998). A better understanding of factors that influence estuary mortality could help to enhance the management of Atlantic Salmon stocks in the face of changing ocean climates (Mills et al. 2013). Despite the importance of estuarine habitats during migration, little is known about Atlantic Salmon smolt behavior and survival in North American estuaries compared with in freshwater and nearshore marine environments (Weitkamp et al. 2014).

The transition of Atlantic Salmon smolts to saltwater is recognized as a period of high mortality in estuaries (Lacroix 2008; Kocik et al. 2009; Dempson et al. 2011) and fjords (Gudjonsson et al. 2005; Svenning et al. 2005; Thorstad et al. 2012b). This period is marked by high predation risk (Hvidsten and Lund 1988; Kocik et al. 2009; Hawkes et al. 2013), physiological stresses (Handeland et al. 1997), and novel environmental conditions (McCormick et al. 1998). In response to these challenges, the process of smolting involves a synchronous suite of changes in physiology, morphology, and behavior that, in concert, enhance the probability of survival following successful saltwater entry (McCormick et al. 1998). Therefore, the seasonal timing of smolt runs is important (McCormick et al. 1998; Thorstad et al. 2012b).

Smolt survival during estuary passage is theoretically maximized by matching migration timing and animal characteristics to predictable annual environmental conditions, along with the presence or absence of predators (Kocik et al. 2009) and sympatric prey buffers (Svenning et al. 2005) during a brief annual period of several weeks, characterizing the "ecological smolt window" (McCormick et al. 1998). To match this ecological window, smolts experience a concurrent period of peak physiological preparedness for saltwater entry, the "physiological smolt window" (McCormick et al. 1998). Gill Na^{+}, K^{+}-ATPase (gill NKA; enzyme code 3.6.3.9; IUBMB 1992) activity in Atlantic Salmon is one enzyme measurement that has been found to be a useful indicator of smolt development and preparedness for saltwater entry (Zaugg 1982; McCormick et al.1987, 1989). During smoltification, gill NKA activity peaks during the spring, resulting in a period of increased saltwater tolerance (Duston and Saunders 1990; McCormick 2013). Evidence relating smolt survival in the wild to gill NKA activity is conspicuously absent from this literature, and long-term performance (e.g., growth) is not clearly linked to higher gill NKA activity (Zydlewski and Zydlewski 2012).

The natural timing of estuary arrival, movement rates through estuaries, and ultimately smolt survival during estuary migration are likely affected by the presence of dams in freshwater along smolt migration routes. Estuarine mortality that occurs naturally from causes such as predation and physiological challenges (Blackwell et al. 1997; Handeland et al. 1997; Halfyard et al. 2013) can be exacerbated by anthropogenic influences, such as passage through dams in freshwater. These influences might occur through migratory delay (Keefer et al. 2012), increased predation (Poe et al. 1991; Blackwell and Juanes 1998), physical injuries (Stier and Kynard 1986; Mathur et al. 2000) that can result in physiological impairment (Zydlewski et al. 2010), and reduced survival during estuary passage. However, the presence and magnitude of dam-related estuary mortality (i.e., delayed or indirect effects of dams) during estuary passage by Atlantic Salmon remains uncertain (Stich et al. 2014). Information about dam-related estuary mortality could have important implications for recovery activities involving Atlantic Salmon stocks, including proposals or actions related to dam removals, such as those occurring in the Kennebec and Penobscot rivers in Maine (Day 2006).

The goal of this study was to quantify movement behavior and survival of Atlantic Salmon smolts in the Penobscot River estuary and their potential drivers in the 2005–2013 period. We classified these drivers into three main categories: (1) individual fish characteristics (fork length, mass, condition factor, and rearing history), (2) migratory history (number of dams passed, release distance from ocean, migratory route), and (3) environmental variability (cumulative temperature experienced, discharge, and photoperiod). Our first objective was to relate estuary arrival date and movement rate of smolts to individual fish characteristics, migratory history, and environmental variability from 2005 through 2013. The second objective of this study was to estimate survival of Atlantic Salmon smolts during migration through the estuary using Cormack–Jolly–Seber (CJS) mark–recapture models. These estimates were used to quantify the effects of fish behavior (estuary arrival date and movement rate), individual fish characteristics, migratory history of fish, and environmental variability on survival.

STUDY SITE

The Penobscot River (Figure 1) is the largest watershed within Maine, encompassing approximately 22,000 km^2.

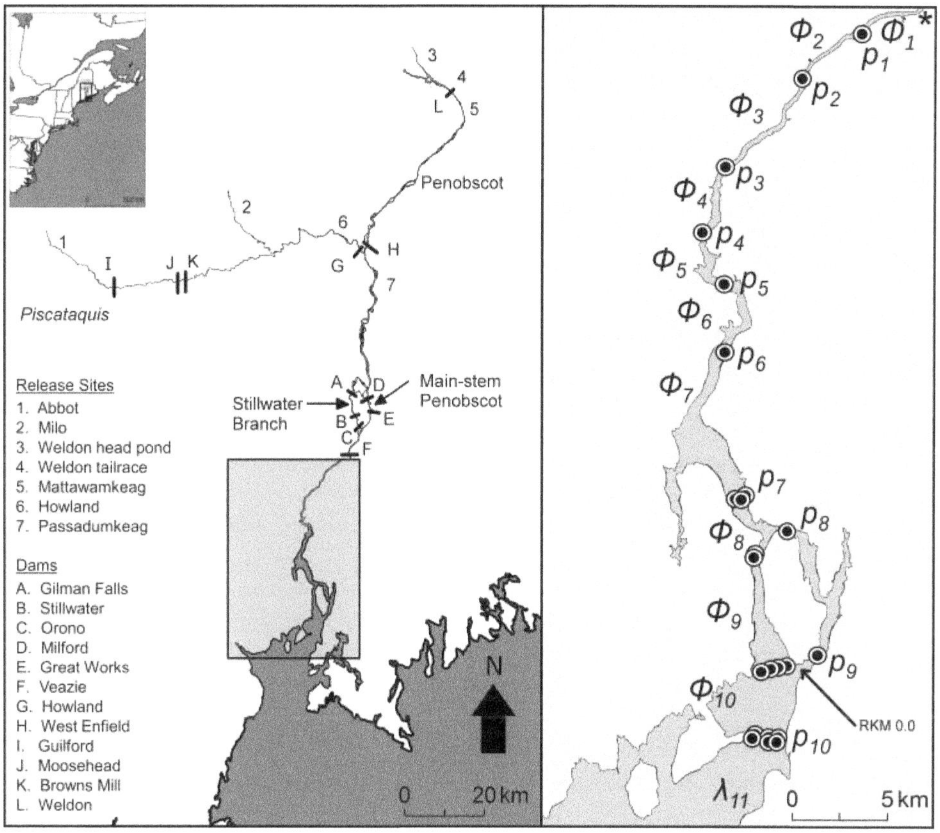

FIGURE 1. Map of the Penobscot River watershed in Maine, showing the locations of tributaries, dams, and release sites in the Penobscot River (left panel). The right panel shows the locations of acoustic receivers used to detect Atlantic Salmon smolts in the estuary. Parameters associated with Cormack–Jolly–Seber survival models are as follows: detection probability at each location following release (p_i), apparent survival within reaches between locations (ϕ_i), and λ_{11} (product of ϕ_{11}in the final reach and p_{11} at the final receiver location). There were 40 receivers in the bay that are not shown. The asterisk indicates the point of virtual release in the estuary for Cormack–Jolly–Seber models.

Atlantic Salmon have been stocked throughout the watershed at egg, fry, parr, and smolt life stages during the last several decades, and limited wild spawning occurs (USASAC 2014). The proportion of the Penobscot River smolt run made up of stocked fish is unknown, but sampling in the Penobscot River bay suggests that more than 90% of the run results from smolt stocking (Sheehan et al. 2011). The majority of the annual adult run (about 83%) is also predominantly composed of hatchery-stocked smolts (USASAC 2012). Hatchery stocking generally occurs less than 160 km above the mouth of the estuary in the Penobscot River and its tributaries (see Figure 1). The peak of the smolt emigration from wild rearing sites occurs between late April and early May most years (USASAC 2014).

The migratory pattern of individual smolts in the Penobscot River depends on rearing history, stocking locations, migratory routing through the lower river, and interannually varying hydropower operations throughout the catchment. All smolts stocked or reared in the upper reaches of the catchment enter the main stem of the Penobscot River at river kilometer (rkm) 100 (measured from the mouth of the estuary; rkm 0 is shown in the right panel of Figure 1), either by passing Howland Dam (G in Figure 1) from the Piscataquis River or by passing West Enfield Dam (H in Figure 1) from the upper Penobscot River. Upon reaching the lower freshwater portion of the Penobscot River (rkm 60), smolts can use one of two migratory paths around a large island: the main-stem Penobscot River to the east (88% of smolts) or the Stillwater Branch (12% of smolts) to the west (Figure 1). On the eastern side of the island (main-stem Penobscot), smolts passed two dams until the removal of the Great Works Dam (E in Figure 1) in 2012. Concurrently, hydropower production was increased at two dams (Stillwater and Orono) in the Stillwater Branch to the west. The details of changes to operations of the Stillwater (B in Figure 1) and Orono (C in Figure 1) dams are described in Stich et al. (2014). Briefly, hydropower generation was approximately doubled at each of these two dams following the addition of a second powerhouse at each facility in spring 2014. On the western side of the island (Stillwater Branch), smolts passed three dams through which survival was high relative to the dams in the main stem during 2005–2013 prior to changes in hydroelectric generation. This resulted in a cumulative 10%

higher survival for smolts moving through the Stillwater Branch compared with the main-stem route in this area (Holbrook et al. 2011; Stich et al. 2014). Finally, smolts traveling either route would then pass the Veazie Dam, at the head of tide, until 2013, when that dam was removed. Dependent upon stocking location, migratory route through the lower Penobscot River, and year of stocking (because of Great Works Dam removal), smolts stocked in freshwater may have passed two to nine dams before entering the estuary. This study occurred prior to the removal of Veazie Dam (F in Figure 1).

The Penobscot River estuary spans 45 km from the mouth of the estuary to the head of tide, which coincides with the former location of Veazie Dam (F in Figure 1). The upper estuary (from p_1 to p_6 in Figure 1) is tidally influenced but uniformly freshwater (Imhoff and Harvey 1972), while the middle estuary (between p_6 and p_{10} in Figure 1) is characterized by mixing of fresh and salt water (Seiwell 1932; Stich et al., in review), and the lower estuary is physically and chemically stratified (Imhoff and Harvey 1972) with low-salinity (10‰) water occupying only the first 1–2 m of the water column (Haefner 1967).

METHODS

Acoustic receiver array.—An array of stationary VR2 and VR2-W acoustic receivers (Amirix Vemco, Halifax, Nova Scotia) was deployed in the river, estuary, and bay of the Penobscot River prior to the start of the Atlantic Salmon smolt run each year of the study (Figure 1). All receivers monitored continuously on a frequency of 69 kHz using omnidirectional hydrophones and were moored to the bottom of the estuary on cement anchors. Where deemed necessary (wide river sections or reaches containing obstructions such as islands), multiple receivers were deployed in a single location to achieve coverage over the entire width of the channel. Detections at all receivers within a discrete cross section were pooled as a single site for data analysis. A total of 11 acoustic receiver locations within the estuary (and several in the bay) were common to all arrays of the present study (2005–2006, 2009–2013). During all years, fish detections at all (about 40) receivers in the bay (downstream of Fort Point: p_{10} in Figure 1) were pooled as a single, final detection event for all analyses.

Fish measurements, acoustic tagging, and releases.—From 2005 through 2013, 1,824 Atlantic Salmon smolts were acoustically tagged (Table 1) and released throughout the Penobscot River watershed in freshwater (Figure 1). Of these, 941 fish were later relocated during estuary migration and used in this study, including 800 hatchery-reared smolts from the U.S. Fish and Wildlife Service Green Lake National Fish Hatchery and 141 wild-reared smolts. Wild smolts released in the Piscataquis River were captured in rotary screw traps by the Maine Department of Marine Resources, tagged, and released immediately downstream. Wild smolts released in the Penobscot River above rkm 100 were captured by Brookfield Renewable Power at the Weldon Dam smolt bypass facility and transported using a 250-L tank with air (identical to the tank used to transport hatchery-reared smolts).

The acoustic tagging methods we used are described in detail by Holbrook et al. (2011) and Stich et al. (2014). Identical procedures were used in all years of the present study (2005–2013). Briefly, individual smolts were anaesthetized using a 100-mg/L solution of MS-222 (tricaine methanesulfonate) buffered to pH 7.0 (using 20-mmol $NaHCO_3$), and fork length (mm) and mass (g) were measured. For each smolt, a 1-cm incision was made offset from the ventral line and 1-cm posterior to the pectoral fin girdle. An acoustic tag was inserted intraperitoneal and the incision was closed with two simple, interrupted knots using 4-0 absorbable vicryl sutures (Ethicon, Somerville, New Jersey). Model V7-2 L (Amirix Vemco, Halifax, Nova Scotia) tags were used in 2005, as well as for wild-origin fish tagged in 2011. Expected battery life of V7-2 L tags was 80 d during 2005 and 69 d in 2011. In all other years, we used model V9-6 L acoustic transmitters (Amirix Vemco, Halifax, Nova Scotia) that had an expected battery life of 82 d (except during 2006 when battery life of V9-6 L transmitters was 80 d). Model V7 tags were 7 mm in diameter, 18.5 mm long, and weighed 1.6 g in air (0.75 g in water), while model V9 tags were 9 mm in diameter, were 20 mm long, and weighed 3.3 g in air (2.0 g in water).

Wild and hatchery-reared smolts were released at up to five locations during any single year. The numbers of fish and release sites varied among years (Table 1), and all fish were released on the day of tagging. To determine the potential for direct mortalities due to handling and surgery, samples of 29–120 smolts were dummy-tagged during several years (2005, 2012, and 2013) concurrent with this study and held in an artificial-stream tank for 1–3 weeks (G. B. Zydlewski, unpublished data). No mortality was observed in any of those fish. Previous studies observed elevated mortality immediately following release (up to 50% in wild smolts but generally less than 5% in hatchery-reared fish) associated with the handling, transport, and release of acoustically tagged smolts in the Penobscot River (Holbrook et al. 2011). All smolts used for analyses in the present study were released a minimum of 50 km above the head of tide (rkm 45), so we assumed that residual effects of tagging were minimal during estuary passage. We included all individuals that were ever detected in the estuary or beyond in the ocean, a total of 941 fish, in analyses of estuary arrival date, movement rate, and survival during the present study.

Covariates of estuary arrival date, movement rate, and survival.—We collected data about several hypothesized covariates of estuary arrival date, movement rate, survival, or detection probability of Atlantic Salmon smolts during estuary migration. These covariates were combined to create a priori hypotheses to test effects of potentially important factors on arrival date, movement rate, and survival. Broadly, we categorized these covariates as characteristics of individual fish,

TABLE 1. Release sites, river kilometer of release sites (rkm), rearing history (Origin), and number of fish used in the study (*n*; numbers in parentheses indicate the total number of fish originally released in each group) for Atlantic Salmon smolts acoustically tagged and released throughout the Penobscot River and its estuary in 2005–2013; twp. = township. Summary statistics were calculated with only fish that were used in the current study and include the mean (SD in parentheses) of fork length (FL; mm), gill Na^+, K^+-ATPase (gill NKA) activity (μmol ADP · mg protein^{-1} · h^{-1}), and mass (g).

Year	Release site	rkm	Origin	FL	Mass	Gill NKA activity	*n*
2005	Howland Dam	99	Hatchery	189 (11)	75 (15)	6.58 (1.95)	90 (150)
	Mattawamkeag twp.	144	Hatchery	185 (12)	69 (15)	6.07 (1.83)	3 (40)
	Milo twp.	142	Hatchery	191 (11)	77 (14)	8.15 (1.94)	44 (85)
	Weldon Dam tailrace	149	Wild	178 (18)	52 (16)	9.08 (1.85)	24 (60)
2006	Milo twp.	142	Hatchery	196 (11)	87 (18)	4.86 (1.24)	38 (72)
	Weldon Dam	149	Hatchery	199 (15)	87 (19)	4.78 (1.85)	53 (146)
	Weldon Dam tailrace	149	Wild	189 (9)	62 (10)	4.12 (1.10)	14 (73)
2009	Milo twp.	142	Hatchery	180 (8)	62 (9)	3.29 (1.38)	73 (100)
	Passadumkeag twp.	92	Hatchery	180 (9)	63 (9)	3.02 (0.82)	77 (100)
2010	Abbot twp.	187	Wild	169 (8)	45 (7)	4.68 (1.10)	19 (75)
	Weldon Dam head pond	162	Wild	180 (14)	55 (13)	4.81 (0.92)	15 (74)
	Milo twp.	142	Hatchery	189 (11)	72 (13)	4.53 (1.04)	63 (100)
	Passadumkeag twp.	92	Hatchery	186 (11)	69 (13)	4.63 (1.13)	77 (100)
2011	Abbot twp.	187	Wild	146 (8)	29 (5)	2.61 (1.17)	55 (75)
	Weldon Dam head pond	162	Wild	163 (19)	42 (16)	3.30 (1.41)	14 (60)
	Milo twp.	142	Hatchery	191 (13)	75 (18)	4.94 (1.47)	55 (100)
	Passadumkeag twp.	92	Hatchery	194 (13)	76 (17)	5.42 (1.49)	74 (100)
2012	Abbot twp.	187	Hatchery	199 (10)	84 (14)	3.35 (1.33)	54 (72)
	Weldon Dam head pond	162	Hatchery	200 (11)	85 (14)	3.48 (1.59)	46 (85)
2013	Abbot twp.	187	Hatchery	185 (11)	70 (13)	2.80 (2.13)	20 (75)
	Weldon Dam head pond	162	Hatchery	185 (9)	71 (11)	2.75 (2.04)	33 (82)
All				187 (16)	71 (19)	4.39 (1.92)	941 (1,824)

characteristics of the migratory history of a fish, and environmental covariates. All continuous covariates were standardized across years and rearing histories prior to each analysis (i.e., timing, movement, and survival analyses) to facilitate ease of interpretation among covariates. Covariate effects were considered statistically significant if the 95% confidence interval of the estimated coefficient for the covariate did not include 0.

Characteristics of individual fish.—Five covariates were used in our models to represent characteristics of individual fish: fork length (mm), Fulton condition factor (K), gill NKA activity, model of acoustic tag that was used (V7 or V9), and rearing history (wild or hatchery). To measure gill NKA activity, a nonlethal gill biopsy (4–6 filaments) was taken from the front, left gill arch of each fish prior to tagging. Individual biopsies were stored at –80°C in 100 μL SEI buffer (250 mM sucrose, 10 mM Na_2-EDTA, 50 mM imidazole) for later analysis of gill NKA activity (expressed as μmol ADP·mg protein^{-1}·h^{-1}) using the method of McCormick (1993). The concentration of NADH at 25°C and 340 nm was used to measure kinetic rate of ouabain-inhibitable ATP hydrolysis, and protein concentration in gill samples was determined using the bicinchoninic acid method (Smith et al. 1985). Gill samples were analyzed in triplicate for gill NKA activity and protein concentration and averaged.

Migratory history of individuals.—We used five covariates to represent the migratory history of each fish: (1) release date, (2) location (in rkm) of release, (3) migratory routing through the lower Penobscot River, (4) number of dams passed during migration (to evaluate dam-related estuary mortality), and (5) median movement rate through the estuary (only in survival models). We used the release date and location of release to represent biological phenomena related to release sites because this allowed us to assess the effects of release locations and fish characteristics at the individual level rather than within artificially structured release groupings. This allowed us to test hypotheses about variation in estuary arrival date, movement rate, and survival over a continuum of release practices rather than within or among groups.

Based on differential survival through the freshwater migration routes in the lower river (Holbrook et al. 2011; Stich et al. 2014), we hypothesized that differential use of the Stillwater Branch or main stem might result in differences in estuary arrival date, movement rates, or dam-related estuary mortality based on the number of dams in each route. Because of imperfect detection at acoustic receivers in the lower river, the use of the Stillwater Branch by smolts was included as a binary covariate in statistical analyses, and fish with an unknown migratory route were assigned the mean value of the covariate (Stillwater Branch = 1, main stem = 0). This assignment allowed us to use the fish for analysis without biasing estimated effects of migratory route. Statistically, this also results in reduced precision in the estimated regression coefficient (i.e., increases type II error) for this covariate.

Smolts released in the freshwater reaches of the Penobscot River passed two to nine dams from their stocking locations to the head of tide during the years of this study (2005–2013). To test whether or not smolts showed behavioral effects or experienced delayed mortality in the estuary due to the passage of dams (hereafter "dam-related estuary mortality"), we examined the relationship between the number of dams a fish passed and estuary arrival date, movement rate, and survival in the estuary. The number of dams passed by each smolt was conditional on the individual's stocking location and migratory route in the lower river (main stem or Stillwater Branch) because the number of dams differed between routes. The potential number of dams passed also was conditioned by year, as in the summer of 2012, after smolt migration was complete, the main-stem Great Works Dam was removed.

Environmental covariates.—We collected information about environmental covariates for estuary arrival date, movement rate, and survival that included temperature and discharge throughout the catchment, as well as photoperiod at the head of tide. These data were collected because of their previously demonstrated influences on the response variables in this study. In-river temperature data (R. Spencer, Maine Department of Marine Resources, unpublished data; and USGS gauge station 01036390) were used to calculate the accumulated thermal units (ATU) experienced by wild smolts in the watershed from January 1 to the date of tagging. Temperature data from outdoor rearing pools at the Green Lake National Fish Hatchery (A. Firmenich, U.S. Fish and Wildlife Service, unpublished data) were used to calculate ATU for hatchery-reared smolts from January 1 to the date of tagging. Photoperiod was calculated from the latitude at the head of tide in the estuary and the ordinal date using the package "geosphere" in R (R Development Core Team 2014). Discharge data were obtained for the U.S. Geological Survey gauge at the West Enfield Dam (H in Figure 1) for each day during all years from 2005 through 2013 and used to characterize discharge in the main stem of the Penobscot River during the period of smolt migration.

Models of estuary arrival date.—Estuary arrival date was assigned as the first detection in the estuary for each smolt detected below Veazie Dam (rkm 45). We estimated the effects of covariates on estuary arrival date using generalized linear models with a $\log_e$ (Poisson family) link function (Montgomery et al. 2006) in R. We used an information-theoretic approach to model selection to test hypotheses about the relative influences of fish characteristics, migratory history, and environmental variables on estuary arrival date by comparing a priori combinations of covariates. We thought that it was important to account for release date regardless of what other covariates were included in the timing models; therefore, photoperiod was included in all models of estuary arrival date. Results were plotted with corresponding calendar dates to facilitate interpretation. We note that the entire smolt run occurred prior to the vernal equinox each year such that photoperiod only ever increased with progressively later calendar dates (i.e., no two dates had the same photoperiod). The measure of discharge used in models of estuary arrival date was the mean of discharge experienced from the date of tagging to the estuary arrival date.

We constructed models containing a single variable that we classified as a "fish characteristic" in any given model to reduce the potential for spurious effects, simplify the model set, and facilitate comparison between competing explanations for factors affecting estuary arrival date. We did not consider models that contained dams passed in addition to the rkm of release or use of Stillwater Branch because (1) there was a strong correlation between rkm of release and dams passed and (2) dams passed was conditional on whether fish migrated through the Stillwater Branch or the main stem in the lower river. We evaluated the relative support for candidate models using Akaike information criterion corrected for sample size (AIC_c; Burnham and Anderson 2002). Approximation of a variance inflation factor ($\hat{c}$) for the most parameterized model prior to analysis indicated reasonable model fit, so model selection was not adjusted.

Movement rate.—Individual movement rates through the estuary were calculated from consecutive seaward detections of Atlantic Salmon smolts for which detections at more than one receiver location in the estuary were recorded. Movement rate (R) of individual fish (i) through estuary reaches (j) was calculated as kilometers per hour (R_{ij}, km/h) based on the distance between consecutive relocations D_{ij} and the amount of time elapsed between the first detections at consecutive locations (T_{ij}) for each fish using the following equation:

$$R_{ij} = \left(\frac{D_{ij}}{T_{ij}}\right). \quad (1)$$

We used linear mixed-effects models (Zuur et al. 2009) in R (R Development Core Team 2014) to estimate the relative effects of (1) fish characteristics, (2) migratory history, (3) detection location in the estuary, (4) and estuary arrival date on the rate of individual fish movements through the estuary. We included an individual-based random effect on the intercept term in all models to account for repeated, unequal numbers of measurements of movement rate for each fish. Movement rate was constrained to be greater than 0 (i.e., predicted movement rate cannot be negative); therefore, we $\log_e$ transformed movement rate prior to analysis.

We included estuary location (in rkm), movement timing (photoperiod), and freshwater discharge in all models of movement rate because we had strong a priori expectations that movement rate changed in the estuary during the smolt window and with discharge. The measure of discharge used in movement models was the mean daily discharge on the day during which a movement was initiated. Similarly, photoperiod was calculated on the dates of individual detections. A second-order term was included for photoperiod (photoperiod^2) because we expected that movement rate would peak during the middle of the smolt run. Because we also hypothesized that movement rate would be faster at the head of tide and at the mouth of the estuary than in the primary mixing zone in the middle estuary, we compared models with only a linear term for rkm to models containing a second-order (quadratic) term for receiver rkm. Allowing for these modifications, we followed the same process for a priori model construction and hypothesis testing based on model selection as was used in the analysis of estuary arrival date. Approximation of $\hat{c}$ for the most parameterized model indicated reasonable model fit, so model selection was not adjusted.

Survival analysis.—The detections at receiver locations were used to develop individual recapture histories (located = 1, not located = 0) for each fish during estuary passage (Figure 1). We used encounter histories to estimate apparent survival (hereafter "survival"; ϕ) and detection probability (p) of smolts in the estuary using a spatially explicit form of the Cormack–Jolly–Seber (CJS) mark–recapture modeling framework (see Figure 1). While the term survival is used throughout this study for simplicity, estimates reflect only the apparent survival and not the true survival of smolts as information about whether fish were alive or dead was not available. We conducted the survival analysis using maximum-likelihood estimation in MARK (version 7.1; White and Burnham 1999) via the RMark package (Laake 2013) in program R (version 3.1.0, R Development Core Team 2014). All parameters were estimated using the logit link function.

We used 12 encounter occasions (i.e., receiver locations) in the CJS models, which included for a "virtual release," or starting location, at the head of tide (Figure 1). We accounted for differences in reach lengths (i.e., distances between receiver locations) by including reach lengths in the model framework so that survival within reaches could be compared directly as per-kilometer rates and appropriate standard errors could be derived within MARK. As a result, the estuary array provided for estimates of survival through 11 reaches of the Penobscot River estuary, each standardized as a per-kilometer rate within those reaches. However, estimates of survival and detection probability are confounded in the final interval of CJS models and so the joint probability of recapture and survival (λ) was estimated in the 11th and final interval (Penobscot Bay), whereas survival was estimated separately in the first 10 reaches.

We investigated the relationships between survival and (1) fish characteristics, (2) migratory history, and (3) environmental effects by comparing models that represented specific a priori hypotheses. We included five covariates to represent the characteristics of individual fish in CJS models: (1) fork length, (2) K, (3) gill NKA activity, (4) rearing history, and (5) tag type. Based on prior work (Zydlewski, unpublished data) and the difference in signal power between tag models, we expected the smaller (V7; 136 dB) tags to have a lower probability of being detected than the larger (V9; 151 dB) tags. Therefore, tag type was included as a covariate of detection probability (V7 = 0, V9 = 1).

We included five covariates to represent migratory history: (1) photoperiod (to represent timing), (2) median movement rate for each fish during estuary migration, (3) migratory route (Stillwater Branch or main stem) in the lower Penobscot River, (4) release rkm, and (5) number of dams passed. We included year, estuary reach, photoperiod, and discharge in the main stem of the river to explain spatial and temporal variation in survival. Year was included as a categorical variable to account for otherwise unexplained annual variation in survival. Each of the environmental covariates was measured for the first detection of each fish at each location. If fish were not detected, the mean value of the covariate was assigned to avoid bias in estimates and retain sample sizes for remaining covariates. The effects of explanatory variables on survival were considered in an additive framework such that fixed effects were interpreted across years rather than within years to avoid the confounding of year with environmental conditions. Year, estuary interval, and discharge were used to model heterogeneity detection probability. The discharge measurement used in the detection model was the mean discharge experienced by each smolt during migration through the estuary such that all fish had some observation for discharge that was not contingent upon detection.

Goodness of fit and model selection for survival analysis.—We used an information-theoretic approach to model selection in order to compare competing hypotheses about the factors affecting survival and detection probability of acoustically tagged Atlantic Salmon smolts in the Penobscot River estuary. We assessed goodness of fit for our most general model using the median $\hat{c}$ approximation in MARK (White and Burnham 1999) to estimate overdispersion. The result of this test indicated that the most general model was slightly overdispersed ($\hat{c} \approx 1.265$); therefore, we adjusted the model selection and the variances of estimated regression coefficients for overdispersion and we used quasi-Akaike information criterion corrected for small sample size ($QAIC_c$) for model selection (Burnham and Anderson 2002).

Our model-selection approach had two hypothesis-driven phases. First, we identified the best spatial and temporal model structure for survival (year, reach, and photoperiod) while allowing detection probability to vary according to year, reach, acoustic tag model, and discharge. We compared a priori combinations of survival and detection parameterizations to determine our model for subsequent hypothesis testing. Based on our experience, we did not consider null (i.e., static or constant) models of survival or detection probability. For detection probability we only considered combinations of explanatory variables that included group (year) and reach effects. Due to limitations of sample sizes, we considered only additive group-by-reach effects (i.e., different intercepts for years in each reach but the same slopes) for both survival and detection probability. Our justification for this decision was that if survival or detection probability varied among sites, then it was likely to have either uniformly improved or decreased but would probably vary in magnitude.

After we identified the best spatial and temporal model structure for survival and accounted for factors influencing detection probability, we constructed a second a priori model set, building on this base model, to test hypotheses based on fish characteristics and migratory history. We did not include more than a single covariate representing migratory history of fish in a given model because of known dependencies between some variables (migratory route and dams passed) and colinearity between others (e.g., release rkm and number of dams passed).

RESULTS

Estuary Arrival Date

The best model used to describe estuary arrival date explained 50% of the variation observed during the past decade (McFadden's pseudo $R^2 = 0.50$; Faraway 2005). The mean arrival date in the Penobscot River estuary for Atlantic Salmon smolts released in freshwater was May 9 (SD, 8 d). Smolts released earlier in the year (modeled as photoperiod at release) arrived in the estuary at an earlier date than those fish released later in the season (Table 2). Over the range of release dates used in the past decade (range = April 12–May 29) arrival date was 24 d earlier on average for the earliest release dates than for the latest release dates.

Smolts released further upstream of the estuary arrived in the estuary at a later date than those fish that were released further downstream. The difference in arrival dates between the furthest upstream and furthest downstream release sites, located approximately 90 km apart (range = rkm 92–rkm 187), was 3 d (Figure 2a), and the effect was not statistically significant at $\alpha = 0.05$ (Table 3) unless discharge was not included in the same model. Observed discharge during the period between release dates and estuary arrival dates in 2005–2013 ranged from a minimum of 175 m^3/s to 2,500 m^3/s. Fish experiencing the greatest discharge between release date and estuary arrival date arrived in the estuary 10 d later than fish experiencing the least discharge (Figure 2b; Table 3).

Atlantic Salmon smolts that were wild reared (i.e., progeny of wild spawning, egg planting, or fry stocking) arrived in the estuary later than smolts that were reared in the hatchery (Table 3). Mean estuary arrival date for wild-reared fish was May 20 (SD = 7 d), whereas mean estuary arrival date for hatchery-reared smolts was May 7 (6 d), a difference of 2 weeks. The majority of this difference results directly from the later tagging dates of wild-reared smolts (mean = May 13) compared with the release dates of hatchery-reared smolts (mean = April 24).

Fish experiencing a warmer thermal history (higher ATU) prior to tagging and release arrived in the estuary earlier than fish experiencing a lower ATU prior to release when the effect of release date was accounted for (Table 2). This relationship indicated that fish experiencing the greatest ATU (550) arrived 8 d earlier than those fish experiencing the lowest ATU (220) prior to release (Figure 2c). However, ATU covaried with discharge, and discharge was a better predictor of estuary arrival date than ATU because ATU was not included in the best model and was not statistically significant when discharge was included in the same model.

Estuary arrival date was earlier with increasing gill NKA activity, until gill NKA activity reached 6 μmol

TABLE 2. Model-selection statistics for generalized linear models that were used to quantify the relationships between the date of arrival by acoustically tagged Atlantic Salmon smolts in the Penobscot River estuary in 2005–2013 and fish characteristics (rearing history [rearing], gill NKA activity), migratory history (photoperiod at release date [PP], release rkm [release], migratory route through the lower river [SW]), and environmental conditions (accumulated thermal units [ATU], discharge experienced from release to arrival date [discharge]). The number of parameters estimated in each model is k, AIC is the Akaike information criteria for each model, ΔAIC_c is the difference between the AIC_c of each model and the best model in the candidate set, and w is the relative probability that each model is the best in the candidate set. Only models with substantial support ($\Delta AIC_c < 2.0$) are shown.

Model	k	AIC_c	ΔAIC_c	w
Discharge + PP + rearing + release	5	6,556.57	0.00	0.16
Discharge + PP + rearing + release + SW	6	6,556.98	0.42	0.13
Discharge + PP + rearing	4	6,558.03	1.47	0.08
ATU + gill NKA activity + PP + release + SW	7	6,558.16	1.59	0.07

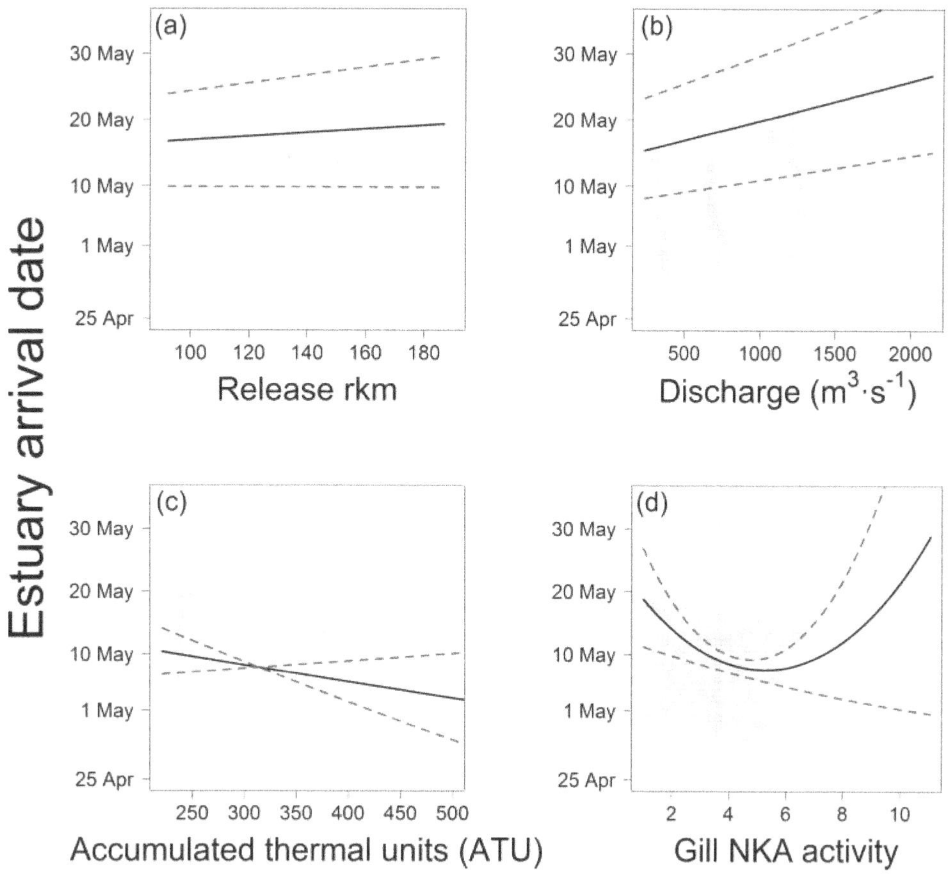

FIGURE 2. Predictions from the generalized linear model relating arrival date of acoustically tagged Atlantic Salmon smolts in the Penobscot River estuary in 2005–2013 to **(a)** release rkm (upstream of estuary), **(b)** discharge experienced from tagging to estuary arrival date, **(c)** accumulated thermal units experienced prior to release, and **(d)** gill NKA activity. The black lines indicate means and the dashed gray lines indicate the 95% CIs.

TABLE 3. Standardized regression coefficients, standard error (SE), and 95% confidence limits (CLs) for the generalized linear model used to model estuary arrival date (as a function of photoperiod at release date [PP], release rkm, rearing history, and discharge). Also shown is the linear mixed-effects model used to model movement rates as a function of location (estuary rkm), discharge, release rkm, and photoperiod for date of movement (PP, PP^2). Covariate effects were significant at $\alpha = 0.05$ if the range of the 95% CLs did not overlap 0.

Model	Parameter	Estimate	SE	Lower 95% CL	Upper 95% CL
Arrival date	Intercept	4.861	0.004	4.854	4.868
	Discharge	0.017	0.003	0.011	0.022
	PP	0.022	0.005	0.013	0.031
	Rearing (wild)	0.040	0.015	0.012	0.069
	Release rkm	0.007	0.004	0.000	0.014
Movement rate	Intercept	0.306	0.019	0.270	0.343
	Discharge	0.034	0.018	0.000	0.069
	Estuary rkm	0.174	0.016	0.144	0.205
	PP	0.057	0.018	0.022	0.092
	PP^2	−0.012	0.011	−0.034	0.009
	Release rkm	0.081	0.016	0.049	0.113

ADP·mg protein^{-1}·h^{-1}, after which estuary arrival date began to increase with gill NKA activity (Figure 2d). As a result, Atlantic Salmon smolts that had very low or very high gill NKA activity arrived in the estuary later than smolts near the mean gill NKA activity. Similar to ATU, gill NKA activity covaried with discharge and was neither included in the best model nor significant when discharge was included in the same model.

Movement Rate

The mean movement rate of smolts through all reaches of the estuary and among all fish was 2.27 km/h (SD = 1.88). The only model of movement rate that received any support was that which included the effects of discharge, date of movement (photoperiod, including quadratic effect), distance released from the ocean, and location at which movement was measured in the estuary. We observed evidence of a quadratic relationship between date of movement and individual movement rate. There was an increased movement rate until the middle of the migration period (early May), after which the movement rate became more variable but appeared to asymptote or even decrease (Figure 3a; Table 3).

The movement rate of smolts decreased in the Penobscot River estuary from the head of tide to the mouth of the estuary (Figure 3b; Table 3) but a second-order term for estuary rkm did not. From the head of tide to the mouth of the estuary, a distance of 50 km, the movement rate of smolts decreased from a mean of 1.8 km/h to 1.1 km/h (Figure 3b).

Smolts released further upstream in the Penobscot River moved faster than tagged smolts released at downstream sites (Figure 3c; Table 3). However, the effect of release rkm on movement rate appeared to be minor in comparison to the effects of arrival date and estuary location: over the range of release rkms used in this study (rkm 92–rkm 187), maximum movement rate increased by less than 0.5 km/h. Movement rate through the estuary increased with increasing discharge (Figure 3d). As with release rkm, the effect of discharge was minimal, resulting in a change of less than 0.3 km/h over the range of discharge observed (217–1,957 m^3/s).

Survival

The survival and detection probability of smolts varied spatially, temporally within years, and among years. In our first phase of hypothesis testing, only the model that allowed

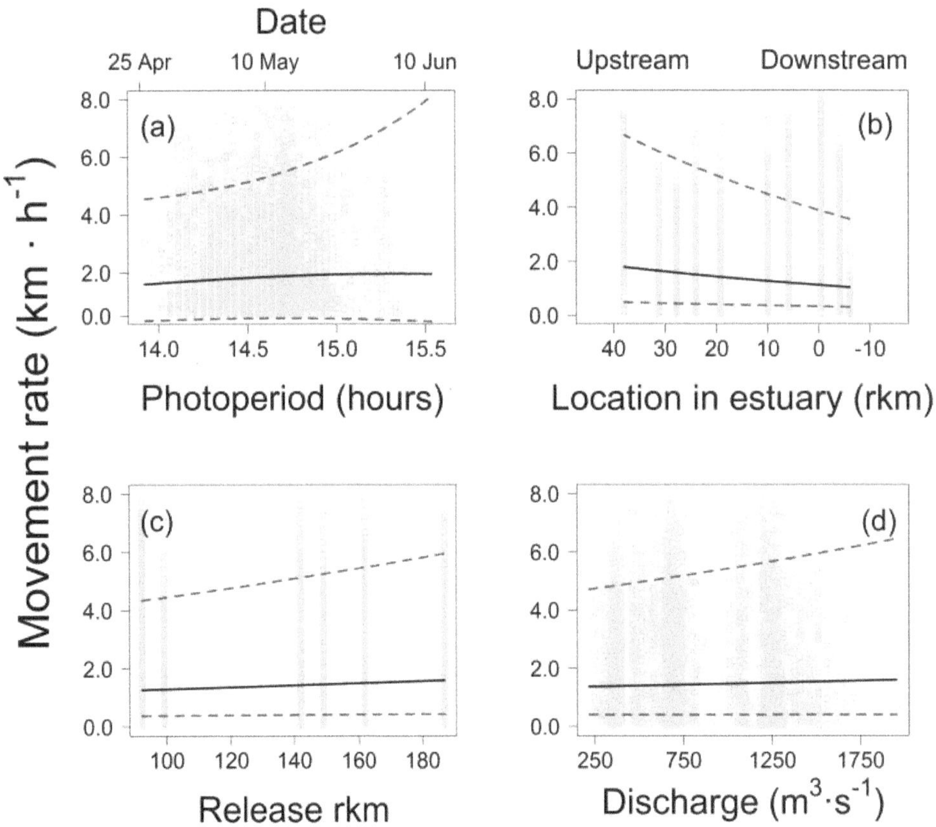

FIGURE 3. Mean (solid line) and 95% CI (dashed lines) predictions from parameters of the generalized linear mixed-effects model used to relate movement rate of acoustically tagged Atlantic Salmon smolts in the Penobscot River estuary in 2005–2013 to **(a)** photoperiod, **(b)** location in the estuary (rkm), **(c)** release distance upstream of the estuary (rkm), and **(d)** discharge experienced during estuary migration.

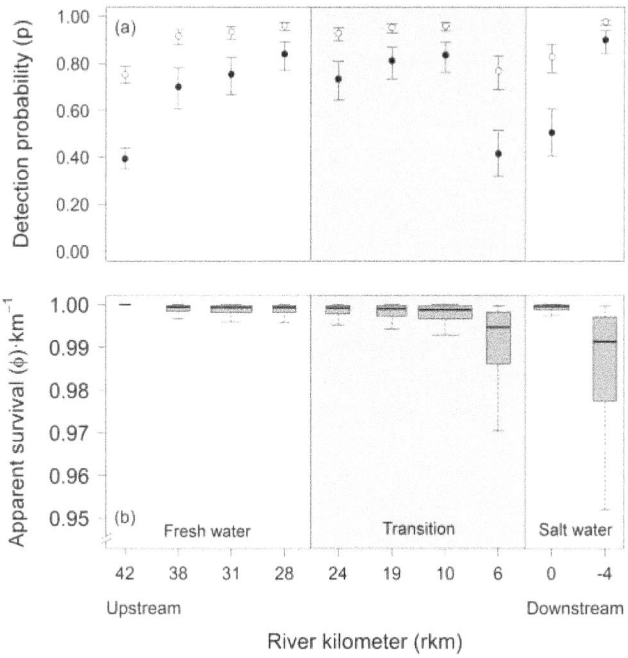

FIGURE 4. Estimates of (**a**) mean (whiskers indicate 95% CI) detection probability for V9 (white) and V7 (black) acoustic tag models in each reach across years 2005–2013 and (**b**) mean apparent survival per kilometer of acoustically tagged Atlantic Salmon smolts in the Penobscot River estuary in 2005–2013, estimated using Cormack–Jolly–Seber models. The solid line in each box in panel (b) indicates median annual survival, box-ends indicate the inner quartile range, and whiskers indicate the 95% CI. The box widths are proportional to the estuary reach lengths to which the estimates apply. From left to right, the three divisions in each panel represent freshwater, the zone of transition from low salinity to high salinity, and salt water within the Penobscot River estuary.

survival to vary among reaches, among years, and with photoperiod (including a quadratic effect) was supported by the data. Survival was highest near the head of tide (>0.99) and the mouth of the estuary and was lowest in the middle estuary (0.98), where tidal influences are strongest (Figure 4). Smolt survival was highest during the middle of the smolt run, peaking in early May, when survival was as much as 70% greater than in the early or late run. Fish arriving in the estuary very early or very late had poor survival (near 0) by comparison to those arriving during the middle of the run (Figure 5a).

As expected, the probability of detecting smolts was higher for fish tagged with large (model V9) acoustic transmitters than for those tagged with smaller (model V7) transmitters (Table 4; Figure 4). The probability of detection also was affected by the discharge experienced by individual smolts during estuary migration (Table 4). Over the range of flows observed (217–1,957 m^3/s), the mean probability of detection was reduced by 0.18 (95% CI = 0.06–0.33) during periods of greatest freshwater discharge compared with periods of lowest discharge. Based on these results, year, estuary reach, and photoperiod all were included in the final parameterization for survival. The final parameterization for the detection model included year, estuary reach, tag model, and discharge experienced by individual fish during migration. We tested all other hypotheses about effects of fish characteristics and migratory history using the parameterization above as a "base model."

After accounting for these effects, the survival of smolts in the Penobscot River estuary was dependent on physiological development (measured as gill NKA activity), dams passed, and the rate at which fish moved through the estuary (Table 5). We found strong evidence for an optimal timing of estuary passage related to survival (Figure 5a; Table 4). Survival increased until mid-May, after which survival became variable but tended to decrease.

The number of dams passed by individual smolts had a strong, negative effect on fish survival in the estuary (Figure 5b; Table 4). The survival of smolts that passed the greatest number of dams (nine) was reduced by 40% compared with those fish that passed only two dams (the minimum). The model including dams passed outperformed the corresponding model with release rkm by nearly 2 $QAIC_c$, suggesting that the majority of the effect of release location on survival was explained by the number of dams passed during migration and not merely by the distance fish had migrated.

The physiological preparedness of smolts for saltwater entry (measured as gill NKA activity) at tagging was positively related to smolt survival during estuary passage (Figure 5c; Table 4). Atlantic Salmon smolts with the highest gill NKA activity had a 25% greater probability of surviving the estuary migration than those fish with the lowest enzyme activity. The median movement rate of smolts was included in the best model of survival, and survival of smolts decreased slightly with increasing movement rate over the observed range of movement rates; however, the effect of this covariate was not statistically significant (Table 4), and the corresponding model that excluded the effects of movement on survival had virtually identical support in the data.

DISCUSSION

By linking the arrival date, movement rate, physiological preparedness, and survival of smolts over nearly a decade, we were able to improve our understanding of the complex relationships and interactions between behavior and survival during estuary migration. Although a number of other studies have explored the effects of factors influencing estuary arrival date, movement rate, and survival of Atlantic Salmon smolts and postsmolts during early marine migration (see Thorstad et al. 2012b), few (if any) studies have had the opportunity to examine all of these processes together in a single population from distant upriver release sites all the way to the ocean. Furthermore, this study provides an unprecedented link between physiological preparedness (gill NKA activity) and performance in the wild. Similarly, this study is the first to clearly

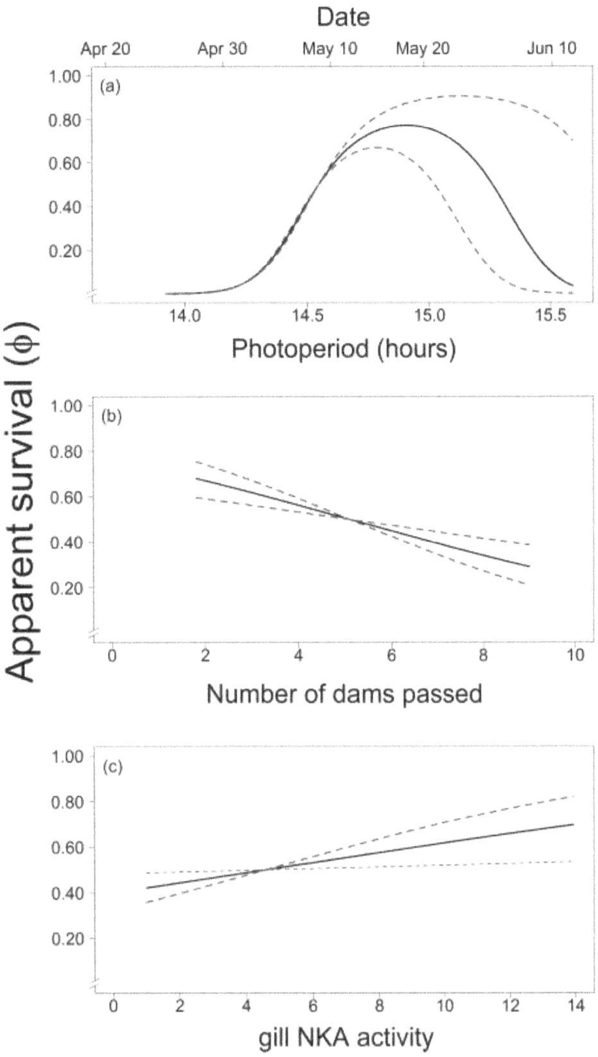

FIGURE 5. Means (solid line) and 95% CIs (dashed lines) of predictions from parameters of the Cormack–Jolly–Seber mark–recapture models used to estimate apparent survival (ϕ) of acoustically tagged Atlantic Salmon smolts through the Penobscot River estuary in 2005–2013. The panels show the effects of **(a)** individual timing, **(b)** number of dams passed by individuals, and **(c)** gill NKA activity of individuals on estimated survival.

relate the estuary survival of Atlantic Salmon directly to the delayed effects of dam passage.

Estuary Arrival Date

Smolt arrival in the Penobscot River estuary was overwhelmingly driven by environmental conditions. We found that estuary arrival date was significantly earlier for smolts subjected to a warmer thermal history (ATU) prior to tagging and for those released in higher freshwater discharges between tagging and arrival date. Migratory behavior in Atlantic Salmon smolts has been shown in multiple laboratory studies to be driven by environmental effects, especially photoperiod (Zydlewski et al. 2014) and temperature (Zydlewski et al. 2005). Those studies have shown that smolts experiencing cooler temperatures during development initiate migration at a later date. Our field studies are consistent with that laboratory work. Sykes et al. (2009) reported that wild Chinook Salmon *Oncorhynchus tshawytscha* smolts initiated migration later when they experienced cooler temperatures during development and during high discharge. Similarly, global trends in the timing of Atlantic Salmon smolt migrations show that smolts experiencing higher discharge migrate later than those experiencing low discharge (Otero et al. 2014). Thus, our study is consistent with previous studies and the understanding that environmental factors drive the onset of migration in smolts.

The timing of estuary arrival was related to the physiological development of Atlantic Salmon smolts in the Penobscot River. Fish with the lowest or greatest gill NKA activity arrived in the estuary later than fish that had intermediate gill NKA activity at release. This is consistent with behavioral adherence to the physiological smolt window for migration (McCormick et al. 1998). However, the arrival date for smolts with the greatest gill NKA activity was highly variable. This variability at the peak of physiological smolting may be due to the heightened sensitivity to stress that is associated with tagging and handling (Carey and McCormick 1998).

Rates of Movement

The rate of movement in the estuary by smolts was influenced by the release location, date of individual movements, and location within the estuary. Smolts released further upstream in freshwater moved faster through the estuary than did smolts released further downstream. This could result from either increased migratory speed by fish released upstream (Jokikokko and Mäntyniemi 2003) or as a result of station-holding (i.e., staging) behavior by smolts released downstream in the watershed (e.g., Strand et al. 2010). Smolts that arrive in the estuary earliest may delay subsequent migration to synchronize ocean entry with optimal temperatures that favor their ability to avoid predators or access food (Otero et al. 2014). Differences in behavioral priming (Dingle and Drake 2007) or environmental conditions among different regions of the watershed (Whalen et al. 1999; Zydlewski et al. 2005) offer alternative explanations. For example, fish in headwater reaches of the river might experience an increased intensity of exposure to environmental factors that prime migratory behavior compared with fish released downstream in main-stem river reaches due to smaller water volumes that increase exposure to changes in temperature, photoperiod, and discharge.

The smolts decreased their migratory speed throughout the course of estuary emigration. It is likely that reductions in the movement rate during estuary emigration were primarily due to changes in current velocity and tidal influences in the lower

TABLE 4. Standardized regression coefficients (logit) and 95% confidence limits (CLs) for individual covariates included in the top-ranked Cormack–Jolly–Seber mark–recapture model used to estimate apparent survival (ϕ) and detection probability (p) of acoustically tagged Atlantic Salmon smolts during emigration through the Penobscot River estuary in 2005–2013. Tag type was classified as a binary covariate: smaller (V7) tags were assigned to 0 and larger (V9) acoustic tags were assigned to 1. The abbreviations PP and PP^2 indicate photoperiod for the date of movement. Covariate effects were significant at $\alpha = 0.05$ if the range of the 95% CLs did not overlap 0.

Parameter	Covariate	Estimate	SE	Lower 95% CL	Upper 95% CL
ϕ	PP	1.631	0.110	1.414	1.847
	PP^2	−0.545	0.086	−0.713	−0.378
	Gill NKA activity	0.192	0.081	0.032	0.351
	Number of dams passed	−0.376	0.092	−0.557	−0.195
	Movement rate	−0.108	0.064	−0.233	0.018
p	Tag type	0.404	0.033	0.340	0.469
	Discharge	−0.475	0.053	−0.579	−0.371

Penobscot River estuary (Fried et al. 1978; Moore et al. 1995). The reduction in movement rates through estuaries might also result from the reversal of migratory direction during the passage of tidal estuaries and bays (Kocik et al. 2009; Dempson et al. 2011; Halfyard et al. 2013). These behaviors appear to be related to tidal cycles in both estuarine (McCleave 1978; Martin et al. 2009) and coastal systems (Lacroix et al. 2005). This behavior previously has been hypothesized to relate to saltwater acclimation (Gudjonsson et al. 2005; Dempson et al. 2011; Halfyard et al. 2013). However, if it is related to saltwater acclimation, then the behavior may not provide an actual fitness benefit because survival is related to gill NKA activity at the time of release in freshwater, which implies that competence for saltwater entry is developed prior to estuary arrival. This finding is corroborated by results of studies in the northeastern Atlantic Ocean, where smolts acclimated in net-pens within an estuary showed no improvement in survival over fish released directly into the mouth of the river (Thorstad et al. 2012a). Other laboratory experiments have found no period of acclimation occurred for smolts upon reaching salt water (Moore et al. 1995). It is, however, possible that fish use tidal movements to minimize energetic costs, especially because this is the period of migration during which smolts are thought to transition from passive to active migration (Hedger et al. 2008; Martin et al. 2009). Further investigation of diurnal and tidal factors that influence behavior and survival on localized spatial and temporal scales could provide an improved understanding of those mechanisms and might have implications for strategic stocking of hatchery-reared smolts.

Survival

The estuary arrival date, individual fish characteristics, number of dams passed, and location within the estuary all influenced Atlantic Salmon smolt survival during estuary passage. We found evidence for a strong optimizing effect of estuary arrival date on the survival of Atlantic Salmon smolts through the estuary, emphasizing the importance of timing in determining the success of migrating smolts (McCormick et al. 1998). Although this trend previously was suspected based on narrow windows of estuary passage in many Atlantic Salmon smolt runs (see Thorstad et al. 2012b), this study provides strong, empirically derived evidence linking survival during estuary migration to the variability in timing of individual estuary passage. The shape and spread of the timing–survival relationship in the Penobscot River suggests that normalizing selection may occur on the timing of estuary arrival through a direct link to survival. This relationship

TABLE 5. Model-selection statistics for Cormack–Jolly–Seber mark–recapture models used to estimate survival of acoustically tagged Atlantic Salmon smolts during emigration through the Penobscot River estuary in 2005–2013 and to test hypotheses about fish characteristics and migratory history that influenced survival. Table headings and model-selection statistics are defined as in Table 2, except here we used quasi-Akaike information criterion corrected for small sample size ($QAIC_c$) for model selection. The top 10 candidate models are shown. All candidate models in this set included the covariates representing spatial and temporal variation, ϕ(reach + year + PP + PP^2), in addition to the covariates shown in each model below. Covariates of apparent survival (ϕ) include the gill NKA activity, number of dams passed (dams), median movement rate in the estuary by individual smolts (R), and distance of release from the ocean (release).

Model	k	$QAIC_c$	$\Delta QAIC_c$	w
ϕ(dams + gill NKA activity + R) p(reach + year + tag type + discharge)	41	7,194.11	0.00	0.24
ϕ(dams + gill NKA activity) p(reach + year + tag type + discharge)	40	7,194.28	0.17	0.22
ϕ(gill NKA activity + release) p(reach + year + tag type + discharge)	40	7,195.80	1.69	0.10
ϕ(gill NKA activity + R + release) p(reach + year + tag type + discharge)	41	7,195.92	1.81	0.10

likely occurs in response to environmental cues and releasing factors prior to the onset of migration and illustrates the importance of timing for successful estuary passage (McCormick et al. 1987; Hoar 1988). This supposition is supported by the strength of release date in freshwater as a predictor of timing for both wild and hatchery fish and the effect of gill NKA activity on survival in the estuary at a later date.

Estuary arrival date was a stronger predictor of smolt survival than the movement rate of individual smolts. The inclusion of movement rate in the best model of survival was somewhat perplexing, given the lack of significance for the covariate and the unexpected direction of the relationship to survival. An increased movement rate through estuaries is postulated to reduce exposure to predators and environmental stressors, such as pollution (McCormick et al. 1998), but our data suggest that perhaps this may not be the case. Based on the similarity between the best model for survival in the present study and the second-ranked survival model (ΔQAIC < 0.20), it seems that the inclusion of movement rate in the survival models resulted in minimal improvement in model fit and that this variable was included because it covaried with some other important variable of interest, such as the number of dams passed, gill NKA activity, or location within the estuary. Based on the other results of this study, it seems likely that there could be synergistic effects of physiological preparedness and movement rate that result in changes to individual survival, but further investigation of the relationships is needed.

Atlantic Salmon smolt survival in the estuary increased with increasing gill NKA activity at the time of release. This result establishes a critical link between the physiological preparation of smolts in freshwater for osmoregulation in the ocean and survival, which has been long suspected by others (Boeuf 1993; Itokazu et al. 2014). A rich literature exists describing the physiological transformations of diadromous fishes (e.g., Zydlewski and Wilkie 2013), and the study of smolt physiology constitutes a large body of work within that field (McCormick et al. 1998; McCormick 2013). It is well established that gill NKA activity is a useful indicator of osmoregulatory ability (Zaugg and McLain 1972; McCormick et al. 1998, 2009) and migratory readiness in salmonids (Aarestrup et al. 2000). While researchers have demonstrated the performance benefits of smolts up-regulating gill NKA activity in laboratory studies relative to sublethal indicators (McCormick et al. 2009), the present study has provided a direct link between the physiological development of smolts and fitness in a study of actively migrating smolts. Although gill NKA activity is not a strong predictor of long-term growth scope or ocean performance (Zydlewski and Zydlewski 2012), our results underscore the importance of physiological preparedness for successful entrance into the marine environment.

We do not suspect that reduced gill NKA activity led to direct mortality due to inability to osmoregulate in the estuary because mortality is not generally observed in laboratory studies of Atlantic Salmon during the period of smolt migration. However, proximate causes of mortality, such as predation (Järvi 1990; Handeland et al. 1997; Hawkes et al. 2013) or acidosis due to synergies between osmoregulatory stress and other forms of stress (Järvi 1989; Price and Schreck 2003; Berli et al. 2014), have the potential to dramatically increase when the osmoregulatory capacity of smolts is suboptimal (McCormick et al. 2009). Research targeting the relationships between smolt physiology and the sources of direct mortality (e.g., predation) in estuaries might help to further unravel the links between physiology and the proximate causes of mortality in the wild.

Our data strongly implicate a delayed, negative effect of dam passage on survival in the estuary, reducing estuary survival by 6–7% per dam passed (see Figure 5). We demonstrated that the cumulative number of dams passed (ranging from two to nine dams in this study) was an important predictor of smolt survival. This result is consistent with the work of Schaller et al. (2014), who found that the number of powerhouses passed by out-migrating Chinook Salmon affected marine survival. It is, however, notable that our results are the first to demonstrate this trend in Atlantic Salmon migrations. Furthermore, the *delayed* (indirect) dam-related mortality experienced in the 50-km Penobscot River estuary as a result of passing nine dams was comparable in magnitude with the cumulative, *acute* (direct) mortality incurred by smolts passing those same nine dams during the 150-km freshwater migration (Holbrook et al. 2011, this study). This suggests that studies of survival at dams, which are the basis for dam permitting, may drastically underestimate the effects of those dams.

Smolts experience injuries, such as descaling, when passing dams in the Penobscot River (Music et al. 2011). Such injuries can severely impair osmoregulatory ability, and impairment can persist for several days after injury (Zydlewski et al. 2010). All fish entering the Penobscot River estuary passed at least one dam prior to estuary arrival. Many fish passed several dams just 24–48 h prior to estuary arrival. Our results suggest that this experience reduced the survival of smolts during estuary passage. Reduction in osmoregulatory ability during estuary passage has previously been linked to a decreased ability to avoid predators (Handeland et al. 1997; Price and Schreck 2003), which could increase mortality in estuaries. Spatial patterns in survival through the Penobscot River estuary indicate that one possible mechanism of reduced survival in the estuary is the interaction between multiple factors (such as dam-related injury, gill NKA activity, and predators) because the greatest reduction in survival occurred upon reaching the saltwater reaches of the estuary, where osmoregulatory perturbance would have the greatest effect.

Conservation and Management Implications

Atlantic salmon smolts may be particularly susceptible to chronic and acute anthropogenic impacts. Climate change, based on our data, has the potential to squeeze Penobscot

River Atlantic Salmon against the ecological and physiological limits to adaptability. Earlier seasonal warming could result in mismatches between physiological and ecological smolt windows based on run timing (McCormick et al. 1997; Todd et al. 2012; Otero et al. 2014). Although changes in physiological smolt development and the timing of initiation of migratory behavior could shift run times initially (McCormick et al. 1997), these changes are also both entrained by circannual rhythms in photoperiod (Zydlewski et al. 2014), so there may be limits to how early physiological and behavioral smolting can occur (Otero et al. 2014). The resilience of many North American stocks to strong selection pressures imposed by these shifts is unknown (Mills et al. 2013; Friedland et al. 2014; Hayes and Kocik 2014).

Successful and expedient passage through estuaries may become increasingly important based on relations between temperature, physiology, and survival (Figure 6). Migratory delay through dams (Keefer et al. 2012) and physical injuries incurred during dam passage (Music et al. 2011) have the potential to further promote the loss of smolt characteristics (McCormick et al. 1999; Marschall et al. 2011) and impair osmoregulatory ability directly (Zydlewski et al. 2010). Recent modeling suggests the possibility that dams affect smolt migrations through migratory delay and potential mismatch in the timing of estuary arrival (McCormick et al. 2009; Marschall et al. 2011).

In the Penobscot River, indirect dam-related estuary mortality is nearly as great as the cumulative direct mortality incurred as a result of dam passage in freshwater (Holbrook et al. 2011; Stich et al. 2014). Indirect, dam-related estuary mortality resultant from passage of a single dam during migration also has the potential to be greater than the acute mortality incurred during passage of multiple hydropower projects in the lower Penobscot River (Holbrook et al. 2011; Stich et al. 2014). This clearly highlights the importance of considering dam-related estuary mortality within regulatory frameworks used to assess the effects of hydropower projects on fish passage.

The removal of Great Works and Veazie dams will likely increase smolt survival in the Penobscot River estuary by reducing dam-related mortality in the estuary (Figure 5), even though these dams resulted in little acute mortality during freshwater smolt migration (Holbrook et al. 2011; Stich et al. 2014). Indeed, the indirect, dam-related estuary mortality resulting from each of these structures (about 6%; Figure 5) was greater than the direct mortality incurred during passage (0–1%) of the facilities (Holbrook et al. 2011; Stich et al. 2014). Because both of these dams were less than a 12-h travel to the estuary, we hypothesize that their effects also may have been greater than dams located further upstream (see Zydlewski et al. 2010), but this remains to be tested. Prior to the removal of Veazie and Great Works dams, 88% of Atlantic Salmon smolts passed both dams (Stich et al. 2014) and 100% of smolts passed Veazie Dam during our study (2005–2013). As such, we expect that estuary survival will increase as a result of the removal of Veazie Dam alone (6%; Figure 5). Additional increased survival is expected for 88% of the emigrating smolts due to Great Works Dam removal (6%; Figure 5). With the removal of these two dams, 20 km of lotic habitat was restored in the lower main stem of the Penobscot River below the now lowermost dam (Milford Dam). This resulted in the opportunity to stock hatchery-reared smolts in the main stem below any dams during the 2014 smolt run.

Smolts incur relatively high rates (about 10%/km) of acute mortality at main-stem dams compared with the background mortality in the Penobscot River (about 1%/km), resulting in a total loss of 40–60% of fish prior to arrival in the estuary (Holbrook et al. 2011; Stich et al. 2014). Fish passing the most dams (nine) during the present study experienced 40% increased dam-related estuary mortality compared with fish passing the fewest dams (two). Smolts stocked in 2014 passed no dams, eliminating dam-related mortality in the estuary (Figure 6). However, stocking nearer to the estuary may require closer attention to the physiological development of smolts (Figure 6) and environmental conditions (Figure 6). Similarly, there are potential ramifications for adult homing that result from stocking further downstream (Gorsky et al. 2009).

The number of smolts exiting the Penobscot River estuary is expected to increase (by 12%) based on reductions in dam-related estuary mortality following the removal of the two lowermost dams. Through improvements in estuary survival by stocking below dams and monitoring smolt gill NKA activity, we expect that the proportion of fish exiting the estuary each year could approximately double in a best-case scenario. Based on a lack of differential survival in marine habitats after leaving Penobscot Bay (Sheehan et al. 2011), and the fact that patterns in marine mortality are similar among North American stocks (Friedland et al. 2003; Mills et al. 2013), this gain would likely translate directly to increases in returning adults.

ACKNOWLEDGMENTS

Financial support for this research was provided by the National Fish and Wildlife Foundation, the American Recovery and Reinvestment Act, and the National Oceanic and Atmospheric Administration (NOAA) Open Rivers Initiative through the Penobscot River Restoration Trust and the NOAA National Marine Fisheries Service, Northeast Fisheries Science Center. Additional funding was provided by Brookfield Renewable Power (Great Lakes Hydro America), the Maine Department of Marine Resources, The Nature Conservancy, the University of Maine, and the U.S. Geological Survey, Maine Cooperative Fish and Wildlife Research Unit. We acknowledge substantial contributions to this work by Michael Bailey, Cory Gardner, Christopher Holbrook, and Andrew O'Malley (Department of Wildlife, Fisheries, and Conservation Biology, University of Maine).

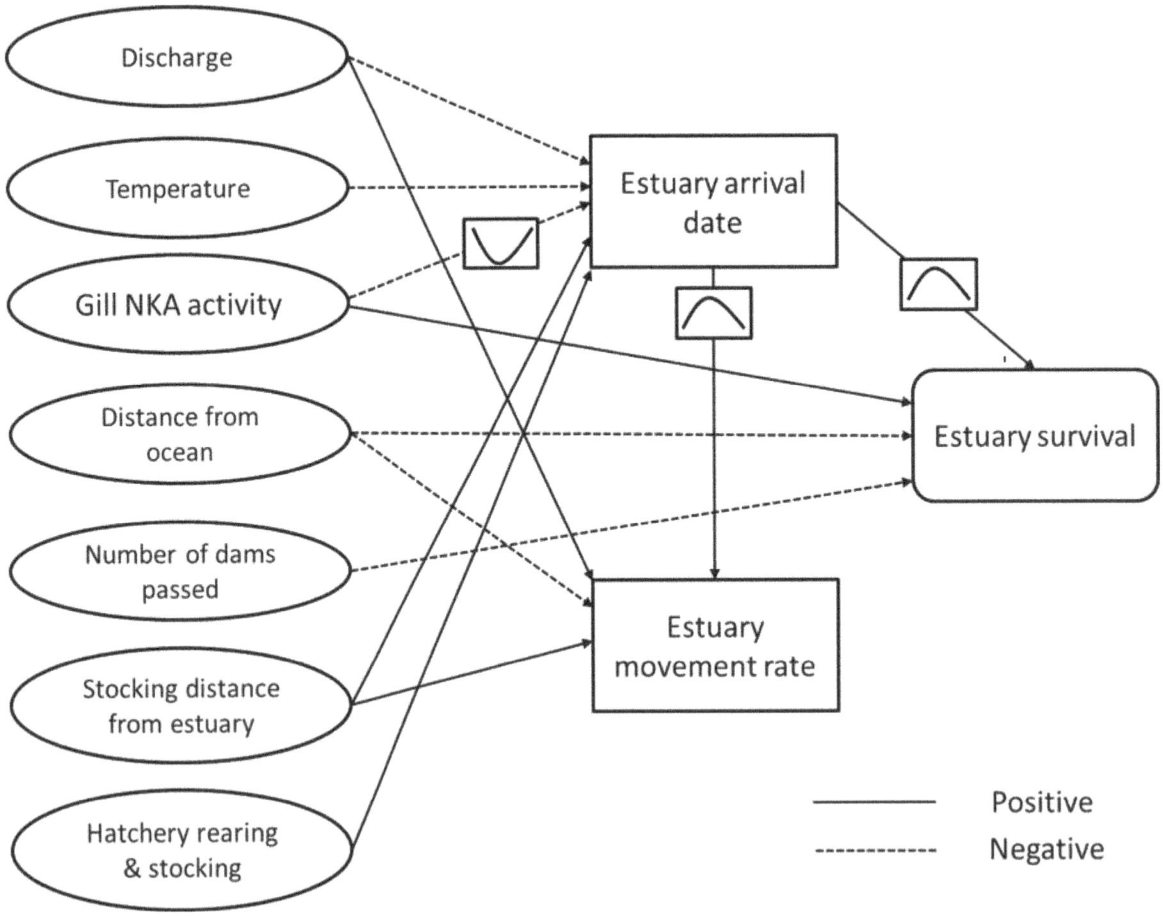

FIGURE 6. Diagram synthesizing the major findings about the factors affecting aspects of estuary migration of Atlantic Salmon smolts during the present study. The dotted lines indicate negative relationships, whereas solid lines indicate positive influences. The assumed causality of relationships is indicated by the direction of the arrows. Nonlinear (quadratic) relationships are indicated by the parabolas in boxes on top of the arrows that describe the relationships.

Acoustic receiver deployment was coordinated collaboratively between the U.S. Geological Survey Maine Cooperative Fish and Wildlife Research Unit, University of Maine, and NOAA. We thank Matthew Altenritter, Phillip Dionne, Stephen Fernandes, and Catherine Johnston (School of Marine Sciences, University of Maine), as well as Graham Goulette and James Hawkes (National Marine Fisheries Service, Northeast Fisheries Science Center), for their efforts in maintaining the acoustic receiver network in the Penobscot River estuary. We thank Michael Bailey (U.S. Fish and Wildlife Service) and Michael Kinnison (University of Maine) for helpful comments on previous versions of this manuscript. The U.S. Fish and Wildlife Service, Green Lake National Fish Hatchery provided Atlantic Salmon smolts and was invaluable in facilitating marking and stocking. This work was done in coordination and cooperation with the NOAA Fisheries Northeast Fisheries Science Center and Maine Department of Marine Resources. Mention of trade names or commercial products does not imply endorsement by the U.S. Government. This work was conducted under the University of Maine Institutional Animal Care and Use Committee protocols 2008-07-01 and A2011-06-06.

REFERENCES

Aarestrup, K., C. Nielsen, and S. S. Madsen. 2000. Relationship between Na^+, K^+-ATPase activity and downstream movement in domesticated and first-generation offspring of wild anadromous Brown Trout (*Salmo trutta*). Canadian Journal of Fisheries and Aquatic Sciences 57:2086–2095.

Berli, B. I., M. J. Gilbert, A. L. Ralph, K. B. Tierney, and P. Burkhardt-Holm. 2014. Acute exposure to a common suspended sediment affects the swimming performance and physiology of juvenile salmonids. Comparative Biochemistry and Physiology Part A Molecular and Integrative Physiology 176:1–10.

Blackwell, B. F., and F. Juanes. 1998. Predation on Atlantic Salmon smolts by Striped Bass after dam passage. North American Journal of Fisheries Management 18:936–939.

Blackwell, B. F., W. B. Krohn, N. R. Dube, and A. J. Godin. 1997. Spring prey use by double-crested cormorants on the Penobscot River, Maine, USA. Colonial Waterbirds 20:77–86.

Boeuf, G. 1993. Salmonid smolting: a pre-adaptation to the oceanic environment. Pages 105–135 *in* J. C. Renkin and F. B. Jensen, editors. Fish ecophysiology. Chapman and Hall, London.

Burnham, K. P., and D. R. Anderson. 2002. Model selection and multimodel inference: a practical information-theoretic approach, 2nd edition. Springer, New York.

Carey, J. B., and S. D. McCormick. 1998. Atlantic Salmon smolts are more responsive to an acute handling and confinement stress than parr. Aquaculture 168:237–253.

Chaput, G. 2012. Overview of the status of Atlantic Salmon (*Salmo salar*) in the North Atlantic and trends in marine mortality. ICES (International Council for the Exploration of the Sea) Journal of Marine Science 69:1538–1548.

Chaput, G., C. M. Legault, D. G. Reddin, F. Caron, and P. G. Amiro. 2005. Provision of catch advice taking account of non-stationarity in productivity of Atlantic Salmon (*Salmo salar* L.) in the Northwest Atlantic. ICES (International Counsel for the Exploration of the Sea) Journal of Marine Science 62:131–143.

Day, L. R. 2006. Restoring native fisheries to Maine's largest watershed: the Penobscot River restoration project. Journal of Contemporary Water Research and Education 134: 29–33.

Dempson, J. B., M. J. Robertson, C. J. Pennell, G. Furey, M. Bloom, M. Shears, L. M. N. Ollerhead, K. D. Clark, R. Hinks, and G. J. Robertson. 2011. Residency time, migration route, and survival of Atlantic Salmon *Salmo salar* smolts in a Canadian fjord. Journal of Fish Biology 78: 1976–1992.

Dingle, H., and V. A. Drake. 2007. What is migration? BioScience 57: 113–121.

Duston, J., and R. L. Saunders. 1990. The entrainment role of photoperiod on hypoosmoregulatory and growth-related aspects of smolting in Atlantic Salmon (*Salmo salar*). Canadian Journal of Zoology 68:707–715.

Faraway, J. J. 2005. Extending the linear model with R: generalized linear, mixed effects and nonparametric regression models. Chapman and Hall, London.

Fried, S. M., J. D. McCleave, and G. W. LaBar. 1978. Seaward migration of hatchery-reared Atlantic Salmon, *Salmo salar*, smolts in the Penosbcot estuary, Maine: riverine movements. Journal of the Fisheries Research Board of Canada 35:76–87.

Friedland, K. 1998. Ocean climate influences on critical Atlantic Salmon (*Salmo salar*) life history events. Canadian Journal of Fisheries and Aquatic Science 55(Supplement 1):119–130.

Friedland, K. D., D. G. Redding, J. R. McMenemy, and K. F. Drinkwater. 2003. Multidecadal trends in North American Atlantic Salmon (*Salmo salar*) and climate trends relevant to juvenile survival. Canadian Journal of Fisheries and Aquatic Sciences 60:563–583.

Friedland, K. D., B. V. Shank, C. D. Todd, P. McGinninty, and J. A. Nye. 2014. Differential response of continental stock complexes of Atlantic Salmon (*Salmo salar*) to the Atlantic multidecadal oscillation. Journal of Marine Systems 133:77–87.

Gorsky, D., J. Trial, J. Zydlewski, and J. McCleave. 2009. The effects of smolt stocking strategies on migratory path selection of adult Atlantic Salmon in the Penobscot River, Maine. North American Journal of Fisheries Management 29:949–957.

Gudjonsson, S., I. R. Jonsson, and T. Antonsson. 2005. Migration of Atlantic Salmon, *Salmo salar*, smolt through the estuary area of River Ellidaar. Environmental Biology of Fishes 74:291–296.

Haefner, P. A. Jr. 1967. Hydrography of the Penobscot River (Maine) estuary. Journal of the Fisheries Research Board of Canada 24: 1553–1571.

Halfyard, E. A., A. J. F. Gibson, M. J. W. Stokesbury, D. E. Ruzzante, and F. G. Whoriskey. 2013. Correlates of estuarine survival of Atlantic Salmon postsmolts from the southern upland, Nova Scotia, Canada. Canadian Journal of Fisheries and Aquatic Sciences 70:452–460.

Handeland, S. O., T. Järvi, A. Fernö, and S. O. Stefansson. 1997. Osmotic stress, antipredator behavior, and mortality of Atlantic Salmon (*Salmo salar*) smolts. Canadian Journal of Fisheries and Aquatic Sciences 53: 2673–2680.

Hawkes, J. P., R. Saunders, A. D. Vashon, and M. S. Cooperman. 2013. Assessing efficacy of non-lethal harassment of double-crested cormorants to improve Atlantic Salmon smolt survival. Northeastern Naturalist 20:1–18.

Hayes, S. A., and J. F. Kocik. 2014. Comparative estuarine and marine migration ecology of Atlantic Salmon and steelhead: blue highways and open plains. Reviews in Fish Biology and Fisheries 24:757–780.

Hedger, R. D., F. Martin, D. Hatin, F. Caron, F. G. Whoriskey, and J. J. Dodson. 2008. Active migration of wild Atlantic Salmon *Salmo salar* smolt through a coastal embayment. Marine Ecological Press Series 355: 235–246.

Hoar, W. S. 1988. The physiology of smolting salmonids. Pages 275–343 *in* W. S. Hoar and D. J. Randall, editors. Fish physiology, volume XIB. Academic Press, New York.

Holbrook, C. M., M. T. Kinnison, and J. Zydlewski. 2011. Survival of migrating Atlantic Salmon smolts through the Penobscot River, Maine: a prerestoration assessment. Transactions of the American Fisheries Society 140:1255–1268.

Hvidsten, N. A., and R. A. Lund. 1988. Predation on hatchery-reared and wild smolts of Atlantic Salmon, *Salmo salar* L., in the estuary of River Orkla, Norway. Journal of Fish Biology 33:121–126.

Imhoff, E. A., and R. L. Harvey. 1972. Penobscot River study. University of Maine, Environmental Studies Center, Technical Report 1, Orono.

Itokazu, Y., R. Käkelä, J. Piironen, X. L. Guan, P. Kiiskinen, and M. Vornanen. 2014. Gill tissue lipids of salmon (*Salmo salar* L.) presmolts and smolts from anadromous and landlocked populations. Comparative Biochemistry and Physiology Part A Molecular and Integrative Physiology 172:39–45.

IUBMB (International Union of Biochemistry and Molecular Biology). 1992. Enzyme nomenclature 1992. Academic Press, San Diego, California.

Järvi, T. 1989. Synergistic effect on mortality in Atlantic Salmon, *Salmo salar*, smolt caused by osmotic stress and presence of predators. Environmental Biology of Fishes 26:149–152.

Järvi, T. 1990. Cumulative acute physiological stress in Atlantic Salmon smolts: the effect of osmotic imbalance and the presence of predators. Aquaculture 89:337–350.

Jokikokko, E., and S. Mäntyniemi. 2003. The survival of stocked Atlantic Salmon smolts during sea run and the timing of migration in the River Simojoki, northern Finland. Aquaculture 219:431–444.

Keefer, M. L., G. A., Taylor, D. F. Garletts, C. K. Helms, G. A. Gauthier, T. M. Pierce, and C. C. Caudill. 2012. Reservoir entrapment and dam passage mortality of juvenile Chinook Salmon in the middle fork Willamette River. Ecology of Freshwater Fish 21:222–234.

Kocik, J. F., J. P. Hawkes, T. F. Sheehan, P. A. Music, and K. F. Beland. 2009. Assessing estuarine and coastal migration and survival of wild Atlantic Salmon smolts from the Naraguagus River, Maine using acoustic telemetry. Pages 293–310 *in* A. Haro, K. L. Smith, R. A. Rulifson, C. M. Moffitt, R. J. Klauda, M. J. Dadswell, R. A. Cunjak, J. E. Cooper, K. L. Beal, and T. S. Avery, editors, Challenges for diadromous fishes in a dynamic global environment. American Fisheries Society, Symposium 69, Bethesda, Maryland.

Laake, J. L. 2013. RMark: an R interface for analysis of capture-recapture data with MARK. National Marine Fisheries Service, Alaska Fisheries Science Center Report 2013-01, Seattle.

Lacroix, G. L. 2008. Influence of origin on migration and survival of Atlantic Salmon (*Salmo salar*) in the Bay of Fundy, Canada. Canadian Journal of Fisheries and Aquatic Sciences 65:2063–2079.

Lacroix, G. L. 2014. Large pelagic predators could jeopardize the recovery of endangered Atlantic Salmon. Canadian Journal of Fisheries and Aquatic Sciences 71:343–350.

Lacroix, G. L., D. Knox, and M. J. W. Stokesbury. 2005. Survival and behaviour of post-smolt Atlantic Salmon in coastal habitat with extreme tides. Journal of Fish Biology 66:485–498.

Marschall, E. A., M. E. Mather, D. L. Parrish, G. W. Allison, and J. R. McMenemy. 2011. Migration delays caused by anthropogenic barriers: modeling dams, temperature, and success of migrating salmon smolts. Ecological Applications 21:3014–3031.

Martin, F., R. D. Hedger, J. J. Dodson, L. Fernandes, D. Hatin, F. Caron, and F. G. Whoriskey. 2009. Behavioural transition during the estuarine migration of wild Atlantic Salmon (*Salmo salar* L.) smolt. Ecology of Freshwater Fish 18:406–417.

Mathur, D., P. G. Heisey, J. R. Skalski, and D. R. Kenney. 2000. Salmonid smolt survival relative to turbine efficiency and entrainment depth in hydroelectric power generation. Journal of the American Water Resources Association 36:737–747.

McCleave, J. D. 1978. Rhythmic aspects of estuarine migration of hatchery-reared Atlantic Salmon (*Salmo salar*) smolts. Journal of fish Biology 12:559–570.

McCormick, S. D. 1993. Methods for non-lethal gill biopsy and measurement of Na^+, K^+-ATPase activity. Canadian Journal of Fisheries and Aquatic Sciences 50:656–658.

McCormick, S. D. 2013. Smolt physiology and endocrinology. Pages 200–237 *in* S. D. McCormick, A. P. Farrell, and C. J. Brauner, editors, Euryhaline fishes. Academic Press, Oxford, UK.

McCormick, S. D., R. A. Cunjak, B. Dempson, M. F. O'Dea, and J. B. Carey. 1999. Temperature-related loss of smolt characteristics in Atlantic Salmon (*Salmo salar*) in the wild. Canadian Journal of Fisheries and Aquatic Sciences 56:1649–1658.

McCormick, S. D., L. P. Hansen, T. P. Quinn, and R. L. Saunders. 1998. Movement, migration, and smolting of Atlantic Salmon (*Salmo salar*). Canadian Journal of Fisheries and Aquatic Sciences 55:77–92.

McCormick, S. D., D. T. Lerner, M. Y. Monette, K. Nieves-Puigdoller, J. T. Kelly, and B. T. Björnsson. 2009. Taking it with you when you go: how perturbations to the freshwater environment, including temperature, dams, and contaminants, affect marine survival of Salmon. Pages 195–214 *in* A. Haro, K. L. Smith, R. A. Rulifson, C. M. Moffitt, R. J. Klauda, M. J. Dadswell, R. A. Cunjak, J. E. Cooper, K. L. Beal, and T. S. Avery, editors. Challenges for diadromous fishes in a dynamic global environment. American Fisheries Society, Symposium 69, Bethesda, Maryland.

McCormick, S. D., and R. L. Saunders. 1987. Preparatory physiological adaptations for marine life in salmonids: osmoregulation, growth and metabolism. Pages 211–229 *in* M. J. Dadswell, R. J. Klauda, C. M. Moffitt, and R. L. Saunders, editors. Common strategies in anadromous and catadromous fishes. American Fisheries Society, Symposium 1, Bethesda, Maryland.

McCormick, S. D., R. L. Saunders, and A. MacIntyre. 1989. Mitochondrial enzyme and Na^+, K^+-ATPase activity, and ion regulation during parr-smolt transformation of Atlantic Salmon (*Salmo salar*). Fish Physiology and Biochemistry 6:231–241.

McCormick, S. D., J. M. Shrimpton, and J. D. Zydlewski. 1997. Temperature effects on osmoregulatory physiology of anadromous fish. Pages 279–301 *in* C. M. Wood and D. G. McDonald, editors. Global warming implications for freshwater and marine fish. University Press, Cambridge, UK.

Miller, A. S., T. F. Sheehan, M. D. Renkawitz, A. L. Meister, and T. J. Miller. 2012. Revisiting the marine migration of US Atlantic Salmon using historical Carlin tag data. ICES (International Council for Exploration of the Sea) Journal of Marine Science 69:1609–1615.

Mills, K. E., A. J. Pershing, T. F. Sheehan, and D. F. Mountain. 2013. Climate and ecosystem linkages explain widespread declines in North American Atlantic Salmon populations. Global Change Biology 19:3046–3061.

Montgomery, D. C., E. A. Peck, and G. G. Vining. 2006. Introduction to linear regression analysis, 4th edition. Wiley, Hoboken, New Jersey.

Moore, A., E. C. E. Potter, N. J. Milner, and S. Bamber. 1995. The migratory behavior of wild Atlantic Salmon (*Salmo salar*) smolts in the estuary of the River Conway, North Wales. Canadian Journal of Fisheries and Aquatic Sciences 52:1923–1935.

Music, P. A., J. P. Hawkes, and M. S. Cooperman. 2011. Magnitude and causes of smolt mortality in rotary screw traps: an Atlantic Salmon case study. North American Journal of Fisheries Management 30:713–722.

Otero, J., J. H. L'Abée-Lund, T. Castro-Santos, K. Leonardsson, G. O. Storvik, B. Jonsson, B. Dempson, I. C. Russell, A. J. Jensen, J.-L. Baglinière, M. Dionne, J. D. Armstrong, A. Romakkaniemi, B. H. Letcher, J. F. Kocik, J. Erkinaro, R. Poole, G. Rogan, H. Lundqvist, J. C. MacLean, E. Jokikokko, J. V. Arnekleiv, R. J. Kennedy, E. Niemelä, P. Caballero, P. A. Music, T. Antonsson, S. Gudjonsson, A. E. Veselov, A. Lamberg, S. Groom, B. H. Taylor, M. Taberner, M. Dillane, F. Arnason, G. Horton, N. A. Hvidsten, I. R. Jonsson, N. Jonsson, S. McKelvey, T. F. Naesje, Ø. Skaala, G. W. Smith, H. Saegrov, N. C. Stenseth, and L. A. Vøllestad. 2014. Basin-scale phenology and effects of climate variability on global timing of initial seaward migration of Atlantic Salmon (*Salmo salar*). Global Change Biology 20:61–75.

Parrish, D. L., R. J. Behnke, S. R. Gephard, S. D. McCormick, and G. H. Reeves. 1998. Why aren't there more Atlantic Salmon (*Salmo salar*)? Canadian Journal of Fisheries and Aquatic Sciences 55(Supplement 1):281–287.

Poe, T. P., H. C. Hansel, S. Vigg, D. E. Palmer, and L. A. Prendergast. 1991. Feeding of predaceous fishes on out-migrating juvenile salmonids in John Day reservoir, Columbia River. Transactions of the American Fisheries Society 120:405–420.

Price, C. S., and C. B. Schreck. 2003. Stress and saltwater-entry behavior of juvenile Chinook Salmon (*Oncorhynchus tshawytscha*): conflicts in physiological motivation. Canadian Journal of Fisheries and Aquatic Sciences 60:910–918.

R Development Core Team. 2014. R: a language and environment for statistical computing. R Foundation for Statistical Computing, Vienna.

Schaller, H. A., C. E. Petrosky, and E. S. Tinus. 2014. Evaluating river management during seaward migration to recover Columbia River stream-type Chinook Salmon considering variation in marine conditions. Canadian Journal of Fisheries and Aquatic Sciences 71:259–271.

Seiwell, H. R. 1932. Some physical characteristics of the water of Penobscot Bay, Maine, especially in relation to the tides and a discussion of the results obtained by duplicate measurements of specific gravity of sea water by the Knudsen and Plummet methods. Internationale Revue der Gesamten Hydrobioligie and Hydrographie 27:315–331.

Sheehan, T. F., M. D. Renkawitz, and R. W. Brown. 2011. Surface trawl survey for US origin Atlantic Salmon *Salmo salar*. Journal of Fish Biology 79:374–398.

Smith, P. K., R. I. Krohn, G. T. Hermanson, A. K. Mallia, F. H. Gartner, M. D. Provenzano, E. K. Fujimoto, N. M. Goeke, B. J. Olson, and D. C. Klenk. 1985. Measurements of protein using bicinchoninic acid. Analytical Biochemistry 150:76–85.

Stich, D. S, M. M. Bailey, J. D. Zydlewski. 2014. Survival of Atlantic Salmon *Salmo salar* smolts through a hydropower complex in the lower Penobscot River, Maine USA. Journal of Fish Biology 85:1074–1096.

Stier, D. J., and B. Kynard. 1986. Use of radio telemetry to determine the mortality of Atlantic Salmon smolts passed through a 17-MW Kaplan turbine at a low-head hydroelectric dam. Transactions of the American Fisheries Society 115:771–775.

Strand, J. E. T., J. G. Davidsen, E. H. Jørgensen, and A. H. Rikardsen. 2010. Seaward migrating Atlantic Salmon smolts with low levels of gill Na^+, K^+-ATPase activity; is sea entry delayed? Environmental Biology of Fishes 90:317–321.

Svenning, M. A., R. Borgstrøm, T. O. Dehli, G. Moen, R. T. Barrett, T. Pedersen, and W. Vader. 2005. The impact of marine fish predation on Atlantic Salmon smolts (*Salmo salar*) in the Tana estuary, north Norway, in the presence of and alternative prey, Lesser Sandeel (*Ammodytes marinus*). Fisheries Research 76:466–474.

Sykes, G. E., C. J. Johnson, and J. M. Shrimpton. 2009. Temperature and flow effects on migration timing of Chinook Salmon smolts. Transactions of the American Fisheries Society 138:1252–1265.

Thorstad, E. B., I. Uglem, B. Finstad, C. M. Chittenden, R. Nilsen, F. Økland, and P. A. Bjørn. 2012a. Stocking location and predation by marine fishes affect survival of hatchery-reared Atlantic Salmon smolts. Fisheries Management and Ecology 19:400–409.

Thorstad, E. B., F. Whoriskey, I. Uglem, A. Moore, A. H. Rikardsen, and B. Finstad. 2012b. A critical life stage of the Atlantic Salmon *Salmo salar*: behaviour and survival during the smolt and initial post-smolt migration. Journal of Fish Biology 81:500–542.

Todd, C. D., K. D. Friedland, J. C. MacLean, B. D. Whyte, I. C. Russell, M. E. Lonergan, and M. B. Morrissey. 2012. Phenological and phenotypic changes in Atlantic Salmon populations in response to a changing climate. ICES (International Council for the Exploration of the Sea) Journal of Marine Science 69:1686–1698.

USASAC (U.S. Atlantic Salmon Assessment Committee). 2012. Annual report of the U.S. Atlantic Salmon Assessment Committee, report 24. Available: http://www.nefsc.noaa.gov/USASAC/Reports/. (August 2014).

USASAC (U.S. Atlantic Salmon Assessment Committee). 2014. Annual report of the U.S. Atlantic Salmon Assessment Committee, report 26. Available: http://www.nefsc.noaa.gov/USASAC/Reports/. (May 2014).

Weitkamp, L. A., G. Goulette, J. Hawkes, M. O'Malley, and C. Lipsky. 2014. Juvenile salmon in estuaries: comparisons between North American Atlantic and Pacific salmon populations. Reviews in Fish Biology and Fisheries 24:713–736.

Whalen, K. G., D. L. Parrish, and S. D. McCormick. 1999. Migration timing of Atlantic Salmon smolts relative to environmental and physiological factors. Transactions of the American Fisheries Society 128:289–301.

White G. C., and K. P. Burnham. 1999. Program MARK: survival estimation from populations of marked animals. Bird Study 46:120–139.

Zaugg, W. S. 1982. Some changes in smoltification and seawater adaptability of salmonids resulting from environmental and other factors. Aquaculture 28:143–151.

Zaugg, W. S., and L. R. McLain. 1972. Changes in gill adenosinetriphosphatase activity associated with parr-smolt transformation in steelhead trout, Coho, and spring Chinook salmon. Journal of the Fisheries Research Board of Canada 29:161–171.

Zuur, A., E. N. Ieno, N. Walker, A. A. Saveliev, and G. M. Smith. 2009. Mixed effects models and extensions in ecology with R. Springer, New York.

Zydlewski, G. B, A. Haro, and S. D. McCormick. 2005. Evidence for cumulative temperature as an initiating and terminating factor in downstream migratory behavior of Atlantic Salmon (*Salmo salar*) smolts. Canadian Journal of Fisheries and Aquatic Sciences 62:68–78.

Zydlewski, G. B., D. S. Stich, and S. D. McCormick. 2014. Photoperiod control of the initiation of downstream migration in Atlantic Salmon smolts. Journal of Fish Biology 85:1074–1096.

Zydlewski, G. B., and J. Zydlewski. 2012. Gill Na^+, K^+-ATPase of Atlantic Salmon smolts in freshwater is not a predictor of long-term growth in seawater. Aquaculture 362–363:121–126.

Zydlewski, J., and M. P. Wilkie. 2013. Freshwater to saltwater transitions in migratory fishes. Pages 254–294 *in* S. D. McCormick, A. P. Farrell, and C. J. Brauner, editors, Euryhaline fishes. Academic Press, Oxford, UK.

Zydlewski, J., G. Zydlewski, and G. R. Danner. 2010. Descaling injury impairs the osmoregulatory ability of Atlantic Salmon smolts entering seawater. Transactions of the American Fisheries Society 138:129–136.

16

Application of Otolith Chemical Signatures to Estimate Population Connectivity of Red Snapper in the Western Gulf of Mexico

Michelle Zapp Sluis*[1]

Department of Oceanography and Coastal Sciences, Louisiana State University, Energy, Coast, and Environment Building, Baton Rouge, Louisiana 70803, USA

Beverly K. Barnett

National Marine Fisheries Service, Southeast Fisheries Science Center, Panama City Laboratory, 3500 Delwood Beach Road, Panama City Beach, Florida 32408, USA

William F. Patterson III

Department of Marine Sciences, University of South Alabama, and Dauphin Island Sea Lab, 101 Bienville Boulevard, Dauphin Island, Alabama 36528, USA

James H. Cowan Jr.

Department of Oceanography and Coastal Sciences, Louisiana State University, Energy, Coast, and Environment Building, Baton Rouge, Louisiana 70803, USA

Alan M. Shiller

Department of Marine Science, University of Southern Mississippi, 1020 Balch Boulevard, Stennis Space Center, Mississippi 39529, USA

Abstract

Otolith chemical signatures of Red Snapper *Lutjanus campechanus* from six nursery regions were used to estimate the sources of recruits to four sampling regions in the western Gulf of Mexico (Gulf) and to estimate whether postsettlement mixing of Red Snapper occurs between the U.S. and Mexican portions of the western Gulf. In a previous study, region-specific otolith signatures (element : Ca ratios: Ba:Ca, Mg:Ca, Mn:Ca, Sr:Ca, and Li:Ca; stable isotope delta values: $\delta^{13}C$ and $\delta^{18}O$) were developed based on age-0 Red Snapper (2005–2007 year-classes) sampled from the six nursery areas. In the present study, subadult and adult Red Snapper (ages 1–3) belonging to those same year-classes were collected from four sampling regions within the western Gulf (two regions in U.S. waters; two regions along the Mexican continental shelf) during summer in 2006–2008. Left sagittal otoliths were used to age subadults and adults to the corresponding nursery year-classes, and right sagittal otoliths were cored for chemical analysis. Off the southwestern U.S. coast, the sampled age-1–3 Red Snapper included locally derived recruits as well as recruits from the northwestern Gulf nursery region. However, analytical results were inconclusive with respect to estimating the connectivity between Red Snapper populations in U.S. and Mexican waters of the western Gulf.

Subject editor: Carl Walters, University of British Columbia, Canada

*Corresponding author: sluism@tamug.edu

[1]Present address: Department of Marine Biology, Texas A&M University at Galveston, 1001 Texas Clipper Road, Galveston, Texas 77554, USA.

The Gulf of Mexico (hereafter, Gulf) fishery for Red Snapper *Lutjanus campechanus* began over 150 years ago off the coast of Pensacola, Florida; however, due to severe overfishing, the stock became depleted by the late 1800s (Camber 1955). The fishery then shifted to the western Gulf from the mouth of the Mississippi River to the southern coast of Texas and even as far south as the Campeche Banks off the coast of Mexico. High landings of Red Snapper continued until the early 1980s, when the U.S. fishing fleet was banned from Mexican waters, thereby restricting the fleet to the western Gulf from Mississippi–Alabama to Texas (Gallaway et al. 1998). Catches continued to decline due to high levels of commercial and recreational exploitation, and bycatch mortality from the shrimp fishery, resulting in Gulf Red Snapper being currently overfished (GMFMC 2010).

Overexploitation of the Red Snapper fishery is also evident in Mexican Gulf waters. The Campeche Banks fishery was initially the national leader in Red Snapper production. However, due to adverse effects from Mexican and Cuban commercial fisheries, and bycatch mortality from the Mexican shrimp fishery, landings of Red Snapper from the Campeche Banks declined by 51.2% between the 1980s and the late 1990s (Monroy-García et al. 2002), and the Mexican stock was estimated to be severely overfished by 2005 (SAGARPA 2006, as cited by Brule et al. 2010). Mexico has established fishing regulations, including commercial finfish permits, hook size restrictions, and an annual catch quota for the Cuban fleet, but there is still a need for stricter regulations.

Management of the U.S. Gulf Red Snapper stock was implemented in November 1984 via the Gulf of Mexico Fishery Management Council's fishery management plan for reef fishes, which was designed to rebuild declining fish stocks. In compliance with regulations set by the Magnuson–Stevens Fishery Conservation and Management Act, several amendments have been adopted to end overfishing and rebuild the Red Snapper stock by 2032. Currently, constraints are placed on both directed fisheries (annual catch limits, bag and minimum size limits, seasonal closures, and reef fish permits) and on the Gulf shrimp fishery (reduction in effort, area closures, and bycatch reduction devices on shrimp trawls; GMFMC 2010).

The Red Snapper population has been categorized into eastern and western substocks (divided by the Mississippi River; SEDAR 2005) based on demographic differences in size at age, maturation rates, and genetic effective population size (N_e) of Red Snapper that occur across the Gulf (Fischer et al. 2004; Saillant and Gold 2006; Jackson et al. 2007). However, plans to rebuild Red Snapper biomass are applied Gulf-wide rather than at the level of individual management subunits. Gold and Saillant (2007) determined that the N_e of Red Snapper off the Louisiana coast was an order of magnitude larger than the N_e off the Alabama and Texas coasts, alluding to spatial differences in the number of viable adults that were able to produce surviving offspring. Uniform reduction of fishing mortality Gulf-wide is expected to result in the western substock recovering faster and to a greater spawning stock biomass level than the eastern substock, since the western substock has a higher biomass and an estimated lower fishing mortality relative to the eastern substock (SEDAR 2013). Thus, without a reconfiguration of the current management approach, the greater size of the western substock is projected to continue.

Demographic differences also exist within the western substock. Studies have shown that Red Snapper collected off Texas are significantly smaller at age and reach a smaller maximum size than those collected off Louisiana (Fischer et al. 2004; Saari 2014). Saari (2014) also reported a higher proportion of older fish collected off north Texas and Louisiana than from all other Gulf regions, which was possibly attributable to the higher stock abundance of the western Gulf. Although differences in Red Snapper growth rates have been linked to increased primary production associated with the Mississippi River plume (Fischer et al. 2004), an understanding of population structure and connectivity could further explain the demographic differences within the western Gulf. Furthermore, the degree of connectivity that exists between the Red Snapper population off south Texas and the population along the northeastern coast of Mexico is unknown. Given that the Mexican stock is severely overfished, high connectivity between Texas and Mexican Red Snapper populations could mean that the Mexican fishery serves as a sink for Texas recruits (sensu Crowder et al. 2000).

The use of otolith chemistry to develop natural tags has become an effective tool for fishery scientists to distinguish juveniles from distinct nursery areas and then estimate the contribution of different nursery areas to adult stocks (Thorrold et al. 1998, 2001; Rooker et al. 2001, 2008). The otolith precipitates as the fish grows and is metabolically inert once formed; thus, the chemical signatures from surrounding seawater accreted onto the growing surface will be permanently retained (as reviewed by Campana 1999). This allows material that is deposited during the juvenile stage to act as a natural marker indicating the nursery of origin. Chemical signatures contained within the core (or juvenile portion) of the otolith can then be used to identify the nursery of origin of adult fish. Barnett and Patterson (2010) determined that the otolith core from an adult Red Snapper could be mechanically extracted and would yield effective results for analyzing nursery chemical signatures. Furthermore, Patterson et al. (2008) and Zapp Sluis et al. (2012) demonstrated that Red Snapper from various nursery regions within the Gulf can be distinguished based on otolith chemical signatures. Employing otolith signatures to examine population connectivity and mixing dynamics is essential to the development of marine population dynamics and the management of fishery stocks (Cowen et al. 2000).

The purpose of this study was to apply the otolith chemical nursery signatures identified by Zapp Sluis et al. (2012) to estimate the population structure and connectivity of Red Snapper in the western Gulf. Specifically, natural tags derived from element : Ca and stable isotope ratios in otoliths of age-0 Red Snapper sampled from six regions throughout the Gulf were compared to element : Ca and stable isotope ratios for the otolith cores in subadult and adult Red Snapper sampled from four western Gulf regions. The objectives were to estimate the sources of Red Snapper recruits to these regions and to examine Red Snapper mixing dynamics between U.S. and Mexican regions of the western Gulf.

METHODS

Sample collection.—Subadult and adult Red Snapper were sampled from the northwestern Gulf (NWG); southwestern Gulf (SWG); southern Gulf shelf between Tampico and Veracruz, Mexico (MEX1); and Campeche Banks, Mexico (MEX2; Figure 1). To correspond to nursery signatures developed for the 2005–2007 year-classes (see Zapp Sluis et al. 2012), Red Snapper of the following ages were targeted during the summer (May–August): age-1 fish were targeted in 2006–2008, age-2 fish were targeted in 2007–2008, and age-3 fish were targeted in 2008. The objective was to sample 50 Red Snapper per year-class in each region over a 3-year period, equaling 1,200 samples total ([50 fish × 1 year-class × 4 regions] + [50 fish × 2 year-classes × 4 regions] + [50 fish × 3 year-classes × 4 regions] = 1,200). Subadult and adult Red Snapper were sampled onboard National Marine Fisheries Service vessels during scientific bottom trawl surveys; from recreational landings around Port Aransas, Texas, and Port Fourchon, Louisiana; and from bycatch in the Mexican shrimp fishery. Due to difficulty in collecting samples from MEX1 and MEX2, sampling in those regions occurred later in the winter (December–March), and no samples were obtained in 2008. Red Snapper TL was measured to the nearest millimeter; both sagittal otoliths were extracted (either in the field or in the laboratory), rinsed free of associated tissue by using double-deionized water ($DDIH_2O$; ultra-pure 18-MΩ/cm water), and stored in individual paper coin envelopes until further laboratory analysis.

Otolith preparation and analysis.—Otoliths were cleaned with a synthetic-bristle brush to remove any adhering tissue, rinsed with $DDIH_2O$, and placed in polyethylene vials to air dry under a class-100 clean hood. The left sagitta was used to determine fish age for each sample. Transverse sections of the otolith were viewed under a dissecting microscope with transmitted light to count opaque zones and accurately determine age via the protocols of Patterson et al. (2001a) and Fischer et al. (2002). Once age was verified, stratified random sampling was used to select the otoliths of up to 50 fish per region per year-class from each summer's sample, and those otoliths were used for coring and chemical analysis.

Right otoliths selected for chemical analysis were embedded in epoxy resin, and a transverse section containing the core was cut with a Buhler Isomet low-speed saw fitted with twin diamond blades separated by a 1.5-mm nylon spacer. Empty sections of epoxy resin from the same block containing the otolith were also cut and affixed to an acid-leached microscope slide with Loctite Super Glue Control Gel. Anterior and posterior ends of the associated epoxy with the embedded transverse otolith section were then affixed to the empty epoxy section with Loctite gel such that the glue did not come into contact with the otolith section. Using the method of Barnett and Patterson (2010), otolith cores were removed from the embedded transverse section with a New Wave MicroMill precision drilling instrument. The empty section of epoxy resin was used to protect the drill bit from possibly hitting the slide and to prevent the otolith core from cracking during the drilling process. A pre-determined path based on the average transverse section perimeters for otoliths from 20 age-0 Red Snapper was programmed into the MicroMill system to extract the age-0 core section from each subadult/adult otolith sample (Figure 2A, B). The drilling process required 24 passes at 75-μm depth per pass with a scan speed of 85 μm/s and at 80% drill speed. Otolith cores were easily extracted from the transverse section with this process (Figure 2C). Extracted cores were placed in clear microcentrifuge tubes for storage until analysis of chemical signatures.

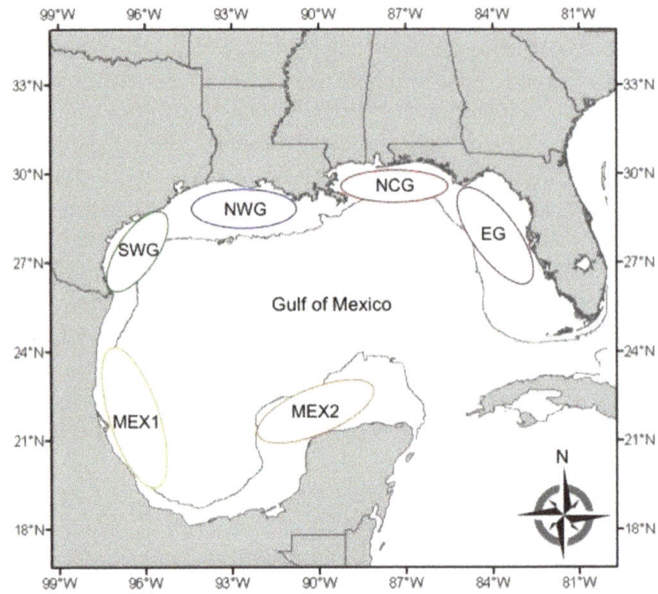

FIGURE 1. Six nursery regions along the continental shelf in the Gulf of Mexico (Gulf), where age-0 Red Snapper were sampled during 2005–2007 (Zapp Sluis et al. 2012). Subadult and adult Red Snapper of those same cohorts were sampled from four of the regions during 2006–2008: the northwestern Gulf (NWG), southwestern Gulf (SWG), Mexico region 1 (MEX1), and Mexico region 2 (MEX2). Additional nursery regions include the eastern Gulf (EG) and north-central Gulf (NCG). The 200-m depth contour line indicates the continental shelf edge.

Prior to elemental analysis or stable isotope analysis, the extracted otolith cores were cleaned under a class-100 clean hood. Dried cores were weighed to the nearest 0.01 mg before

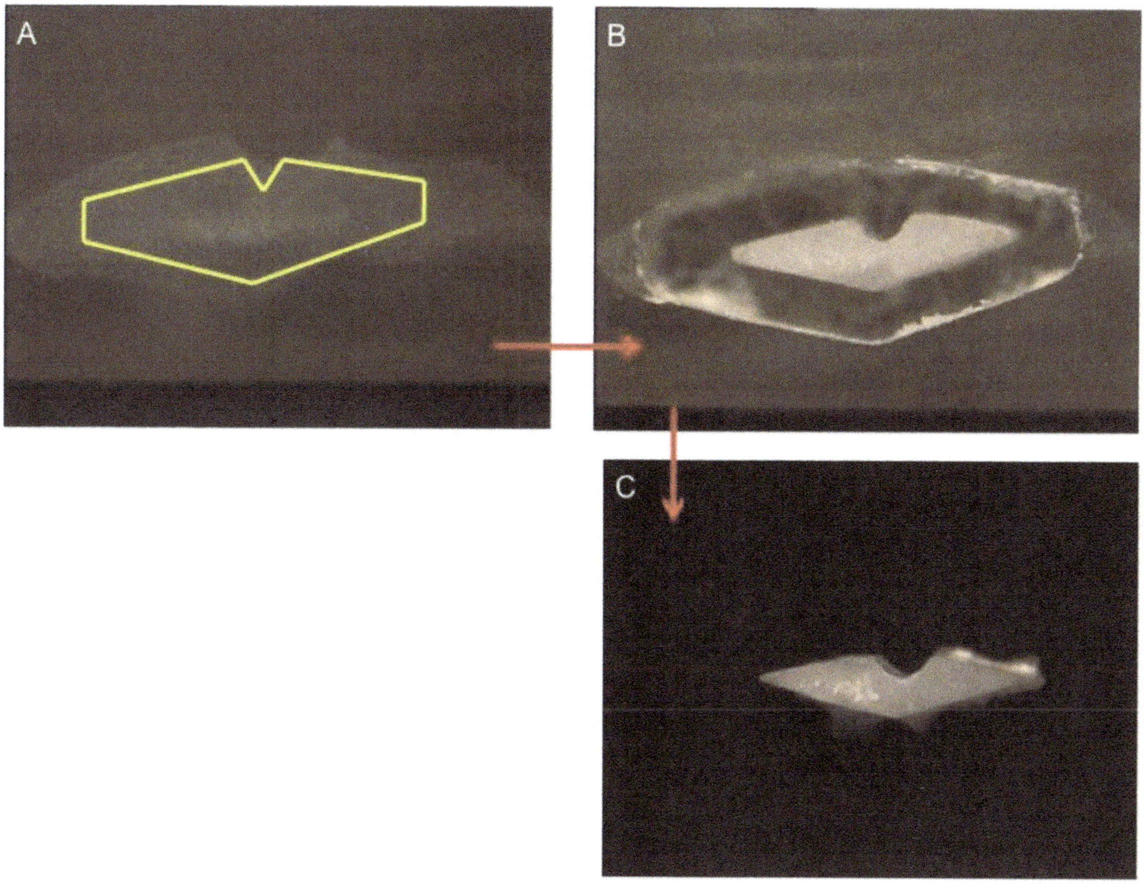

FIGURE 2. Transverse section of a sagittal otolith from an adult Red Snapper, depicting **(A)** a yellow outline of the template pattern, which was used to extract **(B)** the age-0 core with a MicroMill precision drilling instrument. **(C)** The resulting intact extracted core is also shown.

and after cleaning. Whole cores were immersed in 1% ultrapure nitric acid (HNO_3) for 30 s to clean the surface and then were flooded repeatedly with $DDIH_2O$ to remove the acid. Cores were dried under a class-100 clean hood for at least 24 h. Once dried and reweighed, the otolith cores were pulverized with an acid-leached mortar and pestle, and the resulting homogenized powder was divided into two approximately equal proportions. Half of the otolith core powder was weighed to the nearest 0.1 mg and then dissolved in an acid-leached, high-density polyethylene vial by adding 1% ultrapure HNO_3 until a dilution factor of approximately 1,000-fold was achieved. Although total dissolution typically occurred within 1 h, samples were not manipulated for at least 24 h after acid digestion began. Aliquots (5 mL) of the core solutions were sent to the University of Southern Mississippi for trace elemental analysis with a Thermo Fisher Element2 sector field (SF) inductively coupled plasma (ICP) mass spectrometer. Core solutions were spiked with indium at a concentration of 2 ng/mL as an internal standard and then were analyzed for ^{137}Ba, ^{48}Ca, ^{7}Li, ^{55}Mn, ^{25}Mg, and ^{86}Sr. Calibration was achieved using (1) external standards that were made to include approximately the same Ca concentration as the samples and (2) standards that were made without added Ca. Blanks were prepared from 1% ultrapure HNO_3 and were processed through the same sample preparation stages as the sample solutions. Blanks were analyzed concurrently with sample solutions to estimate instrument limits of detection as three SDs of the blank values. Instrument performance and matrix effects were checked by assaying elemental concentrations of an otolith standard reference material that was prepared from adult Red Snapper otoliths (Sturgeon et al. 2005). Solutions of the standard reference material were prepared and analyzed similarly to the otolith core samples. Measured precision (% relative SD; $n = 22$) of the method was 2% for Ba:Ca, 5% for Li:Ca, 6% for Mn:Ca, 16% for Mg:Ca, and 1% for Sr:Ca. Recovery estimates were 102% for Ba:Ca, 79% for Li:Ca, 102% for Mn:Ca, 116% for Mg:Ca, and 94% for Sr:Ca based on comparison with the certified values from Sturgeon et al. (2005). Note that Sturgeon et al.'s (2005) certified values have uncertainties ranging from 2% (Mn:Ca) to 13% (Li:Ca), and thus our results are indicative of satisfactory agreement.

The other half of the powder from each otolith core sample was placed into a 2-mL microcentrifuge tube and was sent to the Stable Isotope Laboratory at the Department of Geology,

University of California–Davis, for stable isotope ($\delta^{13}C$, $\delta^{18}O$) analysis with a Finnigan MAT 251 isotope ratio (IR) mass spectrometer. The instrument was calibrated against the International Atomic Energy Agency's carbonate standard, NBS-19. Accuracy of analytical runs was measured through routine analysis of a check standard that had been stringently calibrated against NBS-19. Method precision based on long-term monitoring of the NBS-19 standard was ±0.02‰ for $\delta^{13}C$ and ±0.06% for $\delta^{18}O$. The isotopic composition of otolith cores are reported in standard delta (δ) notation relative to the Vienna Pee Dee belemnite reference standard:

$$\delta_{\text{sample}}(‰) = \left[\left(R_{\text{sample}}/R_{\text{standard}}\right) - 1\right]\cdot 10^3,$$

where R represents the ratio of heavy isotope to light isotope ($^{13}C/^{12}C$ or $^{18}O/^{16}O$).

Statistical analysis.—Cohort- and year-specific residual values were computed by subtracting mean element : Ca and stable isotope ratios from each respective sample ratio. This process was repeated for the cohort-specific age-0 Red Snapper element : Ca and stable isotope ratios presented in Zapp Sluis et al. (2012). Residuals were computed for otolith chemical signatures from age-0 fish and from subadult/adult fish (core samples) to remove extraneous sources of variance (i.e., ontogenetic effects of disproportionate primordium representation in cored otoliths versus the original three-dimensional structure of age-0 otoliths; instrument drift between sample analysis of age-0 samples and otolith core samples; etc.) when estimating the source regions for subadult and adult samples (Thorrold et al. 2001; Barnett and Patterson 2010).

A Bayesian model was used to estimate the source of recruits to a given region in a given sampling year. Full methodological details on the Bayesian model and accompanying R package (R Development Core Team 2007) used in this study are provided by Smith and Campana (2010). The baseline data set consisted of the residual values of otolith signatures from age-0 Red Snapper that were sampled in nursery regions. Residuals of otolith core signatures from adults and subadults were classified as unknowns (or mixed data) against the age-0 baseline data to estimate their nursery source(s). Even though significant differences among year-classes were evident for the age-0 otolith chemical signatures (Zapp Sluis et al. 2012), they were also combined to determine whether nursery signatures pooled across year-classes could help fill data gaps. Thus, the mixed data for each subadult/adult age-group in each region and each year sampled were classified individually to the year-class-specific baseline data as well as to the baseline data pooled across year-classes. Posterior distributions were used to calculate 95% credible intervals (CIs) for the proportion of mixed samples assigned to each baseline group (i.e., nursery region). An overlap in CIs indicated an ambiguous assignment to the baseline groups.

RESULTS

In total, 1,338 subadult and adult Red Snapper were collected from the four western Gulf sampling regions. Based on the ages estimated for those fish, only 725 individuals corresponded to the designated regions and cohorts of interest, and their otoliths were cored for chemical analysis (Table 1). Few of the samples obtained from MEX1 and MEX2 corresponded to the targeted year-classes, resulting in low sample sizes for those regions. All six elements (Ba, Ca, Li, Mn, Mg, and Sr) were present in concentrations at least two orders of magnitude above detection limits in all samples.

Mean concentrations and natural variability of element : Ca and stable isotope ratios varied across regions and year-classes, as would be expected based upon similar trends in the age-0 baseline data (see Zapp Sluis et al. 2012). The element : Ca and stable isotope ratios also varied among age-groups within a given cohort. For the 2005 cohort, otoliths of age-2 Red Snapper collected from NWG had higher Ba:Ca, Mg:Ca, and Mn:Ca values than the other age-groups. There was also a steady decrease in Li:Ca values with increasing age, and only $\delta^{13}C$ remained constant and within the same range as baseline nursery values (Figure 3). In the 2005-cohort samples from SWG, otolith element : Ca and stable isotope ratios remained constant across age-groups except for a similar increase in otolith Ba:Ca, Mg:Ca, and Mn:Ca values for age-2 fish. In MEX1, otolith ratios for the 2005 cohort remained constant across age-groups except for Li:Ca, $\delta^{13}C$, and $\delta^{18}O$. For MEX2 samples of the 2005 cohort, otolith ratios decreased between age-groups for every element except $\delta^{18}O$, which increased.

For the 2006 cohort, element : Ca and stable isotope ratios in the otoliths of NWG adult/subadult Red Snapper remained fairly constant between age-groups except for a decrease in Mg:Ca and Mn:Ca (Figure 4). Otolith ratios for the 2006 cohort sampled from SWG only remained constant between age-groups for Li:Ca and $\delta^{13}C$ and also exhibited the same decrease in values as the 2006-cohort samples from NWG. For the 2007 cohort, otolith element : Ca and stable isotope ratios in NWG Red Snapper were similar to the corresponding baseline age-0 samples (see Zapp Sluis et al. 2012) except for being more enriched in $\delta^{18}O$ (Figure 5). Otolith ratios for the 2007-cohort samples from SWG were lower in Ba:Ca and Mg:Ca and more enriched in $\delta^{18}O$ relative to the baseline data. Interestingly, in the NWG and SWG samples, $\delta^{18}O$ ratios increased for each age-group within each cohort relative to the baseline age-0 nursery data (see Zapp Sluis et al. 2012).

Classification of subadults and adults (mixed nursery origin) from the 2005 cohort sampled in NWG indicated that the proportion of locally derived fish (i.e., NWG nursery region) increased as age increased (Figure 6). Furthermore, the proportion of NWG-sampled adults assigned to the NWG nursery region differed significantly from the proportions assigned to the other nursery regions, as indicated by the non-overlapping 95% CIs (Figure 6). The secondary source of recruits to NWG was estimated to be the NCG nursery region, with the

TABLE 1. Sample size and TL range of subadult and adult Red Snapper collected from four regions across the Gulf of Mexico (Gulf) during summer in 2006–2008 (NWG = northwestern Gulf; SWG = southwestern Gulf; MEX1 = Mexico region 1; MEX2 = Mexico region 2).

Sampling year	Age	Year-class	Region	Otolith samples (cored and analyzed)	TL (mm) range
2006	1	2005	NWG	51	153–241
			SWG	52	151–226
			MEX1	18	250–280
			MEX2	3	240–250
2007	1	2006	NWG	56	151–235
			SWG	44	153–258
			MEX1	31	230–380
			MEX2	3	240–280
	2	2005	NWG	55	186–443
			SWG	60	232–348
			MEX1	50	240–320
			MEX2	1	480
2008	1	2007	NWG	50	152–209
			SWG	50	151–237
			MEX1		
			MEX2		
	2	2006	NWG	50	220–410
			SWG	50	165–422
			MEX1		
			MEX2		
	3	2005	NWG	50	335–470
			SWG	50	301–457
			MEX1		
			MEX2		

contribution from NCG decreasing as age increased; however, the CIs overlapped with those for the other two nursery regions. Classification of adults from the 2005 cohort sampled in SWG appeared to be locally derived, yet only the age-2 samples differed significantly in the proportion assigned to SWG versus the other nursery regions (Figure 6). Although age-1 and age-2 samples for the 2005 cohort were collected from MEX1 and MEX2, baseline nursery data were not available for these regions; therefore, the samples were not included in Bayesian models for the 2005 cohort. For the 2006 cohort, adult samples from NWG were primarily assigned to the NWG and SWG nursery regions. Among the 2006-cohort adults sampled in SWG, age-1 fish were classified to MEX1 and MEX2 nursery regions, and age-2 fish were primarily assigned to the SWG nursery region. However, the only instance in which CIs did not overlap for the 2006 cohort was for the proportion of MEX2 adult samples that were assigned to the MEX2 nursery region. Although not significantly different at the 95% level, age-1 fish of the 2007 cohort sampled in NWG were classified to both the NWG and SWG nursery regions, while age-1 fish from SWG were typically assigned to the NWG nursery region (Figure 6).

When nursery sources were analyzed by using baseline data that were pooled across year-classes, similar trends emerged. For the 2005 cohort of Red Snapper, adults sampled from NWG were still classified as being primarily locally derived; however, the only case in which CIs did not overlap was for age-3 individuals (Figure 7). Adult samples from SWG were mostly classified to the NWG nursery region rather than being locally derived. Age-1 samples from MEX1 were primarily classified to the MEX1 nursery region, and a large proportion of age-2 samples from MEX1 were assigned to the MEX2 nursery region; however, all of the 95% CIs overlapped. For age-1 fish sampled from MEX2, assignments were mainly to the EG nursery region (although not significantly so); however, this estimate was based on a low sample size ($n = 3$). The age-2 sample from MEX2 could not be analyzed because the sample size was one fish. For the 2006 cohort, age-1 individuals sampled from NWG were assigned to both the NWG and SWG nursery regions, whereas age-2 fish were primarily classified as originating from the SWG nursery region. Adult samples from SWG were predominantly locally derived; the 95% CIs for age-2 fish were nonoverlapping. Adult samples obtained in MEX1 were also assigned to the local MEX1 nursery baseline signature, although there were some instances of CI overlap. The MEX2 adult samples were primarily assigned to the NWG nursery region, but again the CIs greatly overlapped. For the 2007 cohort of Red Snapper, age-1 fish

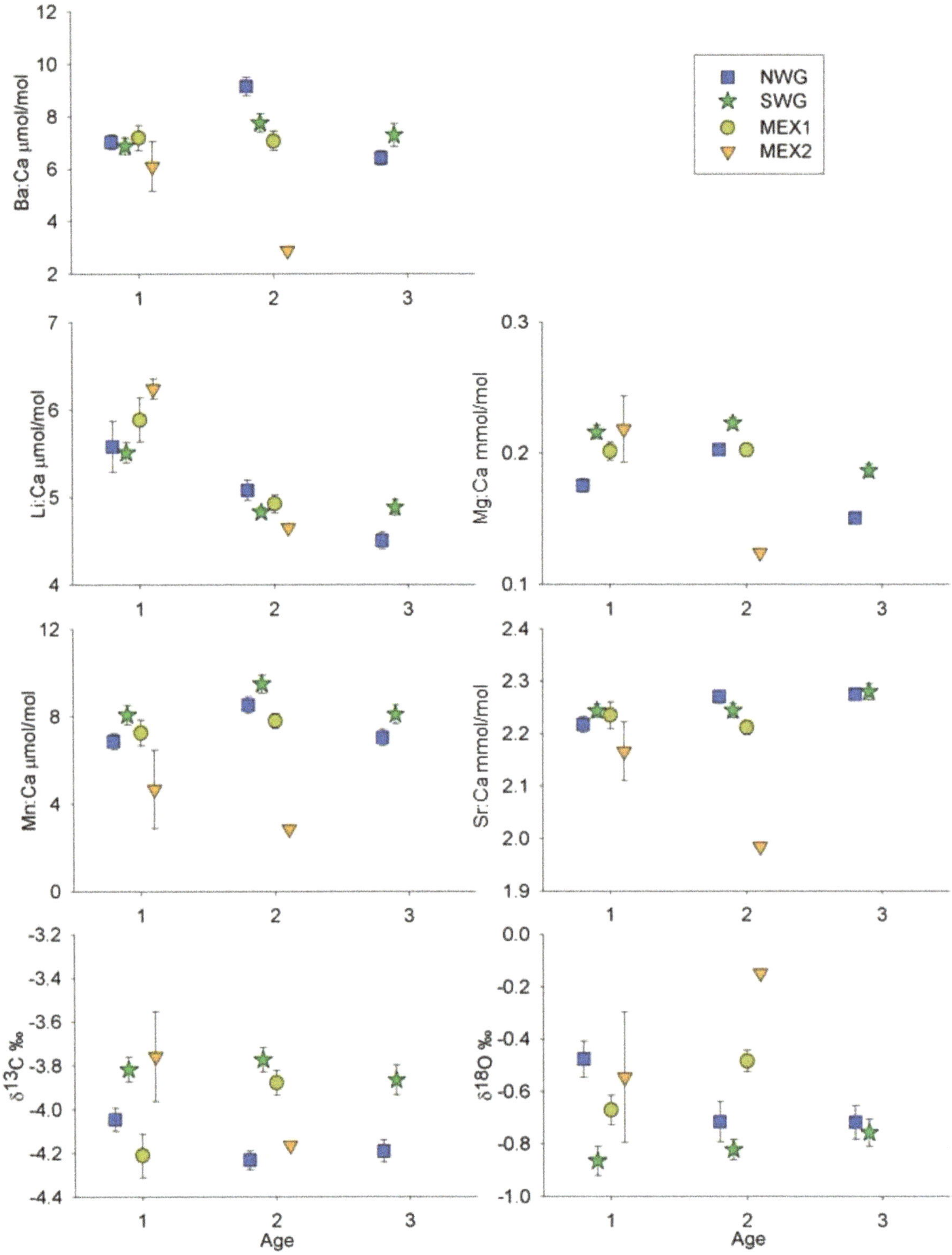

FIGURE 3. Mean (±SE) element : Ca or stable isotope delta ratios for otolith cores from 2005-cohort Red Snapper (subadults and adults, ages 1–3) sampled in four regions of the western Gulf of Mexico (Gulf) during summer in 2006–2008 (NWG = northwestern Gulf; SWG = southwestern Gulf; MEX1 = Mexico region 1; MEX2 = Mexico region 2).

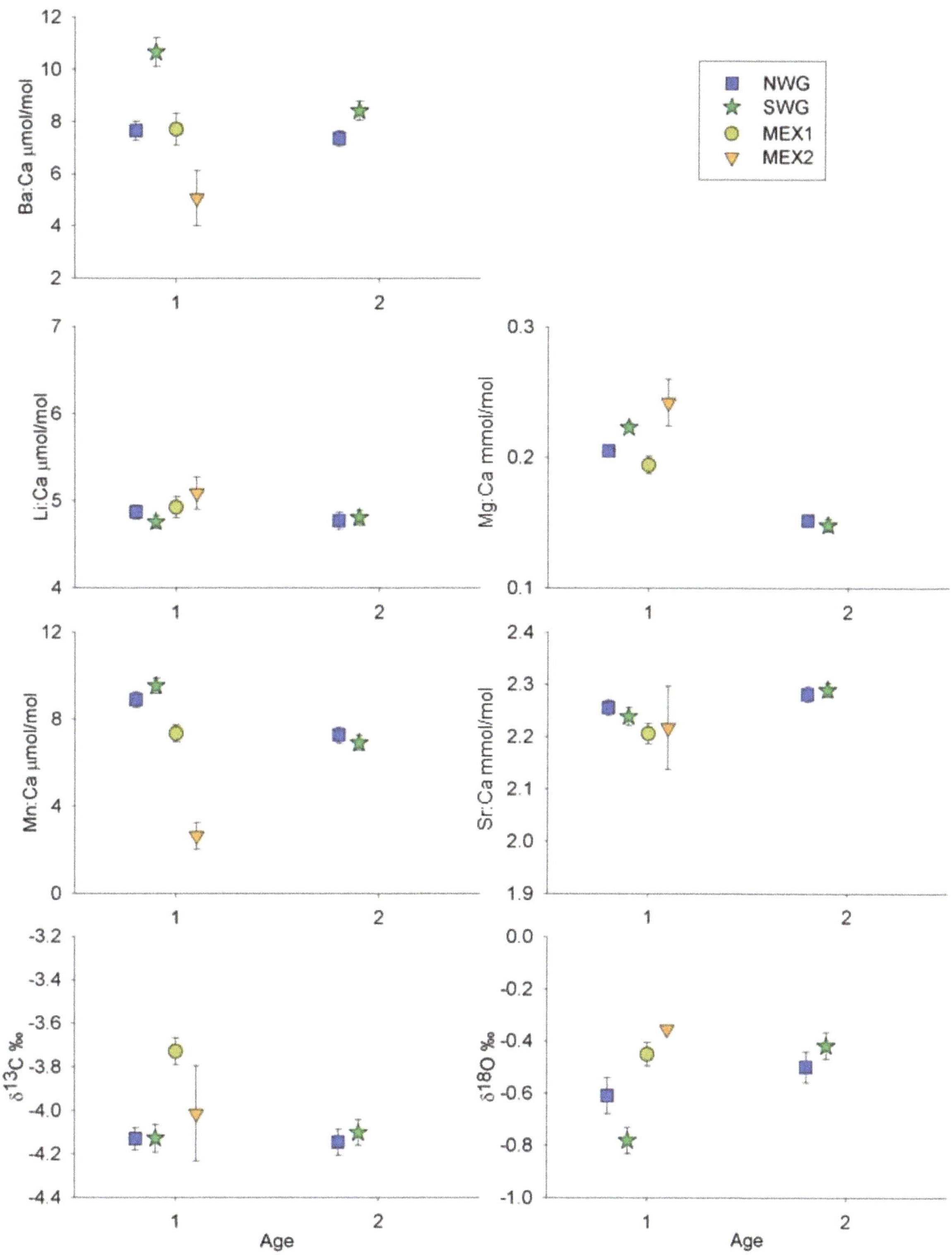

FIGURE 4. Mean (±SE) element : Ca or stable isotope delta ratios for otolith cores from 2006-cohort Red Snapper (subadults and adults, ages 1–2) sampled in four regions of the western Gulf of Mexico during summer in 2007 and 2008 (region codes are defined in Figure 3).

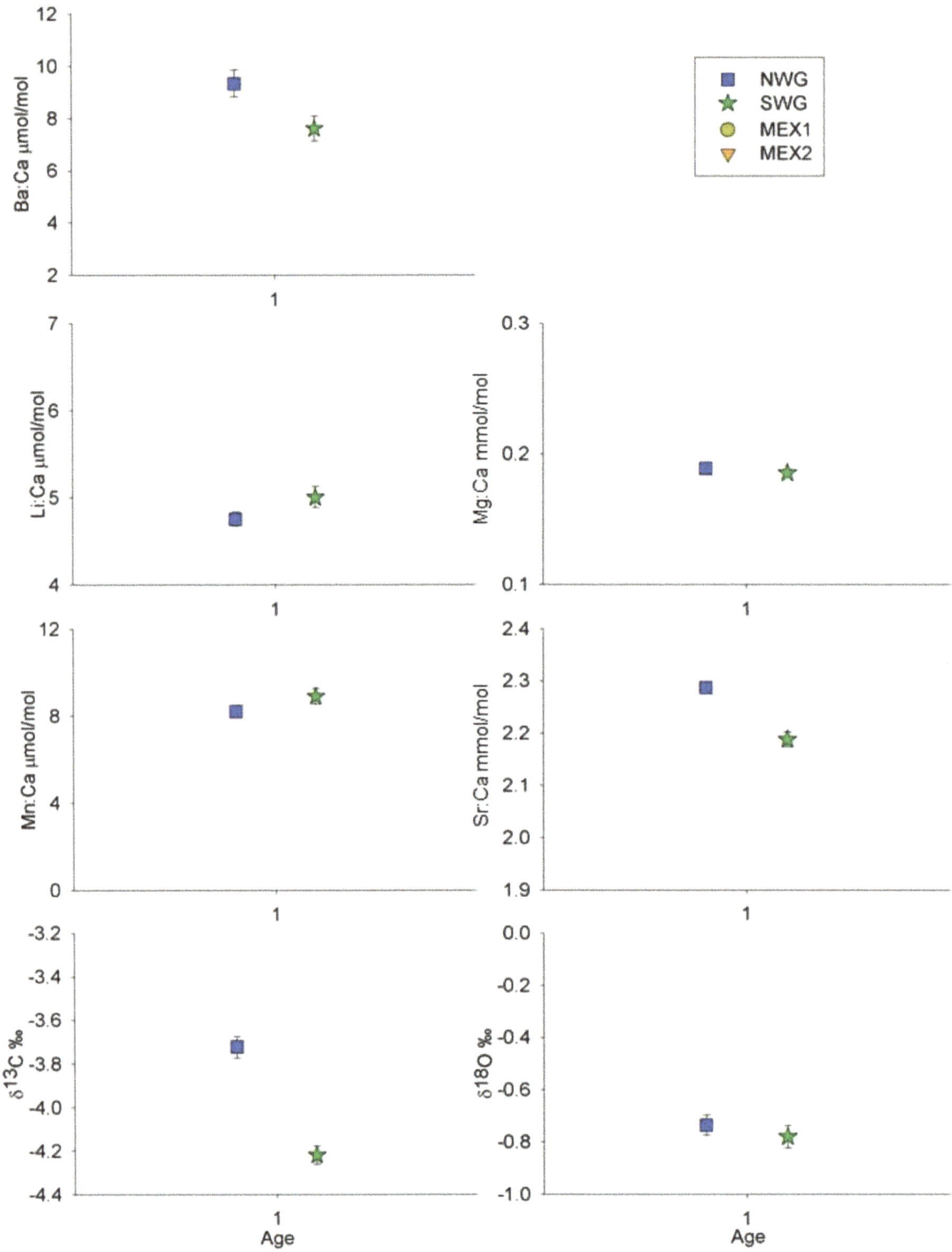

FIGURE 5. Mean (±SE) element : Ca or stable isotope delta ratios for otolith cores from 2007-cohort Red Snapper (subadults and adults, age 1) sampled in two regions of the western Gulf of Mexico during summer 2008 (region codes are defined in Figure 3).

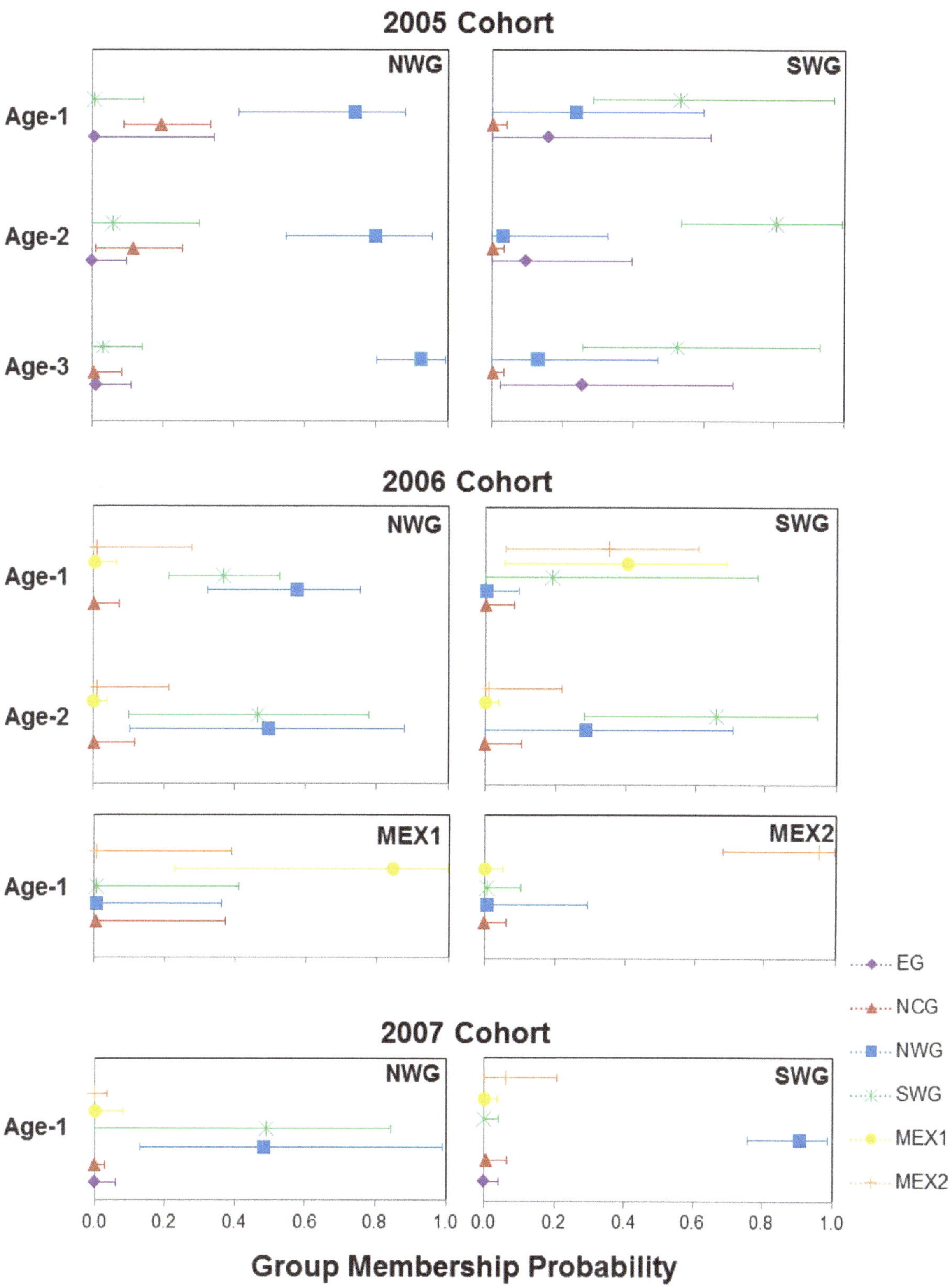

FIGURE 6. Estimated proportions (±95% credible interval) of subadult and adult (ages 1–3) Red Snapper sampled from the western Gulf of Mexico (Gulf) that were assigned to nursery regions based on year-class-specific otolith element : Ca and stable isotope ratio data for age-0 fish from the corresponding (2005–2007) cohorts. The subadult/adult sampling region is specified in the upper-right corner of each panel; symbols represent the assigned nursery regions (EG = eastern Gulf; NCG = north-central Gulf; NWG = northwestern Gulf; SWG = southwestern Gulf; MEX1 = Mexico region 1; MEX2 = Mexico region 2).

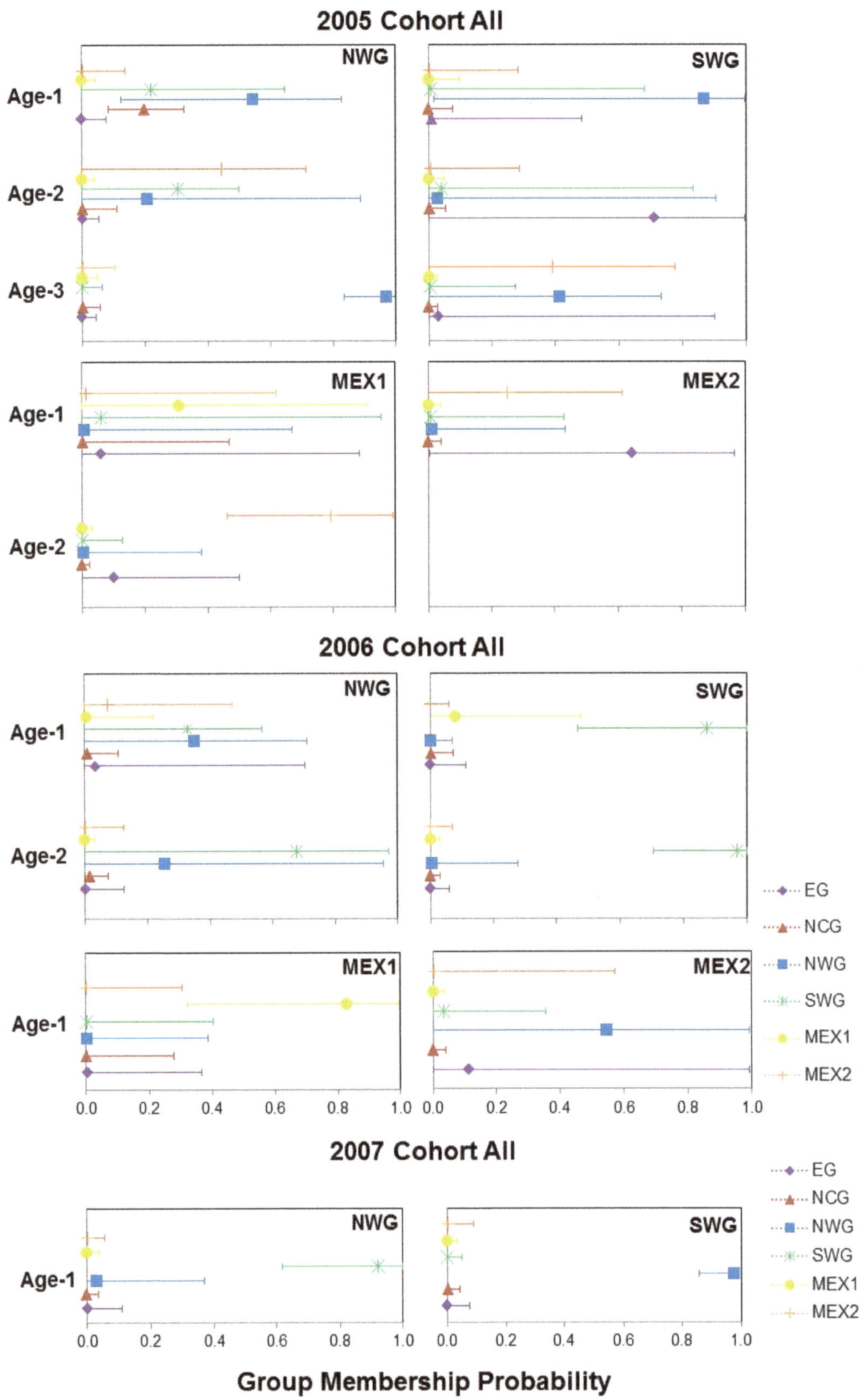

FIGURE 7. Estimated proportions (±95% credible interval) of subadult and adult (ages 1–3) Red Snapper sampled from the western Gulf of Mexico (Gulf) that were assigned to nursery regions based on otolith element : Ca and stable isotope ratio data for age-0 fish pooled across cohorts (2005–2007 year-classes). The subadult/adult sampling region is specified in the upper-right corner of each panel; symbols represent the assigned nursery regions (region codes are defined in Figure 6).

sampled from NWG were classified to the SWG nursery region; likewise, age-1 samples from the SWG were classified to the NWG nursery region.

DISCUSSION

Previous studies of Red Snapper otolith chemistry have indicated that significant postsettlement movement occurs between NWG and SWG (Cowan et al. 2003; Patterson 2007). In the present study, moderate to high percentages of recruits from the NWG nursery were observed among Red Snapper subadults and adults sampled from SWG; in some cases, almost equal proportions of recruits from the NWG and SWG nursery regions were observed among the NWG samples. Therefore, results of this study provide evidence that substantial postsettlement mixing of subadult/adult Red Snapper occurs between U.S. regions within the western Gulf. However, little evidence was detected of Red Snapper mixing between the eastern and western Gulf or between the U.S. and Mexican portions of the Gulf.

Gold and Saillant (2007) estimated that the N_e of Red Snapper was 10-fold higher in NWG than in NCG and SWG. Furthermore, the 2009 Red Snapper stock assessment (SEDAR 2009) indicated that the age distribution in the eastern Gulf was truncated relative to that in the western Gulf; the eastern Gulf population was also projected to have lower productivity than the western substock. The current study demonstrates that (1) the 2005 cohort of Red Snapper in NWG was predominantly composed of locally derived fish; and (2) although some cohorts in SWG were locally derived, the NWG nursery region was an important source of 2007-cohort recruits to SWG. Interestingly, Kulaw (2012) discovered that female Red Snapper in SWG reached 100% maturity faster than females in NWG. This was attributed to signs of juvenescence in the SWG population as it rebuilds from overfishing, whereas the NWG population may have moderate to low fecundity and later maturation due to its higher population size. Therefore, it is possible for the SWG population to be locally derived during strong year-classes (e.g., 2005 cohort; Cowan 2011; SEDAR 2013; Saari 2014) and to receive recruitment from other regions when year-classes are not as strong (e.g., 2007 cohort). Thus, in combination with past research on Red Snapper, the observed classification proportions in this study indicate the NWG's importance as a source of recruits to Red Snapper populations in the western U.S. Gulf.

Previous otolith chemistry studies have indicated that postsettlement movement of Red Snapper does occur but that their movement is limited during the first year of life (Cowan et al. 2003; Patterson 2007; Patterson et al. 2008). However, results of the current study may indicate otherwise. For the 2005 cohort of Red Snapper in NWG, the estimated proportion of locally derived recruits increased as the age of sampled fish increased. If limited movement should occur in the first year of life, with a potential increase in movement as the fish ages, then the trend displayed for locally derived recruits for the NWG 2005 cohort should be reversed. This suggests that Red Snapper are capable of moving over longer distances during the juvenile stage than previously inferred. It could be speculated that the active 2005 hurricane season, which included Hurricanes Katrina and Rita, may be responsible for the estimated large-scale movement of age-1 Red Snapper (Patterson et al. 2001b). Nonetheless, the 2005-cohort age-0 Red Snapper used in the development of nursery signatures were collected after the major hurricane impacts and exhibited the highest classification success, making a hurricane effect less likely. The 2007 cohort showed strong movement in one direction—from NWG to SWG. The 2005 and 2006 cohorts were stronger year-classes than the 2007 cohort (Cowan 2011; SEDAR 2013; Saari 2014), and this may partially explain why higher mixing rates were evident for the 2005 and 2006 cohorts, whereas the 2007 cohort in SWG consisted primarily of recruits from the NWG nursery region. However, much of this is speculation, as a sample size of 30 fish·year-class^{-1}·nursery region^{-1} may be too small to permit accurate discrimination of subadult/adult recruitment sources. Increasing the sample size and the number of age-groups examined may allow for better resolution in understanding the mixing dynamics of Red Snapper populations.

The majority of subadults and adults sampled in SWG were classified as originating from the SWG or NWG nursery region, with only one exception. The 2006-cohort age-1 samples from SWG were proportionately assigned to the MEX1 and MEX2 nursery regions, although the CIs overlapped, indicating a lack of significance at the 95% level. It is perplexing that SWG samples were assigned to and overlapped with the MEX1 and MEX2 nursery regions because (1) the MEX1 and SWG baseline otolith signatures did not overlap for the 2006 cohort; and (2) the MEX2 baseline signatures remained separate from the SWG signatures for both the 2005 and 2006 cohorts (see Zapp Sluis et al. 2012). Although the MEX1 nursery region could be another potential source of Red Snapper recruits for SWG, the transfer of Red Snapper from MEX2 to SWG seems highly unlikely. Prevailing upwelling winds cause circulation on the western Campeche Banks to flow westward along the coast, and during the fall and winter this is met with a down-coast current that extends to the southern Bay of Campeche and generates seasonal offshore transport (Zavala-Hidalgo et al. 2003). These circulation patterns, along with the separation of distance, likely impede the mixing of Red Snapper between SWG and MEX2.

Due to the unbalanced design of the Mexican regional data, only the 2006 cohort age-1 samples could be analyzed unless nursery chemical signatures were pooled across all year-classes. For the 2006 cohort, subadult/adult Red Snapper in MEX2 were estimated to be locally derived, and MEX1 fish were classified to MEX1 baseline samples, although some overlap between 95% CIs was apparent. When examining classification results based on age-0 data pooled across year-classes, the MEX2 Red Snapper were no longer estimated to be locally derived; instead, MEX2 samples from the 2005 and 2006 cohorts were primarily classified to the EG and NWG nursery regions, respectively. For the 2006-cohort samples from MEX1, the classification results based on age-0 data from single year-classes appeared to be the same as the results obtained based on the pooled year-classes. For the 2005 cohort, the pooled classification results indicated that age-1 Red Snapper sampled in MEX1 consisted mainly of recruits from MEX1, although with significant overlap of CIs; the age-2 samples from MEX1 were mainly composed of fish from the MEX2 nursery region. Even though combining the baseline signatures across year-classes resulted in relatively high classification accuracy (70.3%), significant differences in otolith chemical signatures were evident among year-classes for age-0 Red Snapper (Zapp Sluis et al. 2012). Consequently, analyses of classification estimates based on cohort-specific data are more accurate, as evidenced by the notable changes in classification to the NWG and SWG samples when pooled year-class signatures were used. Thus, analytical results were inconclusive regarding the source of recruits to the Mexican Red Snapper populations.

Based on the Red Snapper sampled in this study, a moderate to strong contribution of recruits from the NWG nursery was apparent among adults sampled from NWG and SWG. Unfortunately, connectivity between the western Gulf and Mexican regions is inconclusive at this time, and more data would be required before inferences can be made. Most of the recent increase in spawning stock biomass of Gulf Red Snapper is estimated to have occurred in the western Gulf, and this is projected to continue into the near future (SEDAR 2013). Based on the results of the current study, the center of abundance off the coast in NWG may be expanding outward toward the SWG continental shelf. Furthermore, it appears that some self-recruitment is occurring in SWG. Future work should also determine whether population recovery in the western Gulf is contributing to the relatively recent reappearance of Red Snapper in the far eastern portion of the Gulf. Determining connectivity between eastern and western populations of Red Snapper would be beneficial to the development of efficient regional management for this species.

ACKNOWLEDGMENTS

The present research was supported by the Marine Fisheries Initiative under Grant Number NA05NMF4331072 awarded to W. Patterson and J. Cowan. We thank Steve Tomeny Charters, Fisherman's Wharf Deep Sea Fishing, the National Marine Fisheries Service (fall groundfish surveys and vertical longline surveys), and numerous recreational anglers for allowing us to sample their catch. We are especially grateful to K. Simonsen, M. Hedges Monk, and K. Boswell for assistance in the field; G. Martinez for sampling fish in the MEX1 and MEX2 regions; X. Wang for assistance in running the SF-ICP mass spectrometry samples; D. Winter for running the IR mass spectrometry samples; S. Smith for troubleshooting the program; and two anonymous reviewers for helpful feedback that improved the manuscript. We also thank B. Kline and A. Fischer for their assistance in the laboratory.

REFERENCES

Barnett, B., and W. F. Patterson III. 2010. The effect of coring and pulverizing juvenile Red Snapper, *Lutjanus campechanus*, otoliths on their chemical signatures. Environmental Biology of Fishes 89:463–471.

Brule, T., T. Colas-Marrufo, E. Perez-Diaz, and J. C. Samano-Zapata. 2010. Red Snapper reproductive biology in the southern Gulf of Mexico. Transactions of the American Fisheries Society 139:957–968.

Camber, C. I. 1955. A survey of the Red Snapper fishery of the Gulf of Mexico, with special reference to the Campeche Banks. Florida Board of Conservation Marine Research Laboratory Technical Series 12.

Campana, S. E. 1999. Chemistry and composition of fish otoliths: pathways, mechanisms and applications. Marine Ecology Progress Series 188:263–297.

Cowan, J. H. Jr. 2011. Red Snapper in the Gulf of Mexico and U.S. South Atlantic: data, choices and conclusions. Fisheries 36:319–331.

Cowan, J. H. Jr., M. Woods, W. Patterson III, and D. Nieland. 2003. Otolith microchemistry (and reproductive biology) portion–stock structure of Red Snapper in the northern Gulf of Mexico: is their management as a single stock justified based on spatial and temporal patterns of genetic variation, otolith microchemistry, and growth rates? Marine Fisheries Initiative, Southeast Regional Office, Final Report NA87FF0425, St. Petersburg, Florida.

Cowen, R. K., K. M. M. Lwiza, S. Sponaugle, C. B. Paris, and D. B. Olson. 2000. Connectivity of marine populations: open or closed? Science 287:857–859.

Crowder, L. B., S. J. Lyman, W. F. Figueira, and J. Priddy. 2000. Source-sink population dynamics and the problem of sitting marine reserves. Bulletin of Marine Science 66:799–820.

Fischer, A. J., M. S. Baker Jr., and C. A. Wilson. 2004. Red Snapper (*Lutjanus campechanus*) demographic structure in the northern Gulf of Mexico based on spatial patterns in growth rates and morphometrics. U.S. National Marine Fisheries Service Fishery Bulletin 102:593–603.

Fischer, A. J., C. A. Wilson, and D. L. Nieland. 2002. Age and growth of Red Snapper, *Lutjanus campechanus*, in the northwestern Gulf of Mexico: implications to the unit stock hypothesis. Proceedings of the Gulf and Caribbean Fisheries Institute 53:496–506.

Gallaway, B. J., M. Longnecker, J. G. Cole, and R. M. Meyer. 1998. Estimates of shrimp-trawl bycatch of Red Snapper (*Lutjanus campechanus*) in the Gulf of Mexico. Pages 817–839 *in* F. Funk, T. J. Quinn, J. Heifetz, J. N. Ianelli; J. E. Powers, J. F. Schweigert, P. J. Sullivan, and C. I. Zhang, editors. 15th Lowell Wakefield Fisheries Symposium: fishery stock assessment models for the 21st century. Alaska Sea Grant College Program, AK-SG-98-01, Anchorage.

GMFMC (Gulf of Mexico Fishery Management Council). 2010. Final regulatory amendment to the reef fish fishery management plan to set total allowable catch for Red Snapper. GMFMC, Tampa, Florida.

Gold, J. R., and E. Saillant. 2007. Population structure of Red Snapper in the northern Gulf of Mexico. Pages 201–216 *in* W. F. Patterson III, J. H. Cowan Jr., D. A. Nieland, and G. R. Gitzhugh, editors. Red Snapper ecology and fisheries in the U.S. Gulf of Mexico. American Fisheries Society, Symposium 60, Bethesda, Maryland.

Jackson, M. W., J. H. Cowan Jr., and D. A. Nieland. 2007. Demographic differences in northern Gulf of Mexico Red Snapper reproductive maturation: implication for the unit stock hypothesis. Pages 217–227 *in* W. F. Patterson III, J. H. Cowan Jr., G. R. Fitzhugh, and D. L. Nieland, editors. Red Snapper ecology and fisheries in the U.S. Gulf of Mexico. American Fisheries Society, Symposium 60, Bethesda, Maryland.

Kulaw, D. 2012. Habitat- and region-specific reproductive biology of female Red Snapper (*Lutjanus campechanus*) in the Gulf of Mexico. Master's thesis. Louisiana State University, Baton Rouge.

Monroy-García, C., M. G. Andrade, and J. C. Espinoza. 2002. Análisis de la pesqueria de huachinango (*Lutjanus campechanus*) en el Banco de Campeche. [Analysis of the Red Snapper fishery (*Lutjanus campechanus*) on the Campeche Bank.] Proceedings of the Gulf and Caribbean Fisheries Institute 53:507–515.

Patterson, W. F. III. 2007. A review of Gulf of Mexico Red Snapper movement studies: implications for population structure. Pages 221–236 *in* W. F. Patterson III, J. H. Cowan Jr., G. R. Fitzhugh, and D. L. Nieland, editors. Red Snapper ecology and fisheries in the U.S. Gulf of Mexico. American Fisheries Society, Symposium 60, Bethesda, Maryland.

Patterson, W. F. III, J. H. Cowan Jr., C. A. Wilson, and Z. Chen. 2008. Temporal and spatial variability in juvenile Red Snapper otolith elemental signatures in the northern Gulf of Mexico. Transactions of the American Fisheries Society 137:521–532.

Patterson, W. F. III, J. H. Cowan Jr., C. A. Wilson, and R. L. Shipp. 2001a. Age and growth of Red Snapper, *Lutjanus campechanus*, from an artificial reef off Alabama in the northern Gulf of Mexico. U.S. National Marine Fisheries Service Fishery Bulletin 99:617–627.

Patterson W. F. III, J. C. Watterson, R. L. Shipp, and J. H. Cowan Jr. 2001b. Movement of tagged Red Snapper in the northern Gulf of Mexico. Transactions of the American Fisheries Society 130:533–545.

R Development Core Team. 2007. R: a language and environment for statistical computing. R Foundation for Statistical Computing, Vienna. Available: www.R-project.org. (October 2015).

Rooker, J. R., D. H. Secor, G. DeMetrio, A. J. Kaufman, A. B. Rios, and A. Ticina. 2008. Evidence of trans-Atlantic movement and natal homing of Bluefin Tuna from stable isotopes in otoliths. Marine Ecology Progress Series 368:231–239.

Rooker, J. R., D. H. Secor, V. S. Zdanowicz, and T. Itoh. 2001. Discrimination of northern Bluefin Tuna from nursery areas in the Pacific Ocean using otolith chemistry. Marine Ecology Progress Series 218:275–282.

Saari, C. R. 2014. Regional differences in the age and growth of Red Snapper (*Lutjanus campechanus*) in the U.S. Gulf of Mexico. U.S. National Marine Fisheries Service Fishery Bulletin 112:261–273.

SAGARPA (Secretaría de Agricultura, Ganadería, Desarrollo Rural, Pescay Alimentación). 2006. Carta nacional pesquera. [National fishery chart.] Pages 1–128 *in* A. López-González, editor. Diario official de la Federación, tomo DCXXXV 19, segunda sección). [Official diary of the Federation, volume DCXXXV 19, second section.] Interior Ministry, Mexico City.

Saillant, E., and J. R. Gold. 2006. Population structure and variance effective size of Red Snapper (*Lutjanus campechanus*) in the northern Gulf of Mexico. U.S. National Marine Fisheries Service Fishery Bulletin 104:136–148.

SEDAR (Southeast Data, Assessment, and Review). 2005. Stock assessment report, Gulf of Mexico Red Snapper (SEDAR 7). SEDAR, North Charleston, South Carolina.

SEDAR (Southeast Data, Assessment, and Review). 2009. Update assessment of Red Snapper in the Gulf of Mexico. SEDAR, North Charleston, South Carolina.

SEDAR (Southeast Data, Assessment, and Review). 2013. Stock assessment report, Gulf of Mexico Red Snapper (SEDAR 31). SEDAR, North Charleston, South Carolina.

Smith, S. J., and S. E. Campana. 2010. Integrated stock mixture analysis for continuous and categorical data, with application to genetic-otolith combinations. Canadian Journal of Fisheries and Aquatic Sciences 67:1533–1548.

Sturgeon, R. E., S. N. Willie, L. Yang, R. Greenberg, R. O. Spatz, Z. Chen, C. Scriver, V. Clancy, J. W. Lam, and S. Thorrold. 2005. Certification of a fish otolith reference material in support of quality assurance for trace element analysis. Journal of Analytical Atomic Spectrometry 20:1067–1071.

Thorrold, S. R., C. M. Jones, P. K. Swart, and T. E. Targett. 1998. Accurate classification of juvenile Weakfish *Cynoscion regalis* to estuarine nursery areas based on chemical signatures in otoliths. Marine Ecology Progress Series 173:253–265.

Thorrold, S. R., C. Latkoczy, P. K. Swart, and C. M. Jones. 2001. Natal homing in a marine fish metapopulation. Science 291:297–299.

Zapp Sluis, M., B. K. Barnett, W. F. Patterson III, J. H. Cowan Jr., and A. M. Shiller. 2012. Discrimination of juvenile Red Snapper otolith chemical signatures from Gulf of Mexico nursery regions. Marine and Coastal Fisheries: Dynamics, Management, and Ecosystem Science [online serial] 4:587–598.

Zavala-Hidalgo, J., S. L. Morey, and J. J. O'Brien. 2003. Seasonal circulation on the western shelf of the Gulf of Mexico using a high-resolution numerical model. Journal of Geophysical Research [online serial] 108(C12):3389.

17

Effects of Slotted Water Control Structures on Nekton Movement within Salt Marshes

Matthew E. Kimball*
Baruch Marine Field Laboratory, University of South Carolina, Post Office Box 1630, Georgetown, South Carolina 29442, USA

Lawrence P. Rozas
National Oceanic and Atmospheric Administration, National Marine Fisheries Service, Southeast Fisheries Science Center, Estuarine Habitats and Coastal Fisheries Center, 646 Cajundome Boulevard, Lafayette, Louisiana 70506, USA

Kevin M. Boswell
Marine Sciences Program, Department of Biological Sciences, Florida International University, Biscayne Bay Campus, 3000 Northeast 151st Street, MSB355, North Miami, Florida 33181, USA

James H. Cowan Jr.
Department of Oceanography and Coastal Sciences, Louisiana State University, 2247 Energy, Coast, and Environment Building, Baton Rouge, Louisiana 70803, USA

Abstract

Water control structures (WCSs) restrict hydrological connectivity in salt marshes and thereby impede nekton movement within the greater habitat mosaic. Transient fishery species, which spawn outside salt marshes and must get past these barriers to reach spawning areas or salt-marsh nurseries, are especially vulnerable to these structures. Water control structures incorporating slots (narrow vertical openings spanning most of the water column) are thought to improve nekton passage; however, few studies have directly examined nekton passage through WCS slots. Dual-frequency identification sonar (DIDSON) acoustic imaging was used monthly (April–September 2010) on diurnal flood tides to examine nekton movement through 15-cm-wide slots at two identical WCSs located in Louisiana tidal marsh channels. Nekton behavior was compared between these WCSs and a nearby natural salt-marsh creek. Examination of 12 h of subsampled acoustic data revealed large concentrations of salt-marsh nekton at the WCSs (n = 2,970 individuals total), but passage rates through the slots were low ($\leq$10% of total observed individuals migrated via the slots). Most migrating fish were observed leaving the managed area and swimming against a flood tide. The mean size of migrating individuals (~25 cm TL) did not differ in relation to swimming direction (going into versus exiting the managed marsh) and was similar to that reported from other studies examining similar slot widths. Nekton formed congregations in the WCS channel, but no congregations were observed in the natural salt-marsh creek, even though nekton species composition and sizes were similar among sites. The WCSs in our study appear to function as ecological hot spots, where large individuals may encounter enhanced foraging opportunities but also fishing mortality and where smaller individuals may experience greater predation rates.

Subject editor: Anthony Overton, East Carolina University, Greenville, North Carolina

*Corresponding author: matt@belle.baruch.sc.edu

Salt-marsh habitats serve as critical nursery areas for numerous fish and natant invertebrate species (nekton), including early life history stages. The value of these nurseries depends on the degree of hydrological connectivity with marine spawning areas (Rozas et al. 2013) and the greater mosaic of integral habitats within coastal and estuarine ecosystems (Peterson 2003; Able 2005; Sheaves 2009). Nekton movement through this habitat mosaic—at both small and large spatial and temporal scales—provides an important pathway for the transfer of energy and nutrients among habitats (Kneib 2000; Allen et al. 2013). Movement of nekton between salt marshes and coastal/ocean habitats (i.e., immigration and emigration) provides an important pathway for the trophic transfer of marsh production to the wider estuary and coastal waters (Weinstein et al. 2000). Much of this production supports valuable coastal fisheries (Deegan et al. 2000). Because nekton usage of habitats is largely species specific and varies according to physical and biological factors (Hoese and Moore 1998; Able and Fahay 2010), the unhindered ability of nekton, particularly juveniles, to locate and use suitable estuarine habitat during critical life history stages will determine their success in reaching adulthood.

Many estuaries are degraded and losing critical habitat due to natural and anthropogenic factors (Peterson and Lowe 2009; Kirwan and Megonigal 2013). Conversion of marsh to open water, erosion of marsh edge habitat, and loss (or drastic reduction) of intertidal and subtidal biogenic reefs all potentially alter the function of estuarine ecosystems for plants and animals alike (e.g., Anthony et al. 2009; Peterson and Lowe 2009). Structural marsh management, which incorporates a combination of levees and water control structures (WCSs) to control marsh hydrology, has often been implemented to mitigate such losses and stabilize marsh habitats (Montague et al. 1987; Rogers et al. 1992). Commonly used WCSs include fixed-crest or variable-crest weirs, gated types, and slotted types (Rogers et al. 1994). By regulating water levels, structural marsh management can establish stable hydrological regimes in the managed area but also may restrict life history connectivity by impeding nekton movement (Rogers et al. 1994; Secor and Rooker 2005; Sheaves 2009). Reduced access from WCSs for transient nekton that must migrate to estuarine nursery habitat from nearshore and offshore spawning areas could negatively affect their recruitment into salt marshes during early life history stages and could subsequently limit the emigration of older juveniles or adults to other coastal habitat types.

Water control structures incorporating slots (narrow vertical openings spanning most of the water column) are thought to improve nekton passage (Herke et al. 1992; Rogers et al. 1992; Rulifson and Wall 2006). Importantly, because slots extend the full height of the WCS and permit flow throughout the water column (when open), they may facilitate passage of the entire nekton assemblage rather than only some species. In riverine systems, vertical slots incorporated into dam fishways were found to improve passage (Stuart and Mallen-Cooper 1999; Stuart and Berghuis 2002) and provided access for a large portion of the riverine nekton community (Baumgartner et al. 2010, 2012; Thiem et al. 2013).

Most studies comparing managed and unmanaged salt marshes show that transient species are less abundant in managed areas (Knudsen et al. 1989; McGovern and Wenner 1990; Herke et al. 1992, 1996; Rozas and Minello 1999) and suggest that nekton movement into managed areas is restricted by WCSs. Few studies, however, have directly examined salt-marsh nekton passage through WCSs of any type because collecting unbiased data at these structures is a challenge (Rogers et al. 1994; Hoese and Konikoff 1995). For example, traps and nets may alter nekton behavior by interfering with any back-and-forth movement patterns (assuming general undirected movement). Research opportunities also have been limited by an inability to alter the operation of WCSs or manipulate structure designs and by a lack of replication (i.e., identical WCSs within the same marsh system). Consequently, only three studies have directly examined nekton passage through WCS slots in salt marshes. Rogers et al. (1992) and Rulifson and Wall (2006) used traps to compare unidirectional movement of nekton through WCSs with and without slots. During a prior study within the Breton Sound estuary, southeastern Louisiana, we (Kimball et al. 2010) used high-resolution acoustic imaging to examine the effect of WCS slot width on nekton bidirectional movement, providing unique information on fishes that migrated through the slots as well as those that did not; obtaining such information is not possible with traditional direct-capture techniques. Examining the set of individuals that migrate and the set of individuals that do not provides the context necessary to estimate nekton passage efficiency at a given WCS. Although passage efficiency for riverine nekton at structures has received considerable attention (see recent reviews by Roscoe and Hinch 2010; Bunt et al. 2012; and Noonan et al. 2012), information on salt-marsh nekton passage efficiency is nearly absent from the literature (but see Kimball et al. 2010).

Many marshes along the U.S. Gulf of Mexico and Atlantic coasts are currently under some form of structural marsh management (Knudsen et al. 1985; Montague et al. 1987; Cowan et al. 1988; Robinson and Jennings 2012). Due to the importance of such habitats for numerous fishery species, a better understanding of the impacts of WCSs on nekton movement patterns is critical. Such information may help managers improve the design and operation of WCSs to facilitate rather than impair nekton use of salt marshes. Furthermore, efforts to model the impact of reduced hydrological connectivity (i.e., via WCSs) on salt-marsh nekton population dynamics would also likely benefit from such insights (e.g., Neary 2012; Williams et al. 2012).

In an earlier study examining nekton passage through WCS slots, we (Kimball et al. 2010) used acoustic imaging to overcome the limitations of sampling gear that interfere with nekton movement. Others have also demonstrated the advantages of using this technique (Doehring et al. 2011; Grote et al. 2014). In the present study, we take the next step by incorporating replication and a control in the study design and by using imaging sonar (dual-frequency identification sonar [DIDSON]) to examine nekton movement and behavior at WCS slots. We estimated nekton passage rates through slots by sampling nekton during flood tides at two identical WCSs in a single estuarine system of Louisiana. We also compared the abundance and size of nekton in congregations at these WCSs with those of nekton in a comparable sample volume of an adjacent natural salt-marsh creek (i.e., control) within the same system.

METHODS

Study site.—Our study focused on two WCSs (replicates) and a natural salt-marsh creek (control) located in the southeastern portion of the Calcasieu Lake estuary, Louisiana (Figure 1). The Mangrove Bayou WCS (29°53′37.36″N, 93°13′52.44″W) and the No-Name Bayou WCS (29°50′17.36"N, 93°19′14.06″W) are identical fixed-crest structures. Each WCS consists of four bays (each 2.4 m wide) with a fixed-crest height of 1.4 m (Figure 2). The far-left bay (as viewed from the managed marsh; see Figure 2) contains three vertical slots (0.15 m wide × 1.2 m high). The remaining three bays have no openings. The WCSs are recessed from the lake shoreline in small canals and are positioned perpendicular to the channel. Canals are about 21 m wide at the WCSs and are lined with rip-rap within 10 m and on both sides of each structure (creating a uniform channel profile). Both WCSs control water exchange between the salt marsh and Calcasieu Lake; during our study, the slots remained fully opened. The natural salt-marsh creek site (hereafter, "marsh creek";

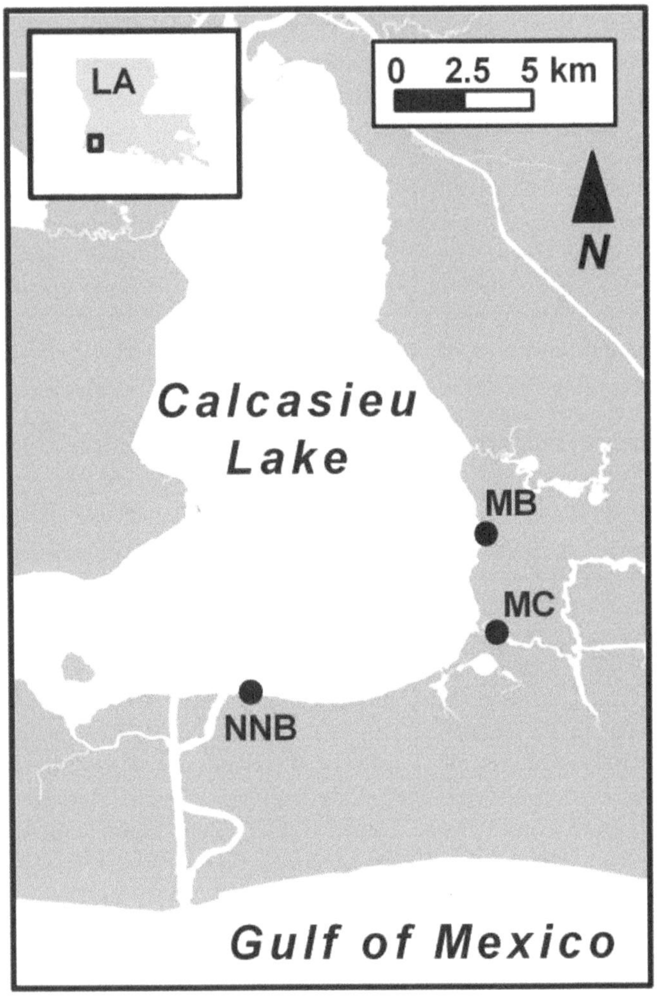

FIGURE 1. Locations of the Mangrove Bayou (MB) and No-Name Bayou (NNB) water control structures and the salt-marsh creek (MC) sample site in the Calcasieu Lake estuary, southwestern Louisiana.

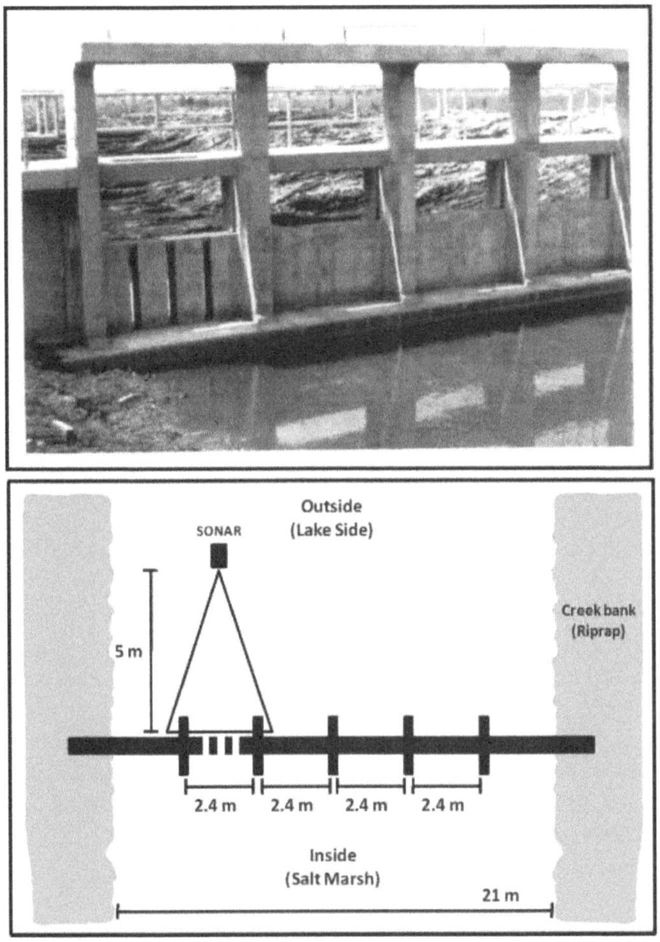

FIGURE 2. Top panel: dewatered view of the No-Name Bayou water control structure (WCS) as seen from the managed marsh (inside), showing the four bays (each 2.4 m wide, with a fixed-crest height of 1.4 m; photo courtesy of Cameron Prairie National Wildlife Refuge, Bell City, Louisiana). The far-left bay contains three vertical slots (0.15 m wide × 1.2 m high). The remaining three bays have no openings. Bottom panel: overhead diagram of the field sampling setup at each WCS, showing the placement of the platform-mounted dual-frequency identification sonar in front of the slotted bay on the lake side (outside).

29°51′33.85″N, 93°13′51.27″W) is located in the managed marsh. The creek at this site is about 25 m wide, with a soft-sediment substrate. The three study sites differed in their approximate water distance from the Gulf of Mexico (Mangrove Bayou WCS: 21.5 km; No-Name Bayou WCS: 10.4 km; marsh creek: 21.4 km).

Field sampling.—Nekton movement through the WCS slots and in the marsh creek was examined during daytime flood tides monthly in April–September 2010 by using a Sound Metrics DIDSON imaging sonar (www.soundmetrics.com). Limiting our observations to flood tides allowed us to focus on nekton movement from Calcasieu Lake into the managed marsh; in addition, our earlier work showed that nekton passage rates peaked during flood tides (Kimball et al. 2010). The DIDSON collects high-resolution acoustic images (4–21 frames/s) and permits sampling in turbid waters and at night. We used the high-frequency mode (1.8 MHz), which uses 96 beams (0.3° horizontal [H] × 14° vertical [V]) resulting in a total field of view of 29° H × 14° V (see Boswell et al. 2008); this mode is most appropriate for collecting high-resolution data at short ranges (<12 m). The DIDSON was mounted on an adjustable platform, which provided a stable yet portable platform suitable for horizontally aimed acoustic monitoring in shallow (~2 m) estuarine habitats (described by Boswell et al. 2007). The platform-mounted DIDSON was adjusted so as to place the sonar in the center of the water column, and the platform was positioned about 5 m from the slots in the lakeside canal. This position optimized data quality, as it minimized the potentially confounding effect of introducing structure immediately in front of the slot opening, yet it still permitted a view of the entire water column at the slots. For the marsh creek, the platform-mounted DIDSON was also centered in the water column and positioned to acquire a similar field of view (i.e., 5-m viewing range, aimed with the flow of water). Once properly positioned, the platform-mounted DIDSON was left in place to continuously record for 4 h during each daytime flood tide sampling period at each site. Power supply and data acquisition were controlled from the walking platform on top of each WCS (Figure 2) or from the creek bank at the marsh creek site, where the DIDSON interfaced with a laptop computer for real-time viewing and data storage.

Acoustic sampling was supplemented with traditional sampling gears—cast nets to target small individuals (<20 cm) and gill nets for targeting large individuals (>20 cm)—to catalog species presence and migration periods in the study area. Cast nets (4.8-mm monofilament mesh; 2.4-m radius) were used to sample nekton in the canal at each WCS, on both sides (i.e., within 5 m of the WCS in the managed and open marshes), and both upstream and downstream of the DIDSON at the marsh creek (i.e., at the shoreline adjacent to the sonar; and 30 and 60 m upstream and downstream). For each flood tide sampling period, casts ($n = 2$) were thrown at each station. A single cast at each WCS was missed in April, so the resulting total of 46 casts is slightly less than the balanced total of 48. All 60 casts were collected at the marsh creek. Gill nets (multiple panels of 25-, 51-, and 76-mm monofilament mesh; 1.8 m deep × 15.5 m long) were used to sample nekton at a single location near each WCS (about 10–30 m away; $n = 12$ sets total). At the marsh creek, gill nets ($n = 12$ sets total) were set diagonally across the creek and were deployed both upstream and downstream of (about 10–30 m away from) the DIDSON. At each location, gill nets were set for 0.5 h once during each flood tide sampling period.

Environmental and physical variables were measured once during each flood tide in the canal near each WCS and at the platform-mounted DIDSON in the marsh creek. Temperature (°C), salinity, and dissolved oxygen (mg/L) were measured at the surface with a YSI Model 85 handheld meter (Yellow Springs Instruments). Samples were collected at the surface once during flood tide and were analyzed for turbidity (NTU) in the laboratory by using a MicroTPW turbidimeter (HF Scientific). Center-channel water depth (m) was measured on both sides of each WCS every 15 min continuously throughout each sampling period with a HOBO water level logger (Onset Computer), and these data were used to determine maximum and minimum center-channel depths during each flood tide sampling period. Due to the uniform bottom profile of the channel, center-channel water depth was representative of the water depth across the channel in front of the WCSs (on both sides). Water depth at the slots, which was measured from the base of each structure (Figure 2), was about 0.5 m less than the center-channel depth. Center-channel water depth (m) in the marsh creek was measured similarly by using a HOBO water level logger located adjacent to the platform-mounted DIDSON.

Water velocity (m/s) at each site was measured with an Aquadopp (right-angle head) acoustic Doppler current profiler (Nortek USA) in October 2010 to examine potential effects on nekton movement. Sampling of flood tide water velocities was performed in October soon after the completion of nekton sampling so as to avoid possible interference between the DIDSON and the current profiler (e.g., acoustic interference from the profiler near the slots). At the WCSs, the profiler was mounted on the underside of a float on the water's surface and was positioned in the open (lakeside) canal with fixed cables about 2.5 m directly in front of the slots. In this position, the profiler measured water velocities at about 10-cm depth intervals throughout the water column every 5 min for a 1-h period. The float-mounted profiler was positioned in the marsh creek at approximately the same channel position where DIDSON sampling occurred, and velocities were measured by the same methods used for the WCSs.

Data analysis.—The DIDSON data for each 4-h flood tide sampling period were divided into two 2-h blocks. The data were then subsampled by randomly selecting six 5-min segments from the total of twenty-four 5-min segments in each 2-h block. Thus, each flood tide sampling period at each site yielded a combined 1 h (twelve 5-min segments) of DIDSON data for analysis (total = 3 sites × 6 months = 18 h of subsampled data). Randomly selecting six 5-min segments from each 2-h block made it more likely that the subsampled data would represent the full span of the 4-h sampling period and lessened the possibility of selecting only segments clustered around a single point in time. Subsampled DIDSON data were analyzed by using the Sound Metrics DIDSON software (version 5.25.41). Output data from DIDSON analyses for each flood tide at each WCS were classified into two categories: migrating individuals and congregating individuals. Migrating individuals were defined as nekton that swam through the WCS slots. All 12 h of subsampled data from the WCSs were examined for migrating individuals. Congregating individuals were defined as those that were observed in front of the WCS but did not swim through the slots. Because congregating individuals were relatively abundant and to reduce the possibility of counting individuals multiple times, we randomly selected and examined 12 frames (one from each 5-min segment) for each hour of subsampled data from both WCSs (n = 144 frames total). For migrating individuals, the number of individuals and length (TL, cm) of each individual observed per 5-min segment were recorded. For congregating individuals, we recorded the number of individuals and length of each individual observed per frame. Swimming direction (i.e., going into the managed marsh or coming from the managed marsh) was determined for each migrant observed at the WCSs. We also calculated an estimate of the relative percentage of migrants at each WCS during each flood tide as follows: (number of migrating individuals per 5-min segment)/(number of congregating individuals per frame). We examined the subsampled data from the marsh creek (n = 72 frames) by following the protocol used for congregating individuals at WCSs; this allowed us to compare the number and size of individuals congregating at each WCS with the number and size of individuals from a natural salt-marsh creek of similar physical characteristics.

Statistical analyses.—Each randomly selected acoustic data unit (i.e., 5-min segment; or frame) was treated as an independent sample in our analyses, in conformance with the procedures used by Kimball et al. (2010). Migrant abundance and relative percentage data were analyzed by using a one-way, randomized complete block (RCB) ANOVA with WCS (n = 2; Mangrove Bayou and No-Name Bayou WCSs) as a factor (GLM procedure in SAS version 9.3). Because our study was conducted under similar environmental conditions during a single extended season (summer), season was not a factor of interest; therefore, the data were blocked based on water temperature (as a more ecologically relevant blocking factor than month) to remove unwanted sources of variation (Potvin 2001). Four water temperature groupings were used based on mean monthly temperature observations as follows (Table 1): 22°C (April), 26°C (May), 30°C (June–July), and 29°C (August–September). To quantify variation in the size of migrants, we used a two-way RCB ANOVA with WCS (n = 2; Mangrove Bayou and No-Name Bayou WCSs) and direction of movement (going into or coming from the managed marsh) as factors and with water temperature as the blocking factor. For the two-way RCB ANOVA of migrant length, interaction terms were included by using a backward stepwise approach in which nonsignificant interactions were excluded from the model. The abundance and size of individuals observed congregating at the WCSs were compared with those of individuals observed at the marsh creek by using a one-way RCB ANOVA with site (n = 3; Mangrove Bayou WCS, No-Name Bayou WCS, and marsh creek) as a factor and with water temperature as the blocking factor. Because some frames at the marsh creek lacked individuals for observation (i.e., July = 5 frames; September = 1 frame), the actual sample size of 210 frames is slightly smaller than the balanced sample size of 216 frames. Prior to analyses, abundance ($\log_{10}[x + 1]$), percentage (arcsine), and length ($\log_{10}$) data were transformed to satisfy the assumptions of ANOVA. Differences in treatment means were examined by using the Tukey–Kramer test for unequal sample sizes (Dunnett 1980; Day and Quinn 1989). We used logit modeling, which is preferred when the response and predictor variables are categorical (Floyd 2001), to independently analyze the effect of WCS

TABLE 1. Mean (SE in parentheses) temperature, salinity, dissolved oxygen (DO), turbidity, and center-channel water depth for each month of nekton sampling (April–September) and water velocity sampling (October) at sites in the Calcasieu Lake estuary, Louisiana. Each variable was measured once per flood tide each month at each sample site (Mangrove Bayou water control structure [WCS], No-Name Bayou WCS, and salt-marsh creek; thus, n = 3 per month). Inside water depth was measured only at the two WCSs and was not measured in April.

Environmental variable	Apr	May	Jun	Jul	Aug	Sep	Oct
Temperature (°C)	22.2 (0.7)	26.1 (1.0)	29.8 (0.8)	31.1 (0.9)	29.5 (0.3)	29.0 (0.3)	25.6 (0.8)
Salinity	17.8 (2.0)	23.6 (0.3)	21.3 (0.3)	20.8 (1.9)	23.1 (3.2)	21.4 (0.2)	22.8 (1.1)
DO (mg/L)	6.2 (0.7)	5.7 (0.4)	5.0 (0.3)	3.3 (0.2)	3.4 (0.4)	3.6 (0.4)	5.0 (0.4)
Turbidity (NTU)	4.7 (2.2)	4.0 (2.1)	3.2 (1.0)	1.2 (0.5)	3.2 (0.6)	24.4 (15.8)	1.5 (0.3)
Outside water depth, maximum (m)	1.38 (0.24)	1.40 (0.15)	1.54 (0.20)	1.43 (0.12)	1.55 (0.16)	1.64 (0.14)	1.41 (0.17)
Outside water depth, minimum (m)	1.32 (0.22)	1.32 (0.12)	1.47 (0.18)	1.40 (0.11)	1.49 (0.14)	1.54 (0.10)	1.36 (0.18)
Inside water depth, maximum (m)		1.36 (0.04)	1.46 (0.01)	1.55 (0.04)	1.51 (0.01)	1.64 (0.03)	1.44 (0.04)
Inside water depth, minimum (m)		1.35 (0.05)	1.42 (0.02)	1.54 (0.04)	1.47 (0.01)	1.60 (0.03)	1.44 (0.04)

on swimming direction (LOGISTIC procedure in SAS). The predictor variable was binary (i.e., WCS = Mangrove Bayou or No-Name Bayou).

Data on the species of nekton collected with traditional direct-capture sampling gears (i.e., cast net and gill net) were for descriptive purposes only and therefore were not statistically analyzed. These data were used to inform the analysis and interpretation of acoustic imaging data by providing the abundances and sizes of the nekton species that were present at the WCSs and the marsh creek during the study period.

Environmental and water quality data were examined with descriptive statistics. Water velocity data were examined for each site separately. Mean water velocities were calculated by using all velocity measurements for each 10-cm depth increment and were plotted to create water column velocity profiles (i.e., velocity [m/s] × depth [m]) for the flood tide measurements at each site.

RESULTS

Environmental and water quality characteristics were similar among the three sites in each month (Table 1). Water temperature, salinity, and dissolved oxygen were consistent among sites during each month and followed general trends associated with summertime in northern Gulf of Mexico estuaries. Turbidity was generally low (<5 NTU) during the study period except in September, when turbidity was high at the marsh creek (55.8 NTU). As expected, the inside water depths observed at the WCSs were similar during each sampling event and never exceeded the fixed-crest height (1.4 m). The marsh creek was included in the mean outside water depth calculations, which were also similar during each sampling event. Outside water depths rarely exceeded the fixed-crest height of the WCSs while the DIDSON was deployed (total of 24 h over 6 months) to collect data: only 15% of the total time at No-Name Bayou (June = 3.5 h) and not at all at Mangrove Bayou. Water velocities were low at all three sites; mean water velocities (averaged over all depths) for the two WCSs were similar (Mangrove Bayou: 0.08 m/s; No-Name Bayou: 0.09 m/s), whereas the marsh creek had a slightly lower mean velocity (0.05 m/s). Maximum velocities were up to three times greater at the WCSs (Mangrove Bayou: 0.13 m/s; No-Name Bayou: 0.16 m/s) than at the marsh creek (0.05 m/s).

Congregating Nekton

During 12 h of monitoring with acoustic imaging, 2,970 individuals were observed congregating at the two WCSs (Mangrove Bayou: n = 1,844 individuals; No-Name Bayou: n = 1,126 individuals; Figure 3B). Although observed in the same discrete time segments, it should be noted that these congregating individuals were observed in 144 image frames rather than from 12 continuous hours of acoustic imaging data (as was done for the migrants). No corrective calculation was

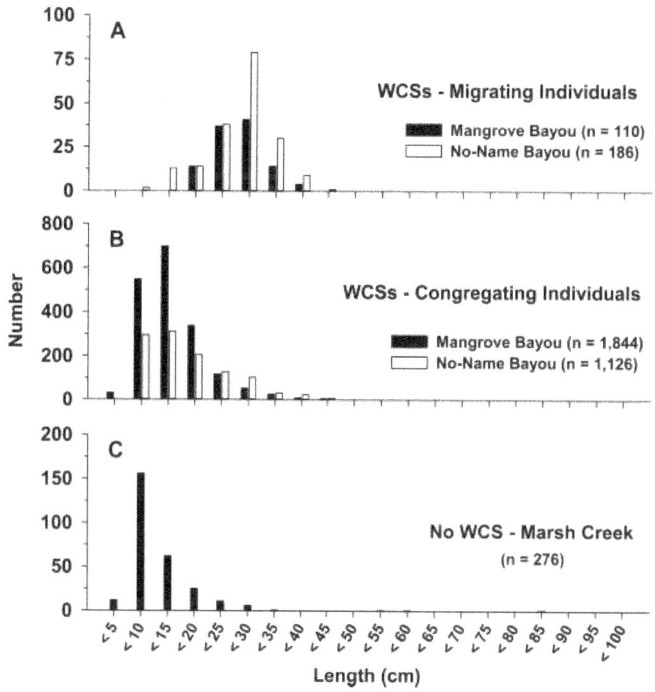

FIGURE 3. Length-frequency distribution for nekton observed with dual-frequency identification sonar at three sites in the Calcasieu Lake estuary: **(A)** migrating individuals (n = 296) at each water control structure (WCS), **(B)** congregating individuals (n = 2,970) at each WCS, and **(C)** individuals (n = 276) at the salt-marsh creek (no WCS). Note that the y-axis scale differs among the panels.

attempted on these congregating nekton data to equalize coverage with the data for migrants. The mean abundance of congregating individuals observed at the Mangrove Bayou WCS was significantly greater than abundance at the No-Name Bayou WCS (P = 0.0382; Figure 4A). Congregating nekton displayed similar size distributions at both of the WCSs (Mangrove Bayou: range = 4–155 cm; No-Name Bayou: range = 5–167 cm), with the majority of individuals (Mangrove Bayou: 86%; No-Name Bayou: 72%) ranging between 5 and 20 cm in length (Figure 3B). The mean length of congregating nekton did not differ significantly between the two WCSs (P = 0.1199; Figure 4B).

Migrant Abundance, Size, and Swimming Direction

In total, 296 individuals were observed migrating through the slots during 12 h of subsampled acoustic data recorded at the two WCSs; this equates to about 25 migrating individuals per hour. Mean migrant abundance was not significantly different between the two WCSs (Table 2; Figure 5A), although more migrating individuals were observed at No-Name Bayou (n = 186 total; 31 migrants/h) than at Mangrove Bayou (n = 110 total; 18 migrants/h). More migrants were also observed swimming out from the managed marsh rather than into the managed marsh at both WCSs (Figure 6A), an

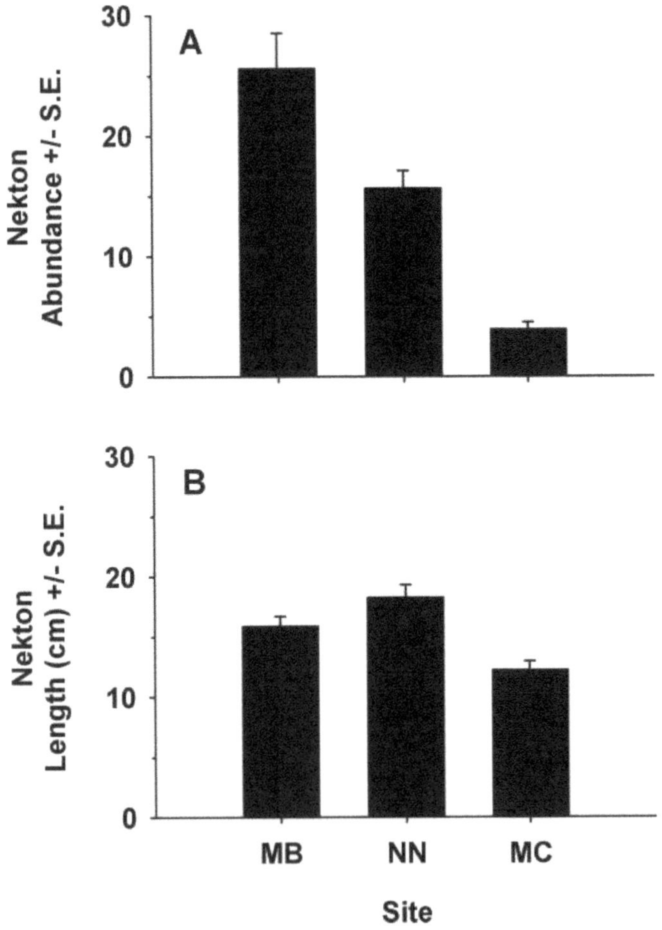

FIGURE 4. Mean (+SE) values of **(A)** nekton abundance (per frame) and **(B)** nekton length (cm TL; per frame) for individuals congregating at each water control structure (WCS; MB = Mangrove Bayou WCS; NN = No-Name Bayou WCS) and individuals in the salt-marsh creek (MC; i.e., no WCS), as observed with dual-frequency identification sonar (n = 3,246 individuals total).

indication that the migration patterns of individuals were not influenced by any particular WCS (Wald $\chi^2 = 3.3056$, df = 1, $P = 0.0690$). Migrating individuals ranged in size from 7 to 44 cm, but most migrants (80%) were 20–35 cm in length, and the size distributions were similar for the WCSs (Figure 3A). Mean migrant length (~25 cm TL) did not vary between WCSs (Table 2; Figure 5C) and did not vary with swimming direction (into or out from the managed marsh; Table 2; Figure 6B). The percentage of migrants was low ($\leq$10% of the observed congregating individuals migrated) at both WCSs (Figure 5B) and did not significantly differ between Mangrove Bayou and No-Name Bayou (Table 2).

Nekton at Water Control Structures versus the Salt-Marsh Creek

Acoustic data revealed that nekton abundance near the WCS slots was three to six times greater than that in comparable sample volumes from the unrestricted marsh creek site ($P < 0.0001$; Table 2; Figures 3B, 3C, 4A), where nekton distributions were likely spread out across the creek. The majority of individuals congregating at WCSs (81%) and observed at the marsh creek (88%) were between 5 and 20 cm in length. Overall size distributions were similar at the WCSs (range = 4–167 cm; Figure 3B) and the marsh creek (range = 4–81 cm; Figure 3C), with some larger individuals (>60 cm) infrequently observed at the WCSs (n = 23 individuals combined) and the marsh creek (n = 1 individual). Mean nekton length, however, was significantly lower at the marsh creek (mean ± SE = 12.19 ± 0.79 cm) than at the WCSs (Mangrove Bayou: 15.89 ± 0.80 cm; No-Name Bayou: 18.25 ± 1.05 cm; $P < 0.0001$; Table 2; Figure 4B).

Direct-Capture Sampling

Thirty-six species of nekton comprising 41,367 individuals were collected with cast nets and gill nets during the study period (Table 3). Abundance was greater at the WCSs (Mangrove Bayou: n = 12,105 individuals; No-Name Bayou: n = 24,832 individuals) than at the marsh creek (n = 4,520 individuals). Species richness was greater at the marsh creek

TABLE 2. Randomized complete block (RCB) ANOVA results (P-values for F-statistics) for the effect of water control structures (WCSs; n = 2; Mangrove Bayou WCS and No-Name Bayou WCS) on migrant abundance (per 5-min segment; n = 144 total segments), the percentage of migrants ([number of migrating individuals per 5-min segment]/[number of congregating individuals per frame]), and individual migrant length (cm TL; n = 296 total individuals) observed with dual-frequency identification sonar. Swimming direction was only included as a factor in the analysis of migrant length; the WCS × swimming direction interaction term was nonsignificant and therefore was excluded from the model. The RCB ANOVA results are also reported for the effect of sample site (n = 3; Mangrove Bayou WCS, No-Name Bayou WCS, and salt-marsh creek) on nekton abundance (per frame; n = 216 total frames) and nekton length (cm TL, per frame; n = 210 total frames). "Nekton" was defined as those individuals observed at the salt-marsh creek as well as those congregating at the WCSs. Factors that were not tested for a given variable were identified as not applicable (na).

Dependent variable	Error df	WCS (df = 1)	Swimming direction (df = 1)	Site (df = 2)
Migrant abundance	139	0.1205	na	na
Migrant percentage	139	0.0862	na	na
Migrant length	290	0.8409	0.4105	na
Nekton abundance	210	na	na	<0.0001
Nekton length	204	na	na	<0.0001

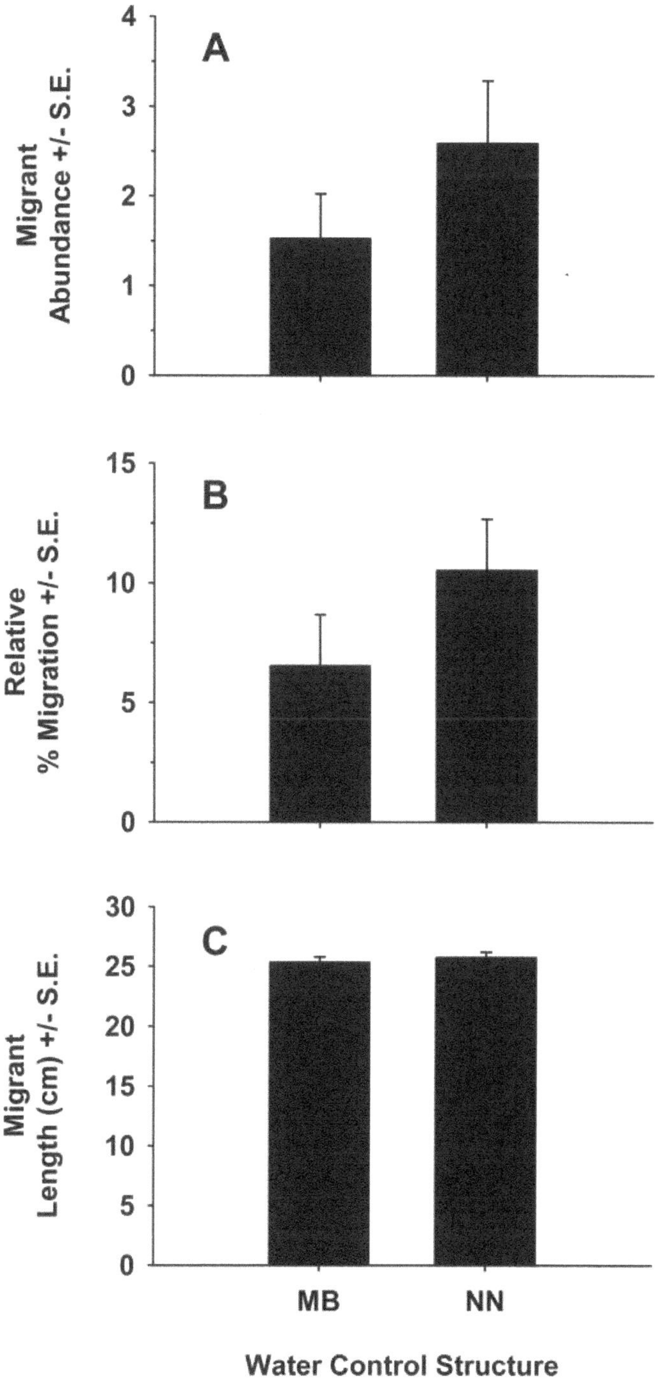

FIGURE 5. Mean (+SE) (**A**) migrant abundance (per 5-min segment), (**B**) relative percentage of migrants (i.e., [number of migrating individuals per 5-min segment]/[number of congregating individuals per frame]), and (**C**) migrant length (cm TL; per 5-min segment), as observed with dual-frequency identification sonar at the Mangrove Bayou (MB) and No-Name Bayou (NN) water control structures ($n = 296$ migrants total).

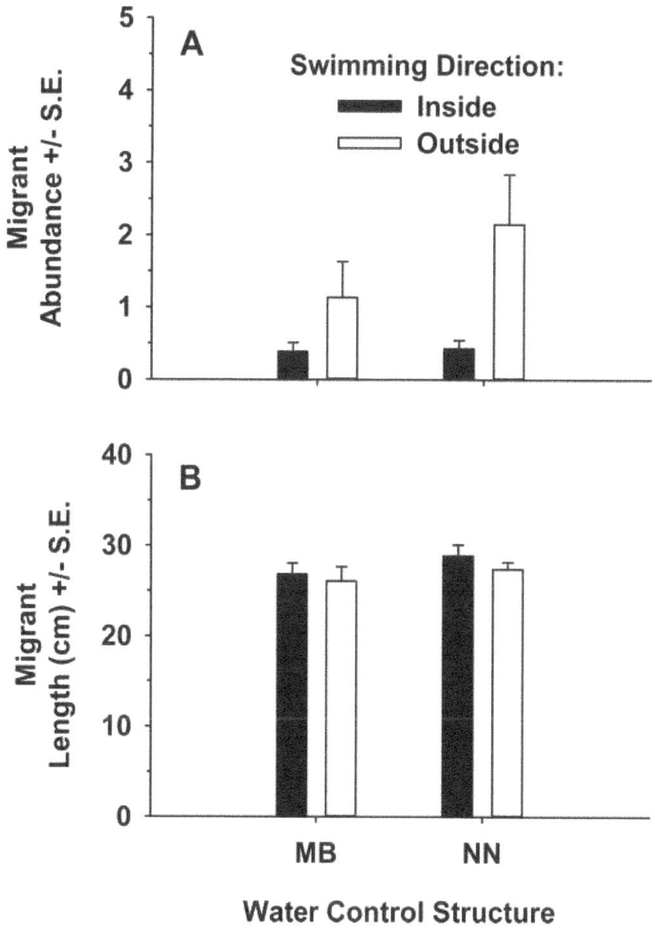

FIGURE 6. Mean (+SE) (**A**) migrant abundance (per 5-min segment) and (**B**) migrant length (cm TL; per 5-min segment), as observed with dual-frequency identification sonar at the Mangrove Bayou (MB) and No-Name Bayou (NN) water control structures ($n = 296$ migrants total). Migrants are classified based on swimming direction (inside = individuals going into the managed salt marsh; outside = individuals coming from the managed marsh).

($n = 30$ species) than at the Mangrove Bayou WCS ($n = 22$ species) and No-Name Bayou WCS ($n = 27$ species). Twice as many species were collected in cast nets ($n = 31$ species) as in gill nets ($n = 15$ species), and almost all individuals were collected in the cast nets (41,299 individuals in cast nets; 68 individuals in gill nets). Gill-net samples consisted primarily of large individuals (15–45 cm; 76%), whereas cast-net samples were mostly small individuals (<15 cm; 96%). The small nekton (<15 cm) we collected were primarily composed of Gulf Menhaden (66%), Atlantic Croakers (16%), and Spot (9%). Larger individuals (>15 cm) were generally less abundant and consisted mostly of Hardhead Catfish, Gizzard Shad, Spotted Seatrout, Striped Mullet, and Black Drum.

DISCUSSION

Passage rates through the WCS slots in our study area were low (i.e., ≤10% of congregating individuals). These passage rate estimates are likely valid because congregating individuals were enumerated from a single frame, which avoided the abundance overestimation that would have been caused by

TABLE 3. Total number of individuals collected (per month) and mean length (mm; SE in parentheses) for each species sampled with cast nets ($n = 106$ casts) and gill nets ($n = 24$ sets) at the three sites combined (Mangrove Bayou and No-Name Bayou water control structures and the salt-marsh creek). Fish and invertebrates were measured in terms of TL (*), FL (#), or carapace width (§).

	Cast net							Gill net						
Species	Apr	May	Jun	Jul	Aug	Sep	Length	Apr	May	Jun	Jul	Aug	Sep	Length
Diamond Killifish *Adinia xenica**	0	1	0	0	0	0	29.0 (0.0)							
Striped Anchovy *Anchoa hepsetus*#	0	0	1	106	0	1	58.3 (1.2)							
Bay Anchovy *Anchoa mitchilli*#	31	23	235	299	120	44	43.0 (0.6)							
Sheepshead *Archosargus probatocephalus**	0	0	0	1	0	0	220.0 (0.0)							
Hardhead Catfish *Ariopsis felis*#								0	0	6	1	4	0	243.2 (19.4)
Alligator Gar *Atractosteus spatula**								0	0	0	0	1	0	947.0 (0.0)
Silver Perch *Bairdiella chrysoura**	0	0	0	0	4	0	86.3 (3.6)							
Gulf Menhaden *Brevoortia patronus*#	10,456	3,221	7,296	6,175	5	16	47.0 (0.5)	1	0	0	0	0	0	150.0 (0.0)
Blue crab *Callinectes sapidus*§	11	2	2	1	0	0	23.8 (6.6)	0	1	0	0	1	2	142.4 (8.2)
Crevalle Jack *Caranx hippos*#	0	0	0	1	2	0	118.7 (17.2)	0	0	0	0	1	0	142.0 (0.0)
Atlantic Spadefish *Chaetodipterus faber**								0	0	0	0	1	0	121.0 (0.0)
Atlantic Bumper *Chloroscombrus chrysurus*#	0	0	0	36	0	0	46.7 (1.0)							
Bay Whiff *Citharichthys spilopterus**	0	1	1	0	0	0	54.0 (1.0)							
Sand Seatrout *Cynoscion arenarius**	0	0	0	5	0	0	153.2 (4.0)							
Spotted Seatrout *Cynoscion nebulosus**								1	1	0	0	1	3	332.8 (40.5)
Sheepshead Minnow *Cyprinodon variegatus**	1	1	0	0	0	0	34.0 (1.0)							
Gizzard Shad *Dorosoma cepedianum*#	0	0	2	1	0	0	198.0 (51.1)	3	2	2	1	4	1	312.0 (13.4)
Ladyfish *Elops saurus*#	0	0	0	9	0	0	223.8 (12.9)							
Northern brown shrimp *Farfantepenaeus aztecus**	0	114	106	6	1	2	65.0 (1.5)							
Gulf killifish *Fundulus grandis**	5	17	1	0	0	0	55.8 (2.2)							
Darter Goby *Ctenogobius boleosoma**	0	1	0	0	0	0	50.0 (0.0)							
Naked Goby *Gobiosoma bosc**	1	1	0	0	0	0	46.0 (1.0)							
Scaled Sardine *Harengula jaguana*#	0	0	0	164	5	235	51.7 (2.6)							
Pinfish *Lagodon rhomboides*#	0	3	9	2	4	1	74.7 (6.3)							
Spot *Leiostomus xanthurus**	248	2,569	211	193	276	9	89.3 (2.1)	0	0	0	0	4	0	174.5 (21.8)
White shrimp *Litopenaeus setiferus**	0	1	0	3	10	10	58.5 (6.5)	0	0	0	0	1	7	131.0 (4.4)
Inland Silverside *Menidia beryllina*#	927	288	9	10	6	0	65.0 (0.9)							
Atlantic Croaker *Micropogonias undulatus**	3,378	2,837	31	40	163	1	68.1 (1.1)	1	1	0	1	0	0	182.0 (0.0)
Striped Mullet *Mugil cephalus*#	14	5	127	7	30	10	164.0 (6.0)	2	1	1	0	2	0	202.0 (6.8)
White Mullet *Mugil curema*#	0	0	406	32	13	9	104.8 (2.1)							
Leatherjack *Oligoplites saurus*#	0	0	0	28	69	4	92.7 (1.8)							
Daggerblade grass shrimp *Palaemonetes pugio**	95	422	12	0	3	0	31.6 (0.3)							
Southern Flounder *Paralichthys lethostigma**								0	1	0	1	0	0	292.0 (3.0)
Black Drum *Pogonias cromis**	0	1	0	0	0	0	940.0 (0.0)	5	0	0	0	0	0	304.2 (22.5)
Red Drum *Sciaenops ocellatus**	0	0	1	0	0	0	217.0 (0.0)	0	1	0	1	0	0	417.5 (27.5)
Atlantic Needlefish *Strongylura marina**	2	0	0	0	0	0	360.0 (40.0)							
Total	15,169	9,508	8,450	7,119	711	342		13	8	9	5	20	13	

counting individuals multiple times. Comparing the number of migrants observed over a continuous period (e.g., 5 min) with discrete "snapshot" estimates of congregating individuals observed at longer time intervals (e.g., 1 frame per 5 min) likely provides the best estimate of passage efficiency. During our previous study (Kimball et al. 2010), we may have underestimated the passage rate (<5%) at a slotted WCS by using continuously observed data from both congregating and migrating individuals.

Water control structures appear to attract and concentrate salt-marsh nekton. Congregating individuals were often observed swimming in a circular pattern or remaining stationary (oriented into the current) near the slots—behaviors also observed in congregating nekton at another salt-marsh WCS (Kimball et al. 2010). In contrast, individuals in the salt-marsh creek displayed primarily random, nondirectional swimming behavior and remained in the DIDSON field of view for only a short time. The congregating nekton was much more abundant (3–6×) at the WCSs than at the salt-marsh creek site even though the assemblages were similar in species composition, dominant species, and individual sizes. High nekton densities at WCSs may indicate ecological hot spots at these critical bottlenecks, where processes such as predation and foraging are enhanced or intensified relative to those in other habitats within the estuary (Sheaves 2009). This concept has already been advanced for structures (e.g., dams) in riverine ecosystems (McLaughlin et al. 2013) and is likely applicable for structures in estuarine ecosystems as well (Sheaves 2009). For some individuals, WCSs appear to be used primarily as a foraging site rather than as an access point for entering or exiting managed marshes, which may account for the low passage rates observed here and elsewhere (Kimball et al. 2010). Turbulence created by water exchange at the slots may concentrate food for planktivores, and the abundance of forage fish (e.g., Gulf Menhaden; Kimball et al. 2010; present study) that are drawn to this food source may in turn attract piscivorous fishes to WCSs. Humans (fishers) take advantage of these nekton aggregations and often target both game fishes (e.g., Spotted Seatrout and Red Drum) and their prey (e.g., penaeid shrimps and blue crabs) at WCSs. The effect of such fishing pressure on these species' populations due to WCS placement in managed marshes is not known. Nekton diet and feeding habits have been examined at structures in rivers, where large nekton aggregations increase competition among predators and lead some species to alter their feeding strategies (Baumgartner 2007). Our lack of understanding about the effect of WCSs on nekton trophic interactions in managed salt marshes warrants further study.

The size of fish observed migrating through WCSs in this region has been consistent in studies of salt marshes and does not appear to be related to slot size. Migrating fish in our study were similar in size to those migrating through a WCS with 15-cm slots (majority of fish = 15–35 cm TL; mean ~ 30 cm TL) located at a salt marsh within Breton Sound estuary (Kimball et al. 2010). Rogers et al. (1992) and Rulifson and Wall (2006) evaluated smaller slot widths (i.e., 10 cm and 4 cm, respectively), but neither study reported the size of migrating nekton. A slot width of 15 cm reportedly limited the migration of large (>100-cm) fish through a dam fishway within a large river of northeastern Australia (Stuart and Berghuis 2002). It is unlikely that slot width limited nekton passage in our study, however, as few individuals (congregating or migrating) larger than 60 cm were observed. Furthermore, increasing the slot width to 45 cm (Stuart and Mallen-Cooper 1999) or 60 cm (Kimball et al. 2010) did not increase passage rates or the size of migrating individuals.

Most of the migrating fish in our study were observed leaving the managed area and swimming against a flood tide. Water velocities at the WCS slots were too low (<0.2 m/s) to limit the migration of most nekton and only exceeded the swimming capabilities of larvae and small (e.g., <3 cm) juveniles (e.g., Mitchell 1989; Luckenbach and Orth 1992; Faria et al. 2009). Individuals migrating out of managed marshes have been reported to be larger than those leaving unmanaged marshes (e.g., penaeid shrimps; Knudsen et al. 1989, 1996), ostensibly due to emigration impedance (resulting in long retention times), reduced predation, or decreased competition in managed marshes (Hoese and Konikoff 1995). Although we observed a similar pattern at another salt marsh (Kimball et al. 2010), some emigrating individuals may have exited the marsh undetected by passing through large flap gates (open only during ebb tide) that could not be monitored. In our study, the similar size of migrants swimming in both directions suggests that salt-marsh nekton—at least the size range of migrants we observed—do not experience long retention times and may transit WCS slots in a more routine, frequent manner.

In addition to examining bidirectional nekton passage, acoustic imaging allowed us to simultaneously observe migrating and congregating individuals at the two WCSs. Without data collected simultaneously from both migrants and congregating individuals, it would be difficult to interpret and compare results from nekton passage studies based on other metrics (e.g., number of migrating individuals per hour). For instance, migrants were observed at a rate of 25 individuals/h in our study, which is consistent with the only other study examining unhindered bidirectional passage of nekton through WCS slots (passage rate = 24 individuals/h; Kimball et al. 2010). Nekton passage rates reported from the only other studies to directly examine movement through slots in a salt marsh were well below (2 individuals/h; Rulifson and Wall 2006) and above (248 individuals/h; Rogers et al. 1992) the rates we observed, but those two studies only assessed nekton passage in one direction. Focusing on passage in a single direction (e.g., upstream or downstream) may be appropriate for riverine nekton exhibiting strong migrational cues (e.g., salmonids); however, salt-marsh nekton at WCSs do not swim unidirectionally most of the time.

Exceptions occur for emigrating life history stages of some nekton species (e.g., penaeid shrimps; Knudsen et al. 1985, 1989, 1996) and for individuals migrating in response to environmental cues (e.g., cold front passage; Herke and Rogers 1984). Observations of such phenomena are more likely to occur with approaches that involve near-continuous (e.g., daily) sampling efforts or that target a single species during critical life history periods (e.g., offshore spawning migration) rather than with efforts focusing on the entire nekton assemblage during specific seasons (e.g., Kimball et al. 2010) or tide stages (present study).

The use of imaging sonars allows a comprehensive examination of nekton movement and behavior at salt-marsh WCSs, but it has some limitations (Kimball et al. 2010; Doehring et al. 2011). Species identification with acoustic imaging is limited to large fishes (e.g., tunas and groupers; Frias-Torres and Luo 2009; Han et al. 2009) or fishes that display unique swimming behaviors (e.g., eels; Mueller et al. 2008; Doehring et al. 2011). Most fishery species recruit to marsh nursery areas as larvae or postlarvae, which are too small to be detected by these acoustic devices. Consequently, we could not examine the effects of WCSs on recruitment processes in our study. Small individuals (<3 cm) of some locally abundant invertebrate species (e.g., penaeid shrimps; Knudsen et al. 1989, 1996) were probably underrepresented in our study, whereas some small juvenile fishes were only identifiable because they formed distinctive schools (e.g., Gulf Menhaden and mullets *Mugil* spp.; Kimball et al. 2010; Doehring et al. 2011). Because WCSs may inhibit immigration into managed marshes (Herke et al. 1992), future work should examine this topic and determine whether WCSs restrict immigration enough to affect fishery production from managed areas. For a more in-depth examination of nekton behavior and species-specific movement patterns at salt-marsh WCSs, techniques other than imaging sonars will be required. For example, PIT technology can be used to collect species-specific data on juvenile fishes (e.g., 12-mm PIT tags; Bass et al. 2012). Currently, a sonar unit is only capable of projecting in one dimension (horizontal or vertical); thus, three-dimensional observation of nekton would require the use of two sonar units simultaneously (K. M. Boswell, unpublished data). In our study, the DIDSON unit was aimed horizontally, so we were unable to determine the position of individuals in the water column. Such information may reveal the depth preferences of migrating nekton at WCSs (e.g., top, middle, and bottom of the slot opening).

Although restricting hydrological connectivity with anthropogenic barriers is known to influence nekton community structure and habitat function in salt marshes (e.g., Rozas and Minello 1999; Rozas et al. 2013), the effects of these barriers on the population dynamics of salt-marsh nekton, particularly transient species, are unknown. To improve our understanding in this area, numerical simulation models would be useful for predicting the effects of WCSs on nekton populations and fishery production from managed marshes. Such models have been used successfully to estimate the effects of spatial configuration and flooding patterns in salt marshes on penaeid shrimp production (Roth et al. 2008). Our estimates of nekton passage efficiency (Kimball et al. 2010 and present study) could be used in such models to estimate the production from managed marshes that is exported to the wider estuary. This type of approach may be an important tool for assessing the effects of levees and storm surge barriers that are currently proposed for coastal states such as Louisiana (CPRA 2012), where large proportions of salt-marsh habitat are already under some form of management (Cowan et al. 1988).

ACKNOWLEDGMENTS

This project was funded by the Gulf States Marine Fisheries Commission and the Louisiana Department of Wildlife and Fisheries. We thank the following for their assistance: H. Finley and M. Harbison (Louisiana Department of Wildlife and Fisheries); G. Harris and R. Gosnell (Cameron Prairie National Wildlife Refuge); M. La Peyre, L. Broussard, and S. Beck (Louisiana State University AgCenter); and T. Mace (University of Louisiana, Lafayette). We are also grateful to H. McCall and M. Thompson for generously providing access to field sites. The suggestions of W. Herke, T. Minello, A. Chester, and several anonymous reviewers improved the manuscript. The findings, conclusions, and recommendations presented in this manuscript are those of the authors and do not necessarily represent the views of the National Oceanic and Atmospheric Administration Fisheries Service.

REFERENCES

Able, K. W. 2005. A re-examination of fish estuarine dependence: evidence for connectivity between estuarine and ocean habitats. Estuarine Coastal and Shelf Science 64:5–17.

Able, K. W., and M. P. Fahay. 2010. Ecology of estuarine fishes, temperate waters of the western North Atlantic. Johns Hopkins University Press, Baltimore, Maryland.

Allen, D. M., S. A. Luthy, J. A. Garwood, R. F. Young, and R. F. Dame 2013. Nutrient subsidies from nekton in salt marsh intertidal creeks. Limnology and Oceanography 58:1048–1060.

Anthony, A., J. Atwood, P. August, C. Byron, S. Cobb, C. Foster, C. Fry, A. Gold, K. Hagos, L. Heffner, D. O. Kellogg, K. Lellis-Dibble, J. J. Opaluch, C. Oviatt, A. Pfeiffer-Herbert, N. Rohr, L. Smith, T. Smythe, J. Swift, and N. Vinhateiro. 2009. Coastal lagoons and climate change: ecological and social ramifications in U.S. Atlantic and Gulf coast ecosystems. Ecology and Society [online serial] 14(1):8.

Bass, A. L., G. R. Giannico, and G. T. Brooks. 2012. Performance of a full-duplex passive integrated transponder (PIT) antenna system in estuarine channels. Marine and Coastal Fisheries: Dynamics, Management, and Ecosystem Science [online serial] 4:145–155.

Baumgartner, L. J. 2007. Diet and feeding habits of predatory fishes upstream and downstream of a low-level weir. Journal of Fish Biology 70:879–894.

Baumgartner, L. J., C. A. Boys, I. G. Stuart, and B. P. Zampatti. 2010. Evaluating migratory fish behavior and fishway performance: testing a combined assessment methodology. Australian Journal of Zoology 58:154–164.

Baumgartner, L. J., T. Marsden, D. Singhanouvong, O. Phonekhampheng, I. G. Stuart, and G. Thorncraft. 2012. Using an experimental in situ fishway to provide key design criteria for lateral fish passage in tropical rivers: a case study from the Mekong River, central Lao PDR. River Research and Applications 28:1217–1229.

Boswell, K. M., M. W. Miller, and C. A. Wilson. 2007. A lightweight transducer platform for use in stationary shallow water horizontal-aspect acoustic surveys. Fisheries Research 85:291–294.

Boswell, K. M., M. P. Wilson, and J. H. Cowan. 2008. A semi-automated approach to estimating fish size, abundance, and behavior from dual-frequency identification sonar (DIDSON) data. North American Journal of Fisheries Management 28:799–807.

Bunt, C. M., T. Castro-Santos, and A. Haro. 2012. Performance of fish passage structures at upstream barriers to migration. Rivers Research and Applications 28:457–478.

Cowan, J. H., R. E. Turner, and D. R. Cahoon. 1988. Marsh management plans in practice: do they work in coastal Louisiana, USA? Environmental Management 12:37–53.

CPRA (Coastal Protection and Restoration Authority of Louisiana). 2012. Louisiana's comprehensive master plan for a sustainable coast. CPRA, Baton Rouge.

Day, R. W., and G. P. Quinn. 1989. Comparisons of treatments after an analysis of variance in ecology. Ecological Monographs 59:433–463.

Dunnett, C. W. 1980. Pairwise multiple comparisons in the homogeneous variance, unequal sample size case. Journal of the American Statistical Association 75:789–795.

Deegan, L. A., J. E. Hughes, and R. A. Rountree. 2000. Salt marsh ecosystem support of marine transient species. Pages 333–365 *in* M. P. Weinstein and D. A. Kreeger, editors. Concepts and controversies in tidal marsh ecology. Kluwer Academic Publishers, Dordrecht, The Netherlands.

Doehring, K., R. G. Young, J. Hay, and A. J. Quarterman. 2011. Suitability of dual-frequency identification sonar (DIDSON) to monitor juvenile fish movement at floodgates. New Zealand Journal of Marine and Freshwater Research 45:413–422.

Faria, A. M., A. F. Ojanguren, L. A. Fuiman, and E. J. Goncalves. 2009. Ontogeny of critical swimming speed of wild-caught and laboratory-reared Red Drum *Sciaenops ocellatus* larvae. Marine Ecology Progress Series 384:221–230.

Floyd, T. 2001. Logit modeling and logistic regression. Pages 197–216 *in* S. M. Scheiner and J. Gurevitch, editors. Design and analysis of ecological experiments. Oxford University Press, New York.

Frias-Torres, S., and J. Luo. 2009. Using dual-frequency sonar to detect juvenile Goliath Grouper *Epinephelus itajara* in mangrove habitat. Endangered Species Research 7:237–242.

Grote, A. B., M. M. Bailey, J. D. Zydlewski, and J. E. Hightower. 2014. Multibeam sonar (DIDSON) assessment of American Shad (*Alosa sapidissima*) approaching a hydroelectric dam. Canadian Journal of Fisheries and Aquatic Sciences 71:545–558.

Han, J., N. Honda, A. Asada, and K. Shibata. 2009. Automated acoustic method for counting and sizing farmed fish during transfer using DIDSON. Fisheries Science 75:1359–1367.

Herke, W. H., E. E. Knudsen, P. A. Knudsen, and B. D. Rogers. 1992. Effects of semi-impoundment of Louisiana marsh on fish and crustacean nursery use and export. North American Journal of Fisheries Management 12:151–160.

Herke, W. H., and B. D. Rogers. 1984. Comprehensive estuarine nursery study completed. Fisheries 9(6):12–16.

Herke, W. H., B. D. Rogers, V. L. Wright, and W. H. Bradshaw. 1996. Postlarval *Penaeus aztecus* and *P. setiferus* transport into, and distribution within, adjacent weired and unweired ponds. Wetlands 16:197–207.

Hoese, H. D., and M. K. Konikoff. 1995. Effects of marsh management on fisheries organisms: the compensatory adjustment hypothesis. Estuaries 18 (1A):180–197.

Hoese, H. D., and R. H. Moore. 1998. Fishes of the Gulf of Mexico: Texas, Louisiana, and adjacent waters, 2nd edition. Texas A&M University Press, College Station.

Kimball, M. E., L. P. Rozas, K. M. Boswell, and J. H. Cowan. 2010. Evaluating the effect of slot size and environmental variables on the passage of estuarine nekton through a water control structure. Journal of Experimental Marine Biology and Ecology 395:181–190.

Kirwan, M. L., and J. P. Megonigal. 2013. Tidal wetland stability in the face of human impacts and sea-level rise. Nature 504:53–60.

Kneib, R. T. 2000. Salt marsh ecoscapes and production transfers by estuarine nekton in the southeastern United States. Pages 267–291 *in* M. P. Weinstein and D. A. Kreeger, editors. Concepts and controversies in tidal marsh ecology. Kluwer Academic Publishers, Dordrecht, The Netherlands.

Knudsen, E. E., R. F. Paille, B. D. Rogers, W. H. Herke, and J. P. Geaghan. 1989. Effects of a fixed-crest weir on brown shrimp *Penaeus aztecus* growth, mortality, and emigration in a Louisiana coastal marsh. North American Journal of Fisheries Management 9:411–419.

Knudsen, E. E., B. D. Rogers, R. F. Paille, and W. H. Herke. 1996. Juvenile white shrimp growth, mortality, and emigration in weired and unweired Louisiana marsh ponds. North American Journal of Fisheries Management 16:640–652.

Knudsen, P. A., W. H. Herke, and E. E. Knudsen. 1985. Emigration of brown shrimp from a low-salinity shallow-water marsh. Proceedings of the Louisiana Academy of Science 48:30–40.

Luckenbach, M. W., and R. J. Orth. 1992. Swimming velocities and behavior of blue crab (*Callinectes sapidus* Rathbun) megalopae in still and flowing water. Estuaries 15:186–192.

McGovern, J. C., and C. A. Wenner. 1990. Seasonal recruitment of larval and juvenile fishes into impounded and non-impounded marshes. Wetlands 10:203–221.

McLaughlin, R. L., E. R. B. Smyth, T. Castro-Santos, M. L. Jones, M. A. Koops, T. C. Pratt, and L. A. Velez-Espino. 2013. Unintended consequences and trade-offs of fish passage. Fish and Fisheries 14:580–604.

Mitchell, C. P. 1989. Swimming performances of some native freshwater fishes. New Zealand Journal of Marine and Freshwater Research 23:181–187.

Montague, C. L., A. V. Zale, and H. F. Percival. 1987. Ecological effects of coastal marsh impoundments. Environmental Management 11:743–756.

Mueller, A. M., T. Mulligan, and P. K. Withler. 2008. Classifying sonar images: can a computer-driven process identify eels? North American Journal of Fisheries Management 28:1876–1886.

Neary, V. S. 2012. Binary fish passage models for uniform and nonuniform flows. River Research and Applications 28:418–428.

Noonan, M. J., J. W. A. Grant, and C. D. Jackson. 2012. A quantitative assessment of fish passage efficiency. Fish and Fisheries 13:450–464.

Peterson, M. S. 2003. A conceptual view of environment–habitat–production linkages in tidal river estuaries. Reviews in Fisheries Science 11: 291–313.

Peterson, M. S., and M. R. Lowe. 2009. Implications of cumulative impacts to estuarine and marine habitat quality for fish and invertebrate resources. Reviews in Fisheries Science 17:505–523.

Potvin, C. 2001. ANOVA: experimental layout and analysis. Pages 63–76 *in* S. M. Scheiner and J. Gurevitch, editors. Design and analysis of ecological experiments. Oxford University Press, New York.

Robinson, K. F., and C. A. Jennings. 2012. Maximizing age-0 Spot export from a South Carolina estuary: an evaluation of coastal impoundment management alternatives via structured decision making. Marine and Coastal Fisheries: Dynamics, Management, and Ecosystem Science [online serial] 4:156–172.

Rogers, B. D., W. H. Herke, and E. E. Knudsen. 1992. Effects of three different water-control structures on the movements and standing stocks of coastal fishes and macrocrustaceans. Wetlands 12:106–120.

Rogers, D. R., B. D. Rogers, and W. H. Herke. 1994. Structural marsh management effects on coastal fishes and crustaceans. Environmental Management 18:351–369.

Roscoe, D. W., and S. G. Hinch. 2010. Effectiveness monitoring of fish passage facilities: historical trends, geographic patterns and future directions. Fish and Fisheries 11:12–33.

Roth, B. M., K. A. Rose, L. P. Rozas, and T. J. Minello. 2008. Relative influence of habitat fragmentation and inundation on brown shrimp *Farfantepenaeus aztecus* production in northern Gulf of Mexico salt marshes. Marine Ecology Progress Series 359:185–202.

Rozas, L. P., C. W. Martin, and J. F. Valentine. 2013. Effects of reduced hydrological connectivity on the nursery use of shallow estuarine habitats within a river delta. Marine Ecology Progress Series 492:9–20.

Rozas, L. P., and T. J. Minello. 1999. Effects of structural marsh management on fishery species and other nekton before and during spring drawdown. Wetlands Ecology and Management 7:121–139.

Rulifson, R. A., and B. L. Wall. 2006. Fish and blue crab passage through water control structures of a coastal bay lake. North American Journal of Fisheries Management 26:317–326.

Secor, H., and J. R. Rooker. 2005. Connectivity in the life histories of fishes that use estuaries. Estuarine Coastal and Shelf Science 64:1–3.

Sheaves, M. 2009. Consequences of ecological connectivity: the coastal ecosystem mosaic. Marine Ecology Progress Series 391:107–115.

Stuart, I. G., and A. P. Berghuis. 2002. Upstream passage of fish through a vertical-slot fishway in an Australian subtropical river. Fisheries Management and Ecology 9:111–122.

Stuart, I. G., and M. Mallen-Cooper. 1999. An assessment of the effectiveness of a vertical-slot fishway for non-salmonid fish at a tidal barrier on a large tropical/subtropical river. Regulated Rivers: Research and Management 15:575–590.

Thiem, J. D., T. R. Binder, P. Dumont, D. Hatin, C. Hatry, C. Katopodis, K. M. Stamplecoskie, and S. J. Cooke. 2013. Multispecies fish passage behavior in a vertical slot fishway on the Richelieu River, Quebec, Canada. River Research and Applications 29:582–592.

Weinstein, M. P., S. Y. Litvin, K. L. Bosley, C. M. Fuller, and S. C. Wainright. 2000. The role of tidal salt marsh as an energy source for marine transient and resident finfishes: a stable isotope approach. Transactions of the American Fisheries Society 129:797–810.

Williams, J. G., G. Armstrong, C. Katopodis, M. Larinier, and F. Trvade. 2012. Thinking like a fish: a key ingredient for development of effective fish passage facilities at river obstructions. River Research and Applications 28:407–417.

18

Estimated Survival of Subadult and Adult Atlantic Sturgeon in Four River Basins in the Southeastern United States

Joseph E. Hightower*
Department of Applied Ecology, North Carolina State University, Campus Box 7617, Raleigh, North Carolina 27695-7617, USA

Michael Loeffler
North Carolina Division of Marine Fisheries, 1367 US-17 South, Elizabeth City, North Carolina 27909, USA

William C. Post
South Carolina Department of Natural Resources, Marine Resources Research Institute, Post Office Box 12559, Charleston, South Carolina 29422, USA

Douglas L. Peterson
Warnell School of Forestry and Natural Resources, University of Georgia, 180 East Green Street, Athens, Georgia 30602, USA

Abstract

Prompted by concerns about the status of Atlantic Sturgeon *Acipenser oxyrinchus oxyrinchus*, in 2012 the National Oceanic and Atmospheric Administration listed one distinct population segment (DPS) as threatened (Gulf of Maine) and listed the remaining four DPSs as endangered (New York Bight, Chesapeake Bay, Carolina, and South Atlantic). To provide information for recovery planning, we estimated the survival of subadult and adult Atlantic Sturgeon in two river basins within the Carolina DPS (Roanoke and Cape Fear rivers, North Carolina) and two basins within the South Atlantic DPS (Ashepoo–Combahee–Edisto rivers [ACE], South Carolina; Altamaha River, Georgia). Estimated detection probability varied strongly by season but was similar among river basins, likely reflecting a winter migration into marine waters with minimal receiver coverage. Apparent monthly survival was very high and precisely estimated for the Roanoke River (0.985; 95% credible interval [CI] = 0.970–0.995), Cape Fear River (0.979; 95% CI = 0.971–0.986), ACE (0.989; 95% CI = 0.979–0.993), and Altamaha River (0.985; 95% CI = 0.973–0.994) basins. A pooled estimate for 87 adults from all four basins was 0.988 (95% CI = 0.982–0.992). The monthly rates implied annual apparent survival rates of 0.839 (Roanoke River basin), 0.778 (Cape Fear River basin), 0.871 (ACE basin), and 0.842 (Altamaha River basin); the pooled estimate for adults was 0.860. Our estimated survival rates were similar to other recent estimates for Atlantic Sturgeon but lower than recent estimates for several populations of Gulf Sturgeon *A. oxyrinchus desotoi*. Recovery of Atlantic Sturgeon in these southeastern rivers will occur more quickly if survival can be increased to a level that is consistent with published estimates of true natural mortality (0.03–0.07; annual survival $\geq$ 0.93).

Subject editor: Donald Noakes, Vancouver Island University, Nanaimo, British Columbia

*Corresponding author: jhightower@ncsu.edu

Atlantic Sturgeon *Acipenser oxyrinchus oxyrinchus* occur along the East Coast of North America, with known spawning populations from the St. Lawrence River, Quebec, to the Altamaha River or Satilla River, Georgia (Smith 1985; Wirgin et al. 2000, 2007). Historically, this subspecies was an important component of coastal ecosystems and supported commercial fisheries, but it has declined dramatically in abundance due to a variety of factors, including overfishing and habitat loss (Smith 1985). The National Oceanic and Atmospheric Administration conducted a status review of the subspecies and classified it into five distinct population segments (DPSs): Gulf of Maine, New York Bight, Chesapeake Bay, Carolina, and South Atlantic (Atlantic Sturgeon Status Review Team 2007). In 2012, the Gulf of Maine DPS was listed as threatened and the remaining DPSs were listed as endangered under the Endangered Species Act of 1973 (NOAA 2012a, 2012b). Issues leading to the designation of southeastern U.S. DPSs (Carolina and South Atlantic) as endangered included low abundance and concerns about survival (e.g., commercial bycatch and poor habitat quality; NOAA 2012a). Reliable information on survival, including temporal and spatial variation, is important for monitoring Atlantic Sturgeon recovery and restoring populations within each DPS.

Relatively little is known about the survival of Atlantic Sturgeon, particularly those belonging to southeastern U.S. populations. Peterson et al. (2008) obtained annual survival estimates of 0.79–0.83 for adult Atlantic Sturgeon in the Altamaha River based on a catch curve analysis of age data. Apparent annual survival of juvenile Atlantic Sturgeon in the Altamaha River was low (0.03–0.34), but true survival was unknown due to the confounding effect of permanent emigration (Schueller and Peterson 2010). Kahnle et al. (2007) used length and age data for Hudson River Atlantic Sturgeon and obtained survival estimates of 0.76 for males (ages 15–31) and 0.92 for females (ages 22–40). Pine et al. (2001) developed population models for the Gulf Sturgeon *A. oxyrinchus desotoi*; based on capture–recapture data and prior studies, they used annual survival estimates of 0.75 for ages 1–3 and 0.84 for ages 4–25. Most recently, annual survival of Gulf Sturgeon was estimated to be 0.92 or higher in three of four geographic areas (Rudd et al. 2014).

In the present study, we used telemetry to estimate the survival rates of Atlantic Sturgeon within several southeastern U.S. rivers. Similar approaches have been used to estimate survival rates for Green Sturgeon *A. medirostris* along the Pacific coast of North America and for Gulf Sturgeon in the Gulf of Mexico and tributary rivers (Lindley et al. 2008; Rudd et al. 2014). Adult and subadult survival rates are key population parameters, so the availability of comparable but independent estimates from multiple rivers should be very useful in the planning of Atlantic Sturgeon recovery efforts. Our methods should also be useful for application to other species with telemetry detections from marine and riverine receiver arrays.

METHODS

Atlantic Sturgeon were captured primarily with set gill nets, and acoustic transmitters were surgically implanted into the fish by using methods similar to those of Fox et al. (2000). Transmitters were VEMCO V16-4H acoustic tags (nominal life span = 1,157 d) or VEMCO V16-6H tags (nominal life span = 1,633 d). The transmitters were detected by using fixed arrays of VEMCO VR2W submersible receivers that were deployed in riverine and estuarine waters of four basins: the Roanoke River, North Carolina; Cape Fear River, North Carolina; Ashepoo–Combahee–Edisto rivers (ACE), South Carolina; and Altamaha River, Georgia. Our transmitters also had the potential to be detected at other receiver arrays deployed from Florida to Maine by 35 research groups participating in the Atlantic Coast Telemetry Network (www.theactnetwork.com) based at Delaware State University.

We analyzed data separately for each basin to detect spatial variation in survival, and we also produced a pooled survival estimate for adult-sized Atlantic Sturgeon (>150 cm TL at tagging; Bain 1997). Each analysis was based on a capture history matrix with rows for individual fish and columns for months. The matrix contained a zero when no detections occurred in a particular month and a 1 if there were one or more detections. To minimize concerns about mortality due to capture, handling, and tag implantation, the analysis period for each individual began 1 month after tagging. We censored a few individuals at the expected end of transmitter battery life. The period of analysis was 46 months (October 2010–July 2014) for the Roanoke River basin (fish tagged in the Roanoke River and Albemarle Sound), 39 months (May 2011–July 2014) for the Cape Fear River basin, 47 months (September 2010–July 2014) for the ACE basin and for the pooled analysis, and 34 months (May 2011–February 2014) for the Altamaha River basin. An additional three fish from the Roanoke River basin, four fish from the Altamaha River basin, and seven fish from the ACE basin were excluded from our analysis because they either were not detected during our period of analysis or were detected only during a single period after tagging.

Capture histories were analyzed using a Cormack–Jolly–Seber model, which is an open capture–recapture model that provides estimates of detection probability (probability that a live fish will be detected on one or more receivers) and survival (Kéry and Schaub 2012). Parameter estimates were obtained through a Bayesian state-space modeling framework (Kéry and Schaub 2012) by using OpenBUGS software (Spiegelhalter et al. 2010). The Cormack–Jolly–Seber model provides estimates of apparent (or "local") survival—that is, a fish's probability of surviving and being in the study area (in this instance, the areas covered by receivers; Kéry and Schaub 2012). Given the short (monthly) interval for detections relative to the period of analysis, we assumed that the bias in estimated survival due to the timing of tagging or detections was negligible. To examine variation in survival among individuals, we also calculated a simple indicator of survival for Atlantic Sturgeon in the Roanoke River basin: the range of months from release to final detection, expressed as a proportion of the number of possible months (from release to the end of battery life or to the end of the study).

Selection of the best model is a difficult and subjective process, and inferences are conditional on the model that is chosen (Link and Barker 2010; Kéry and Schaub 2012). Preliminary modeling indicated (1) little variation in apparent survival among periods but (2) strong seasonality in detection probability. Therefore, our base model for each basin or for the pooled analysis was defined as including time-dependent detection probability and constant survival. Our alternative model added fish TL as a covariate for survival. We chose the preferred model based on whether the 95% credible interval (CI) for the TL slope parameter contained zero. Following Kéry and Schaub (2012), we used uninformative prior distributions for mean survival (uniform, 0–1) and the logit-scale slope for the TL covariate (normal, mean = 0, variance = 1,000; bounds = −10 to 10). Example OpenBUGS code for modeling Atlantic Sturgeon in the Roanoke River basin is provided in the Appendix.

An Atlantic Sturgeon stock-rebuilding target of at least 20 protected mature age-classes for each spawning stock was established by the Atlantic States Marine Fisheries Commission (ASMFC 1990). The equilibrium production of mature females given our (combined-sex) estimates of annual survival was compared with production at an annual survival rate of 0.93 (based on an assumed rate of natural mortality $M = 0.07$; Kahnle et al. 2007). Maturation was assumed to follow the schedule reported by Kahnle et al. (2007). We also estimated the probability that Atlantic Sturgeon in our study rivers had annual survival rates greater than 0.93 (step function in the Appendix), which is a potential target level based on the M of 0.07 (Kahnle et al. 2007).

RESULTS

Sample sizes, the size range of tagged fish, and the analysis period varied among river basins (Table 1; Figure 1). Detection probability varied strongly among seasons, but the patterns were quite similar among river basins (Figure 2). Detection probability was generally low in December–February but rebounded sharply by April. Estimates of detection probability for April–October were relatively high in most cases, although there were occasional low values during summer (e.g., Roanoke River, ACE, and Altamaha River basins in July 2011). Mean monthly detection probability values were relatively similar among basins and were unaffected by inclusion of the TL covariate for survival (Table 1).

Mean monthly estimates of apparent survival were high, precise, and similar among basins (Table 1). The monthly values implied annual apparent survival rates of 0.839 (95% CI = 0.692–0.938) for the Roanoke River basin, 0.778 (95% CI = 0.699–0.846) for the Cape Fear River basin, 0.871 (95%

TABLE 1. Cormack–Jolly–Seber models fitted to capture history matrices for Atlantic Sturgeon that were tagged with acoustic transmitters in four river basins (Roanoke River, Cape Fear River, Ashepoo–Combahee–Edisto rivers [ACE], and Altamaha River) and for all adult-sized individuals (>150 cm TL at tagging) pooled across basins. Preferred models (shown in bold italics) were chosen based on whether the 95% credible interval (CI) for the slope of the TL effect on survival contained zero. Reported estimates are medians. Subadults (<150 cm TL at tagging) comprised 65% of tagged fish in the Roanoke River basin, 93% in the Cape Fear River basin, 12% in the ACE basin, and 5% in the Altamaha River basin (log*L* = log likelihood).

Basin	Number tagged	TL (cm) range	Detection probability	Survival rate	Log*L*	Mean detection probability	Monthly survival: Estimate	Monthly survival: 95% CI	Fish TL slope: Estimate	Fish TL slope: 95% CI
Roanoke	20	72–178	***Time***	***Constant***	***−238.7***	***0.498***	***0.985***	***0.970–0.995***		
			Time	Size dependent	−240.2	0.498	0.988	0.972–0.998	0.02	−0.00 to 0.07
Cape Fear	81	67–175	***Time***	***Constant***	***−920.0***	***0.496***	***0.979***	***0.971–0.986***		
			Time	Size dependent	−922.5	0.495	0.980	0.972–0.989	−0.01	−0.02 to 0.01
ACE	42	141–202	***Time***	***Constant***	***−676.0***	***0.477***	***0.989***	***0.979–0.993***		
			Time	Size dependent	−677.0	0.477	0.989	0.982–0.994	−0.02	−0.06 to 0.01
Altamaha	41	144–231	***Time***	***Constant***	***−442.2***	***0.525***	***0.985***	***0.973–0.994***		
			Time	Size dependent	−444.7	0.525	0.988	0.977–0.995	−0.02	−0.05 to 0.02
Pooled	87	150–231	***Time***	***Constant***	***−1,252.5***	***0.464***	***0.988***	***0.982–0.992***		
			Time	Size dependent	−1,253.5	0.464	0.988	0.982–0.992	−0.00	−0.00 to 0.00

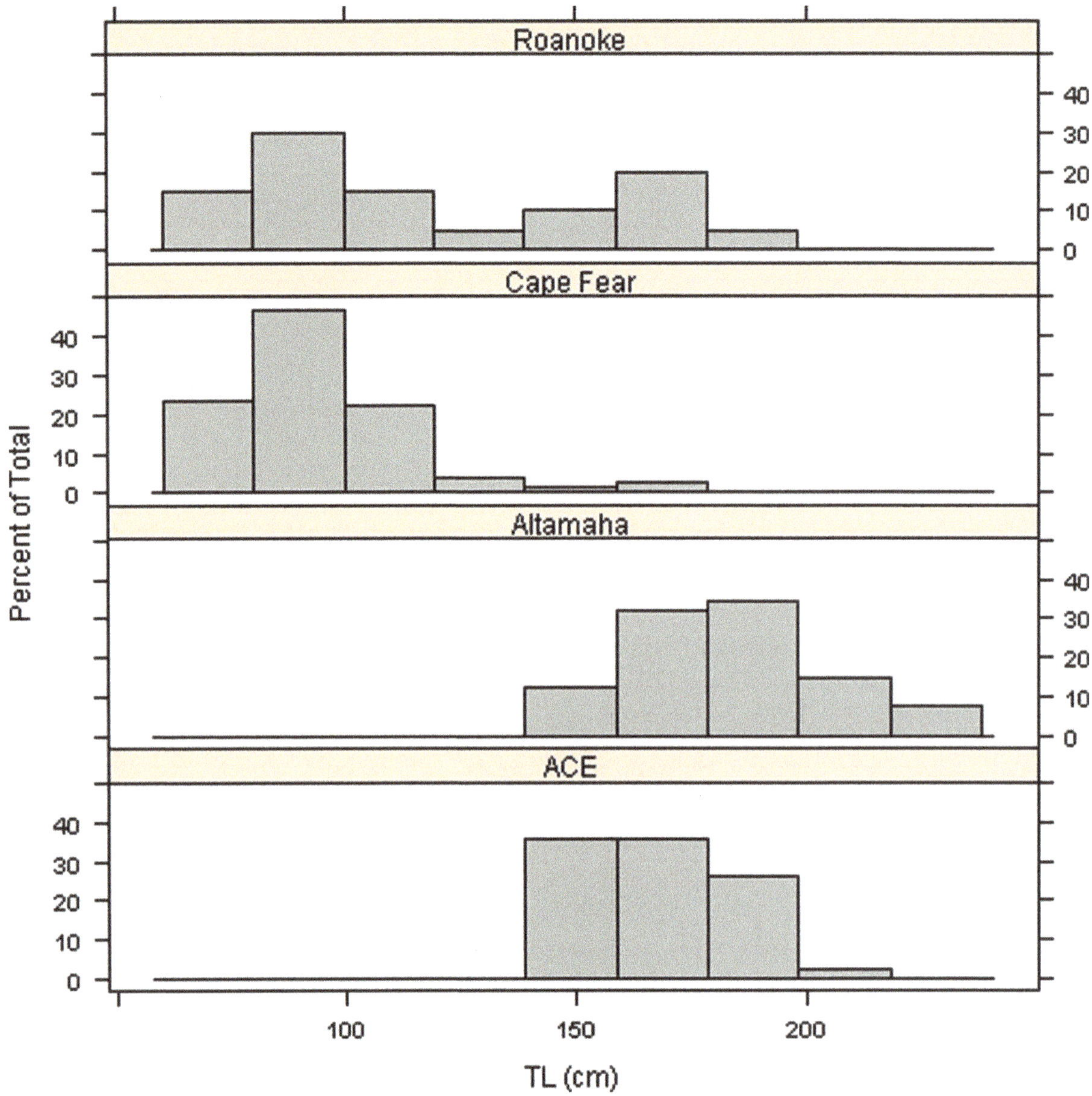

Figure 1. Length frequency distribution (TL, cm) of Atlantic Sturgeon that were tagged with acoustic transmitters in four basins (2010–2014): Roanoke River ($N = 20$), Cape Fear River ($N = 81$), Ashepoo–Combahee–Edisto rivers (ACE; $N = 42$), and Altamaha River ($N = 41$).

CI = 0.796–0.928) for the ACE basin, and 0.842 (95% CI = 0.722–0.932) for the Altamaha River basin. The pooled estimate of annual apparent survival for adults was 0.860 (95% CI = 0.799–0.909). The probability that annual survival was greater than 0.93 ranged from 5.0×10^{-5} to 0.04 for the four basins and was 0.002 for the pooled analysis. A size effect (TL covariate) was not evident for any of the four basins or for the pooled model, although the lower bound of the 95% CI for the Roanoke River basin was close to zero. This result was probably due to four of the smaller Roanoke River fish (78–121 cm TL), which were detected for short periods relative to the total possible number of months at risk (Figure 3). Those four fish were last seen in April, August, September, and November, so there was no obvious pattern in their timing of apparent mortality.

DISCUSSION

Tagging of Atlantic Sturgeon in multiple rivers and monitoring the fish through a collaborative coastwide network of receiver arrays proved to be an effective approach for investigating apparent survival. Seasonal patterns of detection were quite similar among river basins. The low detection probabilities in winter likely reflected an offshore migration to areas with low receiver coverage. The timing was consistent with previous work conducted in two South Carolina rivers by

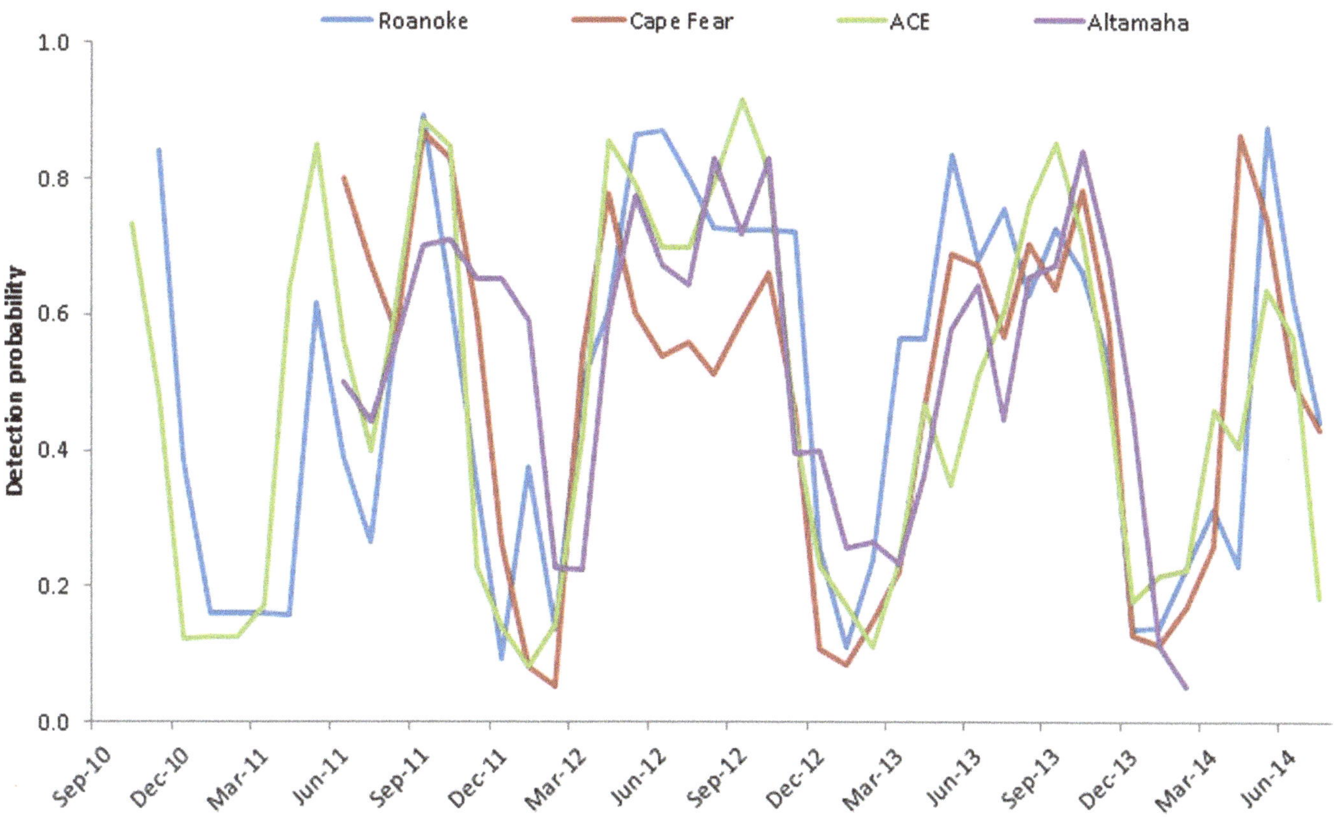

FIGURE 2. Estimated detection probability (2010–2014) for Atlantic Sturgeon that were tagged with acoustic transmitters in four river basins (Roanoke River, Cape Fear River, Ashepoo–Combahee–Edisto rivers [ACE], and Altamaha River); values are based on the preferred model for each basin (see Table 1).

Collins et al. (2000b), who documented an ocean migration of Atlantic Sturgeon in October or November and a return to estuarine or riverine waters in approximately March. During the current study, Roanoke River Atlantic Sturgeon migrated into marine waters during November or December and returned to Albemarle Sound and the Roanoke River during late spring or summer (Flowers 2015). Laney et al. (2007) conducted winter trawl surveys of the North Carolina coast and showed that shallow, nearshore marine waters served as important overwintering habitat for Atlantic Sturgeon. The North Carolina coastal section represents one of several gaps in East Coast receiver coverage (Flowers 2015) that could account for our low number of winter detections. Furthermore, many receiver arrays to the north of Chesapeake Bay are removed during winter, so there were periods in which there was no receiver coverage of northern ocean waters. Another factor contributing to the low number of winter detections may have been a lack of movement. Riverine studies have shown that Shortnose Sturgeon *A. brevirostrum* are highly aggregated and exhibit little movement during winter (O'Herron et al. 1993; Fernandes et al. 2010). Our estimates of apparent survival are unlikely to have been biased by the low detection probabilities in winter, because the transmitters have long life spans and the status of a tagged individual is made clear upon its re-detection in spring or summer.

Monthly estimates of mean apparent survival were 0.979–0.989 for the four river basins, and the pooled adult estimate was 0.988. True survival in this case may be quite similar to apparent survival given the very high survival estimates and the widespread receiver coverage. Atlantic Sturgeon undoubtedly occupy areas without receiver coverage at times, but this temporary emigration should not have biased our long-term survival estimates (Williams et al. 2001). Other potential biases in our estimated survival rates could include transmitter failure or expulsion. We are unaware of any studies indicating that transmitter failure is likely. Tank studies of sturgeons have generally shown that tag expulsion is unlikely (Collins et al. 2002; Neely et al. 2009; although see Boone et al. 2013).

Assuming that our estimates of apparent survival have negligible bias due to permanent emigration or transmitter failure, we obtained annual survival estimates of 0.839 for the Roanoke River basin, 0.778 for the Cape Fear River basin, 0.871 for the ACE basin, and 0.842 for the Altamaha River basin, and we generated a pooled estimate of 0.860 for adults. Given the overlapping marine distributions and shared threats among populations and DPSs, the pooled estimate of adult survival should be useful for large-scale recovery planning (Dunton et al. 2012; Waldman et al. 2013; Wirgin et al. 2015a, 2015b). The survival estimate for the ACE basin is probably low, as nine fish that were not detected during the final study

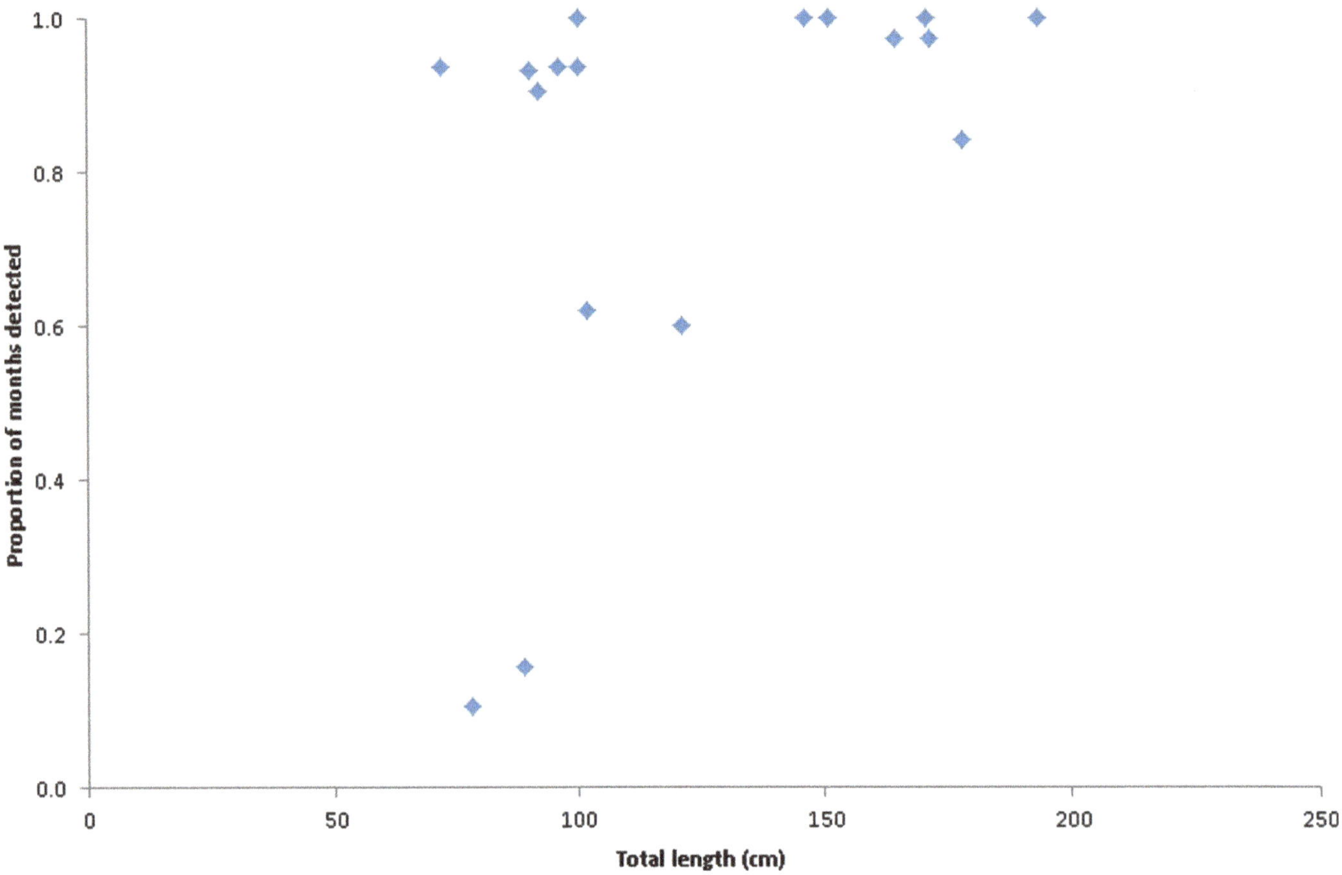

FIGURE 3. Range of months from release to final detection, expressed as a proportion of possible months at large (from release to the end of the study or to the end of transmitter battery life), presented in relation to TL (cm) at tagging for Roanoke River basin Atlantic Sturgeon.

period were detected later (between August and December 2014). Our survival estimates are consistent with other recent estimates for subadult and adult Atlantic Sturgeon (Kahnle et al. 2007; Peterson et al. 2008) but lower than estimates for Gulf Sturgeon in some eastern Gulf of Mexico rivers (Rudd et al. 2014). Our survival estimates imply a range of 0.14–0.25 for the total instantaneous mortality rate ($Z = -\log_e$[survival]). Given that there is no legal harvest of Atlantic Sturgeon, these Z-estimates likely reflect a combination of true M (e.g., due to predation, disease, or senescence) and mortality due to anthropogenic sources, such as degraded habitat, vessel strikes, or bycatch mortality in ocean, estuarine, or inland waters (Collins et al. 1996, 2000a; Secor and Waldman 1999; Stein et al. 2004; Oakley and Hightower 2007; Simpson and Fox 2009; Brown and Murphy 2010; Balazik et al. 2012). The Z-estimates could also be biased upward by permanent emigration from any of the areas with receiver coverage. The extent of this bias is unknown but should be lessened by the long periods of analysis and by the presence of offshore arrays, which can encounter fish that do not return to riverine waters.

Prior modeling studies for Atlantic Sturgeon have assumed an adult M of 0.07 (Boreman 1997; Kahnle et al. 2007), whereas Rudd et al. (2014) obtained a field-based M-estimate of 0.03 for the largest population of Gulf Sturgeon (Suwannee River, Florida). Modeling studies focused on adult Atlantic Sturgeon have suggested a target Z of 0.10–0.12 for a sustainable fishery (Boreman 1997; Kahnle et al. 2007); Pine et al. (2001) suggested that at Z-values greater than 0.17–0.19, the rebuilding of Gulf Sturgeon populations would be unlikely. We would expect a more robust or rapid recovery of Atlantic Sturgeon in our study rivers if total mortality can be decreased to levels that are consistent with true M (0.03–0.07; i.e., an annual survival rate of 0.93 or higher). However, we also note that our survival estimates were similar among the four rivers, whereas Atlantic Sturgeon abundance is likely much higher for the Altamaha River (Peterson et al. 2008). Flowers and Hightower (2015) estimated that the abundance of riverine Atlantic Sturgeon was higher for the Edisto River than for the Cape Fear River, which in turn had substantially higher abundance than the Roanoke River. Thus, abundance differences among rivers may be attributable to other factors, such as survival at early life stages.

One measure of the difference between our current survival estimates and the target levels is the equilibrium production of mature females. Starting from an initial cohort of 100 females at age 11 (the first age with a non-zero proportion of mature individuals) and using a target survival rate of

0.93 ($M = 0.07$), we obtain 50 age-classes (through age 60) with at least one mature individual, resulting in a total of 1,020 mature females (see Supplementary Table S.1 and Figure S.1 online). Our highest annual survival rate (0.87) would result in 34 age-classes containing at least one individual (still exceeding the Atlantic States Marine Fisheries Commission's target) but would only yield 419 mature females—a 59% reduction in equilibrium spawning stock production. Clearly, moderate differences in survival have large impacts over the long life span of Atlantic Sturgeon.

Continued monitoring of these populations is warranted because many of the fish analyzed in this study have transmitter expiration dates of 2015 or later. Future detections will eliminate uncertainty about the status of fish that have not been detected recently and will provide a longer time frame for judging temporal variability in survival. Information about the locations where fish were last detected might prove beneficial in identifying threats to survival. These spatially explicit detections can also be used in multistate models (e.g., Flowers 2015) that allow for spatial and temporal variation in survival. Continued monitoring could also provide feedback about whether regulation changes (e.g., gear restrictions or seasonal closures) are improving survival. An increase in the tagging of subadult Atlantic Sturgeon would aid in judging the size dependence of survival, particularly in the ACE basin, where only large fish were tagged during the present study. Subadults would be expected to spend more time in estuarine waters and less time in the ocean than adults, so threats to survival would differ for the two groups. A larger sample size of subadults would also be beneficial for assessing the timing of apparent mortalities relative to potential risk factors in estuarine waters.

ACKNOWLEDGMENTS

We thank the state, federal, and university biologists who provided assistance in the collection and tagging of adult Atlantic Sturgeon; we also thank the collaborating scientists who provided telemetry detections through the Atlantic Coast Telemetry Network. We are grateful to two anonymous reviewers for their helpful comments on a previous version of this paper. Any use of trade, firm, or product names is for descriptive purposes only and does not imply endorsement by the U.S. Government. This study was performed under the auspices of Permit 13-037-O from the Institutional Animal Care and Use Committee and Scientific Research Permit 16375-01 from the National Oceanic and Atmospheric Administration–Fisheries.

REFERENCES

ASMFC (Atlantic States Marine Fisheries Commission). 1990. Atlantic States Marine Fisheries Commission interstate fishery management plan for Atlantic Sturgeon. ASMFC, Fisheries Management Report 17, Washington, D.C.

Atlantic Sturgeon Status Review Team. 2007. Status review of Atlantic Sturgeon (*Acipenser oxyrinchus oxyrinchus*). Report to the National Marine Fisheries Service, Northeast Regional Office, Gloucester, Massachusetts.

Bain, M. B. 1997. Atlantic and Shortnose sturgeons of the Hudson River: common and divergent life history attributes. Environmental Biology of Fishes 48:347–358.

Balazik, M. T., K. J. Reine, A. J. Spells, C. A. Fredrickson, M. L. Fine, G. C. Garman, and S. P. McIninch. 2012. The potential for vessel interactions with adult Atlantic Sturgeon in the James River, Virginia. North American Journal of Fisheries Management 32:1062–1069.

Boone, S. S., S. J. Divers, A. C. Camus, D. L. Peterson, C. A. Jennings, J. L. Shelton, and S. M. Hernandez. 2013. Pathologic and physiologic effects associated with long-term intracoelomic transmitters in captive Siberian Sturgeon. North American Journal of Fisheries Management 33:869–877.

Boreman, J. 1997. Sensitivity of North American sturgeon and Paddlefish to fishing mortality. Environmental Biology of Fishes 48:399–405.

Brown, J. J., and G. W. Murphy. 2010. Atlantic Sturgeon vessel-strike mortalities in the Delaware estuary. Fisheries 35:72–83.

Collins, M. R., D. W. Cooke, T. I. J. Smith, W. C. Post, D. C. Russ, and D. C. Walling. 2002. Evaluation of four methods of transmitter attachment on Shortnose Sturgeon, *Acipenser brevirostrum*. Journal of Applied Ichthyology 18:491–494.

Collins, M. R., S. G. Rogers, and T. I. J. Smith. 1996. Bycatch of sturgeons along the southern Atlantic coast of the USA. North American Journal of Fisheries Management 16:24–29.

Collins, M. R., S. G. Rogers, T. I. J. Smith, and M. L. Moser. 2000a. Primary factors affecting sturgeon populations in the southeastern United States: fishing mortality and degradation of essential habitats. Bulletin of Marine Science 66:917–928.

Collins, M. R., T. I. J. Smith, W. C. Post, and O. Pashuk. 2000b. Habitat utilization and biological characteristics of adult Atlantic Sturgeon in two South Carolina rivers. Transactions of the American Fisheries Society 129:982–988.

Dunton, K. J., D. Chapman, A. Jordaan, K. Feldheim, S. J. O'Leary, K. A. McKown, and M. G. Frisk. 2012. Genetic mixed-stock analysis of Atlantic Sturgeon *Acipenser oxyrinchus oxyrinchus* in a heavily exploited marine habitat indicates the need for routine genetic monitoring. Journal of Fish Biology 80:207–217.

Fernandes, S. J., G. B. Zydlewski, J. D. Zydlewski, G. S. Wippelhauser, and M. T. Kinnison. 2010. Seasonal distribution and movements of Shortnose Sturgeon and Atlantic Sturgeon in the Penobscot River estuary, Maine. Transactions of the American Fisheries Society 139:1436–1449.

Flowers, H. J. 2015. Methods for monitoring and assessing sturgeon populations using technology. Doctoral dissertation. North Carolina State University, Raleigh.

Flowers, H. J., and J. E. Hightower. 2015. Estimating sturgeon abundance in the Carolinas using side-scan sonar. Marine and Coastal Fisheries: Dynamics, Management, and Ecosystem Science [online serial] 7:1–9.

Fox, D. A., J. E. Hightower, and F. M. Parauka. 2000. Gulf Sturgeon spawning migration and habitat in the Choctawhatchee River system, Alabama–Florida. Transactions of the American Fisheries Society 129:811–826.

Kahnle, A. W., K. A. Hattala, and K. A. McKown. 2007. Status of Atlantic Sturgeon of the Hudson River estuary, New York, USA. Pages 347–363 *in* J. Munro, D. Hatin, J. E. Hightower, K. A. McKown, K. J. Sulak, A. Kahnle, and F. Caron, editors. Proceedings of the symposium on anadromous sturgeon: status and trend, anthropogenic impact, and essential habitat. American Fisheries Society, Symposium 56, Bethesda, Maryland.

Kéry, M., and M. Schaub. 2012. Bayesian population analysis using WinBUGS: a hierarchical perspective. Academic Press, Waltham, Massachusetts.

Laney, R. W., J. E. Hightower, B. R. Versak, M. F. Mangold, W. W. Cole Jr., and S. E. Winslow. 2007. Distribution, habitat use, and size of Atlantic Sturgeon captured during cooperative winter tagging cruises. Pages 167–182 *in* J. Munro, D. Hatin, J. E. Hightower, K. McKown, K. J. Sulak, A. W.

Kahnle, and F. Caron, editors. Anadromous sturgeons: habitats, threats, and management. American Fisheries Society, Symposium 56, Bethesda, Maryland.

Lindley, S. T., M. L. Moser, D. L. Erickson, M. Belchik, D. W. Welch, E. L. Rechisky, J. T. Kelly, J. Heublein, and A. P. Klimley. 2008. Marine migration of North American Green Sturgeon. Transactions of the American Fisheries Society 137:182–194.

Link, W. A., and R. J. Barker. 2010. Bayesian inference with ecological applications. Academic Press, London.

Neely, B. C., K. D. Steffensen, and M. A. Pegg. 2009. A comparison of gastrically and surgically implanted telemetry transmitters in Shovelnose Sturgeon. Fisheries Management and Ecology 16:323–328.

NOAA (National Oceanic and Atmospheric Administration). 2012a. Endangered and threatened wildlife and plants; final listing determination for two distinct population segments of Atlantic Sturgeon (*Acipenser oxyrinchus oxyrinchus*) in the southeast. Federal Register 77:24(February 6, 2012):5914–5982.

NOAA (National Oceanic and Atmospheric Administration). 2012b. Endangered and threatened wildlife and plants; threatened and endangered status for distinct population segments of Atlantic Sturgeon in the northeast region, final rule. Federal Register 77:24(February 6, 2012):5880–5912.

O'Herron, J. C. II, K. W. Able, and R. W. Hastings. 1993. Movements of Shortnose Sturgeon (*Acipenser brevirostrum*) in the Delaware River. Estuaries 16:235–240.

Oakley, N. C., and J. E. Hightower. 2007. Status of Shortnose Sturgeon in the Neuse River, North Carolina. Pages 273–284 *in* J. Munro, D. Hatin, J. E. Hightower, K. McKown, K. J. Sulak, A. W. Kahnle, and F. Caron, editors. Anadromous sturgeons: habitats, threats, and management. American Fisheries Society, Symposium 56, Bethesda, Maryland.

Peterson, D. L., P. Schueller, R. DeVries, J. Fleming, C. Grunwald, and I. Wirgin. 2008. Annual run size and genetic characteristics of Atlantic Sturgeon in the Altamaha River, Georgia. Transactions of the American Fisheries Society 137:393–401.

Pine, W. E. III, M. S. Allen, and V. J. Dreitz. 2001. Population viability of the Gulf of Mexico Sturgeon: inferences from capture–recapture and age-structured models. Transactions of the American Fisheries Society 130:1164–1174.

Rudd, M. B., R. N. M. Ahrens, W. E. Pine III, and S. K. Bolden. 2014. Empirical, spatially explicit natural mortality and movement rate estimates for the threatened Gulf Sturgeon (*Acipenser oxyrinchus desotoi*). Canadian Journal of Fisheries and Aquatic Sciences 71:1407–1417.

Schueller, P., and D. L. Peterson. 2010. Abundance and recruitment of juvenile Atlantic Sturgeon in the Altamaha River, Georgia. Transactions of the American Fisheries Society 139:1526–1535.

Secor, D. H., and J. R. Waldman. 1999. Historical abundance of Delaware Bay Atlantic Sturgeon and potential rate of recovery. Pages 203–216 *in* J. A. Musick, editor. Life in the slow lane: ecology and conservation of long-lived marine animals. American Fisheries Society, Symposium 23, Bethesda, Maryland.

Simpson, P. C., and D. A. Fox. 2009. A contemporary understanding of the Delaware River Atlantic Sturgeon: survival in a highly impacted aquatic ecosystem. Pages 867–870 *in* A. J. Haro, K. L. Smith, R. A. Rulifson, C. M. Moffitt, R. J. Klauda, M. J. Dadswell, R. A. Conjak, J. E. Cooper, K. L. Beal, and T. S. Avery, editors. Challenges for diadromous fishes in a dynamic global environment. American Fisheries Society, Symposium 69, Bethesda, Maryland.

Smith, T. I. J. 1985. The fishery, biology, and management of Atlantic Sturgeon, *Acipenser oxyrinchus*, in North America. Environmental Biology of Fishes 14:61–72.

Spiegelhalter, D., A. Thomas, N. Best, and D. Lunn. 2010. OpenBUGS user manual, version 3.1.2. MRC Biostatistics Unit, Cambridge, UK.

Stein, A. B., K. D. Friedland, and M. Sutherland. 2004. Atlantic Sturgeon marine bycatch and mortality on the continental shelf of the northeast United States. North American Journal of Fisheries Management 24: 171–183.

Waldman, J. R., T. King, T. Savoy, L. Maceda, C. Grunwald, and I. Wirgin. 2013. Stock origins of subadult and adult Atlantic Sturgeon, *Acipenser oxyrinchus*, in a non-natal estuary, Long Island Sound. Estuaries and Coasts 36:257–267.

Williams, B. K., J. D. Nichols, and M. J. Conroy. 2001. Analysis and management of animal populations. Academic Press, San Diego, California.

Wirgin, I., M. W. Breece, D. A. Fox, L. Maceda, K. W. Wark, and T. King. 2015a. Origin of Atlantic Sturgeon collected off the Delaware coast during spring months. North American Journal of Fisheries Management 35:20–30.

Wirgin, I., C. Grunwald, J. Stabile, and J. Waldman. 2007. Genetic evidence for relict Atlantic Sturgeon stocks along the mid-Atlantic coast of the USA. North American Journal of Fisheries Management 27:1214–1229.

Wirgin, I., L. Maceda, C. Grunwald, and T. L. King. 2015b. Population origin of Atlantic Sturgeon *Acipenser oxyrinchus oxyrinchus* by-catch in U.S. Atlantic coast fisheries. Journal of Fish Biology 86:1251–1270.

Wirgin, I., J. R. Waldman, J. Rosko, R. Gross, M. R. Collins, S. G. Rogers, and J. Stabile. 2000. Genetic structure of Atlantic Sturgeon populations based on mitochondrial DNA control region sequences. Transactions of the American Fisheries Society 129:476–486.

Appendix: OpenBUGS Code for the Cormack–Jolly–Seber model

Example of an OpenBUGS code for fitting a Cormack–Jolly–Seber model for Atlantic Sturgeon, with time-dependent detection probability and survival that varies as a function of the TL covariate.

```
model
  {
    # Priors
    mean.S ~ dunif(0, 1)                   # Uninformative on probability scale (Kéry and Schaub 2012:186)
    mu_S <-log(mean.S/(1-mean.S))          # Logit transformation (Kéry and Schaub 2012:186)
    b_S ~ dnorm(0, 0.001) I(-10, 10)       # Prior for logit-scale slope for TL (Kéry and Schaub 2012:190)
  for (i in 1:NumInd)
    {
      for (j in 1:Periods-1)
      {
        logit(S[i,j]) <-mu_S + b_S * (TL[i]-mean_TL)
      }
    }
  mean_p <-mean(p[2:Periods])
  for (j in 1: Periods)
    {
      p[j] ~ dunif(0, 1)
    }
  mean_TL <-mean(TL[])
  # Likelihood
  for (i in 1:NumInd)
    {
  z[i,first[i]] ~dbern(1)         # Known to be alive at the time of entry into this analysis
  for (j in first[i]+1:last[i])
      {
      eff.S[i,j-1] <-z[i,j-1] * S[i, j-1]
      z[i,j] ~ dbern(eff.S[i,j-1])        # Current state depends on state at previous occasion and survival probability
      eff.p[i,j] <-z[i,j] * p[j]     # Detection probability for an individual depends on its state (must be alive) and estimated p
     y[i,j] ~ dbern(eff.p[i,j])
     }
    }
  Annual.S <-pow(mean.S, 12)
  Good.S <-step(Annual.S-0.93)
  }
```

Permissions

All chapters in this book were first published in MCFDMES, by Taylor & Francis; hereby published with permission under the Creative Commons Attribution License or equivalent. Every chapter published in this book has been scrutinized by our experts. Their significance has been extensively debated. The topics covered herein carry significant findings which will fuel the growth of the discipline. They may even be implemented as practical applications or may be referred to as a beginning point for another development.

The contributors of this book come from diverse backgrounds, making this book a truly international effort. This book will bring forth new frontiers with its revolutionizing research information and detailed analysis of the nascent developments around the world.

We would like to thank all the contributing authors for lending their expertise to make the book truly unique. They have played a crucial role in the development of this book. Without their invaluable contributions this book wouldn't have been possible. They have made vital efforts to compile up to date information on the varied aspects of this subject to make this book a valuable addition to the collection of many professionals and students.

This book was conceptualized with the vision of imparting up-to-date information and advanced data in this field. To ensure the same, a matchless editorial board was set up. Every individual on the board went through rigorous rounds of assessment to prove their worth. After which they invested a large part of their time researching and compiling the most relevant data for our readers.

The editorial board has been involved in producing this book since its inception. They have spent rigorous hours researching and exploring the diverse topics which have resulted in the successful publishing of this book. They have passed on their knowledge of decades through this book. To expedite this challenging task, the publisher supported the team at every step. A small team of assistant editors was also appointed to further simplify the editing procedure and attain best results for the readers.

Apart from the editorial board, the designing team has also invested a significant amount of their time in understanding the subject and creating the most relevant covers. They scrutinized every image to scout for the most suitable representation of the subject and create an appropriate cover for the book.

The publishing team has been an ardent support to the editorial, designing and production team. Their endless efforts to recruit the best for this project, has resulted in the accomplishment of this book. They are a veteran in the field of academics and their pool of knowledge is as vast as their experience in printing. Their expertise and guidance has proved useful at every step. Their uncompromising quality standards have made this book an exceptional effort. Their encouragement from time to time has been an inspiration for everyone.

The publisher and the editorial board hope that this book will prove to be a valuable piece of knowledge for researchers, students, practitioners and scholars across the globe.

List of Contributors

Yasutoki Shibata
Tohoku National Fisheries Research Institute, Fisheries Research Agency, Samemachi, Aomori 031 0841, Japan

Manabu Yamada and Toshihiro Wada
Fukushima Fisheries Experiment Station, Iwaki, Fukushima 970 0316, Japan

Masaki Itou
Tohoku National Fisheries Research Institute, Fisheries Research Agency, Samemachi, Aomori 031 0841, Japan

Harumi Yamada
Seikai National Fisheries Research Institute, Fisheries Research Agency, Taira-machi, Nagasaki 851 2213, Japan

Tadahiro Sohtome, Takashi Iwasaki, Tooru Sakuma and Takuji Mizuno
Fukushima Fisheries Experiment Station, Iwaki, Fukushima 970 0316, Japan

Joseph H. Tarnecki and William F. Patterson III
Dauphin Island Sea Lab, Department of Marine Sciences, University of South Alabama, 101 Bienville Boulevard, Dauphin Island, Alabama 36528, USA

Jason M. Morson and Eleanor A. Bochenek
Haskin Shellfish Research Laboratory–Rutgers, The State University of New Jersey, 6959 Miller Avenue, Port Norris, New Jersey 08349, USA

Eric N. Powell
Gulf Coast Research Laboratory, University of Southern Mississippi, 703 East Beach Drive, Ocean Springs, Mississippi 39564, USA

Emerson C. Hasbrouck
Cornell University Cooperative Extension, Marine Program, 423 Griffing Avenue, Riverhead, New York 11901, USA

Charles F. Cotton
Virginia Institute of Marine Science, 1208 Greate Road, Gloucester Point, Virginia 23062, USA

Jennifer E. Gius
Haskin Shellfish Research Laboratory–Rutgers, The State University of New Jersey, 6959 Miller Avenue, Port Norris, New Jersey 08349, USA

Kristin Gerbino and Tara Froehlich
Cornell University Cooperative Extension, Marine Program, 423 Griffing Avenue, Riverhead, New York 11901, USA

Stephen P. Geiger
Florida Fish and Wildlife Conservation Commission, Fish and Wildlife Research Institute, 100 8th Avenue Southeast, St. Petersburg, Florida 33701, USA

William S. Arnold
National Oceanic and Atmospheric Administration, Southeast Regional Office, 263 13th Avenue South, St. Petersburg, Florida 33701, USA

Sarah Stephenson and Keith Fischer
Florida Fish and Wildlife Conservation Commission, Fish and Wildlife Research Institute, 100 8th Avenue Southeast, St. Petersburg, Florida 33701, USA

David J. Teel
National Oceanic and Atmospheric Administration, National Marine Fisheries Service, Northwest Fisheries Science Center, Conservation Biology Division, Post Office Box 130, Manchester, Washington 98353, USA

Brian J. Burke
National Oceanic and Atmospheric Administration, National Marine Fisheries Service, Northwest Fisheries Science Center, Fish Ecology Division, 2725 Montlake Boulevard East, Seattle, Washington 98112, USA

David R. Kuligowski
National Oceanic and Atmospheric Administration, National Marine Fisheries Service, Northwest Fisheries Science Center, Conservation Biology Division, Post Office Box 130, Manchester, Washington 98353, USA

Cheryl A. Morgan
Cooperative Institute for Marine Resources Studies, Hatfield Marine Science Center, Oregon State University, 2030 South Marine Science Drive, Newport, Oregon 97365, USA

Donald M. Van Doornik
National Oceanic and Atmospheric Administration, National Marine Fisheries Service, Northwest Fisheries Science Center, Conservation Biology Division, Post Office Box 130, Manchester, Washington 98353, USA

Anne E. McElroy, Robert M. Cerrato and Michael G. Frisk
School of Marine and Atmospheric Sciences, Stony Brook University, Stony Brook, New York 11794-5000, USA

Brian K. Gallagher
Chesapeake Biological Laboratory, University of Maryland Center for Environmental Sciences, Post Office Box 38, Solomons, Maryland 20688, USA

Lyndie A. Hice
Delaware National Estuarine Research Reserve, Dover, Delaware 19901, USA

J. Kevin Craig
National Marine Fisheries Service, Southeast Fisheries Science Center, Beaufort Laboratory, Beaufort, North Carolina 28516, USA

William E. Smith
North Carolina Division of Marine Fisheries, 3441 Arendell Street, Morehead City, North Carolina 28557, USA
U.S. Fish and Wildlife Service, Bay-Delta Fish and Wildlife Office, 650 Capitol Mall, Sacramento, California 95814, USA

Frederick S. Scharf
Department of Biology and Marine Biology, University of North Carolina Wilmington, 601 South College Road, Wilmington, North Carolina 28403, USA

James P. Monaghan
North Carolina Division of Marine Fisheries, 3441 Arendell Street, Morehead City, North Carolina 28557, USA

C. E. Porch
National Marine Fisheries Service, Southeast Fisheries Science Center, Sustainable Fisheries Division, 75 Virginia Beach Drive, Miami, Florida 33149-1099, USA

G. R. Fitzhugh
National Marine Fisheries Service, Southeast Fisheries Science Center, Panama City Laboratory, 3500 Delwood Beach Road, Panama City, Florida 32408, USA

E. T. Lang and H. M. Lyon
Riverside Technology Contracting for National Marine Fisheries Service, Panama City Laboratory, 3500 Delwood Beach Road, Panama City, Florida 32408, USA
Louisiana Department of Wildlife and Fisheries, 2000 Quail Drive, Baton Rouge, Louisiana 70808, USA

B. C. Linton
National Marine Fisheries Service, Southeast Fisheries Science Center, Sustainable Fisheries Division, 75 Virginia Beach Drive, Miami, Florida 33149-1099, USA

Shang-Yin Vanson Liu
Department of Oceanography and Department of Marine Biotechnology and Resources, National Sun Yat-Sen University, Kaohsiung 80424, Taiwan

I-Hsiang Huang
Department of Oceanography, National Sun Yat-Sen University, Kaohsiung 80424, Taiwan

Min-Yun Liu
Taiwan Ocean Research Institute, National Applied Research Laboratories, Kaohsiung 852, Taiwan

Hung-Du Lin
Affiliated School of National Tainan First Senior High School, Tainan 701, Taiwan

Feng-Yu Wang
Taiwan Ocean Research Institute, National Applied Research Laboratories, Kaohsiung 852, Taiwan

Te-Yu Liao
Department of Oceanography, National Sun Yat-Sen University, Kaohsiung 80424, Taiwan

Jinn-Shing Weng
Executive Yuan, Council of Agriculture, Fisheries Research Institute, Coastal and Offshore Resources Research Center, Kaohsiung 80627, Taiwan; and Department of Environmental Biology and Fisheries Science, National Taiwan Ocean University, Keelung 20224, Taiwan

Ming-An Lee
Department of Environmental Biology and Fisheries Science, National Taiwan Ocean University, Keelung 20224, Taiwan

Kwang-Ming Liu
Institute of Marine Affairs and Resource Management, National Taiwan Ocean University, Keelung 20224, Taiwan; and George Chen Shark Research Center, National Taiwan Ocean University, Keelung 20224, Taiwan

Ming-Shu Hsu, Mine-Kune Hung and Long-Jing Wu
Executive Yuan, Council of Agriculture, Fisheries Research Institute, Coastal and Offshore Resources Research Center, Kaohsiung 80627, Taiwan

H. Jared Flowers.
North Carolina Cooperative Fish and Wildlife Research Unit, Department of Applied Ecology, North Carolina State University, 127 David Clark Labs, Campus Box 7617, Raleigh, North Carolina 27695-7617, USA

Joseph E. Hightower
U.S. Geological Survey, North Carolina Cooperative Fish and Wildlife Research Unit, Department of Applied Ecology, North Carolina State University, 127 David Clark Labs, Campus Box 7617, Raleigh, North Carolina 27695-7617, USA

Keith J. Dunton
School of Marine and Atmospheric Sciences, Stony Brook University, Stony Brook, New York 11794-5000, USA

Adrian Jordaan
Department of Environmental Conservation, University of Massachusetts Amherst, Holdsworth Hall, 160 Holdsworth Way, Amherst, Massachusetts 01003, USA

David O. Conover
School of Marine and Atmospheric Sciences, Stony Brook University, Stony Brook, New York 11794-5000, USA

Michael G. Frisk
School of Marine and Atmospheric Sciences, Stony Brook University, Stony Brook, New York 11794-5000, USA

Kim A. McKown and Lisa A. Bonacci
New York State Department of Environmental Conservation, Division of Fish, Wildlife and Marine Resources, Bureau of Marine Resources, 205 North Belle Mead Road, Suite 1, East Setauket, New York 11733, USA

Márcia Regina Denadai
Centro Universitário Módulo, Rua Frei Pacífico Wagner 653, 11660-903 Caraguatatuba, São Paulo, Brazil; and Departamento de Oceanografia Biológica, Instituto Oceanográfico da Universidade de São Paulo, Praça do Oceanográfico 191, 05508-120 São Paulo, São Paulo, Brazil

Máıra Pombo
Departamento de Oceanografia Biológica, Instituto Oceanográfico da Universidade de São Paulo, Praça do Oceanográfico 191, 05508-120 São Paulo, São Paulo, Brazil

Ligia Coletti Bernadochi
Centro Universitário Módulo, Rua Frei Pacífico Wagner 653, 11660-903 Caraguatatuba, São Paulo, Brazil

Alexander Turra
Departamento de Oceanografia Biológica, Instituto Oceanográfico da Universidade de São Paulo, Praça do Oceanográfico 191, 05508-120 São Paulo, São Paulo, Brazil

Melissa A. Yencho
School of Marine and Atmospheric Sciences, Stony Brook University, Stony Brook, New York 11794-5000, USA

Adrian Jordaan
Department of Environmental Conservation, University of Massachusetts Amherst, Holdsworth Hall, Amherst, Massachusetts 01003, USA

Robert M. Cerrato
School of Marine and Atmospheric Sciences, Stony Brook University, Stony Brook, New York 11794-5000, USA

Hannes Baumann
Department of Marine Sciences, University of Connecticut, 1080 Shennecossett Road, Groton, Connecticut 06340-6048, USA

Michael G. Frisk
School of Marine and Atmospheric Sciences, Stony Brook University, Stony Brook, New York 11794-5000, USA

Daniel S. Stich
Department of Wildlife, Fisheries, and Conservation Biology, University of Maine, 5755 Nutting Hall, Orono, Maine 04469, USA

Gayle B. Zydlewski
School of Marine Sciences, University of Maine, 5706 Aubert Hall, Orono, Maine 04469, USA

John F. Kocik
National Oceanic and Atmospheric Administration, Northeast Fisheries Science Center, 17 Godfrey Drive, Suite 1, Orono, Maine 04473, USA

Joseph D. Zydlewski
U.S. Geological Survey, Maine Cooperative Fish and Wildlife Research Unit, University of Maine, 5755 Nutting Hall, Orono, Maine 04469, USA

Michelle Zapp Sluis
Department of Oceanography and Coastal Sciences, Louisiana State University, Energy, Coast, and Environment Building, Baton Rouge, Louisiana 70803, USA
Department of Marine Biology, Texas A&M University at Galveston, 1001 Texas Clipper Road, Galveston, Texas 77554,USA

Beverly K. Barnett
National Marine Fisheries Service, Southeast Fisheries Science Center, Panama City Laboratory, 3500 Delwood Beach Road, Panama City Beach, Florida 32408, USA

William F. Patterson III
Department of Marine Sciences, University of South Alabama, and Dauphin Island Sea Lab, 101 Bienville Boulevard, Dauphin Island, Alabama 36528, USA

James H. Cowan Jr.
Department of Oceanography and Coastal Sciences, Louisiana State University, Energy, Coast, and Environment Building, Baton Rouge, Louisiana 70803, USA

Alan M. Shiller
Department of Marine Science, University of Southern Mississippi, 1020 Balch Boulevard, Stennis Space Center, Mississippi 39529, USA

Matthew E. Kimball
Baruch Marine Field Laboratory, University of South Carolina, Post Office Box 1630, Georgetown, South Carolina 29442, USA

Lawrence P. Rozas
National Oceanic and Atmospheric Administration, National Marine Fisheries Service, Southeast Fisheries Science Center, Estuarine Habitats and Coastal Fisheries Center, 646 Cajundome Boulevard, Lafayette, Louisiana 70506, USA

Kevin M. Boswell
Marine Sciences Program, Department of Biological Sciences, Florida International University, Biscayne Bay Campus, 3000 Northeast 151st Street, MSB355, North Miami, Florida 33181, USA

James H. Cowan Jr.
Department of Oceanography and Coastal Sciences, Louisiana State University, 2247 Energy, Coast, and Environment Building, Baton Rouge, Louisiana 70803, USA

Joseph E. Hightower
Department of Applied Ecology, North Carolina State University, Campus Box 7617, Raleigh, North Carolina 27695-7617, USA

Michael Loeffler
North Carolina Division of Marine Fisheries, 1367 US-17 South, Elizabeth City, North Carolina 27909, USA

William C. Post
South Carolina Department of Natural Resources, Marine Resources Research Institute, Post Office Box 12559, Charleston, South Carolina 29422, USA

Douglas L. Peterson
Warnell School of Forestry and Natural Resources, University of Georgia, 180 East Green Street, Athens, Georgia 30602, USA